Lecture Notes in Computer Science   12701

More information about this subseries at http://www.springer.com/series/7407

Tiziana Calamoneri · Federico Corò (Eds.)

Algorithms and Complexity

12th International Conference, CIAC 2021
Virtual Event, May 10–12, 2021
Proceedings

 Springer

Editors
Tiziana Calamoneri 🆔
Sapienza University of Rome
Rome, Italy

Federico Corò 🆔
Sapienza University of Rome
Rome, Italy

ISSN 0302-9743   ISSN 1611-3349   (electronic)
Lecture Notes in Computer Science
ISBN 978-3-030-75241-5   ISBN 978-3-030-75242-2   (eBook)
https://doi.org/10.1007/978-3-030-75242-2

LNCS Sublibrary: SL1 – Theoretical Computer Science and General Issues

This Springer imprint is published by the registered company Springer Nature Switzerland AG
The registered company address is: Gewerbestrasse 11, 6330 Cham, Switzerland

Preface

This volume contains the proceedings of CIAC 2021: the 12th International Conference on Algorithms and Complexity; it includes all contributed papers and information on the invited lectures delivered at the conference. This conference series presents original research contributions in the theory and applications of algorithms and computational complexity.

The conference was supposed to be held in Larnaca, Cyprus, during May 10–12, 2021, but, due to the COVID-19 pandemic, it was held remotely over the same period.

The volume begins with the descriptions of the invited lectures. We are grateful to Henning Fernau (Trier University, Germany), Katharina Huber (University of East Anglia, UK), and Seffi Naor (Technion Haifa, Israel) for kindly accepting our invitation to give plenary lectures.

Then, the volume continues with all the contributed papers, arranged alphabetically by the last names of their first author. In response to a Call for Papers, the Program Committee received 78 submissions handled through the EasyChair system. Each submission was reviewed by three Program Committee members and 27 papers were selected for inclusion in the scientific program, on the basis of originality, quality, and relevance to theoretical computer science. Among these papers, the Program Committee selected as recipient of the best paper award the work entitled "The Weisfeiler-Leman Algorithm and Recognition of Graph Properties" co-authored by Frank Fuhlbrück, Johannes Koebler, Ilia Ponomarenko, and Oleg Verbitsky. The reason lies in its potential future applications and in the novelty of the results and constructions. We congratulate the authors and are grateful to Springer for sponsoring this award.

A special issue of *Theoretical Computer Science* will be devoted to publishing an extended version of a selection of the papers presented at the conference.

We wish to thank all authors who submitted papers for consideration, the Program Committee for its hard work, and the external reviewers who assisted the Program Committee in the evaluation process.

Last but not least, for their very dedicated work, special thanks go to Chryssis Georgiou and Anna Philippou, from the University of Cyprus, who served in the organizing committee.

March 2021

Tiziana Calamoneri
Federico Corò

Organization

Steering Committee

Giorgio Ausiello	Sapienza University of Rome, Italy
Pinar Heggernes	University of Bergen, Norway
Vangelis Paschos	Paris Dauphine University, France
Rossella Petreschi	Sapienza University of Rome, Italy
Peter Widmayer	ETH Zurich, Switzerland

Program Committee

Cristina Bazgan	Paris Dauphine University, France
Tiziana Calamoneri (Chair)	Sapienza University of Rome, Italy
Jianer Chen	Texas A&M University, USA
Peter Damaschke	Chalmers University of Technology, Sweden
Thomas Erlebach	University of Leicester, UK
Guillaume Fertin	Nantes University, France
Jiří Fiala	Charles University, Czech Republic
Paola Flocchini	University of Ottawa, Canada
Dimitris Fotakis	NTU Athens, Greece
Paolo Franciosa	Sapienza University of Rome, Italy
Leszek Antoni Gąsienie	University of Liverpool, UK
Mordecai Golin	Hong Kong University of Science and Technology, Hong Kong
Jan Kratochvíl	Charles University, Czech Republic
Vadim Lozin	University of Warwick, UK
Bodo Manthey	University of Twente, Netherlands
Marios Mavronicolas	University of Cyprus, Cyprus
Cécile Murat	Paris Dauphine University, France
Gaia Nicosia	Roma Tre University, Italy
Hirotaka Ono	Nagoya University, Japan
Maurizio Patrignani	Roma Tre University, Italy
Tomasz Radzik	King's College London, UK
Laura Sanità	University of Waterloo, Canada
Charles Semple	University of Canterbury, New Zealand
Blerina Sinaimeri	Claude Bernard Lyon 1 / Inria, France
Csaba D. Tóth	California State University, Northridge, USA
Luca Trevisan	Bocconi University, Italy
Ryuhei Uehara	JAIST, Japan

Organizing Committee

Federico Corò   Sapienza University of Rome, Italy
  (Publicity)
Chryssis Georgiou   University of Cyprus, Cyprus
  (Co-chair)
Anna Philippou   University of Cyprus, Cyprus
  (Co-chair)

Additional Reviewers

Bogdan Alecu	Tesshu Hanaka	Fukuhito Ooshita
Carlos Alegría	Marc Heinrich	Giacomo Ortali
Ei Ando	Adam Hesterberg	Yota Otachi
Antonios Antoniadis	Shahid Hussain	Vangelis Paschos
Nicola Apollonio	Vít Jelínek	Panagiotis Patsilinakos
Lorenzo Balzotti	Alkis Kalavasis	Daniel Paulusma
Evripidis Bampis	Martin Klazar	Christophe Picouleau
Valentin Bartier	Kim-Manuel Klein	Andrea Ribichini
René Van Bevern	Dušan Knop	Simona E. Rombo
Nicolas Boria	Grigorios Koumoutsos	Andre Rossi
Manuel Borrazzo	Martin Koutecky	Toshiki Saitoh
Laurent Bulteau	Rastislav Kralovic	Richard Santiago
Katarina Cechlarova	Dmitry Kravchenko	Jayalal Sarma
Keren Censor-Hillel	Temur Kutsia	Arseny Shur
Dimitris Christou	O-Joung Kwon	Florian Sikora
Matteo Cosmi	Pierre L'Ecuyer	Stratis Skoulakis
Andrzej Czygrinow	Michael Lampis	Eric Tannier
Fabrizio D'Amore	Isabella Lari	Alessandra Tappini
Samir Datta	Massimo Lauria	Sonia Toubaline
Riccardo Dondi	Philip Lazos	Pavel Veselý
Wolfgang Dvořák	Jean-Guy Mailly	Stéphane Vialette
Christoph Dürr	Martin Mares	Kunihiro Wasa
Till Fluschnik	Andrea Marino	Mingyu Xiao
Florent Foucaud	Ian McQuillan	Peter Zeman
Fabrizio Frati	Takaaki Mizuki	Philipp Zschoche
Claude Gravel	Roman Nedela	
Niels Grüttemeier	Jana Novotna	

Contents

x   Contents

Invited Lecture

Invited Talks

Henning Fernau[1]([⊠]), Katharina T. Huber[2], and Joseph (Seffi) Naor[3]

[1] Universität Trier, Fachbereich 4, Informatikwissenschaften, CIRT,
54286 Trier, Germany
`fernau@informatik.uni-trier.de`
[2] School of Computing Sciences, University of East Anglia, Norwich, UK
`k.huber@uea.ac.uk`
[3] Computer Science Department, Technion, 32000 Haifa, Israel
`naor@cs.technion.ac.il`
`https://people.uea.ac.uk/k_huber`

Abstract. This document contains the summaries of the invited talks that have been delivered at CIAC 2021. A detailed description of the lecture by Henning Fernau (titled *Abundant Extensions*) is on pages 1–15, while the abstracts of the lectures by Katharina T. Huber (*Phylogenetic networks, a way to cope with complex evolutionary processes*) and Joseph (Seffi) Naor (*Recent Advances in Competitive Analysis of Online Algorithms*) are on pages 16 and 17, respectively.

Abundant Extensions

by Katrin Casel, Henning Fernau, Mehdi Khosravian Ghadikolaei, Jérôme Monnot, Florian Sikora[1]

Most algorithmic techniques dealing with constructing solutions to combinatorial problems build solutions in an incremental fashion. Often, the whole procedure could be visualized by means of a search tree. To the leaves of such a search tree, we can associate solutions, while to the inner nodes, only pre-solutions can be associated. By looking at all leaves of the search tree, an optimum solution can be found if desired. In particular for reasons of speed, it is crucial to prune off potential branches of the search tree by deciding at an early stage if a certain pre-solution can be ever extended to a solution that is optimal. But, how easy is it to tell if such pruning is possible?

In this survey paper, we first present a motivating example as an introduction, followed by a general framework for extension problems. Then we show that such problems are really abundant, and these examples also prove that the complexity of these extension problems can be the same as or different from that of the original combinatorial question.

---

[1] Katrin Casel is with Hasso Plattner Institute, University of Potsdam and Universität Trier; Henning Fernau is with Universität Trier, Mehdi Khosravian Ghadikolaei, Jérôme Monnot, Florian Sikora are with Université Paris-Dauphine, PSL University.

© Springer Nature Switzerland AG 2021
T. Calamoneri and F. Corò (Eds.): CIAC 2021, LNCS 12701, pp. 3–19, 2021.
https://doi.org/10.1007/978-3-030-75242-2_1

1   Introduction

Let us start motivating our considerations by looking at one of the best-known combinatorial graph notions, namely that of a vertex cover in an undirected graph G, i.e., a set C of vertices such that each edge of G has at least one vertex from C incident to it. To be more concrete, consider the task of finding a smallest vertex cover in the graph $G = (V, E)$, where $V = \{x, y_1, y_2, y_3, y_4\} \cup \{z_{i,j} \mid 1 \le i \le 4, 1 \le j \le 2\}$. The edge set E is given by $E = \{xy_i \mid 1 \le i \le 4\} \cup \{y_i z_{i,j} \mid 1 \le i \le 4, 1 \le j \le 2\}$ (see Fig. 1). How to approach this problem? As we know that the corresponding general optimization problem MINIMUM VERTEX COVER cannot be solved in polynomial time, unless P = NP, we can follow different strategies. (Clearly, for the concrete example, we could find a smallest vertex cover even in linear time, as our example graph is a tree. We ignore this possibility in the following and use the example for illustrative purposes only.)

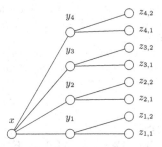

Fig. 1. Toy graph for our running example.

Design a Heuristics. A very natural greedy strategy is to recursively select a vertex of highest degree to be put in a pre-solution. As after selecting a highest-degree vertex v, all incident edges have been covered, it is clear that in the recursion, we remove v and its incident edges from the graph. In our example, this would mean that we first put x into the pre-solution and then, one after the other, y_1, y_2, y_3, and y_4 (in any order). We are left with an edge-less graph that has $\emptyset$ as a vertex cover. Hence, altogether we computed the vertex cover $\{x, y_1, y_2, y_3, y_4\}$. However, notice that this cover is not inclusion-wise minimal, as we can safely remove x from this solution to produce another (smaller) vertex cover. Incidentally, this is also a smallest vertex cover for this example.

Of course, we could walk through our solution again and check if it is inclusion-wise minimal. More precisely, whenever we detect a vertex such that all its neighbors are contained in the solution, we can remove it from the solution. Yet, it feels a bit awkward to make such amendments with hindsight, it would look more natural to avoid producing non-minimal vertex covers from the very beginning. We could try to integrate a minimality check in the greedy step itself: Pick a vertex of largest degree, unless this would mean that a previously selected vertex (put into the pre-solution) would be 'completely encircled'. In our example, this could mean that we avoid putting y_4 into the solution after having put x, y_1, y_2, y_3 there. Then, we would necessarily arrive at the vertex cover $\{x, y_1, y_2, y_3, z_{4,1}, z_{4,2}\}$, which is indeed a minimal solution, but clearly not the smallest one.

Design a Branching Algorithm. This means that we are really looking for a smallest vertex cover in any input graph. Assuming $\mathsf{P} \neq \mathsf{NP}$, we must be content with an algorithm potentially using exponential time. Again, most such algorithms would pick a vertex of highest degree and branch into two cases: to put this vertex into a pre-solution or not, and then working further on recursively. Clearly, we cannot hope for a good algorithmic solution of the question if there is a solution of optimum size (or not) in a certain branch, as this is exactly the same NP-hard question that we started out with. Therefore, we have to relax our question. A natural relaxation would be to ask if (in a certain branch) there is still an inclusion-wise minimal vertex cover. In a certain sense, this question does sound easier, because we can clearly compute an inclusion-wise minimal vertex cover of the whole graph by a sort of greedy strategy as described in the previous paragraph. However, remember that along with running our branching algorithm, we already put some vertices in a pre-solution. As we do not want to undo this decision (because we even know that the case when a certain vertex is not put into the cover is handled in a different branch), we like to know if there is a minimal vertex cover extending the currently considered pre-solution. As we will see, this type of question is indeed quite difficult to handle in general.

Approximation and Parameterized Algorithms. MINIMUM VERTEX COVER is one of the standard problems both concerning approximation and parameterized algorithms. Often, heuristics as the one discussed above can be used or tweaked to achieve provable (reasonable) approximation ratios. With our greedy idea, we have to be somewhat cautious, but there is a randomized algorithm that implements the intuition that high-degree vertices should be put with preference into the vertex cover. Namely, we can associate a probability proportional to the degree of each vertex to select it into a pre-solution in the next step of the algorithm. More formally, the probability of $v \in V$ in $G = (V, E)$ would be $p_v = \frac{|N(v)|}{2|E|}$. With our example, it is interesting to note that at the very beginning, $p_x = \frac{4}{24} = \frac{1}{6}$ is clearly highest, but $p_{y_i} = \frac{3}{24} = \frac{1}{8}$ is only slightly less, and as it does not really matter which of the y_i we pick, the probability to choose any of them is $\frac{1}{2}$, much higher than the probability of choosing x. Assuming that we chose y_4 in the first step, after the first recursion step, the probability of choosing x is still $p'_x = \frac{3}{18} = \frac{1}{6}$, but the probability to choose any of y_1, y_2, y_3 is also $\frac{1}{6}$. Developing this example further shows that we actually arrive at a smallest vertex cover with quite some probability. Conversely, the strategy that we designed for a branching algorithm would indeed be also a standard strategy in parameterized algorithmics. We only refer to the textbook [14]. As a side-note, another strategy predominant in parameterized algorithmics would work out perfectly on our example, namely the strategy of using reduction rules.

In our example, a well-known rule says to always select the neighbor vertex of a vertex of degree one. This strategy alone would solve our example to optimality. Furthermore, in particular for MINIMUM VERTEX COVER, algorithms have been developed that combine both approaches, see [8,16].

More Scenarios: Enumeration and Counting. Recently, the task of enumerating all inclusion-wise minimal vertex covers (or similar questions, related to other combinatorial problems) has been given quite some considerations in the literature. A related task is to count all inclusion-wise minimal vertex covers. We refer to [5–7,15,17,19–21,26,27,33,35]. Note that building up from only relevant pre-solutions yields the possibility to define an order on solutions and hence enumerate without repetition. It was this scenario of enumeration where the question of extension was asked for VERTEX COVER EXTENSION in [15].

Maximin and Minimax Problems. Especially greedily solving the task of finding (and also approximating) a minimal vertex cover lower-bounded by a given number k (a minimal vertex cover of maximum size) requires knowing that (at least a significant part of) the pre-solution remains in a minimal solution. In fact it is the difficulty of deciding extendability of pre-solutions that shows why all approximation strategies that work for minimum vertex cover fail to be useful for finding a minimal vertex cover of maximum size. The second part of the title of this paragraph already indicates a further definitorial problem: What should "extension" mean in the context of problems that maximize a certain value? We will answer this question below by introducing a framework based on a partial order approach, somewhat along the lines of Manlove's work [31].

On a more general note, the following question was already asked in 1956 by Kurt Gödel in a famous letter to Joh(an)n von Neumann [36]: *It would be interesting to know [. . .] how strongly in general the number of steps in finite combinatorial problems can be reduced with respect to simple exhaustive search.* The mentioned pruning of search branches and hence the question of finding (pre-)solution extensions lies at the heart of this question.

2   A General Framework of Extension Problems

In order to formally define our concept of minimal extension, we define what we call *monotone problems* which can be thought of as problems in NPO with the addition of a set of pre-solutions (which includes the set of solutions) together with a partial ordering on this new set. Formally, we define such monotone problems as 5-tuples $\Pi = (\mathcal{I}, presol, sol, \preceq, m)$ (where $\mathcal{I}, sol, m$ with an additional $goal \in \{\min, \max\}$ yields an NPO problem) defined by:

- $\mathcal{I}$ is the set of instances, recognizable in polynomial time.
- For $I \in \mathcal{I}$, $presol(I)$ is the set of *pre-solutions* and, in a reasonable representation of instances and pre-solutions, the length of the encoding of any $y \in presol(I)$ is polynomially bounded in the length of the encoding of I.
- For $I \in \mathcal{I}$, $sol(I)$ is the *set of solutions*, which is a subset of $presol(I)$.

- There exists an algorithm which, given (I, U), decides in polynomial time if $U \in presol(I)$; similarly there is an algorithm which decides in polynomial time if $U \in sol(I)$.
- For $I \in \mathcal{I}$, $\preceq$ is a partial ordering on $presol(I)$ and there exists an algorithm that, given an instance I and $U, U' \in presol(I)$, can decide in polynomial time if $U' \preceq U$.
- For each $I \in \mathcal{I}$, the set of solutions $sol(I)$ is upward closed with respect to $\preceq$, i.e., $U \in sol(I)$ implies $U' \in sol(I)$ for all $U, U' \in presol(I)$ with $U \preceq U'$.
- m is a polynomial-time computable function which maps pairs (I, U) with $I \in \mathcal{I}$ and $U \in presol(I)$ to non-negative rational numbers; $m(I, U)$ is the *value of* U.
- For $I \in \mathcal{I}$, $m(I, \cdot)$ is *monotone* with respect to $\preceq$, meaning that the property $U' \preceq U$ for some $U, U' \in presol(I)$ either always implies $m(I, U') \leq m(I, U)$ or $m(I, U') \geq m(I, U)$.

Given a monotone problem $\Pi = (\mathcal{I}, presol, sol, \preceq, m)$, we denote by $\mu(sol(I))$ the set of *minimal feasible solutions of* I, formally given by

$$\mu(sol(I)) = \{S \in sol(I) \colon ((S' \preceq S) \wedge (S' \in sol(I))) \to S' = S\}.$$

Further, given $U \in presol(I)$, we define

$$ext(I, U) = \{U' \in \mu(sol(I)) \colon U \preceq U'\}$$

to be the set of *extensions* of U. Sometimes, $ext(I, U) = \emptyset$, which makes the question of the existence of such extensions interesting. Hence, finally, the *extension problem* for Π, written EXT Π, is defined as follows: An instance of EXT Π consists of an instance $I \in \mathcal{I}$ together with some $U \in presol(I)$, and the associated decision problems asks if $ext(I, U) \neq \emptyset$.

The reader is encouraged to work through all these definitions having VERTEX COVER in mind. Here, the partial ordering is simply set inclusion.

With these formal definitions, we try to capture aspects of extension that could be used to transfer properties among different specific extension problems. The requirement that the set of solutions is upward closed with respect to the partial ordering relates to *independence systems*, see [34]. This choice also models greedy strategies that attempt to build up solutions gradually by stepwise improvements towards feasibility. Note that such greedy approaches usually do not employ steps that transform a solution back into a pre-solution that is not feasible.

Adding the function m to the formal description of a monotone problem is on the one hand reminiscent of the problem class NPO, on the other hand it also allows us to study approximate extension as follows. For a monotone problem Π, one might ask for input (I, U) not to extend exactly U (if this is not possible) but to find a pre-solution U' as close as possible to U so that (I, U') has an extension. Such optimization formulations were studied for extension versions of the vertex cover and independent set problems under the notion *price of extension* (PoE) in [11], where the partial order is the subset or superset relation.

However, this idea was not yet studied in the general scenario presented so far. Further discussions on this are contained in Sect. 4.

Further, the function m allows to discuss the parameterized complexity of extension problems, where we define the *standard parameter* for an extension problem EXT Π for a monotone problem $\Pi = (\mathcal{I}, presol, sol, \preceq, m)$ to be the value of the given pre-solution, i.e., the parameter for instance (I, U) of EXT Π is $m(I, U)$. The *dual* parameterizations as discussed in [11] and [10] to derive the results summarized in Table 1, can be modelled as follows in this framework. The dual parameter is given by the difference of the value of the given pre-solution to the maximum $m_{max}(I) := \max\{m_I(y) : y \in presol(I)\}$, so the parameter for instance (I, U) of EXT Π is $m_{max}(I) - m(I, U)$. Note that this can only be properly defined for monotone problems Π where $m_{max}(I) := \max\{m_I(I, U) : U \in presol(I)\}$ exists for all $I \in \mathcal{I}$. In this case we say that Π *admits a dual parameterization*, and observe the following.

Proposition 1. *Let* $\Pi = (\mathcal{I}, presol, sol, \preceq, m)$ *be a monotone problem that admits a dual parameterization. If, for all $I \in \mathcal{I}$ and $U \in presol(I)$, $Above(U) = \{V : V \in sol(I), U \preceq V\}$ can be enumerated in FPT-time, parameterized by $m_{max}(I) - m(I, U)$, then EXT Π with dual parameterization is in FPT.*

In order to enumerate $Above(U)$, it is often easiest to actually enumerate (if possible in FPT-time) the superset $\{V : V \in presol(I), U \preceq V\}$ instead and then check if the enumerated pre-solution is a solution, which can be done in polynomial time in our framework.

Although we strongly linked the definition of monotone problems to NPO, the corresponding extension problems do not generally belong to NP (in contrast to the canonical decision problems associated to NPO problems), as we will see in Example 5.

Still, we can prove a general upper bound as follows. Recall that $\Sigma_1^p = $ NP and that co-NP $\subseteq \Sigma_2^p$ in the usual terminology regarding the first levels of the polynomial-time hierarchy, see for example the textbook [1] for more definitions.

Proposition 2. *If Π is a monotone problem, then EXT Π can always be solved within Σ_2^p.*

One of the consequences is that we cannot expect to obtain PSPACE-hard extension problems within our framework.

Under certain circumstances, as described in the following, we can prove membership of EXT Π in $\Sigma_1^p = $ NP. To this end, consider the finer structure of the ordering $\preceq$ defined on $presol(I)$ for an instance I of Π. For $U, U' \in presol(I)$, call U' an *immediate predecessor* of U if $U' \preceq U$ and U' is a maximal element in $Below(U) = \{X \in presol(I) : X \preceq U \land X \neq U\}$, i.e., there exists no $U'' \neq U'$ with $U'' \in Below(U)$ and $U' \preceq U''$. We say that a monotone problem Π *admits polynomial enumeration of predecessors* if $Below(U)$ never contains infinite chains for any pre-solution U and if there exists a polynomial-time algorithm that, given any instance I of Π and $U \in presol(I)$, enumerates all immediate predecessors of U in polynomial time.

Proposition 3. *If Π is a monotone problem that admits polynomial enumeration of predecessors, then* EXT Π *can be solved within* $\Sigma_1^p = $ NP.

For the proofs of the previous two propositions, we refer to [12].

3   Some Examples

We now discuss quite a number of examples. This should help the reader understand how abundant extension problems are and how different they look. This section also serves as a sort of literature survey. Let us start with a problem from logic that seems to be one of the first ones that really fits into our framework.

Example 4. The question of finding extensions to minimal solutions was encountered in the context of proving hardness results for (efficient) enumeration algorithms for Boolean formulae, in the context of matroids and similar situations; see [6,27]. More precisely, it is NP-hard to decide if a pre-solution can be extended for the problem of computing prime implicants of the dual of a Boolean function; a problem which can also be seen as finding a minimal hitting set for the set of prime implicants of the input function. Interpreted in this way, the proof of [6] yields NP-hardness for an extension version of 3-HITTING SET.   □

Logical problems also allow for constructing extension problems with very specific features.

Example 5. Consider the monotone problem $\Pi_\tau = (\mathcal{I}, presol, sol, \preceq, m)$ with:

- $\mathcal{I} = \{F \colon F \text{ is a Boolean formula}\}$.
- $presol(F) = sol(F) = \{\phi \mid \phi \colon \{1, \dots, n\} \to \{0,1\}\}$ for a formula $F \in \mathcal{I}$ on n variables.
- For $\phi, \psi \in presol(F)$, $\phi \preceq \psi$ if either $\phi = \psi$, or assigning variables according to ψ satisfies F while an assignment according to ϕ does not.
- $m \equiv 1$ (plays no role for the extension problem).

The associated extension problem EXT Π_τ corresponds to the co-NP-complete problem TAUTOLOGY in the following way: Given a Boolean formula F which, w.l.o.g., is satisfied by the all-ones assignment $\psi_1 \equiv 1$, it follows that (F, ψ_1) is a yes-instance for EXT Π_τ if and only if F is a tautology, as ψ_1 is in $\mu(sol(F))$ if and only if there does not exist some $\psi_1 \neq \phi \in sol(F)$ with $\phi \preceq \psi_1$, so, by definition of the partial ordering, an assignment ϕ which does not satisfy F. Consequently, EXT Π_τ is not in NP, unless co-NP = NP.   □

Let us mention some well-known graph problems that can quite naturally be modelled as monotone problems with $\mathcal{I}$ always as the set of undirected graphs.

Example 6. Denoting instances by $G = (V, E)$, and the simple cardinality function as objective, i.e., $m(G, U) = |U|$ for all $U \in presol(G)$:

- VERTEX COVER (VC): $\preceq\, = \,\subseteq$, $presol(G) = 2^V$, $C \in sol(G)$ iff each $e \in E$ is incident to at least one $v \in C$;

- EDGE COVER (EC): $\preceq = \subseteq$, $presol(G) = 2^E$, $C \in sol(G)$ iff each $v \in V$ is incident to at least one $e \in C$;
- INDEPENDENT SET (IS): $\preceq = \supseteq$, $presol(G) = 2^V$, $S \in sol(G)$ iff $G[S]$ contains no edges;
- EDGE MATCHING (EM): $\preceq = \supseteq$, $presol(G) = 2^E$, $S \in sol(G)$ iff none of the vertices in V is incident to more than one edge in S;
- DOMINATING SET (DS): $\preceq = \subseteq$, $presol(G) = 2^V$, $D \in sol(G)$ iff $N[D] = V$;
- EDGE DOMINATING SET (EDS): $\preceq = \subseteq$, $presol(G) = 2^E$, $D \in sol(G)$ iff each edge belongs to D or is adjacent to some $e \in D$.

When studying monotone graph problems restricted to some particular graph classes, this formally means that the instance set $\mathcal{I}$ contains only graphs that fall into the graph class under consideration. We hence arrive at problems like EXT VC (or EXT IS, resp.), where the instance is specified by a graph $G = (V, E)$ and a vertex set U, and the question is if there is some *minimal* vertex cover $C \supseteq U$ (or some *maximal* independent set $I \subseteq U$). Notice that the instance (G, V) of EXT IS can be solved by the exhaustive greedy approach that, starting from $\emptyset$, gradually adds vertices and deletes their closed neighborhood. Note that this gives an independent set U that trivially satisfies $U \subset V$ so $V \preceq U$. Further, for any $w \in V \setminus U$, the set $U \cup \{w\}$ is not an independent set by the construction of U, which means that there exists no independent set $U' \neq U$ with $U \subset U'$ (i.e., $U' \preceq U$), as $\{U \cup \{w\} \mid w \in V \setminus U\}$ is the set of immediate predecessors of U. Similarly, $(G, \emptyset)$ is an easy instance of EXT VC. We will show that this impression changes for other instances.

EXT VC and EXT IS were studied in [11]. While these problems are NP-complete on planar bipartite sub-cubic graphs, they are polynomial-time decidable in chordal and in circular-arc graphs. Also, EXT VC remains NP-hard, even restricted to planar cubic graphs, see [2]. EXT DS is treated in [12] and shown to be NP-complete on planar bipartite sub-cubic graphs, also cf. [3]. In [12], it is also explained why, under the standard parameterization, we obtain a W[3]-complete problem, quite a peculiar thing in parameterized complexity theory; see the comments in [13].

In [29], Khosravian et al. studied the following extension variant of the CONNECTED VERTEX COVER problem, denoted by EXT CVC: given a connected graph $G = (V, E)$ together with a subset $U \subseteq V$ of vertices, the goal is to decide whether there exists a minimal connected vertex cover of G containing U. It is shown that EXT CVC is polynomial-time decidable in chordal graphs while it is NP-complete on bipartite graphs of maximum degree 3 even if U is an independent set.

Extension variants of three edge graph problems, namely EDGE COVER, EDGE MATCHING and EDGE DOMINATING SET, (here denoted by EXT EC, EXT EM and EXT EDS, respectively) were studied in [10]; it is shown that all these problems are NP-hard in planar bipartite graphs of maximum degree 3. Further, EXT EM is polynomial-time decidable when the forbidden edges $\overline{U} = E \setminus U$ form an induced matching.

In [11] and [10], extension variants of some classical graph problems were also studied from a parameterized complexity point of view, in particular under the standard or dual parameters. A summary of these parameterized results is presented in Table 1. Further, [11] contains some complexity lower bounds for extension problems assuming the Exponential Time Hypothesis (ETH)[2].

The paragraph after Proposition 1 shows that, with the dual parameterization, EXT DS belongs to FPT.   □

Notice that, as illustrated for EXT IS, each of the above monotone graph problems admits polynomial enumeration of predecessors. Therefore, the corresponding extension problems all lie in NP. It is instructive to have another look at the monotone problem Π_τ from Example 5 whose extension variant corresponds to TAUTOLOGY. Here, the partial order $\preceq$ on $\{1, \ldots, n\}^{\{0,1\}}$ can be also described as follows (with respect to a given Boolean formula F):

- All assignments that do not satisfy F are mutually incomparable, while
- each of them is strictly smaller (with respect to $\preceq$) than any assignments that satisfy F,
- which are again incomparable amongst themselves.

As a formula may possess exponentially many non-satisfying assignments, Π_τ does not admit polynomial enumeration of predecessors. In view of our earlier findings, this is a pre-requisite to prove co-NP-hardness of the extension variant.

In actual fact, a sort of generalization of EXT IS has been considered much before as an extension problem, as discussed in the following. Supposedly, this is really the first problem from the literature that fits into our framework.

Example 7. The extension problem EXT IND SYS (also called FLASHLIGHT)[3] for independent systems was proposed in [30]. An *independent system* is a set system $(V, \mathcal{E})$, $\mathcal{E} \subseteq 2^V$, that is hereditary under inclusion. In the extension problem EXT IND SYS, given as input $X, Y \subseteq V$, one asks for the existence of a maximal independent set including X that does not intersect with Y. Lawler et al. [30] proved that EXT IND SYS is NP-complete, even when $X = \emptyset$. In order to enumerate all (inclusion-wise) minimal dominating sets of a given graph, Kanté et al. studied a restriction of EXT IND SYS: the problem of deciding if there exists a minimal dominating set containing X (denoted by EXT DS here). They proved that EXT DS is NP-complete, even in special graph classes like split graphs, chordal graphs and line graphs [24,25]. Moreover, they proposed a linear algorithm for split graphs when X, Y is a partition of the clique part [23]. Further discussions are contained in Example 6. EXT DS also links to the following

---

[2] ETH is a conjecture asserting that there is no $2^{o(n)}$ (i.e., no sub-exponential) algorithms for solving 3-SAT, where n is the number of variables; the number of clauses is somehow subsumed into this expression, as this number can be assumed to be sub-exponential in n (after applying the famous sparsification procedure); cf. [22].

[3] Presumably, this is the origin of the naming of "flashlight" for an algorithm that tries to prune branches of a search tree when enumerating minimal sets with a certain property, e.g., minimal vertex covers.

open question in the area of enumeration algorithms: Is it possible to design an exact algorithm for UPPER DOMINATION (the task to find a minimal dominating set of maximum cardinality) that avoids enumerating all minimal dominating sets? □

So far, it might appear that every classical decision problem yields exactly one corresponding extension problem. However, different algorithmic (greedy) strategies for a classical problem result in different corresponding sets of pre-solutions and orderings, hence different extension problems. We explain this point by discussing GRAPH COLORING.[4]

Example 8. Consider for example the following two greedy strategies of finding a proper vertex coloring. Formally, vertex colorings of a graph $G = (V, E)$ are functions $c \colon V \to \{1, \ldots, k\}$ for some $k \in \mathbb{N}$, and they are proper (hence a solution to the graph coloring problem) if $c(u) \neq c(v)$ for all edges $uv \in E$. Starting from a base coloring c that assigns the same color to all vertices, formally $c \colon V \to \{1\}$, consider the following two options as greedy improvement strategies for a coloring $c \colon V \to \{1, \ldots, k\}$:

(a) Pick an index $i \in \{1, \ldots, k\}$ and a subset C_i of $\{v \in V \mid c(v) = i\}$. Define the improved coloring c' on $\{1, \ldots, k\}$ by $c'(v) = k + 1$ for all $v \in C_i$, and $c'(v) = c(v)$ for all $v \in V \setminus C_i$. *(split one color class into two)*
(b) Pick an independent set $C \subseteq V$ and define the improved coloring c' on $\{1, \ldots, k\}$ by $c'(v) = k + 1$ for all $v \in C$, and $c'(v) = c(v)$ for all $v \in V \setminus C$. *(recolor an independent set)*

These two ideas, expressed as partial orderings towards and among feasible solutions, yield the partial orderings denoted *a-chromatic* and *b-chromatic* in [31], formally defined as follows. Two colorings c_1, c_2 for a graph G satisfy $c_1 \preceq c_2$ iff c_2 uses exactly one color more than c_1, i.e., $c_1 \colon V \to \{1, \ldots, k\}$ and $c_2 \colon V \to \{1, \ldots, k+1\}$ with the following conditions:

a-chromatic there exists a color i such that $c_1(v) \neq c_2(v)$ only for v with $c_1(v) = i$ and $c_2(v) = k + 1$ (split one color into two),
b-chromatic $c_1(v) \neq c_2(v)$ only for v with $c_2(v) = k + 1$ AND the color class $k + 1$ forms an independent set (recolor an independent set).

This would lead to problems like EXT A-CHROM or EXT B-CHROM. The complexities of these extension problems seem to be completely **unexplored**. □

The reader is also referred to [31] as a rich source of other examples for instance orderings. In actually fact, all these are barely explored from the viewpoint of extension problems.

As a non-graph example for extension that fits within our framework, we will now discuss an extension version of BIN PACKING.

---

[4] The reader might have expected us talking on PRECOLORING EXTENSION and similar problems known from the literature (see, e.g., [4,32]); however, these types of extension problems do not really fit into our framework due to the lack of a suitable notion of a partial order.

Example 9. We consider as ordering the so-called partition ordering. Bin packing can be modelled as monotone problem as follows. Instances in $\mathcal{I}$ are sets $X = \{x_1, \ldots, x_n\}$ of items and a weight function w that associates rational numbers $w(x_i) \in (0,1)$ to items, $presol(X)$ contains all partitions of X, and a partition π of X is in $sol(X)$, if for each set $Y \in \pi$, $\sum_{y \in Y} w(y) \leq 1$. For two partitions π_1, π_2 of X, we define the partial ordering $\preceq$ by $\pi_1 \preceq \pi_2$ iff π_2 is a refinement of π_1, i.e., π_2 can be obtained from π_1 by splitting up its sets into a larger number of smaller sets. The traditional aim of the bin packing problem is to find a feasible π such that $|\pi|$, the number of bins, is minimized, hence we set $m(X, \pi) = |\pi|$.

Notice that $\{X\}$ is the smallest partition with respect to $\preceq$. Clearly, the set of solutions is upward closed. Now, a solution is minimal if merging any two of its sets into a single set yields a partition π such that there is some $Y \in \pi$ with $w(Y) := \sum_{y \in Y} w(y) > 1$. Aside from modeling the greedy strategy that gradually splits up bins that are too large, fixing a pre-solution can be interpreted

Table 1. Survey on parameterized complexity results for extension problems

Param.	Ext. of						
	EC	EM	EDS	IS	VC	DS	BP
Standard	FPT	FPT	W[1]-hard	FPT	W[1]-compl	W[3]-compl	para-NP
Dual	FPT	FPT	FPT	W[1]-compl	FPT	FPT (Example 6)	FPT (Example 9)

as encoding knowledge about which items should not be put together in one bin. This describes the problem EXT BP, which takes as input a set of items X with weight function w, a partition π_U of X (i.e., $\pi_U \in presol(X)$), and asks if $ext(X, \pi_U) \neq \emptyset$. In [12], it is shown that EXT BP is NP-complete, even if the pre-solution π_U contains only two sets. This also proves para-NP-hardness with respect to the standard parameter.

Dual parameterization easily yields membership in FPT by kernelization. Consider the reduction rule that, for a partition π_U of X given by sets $X_1, \ldots, X_k$, removes for all i with $X_i = \{x_i\}$ the elements x_i from X and X_i from π_U, leaving the dual parameter $k_d = n - k$ unaffected. An irreducible instance is then a partition $\pi_U = \{X_1, \ldots, X_k\}$ of X, $|X| = n$, with $|X_i| \geq 2$ and hence $2k \leq n = |X| = \sum_{i=1}^k |X_i|$, so that $k_d = n - k \geq \frac{1}{2}n$. It is known that the number of partitions of an m-element set is given by the mth Bell number, which again is upper-bounded by $\mathcal{O}(m^m)$. Hence, by simple brute-force, an instance (X, π_U) can be solved in time $\mathcal{O}^*(k_d^{k_d})$. □

Our last example comes from the area of formal languages.

Example 10. In [18], extension variants of one of the most famous combinatorial problems in automata theory, namely the SYNCHRONIZING WORD problem for deterministic finite automata (DFA), were considered. This means that we are given a deterministic finite automaton A and an input word u, and the task is to find a minimal extension w of u such that w is synchronizing A, where minimality is defined with respect to a partial order $\preceq$ on the set Σ^* of input words. A word $w \in \Sigma^*$ is *synchronizing* A if there is some state q_{sync} such that, wherever

A starts processing w, it will end in the synchronizing state q_{sync}. The complexity status of these extension problems vary with the choice of $\preceq$. For further details, we refer to the discussions in [9, 12, 18]. Let us only mention that the complexity status of this extension problem with respect to the subword ordering is **open**. In particular, no polynomial-time algorithm is known. Conversely, for the prefix- or suffix-orderings, the corresponding extension problem can be solved in polynomial time. Observe that upward closedness of the solution space follows from the fact that if $w \in \Sigma^*$ is a synchronizing word, then for any $x, y \in \Sigma^*$, xwy is also synchronizing.   □

4   Summary and Suggestions for Further Studies

In [10–12], extension variants of some classical graph problems were also studied from a parameterized complexity point of view, in particular under the standard and dual parameterizations. A summary of these parameterized results is presented in Table 1. All these problems are NP-hard, often for quite restrictive conditions, as also discussed above. Recall that the combinatorial problems underlying EXT EC and Ext EM, namely those of finding minimum edge covers or maximum matchings, are solvable in polynomial time; still, the extension variations are much harder. Example 10 shows the converse possibility, that from NP-hard combinatorial questions (finding smallest synchronizing words), we may result in polynomial-time solvable extension problems. The focus on this paper is to introduce a general framework of this type of problems and to illustrate it presenting several quite different examples. More examples are contained in [28].

Extension problems were also studied under the lens of approximability, leading to the notion *price of extension* already mentioned before. More precisely, this notion was defined differently for the subset and superset orders in [10, 11].

- For the subset order, given some $I \in \mathcal{I}$ and some $U \in presol(I)$ of a monotone problem $\Pi = (\mathcal{I}, presol, sol, \subseteq, m)$, with $m = | \cdot |$, the task is to find an $S \in \mu(sol(I))$ that maximizes $|S \cap U|$. This variation is called $\text{EXT}_{\max}\Pi$. An example would be VERTEX COVER.
- For the superset order, given some $I \in \mathcal{I}$ and some $U \in presol(I)$ of a monotone problem $\Pi = (\mathcal{I}, presol, sol, \supseteq, m)$, with $m = | \cdot |$, the task is to find an $S \in \mu(sol(I))$ that minimizes $|U| + |S \cap \overline{U}| = |U \cup S|$. This variation is called $\text{EXT}_{\min}\Pi$. An example would be INDEPENDENT SET.

One suggestion for further studies would be to define and study the *price of extension* in the general scenario. We formulate some suggestions in the sequel. Let $\Pi = (\mathcal{I}, presol, sol, \preceq, m)$ be some monotone problem. This means that two cases may happen, discussing some arbitrary $I \in \mathcal{I}$:

(a) For all $U, U' \in presol(I)$, $U' \preceq U$ implies $m(I, U') \leq m(I, U)$.
(b) For all $U, U' \in presol(I)$, $U' \preceq U$ implies $m(I, U') \geq m(I, U)$.

In case (a), we can define $\text{EXT}_{\max}\Pi$ as the optimization problem that, given $I \in \mathcal{I}$ and $U \in presol(I)$, asks to find a pre-solution $U' \in presol(I)$ with $U' \preceq U$

and $ext(I, U') \neq \emptyset$ that maximizes $m(I, U')$. Note that this exactly yields the price of extension of VERTEX COVER introduced in [11] with $\preceq = \subseteq$ and $m = |\cdot|$. To explain this notion with another concrete example, reconsider Example 10. Instances are given by a DFA A with input alphabet Σ and some word $u \in \Sigma^*$. Then, $presol(A) = \Sigma^*$ and $sol(A) = \{w \in \Sigma^* \mid w$ synchronizes $A\}$. Let $\preceq$ denote the subword ordering on Σ^*. This defines the problem EXT SYNC-SUB. Then, m is the length function on words. Now, in $\text{EXT}_{\max}$ SYNC-SUB, we ask to find a subword x of u of maximum length that is a subword of some synchronizing word w such that w has no proper subword that is synchronizing.

In case (b), we can define $\text{EXT}_{\min}\Pi$ as the optimization problem that, given $I \in \mathcal{I}$ and $U \in presol(I)$, asks to find a pre-solution $U' \in presol(I)$ with $U' \preceq U$ and $ext(I, U') \neq \emptyset$ that minimizes $m(I, U')$. This second case generalizes the price of extension for INDEPENDENT SET, with $\preceq = \supseteq$ and $m = |\cdot|$.

Alternatively, and somehow unifying both cases, we could also consider the task to find a pre-solution $U' \in presol(I)$ with $U' \preceq U$ and $ext(I, U') \neq \emptyset$ that minimizes $|m(I, U') - m(I, U)|$. This variation has the caveat that the optimization function can take the value zero, which happens especially in case $ext(I, U) \neq \emptyset$. In fact, for all examples discussed here, $|m(I, U') - m(I, U)| = 0$ and $U' \preceq U$ hold if and only if $U = U'$ which means that the optimization function takes value zero if and only if (I, U) is a yes-instance of the extension problem.

Let us mention another research direction connected to what we defined above: one could also consider the natural parameterizations of all variants of optimization problems. For instance, in the last case, this parameterized decision problem (with natural parameter k) would read as follows, referring to a monotone problem $\Pi = (\mathcal{I}, presol, sol, \preceq, m)$:
Given $I \in \mathcal{I}$, $U \in presol(I)$ and $k \in \mathbb{N}$, does there exists a pre-solution $U' \in presol(I)$ with $U' \preceq U$ and $ext(I, U') \neq \emptyset$ such that $|m(I, U') - m(I, U)| \leq k$?

Especially considering the hardness often already encountered for $k = 0$, one could (alternatively) consider $|U|$ or its dual as a parameter for this problem, possibly also touching the idea of parameterized approximation.

This whole area of optimization variants to extension problems is widely open. Yet, there is so much more to explore in this area. Probably, we only saw the peak of an iceberg so far. To mention one more research direction: Relatively little is known about extension problems when restricting the range of instances. For instance, what about extensions of graph problems restricted to certain graph classes? Often, some kind of hardness results are derived when looking at the corresponding enumeration problem; we only refer to [19,20,23,25] and the discussions on dominating set in [28]. Yet, a more systematic discussion of this topic is lacking. We finally recall that several concrete open questions have been mentioned throughout this paper. We hope that this serves as an appetizer to extension problems.

References

1. Arora, S., Barak, B.: Computational Complexity - A Modern Approach. Cambridge University Press, Cambridge (2009)

2. Bazgan, C., Brankovic, L., Casel, K., Fernau, H.: On the complexity landscape of the domination chain. In: Govindarajan, S., Maheshwari, A. (eds.) CALDAM 2016. LNCS, vol. 9602, pp. 61–72. Springer, Cham (2016). https://doi.org/10.1007/978-3-319-29221-2_6

3. Bazgan, C., et al.: The many facets of upper domination. Theoret. Comput. Sci. **717**, 2–25 (2018)

4. Biró, M., Hujter, M., Tuza, Z.: Precoloring extension. I. Interval graphs. Discrete Math. **100**(1–3), 267–279 (1992)

5. Boros, E., Elbassioni, K., Gurvich, V., Khachiyan, L., Makino, K.: On generating all minimal integer solutions for a monotone system of linear inequalities. In: Orejas, F., Spirakis, P.G., van Leeuwen, J. (eds.) ICALP 2001. LNCS, vol. 2076, pp. 92–103. Springer, Heidelberg (2001). https://doi.org/10.1007/3-540-48224-5_8

6. Boros, E., Gurvich, V., Hammer, P.L.: Dual subimplicants of positive Boolean functions. Optim. Methods Softw. **10**(2), 147–156 (1998)

7. Boros, E., Gurvich, V., Khachiyan, L., Makino, K.: Dual-bounded generating problems: partial and multiple transversals of a hypergraph. SIAM J. Comput. **30**(6), 2036–2050 (2000)

8. Brankovic, L., Fernau, H.: A novel parameterised approximation algorithm for minimum vertex cover. Theoret. Comput. Sci. **511**, 85–108 (2013)

9. Bruchertseifer, J., Fernau, H.: Synchronizing words and monoid factorization: a parameterized perspective. In: Chen, J., Feng, Q., Xu, J. (eds.) TAMC 2020. LNCS, vol. 12337, pp. 352–364. Springer, Cham (2020). https://doi.org/10.1007/978-3-030-59267-7_30

10. Casel, K., Fernau, H., Khosravian Ghadikolaei, M., Monnot, J., Sikora, F.: Extension of some edge graph problems: standard and parameterized complexity. In: Gąsieniec, L.A., Jansson, J., Levcopoulos, C. (eds.) FCT 2019. LNCS, vol. 11651, pp. 185–200. Springer, Cham (2019). https://doi.org/10.1007/978-3-030-25027-0_13

11. Casel, K., Fernau, H., Ghadikoalei, M.K., Monnot, J., Sikora, F.: Extension of vertex cover and independent set in some classes of graphs. In: Heggernes, P. (ed.) CIAC 2019. LNCS, vol. 11485, pp. 124–136. Springer, Cham (2019). https://doi.org/10.1007/978-3-030-17402-6_11

12. K. Casel, H. Fernau, M. Khosravian G., J. Monnot, and F. Sikora. On the complexity of solution extension of optimization problems. Manuscript submitted to Theoretical Computer Science (2020)

13. Chen, J., Zhang, F.: On product covering in 3-tier supply chain models: natural complete problems for W[3] and W[4]. Theoret. Comput. Sci. **363**(3), 278–288 (2006)

14. Cygan, M., Fomin, F.V., Kowalik, L., Lokshtanov, D., Marx, D., Pilipczuk, M., Pilipczuk, M., Saurabh, S.: Lower bounds for kernelization. Parameterized Algorithms, pp. 523–555. Springer, Cham (2015). https://doi.org/10.1007/978-3-319-21275-3_15

15. Damaschke, P.: Parameterized enumeration, transversals, and imperfect phylogeny reconstruction. Theoret. Comput. Sci. **351**(3), 337–350 (2006)

16. Fellows, M.R., Kulik, A., Rosamond, F.A., Shachnai, H.: Parameterized approximation via fidelity preserving transformations. J. Comput. Syst. Sci. **93**, 30–40 (2018)

17. Fernau, H.: On parameterized enumeration. In: Ibarra, O.H., Zhang, L. (eds.) COCOON 2002. LNCS, vol. 2387, pp. 564–573. Springer, Heidelberg (2002). https://doi.org/10.1007/3-540-45655-4_60

18. Fernau, H., Hoffmann, S.: Extensions to minimal synchronizing words. J. Automata Lang. Comb. **24**, 287–307 (2019)
19. Golovach, P.A., Heggernes, P., Kanté, M.M., Kratsch, D., Villanger, Y.: Enumerating minimal dominating sets in chordal bipartite graphs. Discrete Appl. Math. **199**, 30–36 (2016)
20. Golovach, P.A., Heggernes, P., Kanté, M.M., Kratsch, D., Villanger, Y.: Minimal dominating sets in interval graphs and trees. Discrete Appl. Math. **216**, 162–170 (2017)
21. Golovach, P.A., Heggernes, P., Kratsch, D., Villanger, Y.: An incremental polynomial time algorithm to enumerate all minimal edge dominating sets. Algorithmica **72**(3), 836–859 (2015)
22. Impagliazzo, R., Paturi, R., Zane, F.: Which problems have strongly exponential complexity? J. Comput. Syst. Sci. **63**(4), 512–530 (2001)
23. Kanté, M.M., Limouzy, V., Mary, A., Nourine, L.: On the enumeration of minimal dominating sets and related notions. SIAM J. Discrete Math. **28**(4), 1916–1929 (2014)
24. Kanté, M.M., Limouzy, V., Mary, A., Nourine, L., Uno, T.: Polynomial delay algorithm for listing minimal edge dominating sets in graphs. In: Dehne, F., Sack, J.-R., Stege, U. (eds.) WADS 2015. LNCS, vol. 9214, pp. 446–457. Springer, Cham (2015). https://doi.org/10.1007/978-3-319-21840-3_37
25. Kanté, M.M., Limouzy, V., Mary, A., Nourine, L., Uno, T.: A polynomial delay algorithm for enumerating minimal dominating sets in chordal graphs. In: Mayr, E.W. (ed.) WG 2015. LNCS, vol. 9224, pp. 138–153. Springer, Heidelberg (2016). https://doi.org/10.1007/978-3-662-53174-7_11
26. Khachiyan, L., Boros, E., Elbassioni, K.M., Gurvich, V.: On enumerating minimal dicuts and strongly connected subgraphs. Algorithmica **50**(1), 159–172 (2008)
27. Khachiyan, L.G., Boros, E., Elbassioni, K.M., Gurvich, V., Makino, K.: On the complexity of some enumeration problems for matroids. SIAM J. Discrete Math. **19**(4), 966–984 (2005)
28. Khosravian Ghadikolaei, M.: Extension of NP Optimization Problems. Ph.D. thesis, Université Paris Dauphine, France, July 2019
29. Khosravian Ghadikoalei, M., Melissinos, N., Monnot, J., Pagourtzis, A.: Extension and its price for the connected vertex cover problem. In: Colbourn, C.J., Grossi, R., Pisanti, N. (eds.) IWOCA 2019. LNCS, vol. 11638, pp. 315–326. Springer, Cham (2019). https://doi.org/10.1007/978-3-030-25005-8_26
30. Lawler, E.L., Lenstra, J.K., Kan, A.H.G.R.: Generating all maximal independent sets: NP-hardness and polynomial-time algorithms. SIAM J. Comput. **9**, 558–565 (1980)
31. Manlove, D.F.: Minimaximal and maximinimal optimisation problems: a partial order-based approach. Ph.D. thesis, University of Glasgow, Computing Science (1998)
32. Marx, D.: NP-completeness of list coloring and precoloring extension on the edges of planar graphs. J. Graph Theory **49**(4), 313–324 (2005)
33. Moon, J.W., Moser, L.: On cliques in graphs. Israel J. Math. **3**, 23–28 (1965)
34. Schrijver, A.: Combinatorial Optimization: Polyhedra and Efficiency. Springer, Heidelberg (2003)
35. Schwikowski, B., Speckenmeyer, E.: On enumerating all minimal solutions of feedback problems. Discrete Appl. Math. **117**, 253–265 (2002)
36. Sipser, M.: The history and status of the P versus NP question. In: Kosaraju, S.R., Fellows, M., Wigderson, M.A., Ellis, J.A. (eds.) Proceedings of the 24th Annual ACM Symposium on Theory of Computing, STOC, pp. 603–618. ACM (1992)

Phylogenetic Networks, A Way to Cope with Complex Evolutionary Processes

by Katharina T. Huber[5]

Understanding how pathogens such as Covid-19, birdflu, or ash dieback might have arisen is among some of the most challenging scientific questions of today. Phylogenetics is a burgeoning area at the interface of Computer Science, Mathematics, Statistics, Evolutionary Biology and also Medicine concerned with developing powerful algorithms and mathematical methodology to help with this. Going back to at least the beginning of the 19th century, treelike structures (now formalized as phylogenetic trees) have been used to visualize and model the evolution of a set of organisms of interest. Similar to a genealogy, such a tree is a certain rooted or unrooted graph-theoretical tree whose leaf set is the set of organisms of interest. In the rooted case, the unique root represents the last common ancestor of the organisms under consideration and the interior vertices correspond to hypothetical speciation events.

Growing evidence from the tsunami-like amounts of data generated by modern sequencing technologies however suggests that for certain organisms the model of a phylogenetic tree might be too simplistic to explain their complex evolutionary past (e.g. recombination in viruses or hybridization in plants). This has led to the introduction of phylogenetic networks as a tool to model and visualise evolutionary relationships between organisms. Introduced in rooted and unrooted form, these graphs naturally generalize phylogenetic trees in terms of a rooted directed acyclic graph (rooted case) or as a splits graph (unrooted case).

Although deep algorithmic and mathematical results concerning phylogenetic networks have been established over the years, numerous questions (including some very fundamental ones) have remained open so far. These include

(i) What kind of data do we require to be able to uniquely reconstruct the evolutionary scenario that gave rise to it?

(ii) How can we combine potentially conflicting gene trees (i.e. phylogenetic trees supported by a gene or a genomic region) into an overall evolutionary scenario for a set of organisms of interest?

(iii) How many potential phylogenetic networks can a set of organisms of interest support and what can we say about their space of phylogenetic networks?

(iv) How are rooted and unrooted phylogenetic networks related?

In this talk, we first give a brief introduction to phylogenetics in general and phylogenetic networks in particular and then discuss recent developments regarding some of the questions above. This will also include pointing out potential further directions of research.

---

[5] Katharina T. Huber is with School of Computing Sciences, University of East Anglia, UK.

Recent Advances in Competitive Analysis of Online Algorithms

by Joseph (Seffi) Naor[6]

This talk will survey recent advances in competitive analysis of online algorithms. I will discuss recent work on deriving online algorithms for several problems from Bregman projections and its connections to previous work on online primal-dual algorithms. A primal-dual approach to the k-taxi problem, a generalization of the k-server problem, will be discussed, as well as non-standard caching models such as writeback-aware caching.

---

[6] Joseph (Seffi) Naor is with Computer Science Dept., Technion, Haifa 32000, Israel.

Contributed Papers

Three Problems on Well-Partitioned Chordal Graphs

Jungho Ahn[1,2] , Lars Jaffke[3(✉)] , O-joung Kwon[2,4] , and Paloma T. Lima[3]

[1] Department of Mathematical Sciences, KAIST, Daejeon, South Korea
junghoahn@kaist.ac.kr
[2] Discrete Mathematics Group, IBS, Daejeon, South Korea
[3] Department of Informatics, University of Bergen, Bergen, Norway
{lars.jaffke,paloma.lima}@uib.no
[4] Department of Mathematics, Incheon National University, Incheon, South Korea
ojoungkwon@inu.ac.kr

Abstract. In this work, we solve three problems on well-partitioned chordal graphs. First, we show that every connected (resp., 2-connected) well-partitioned chordal graph has a vertex that intersects all longest paths (resp., longest cycles). It is an open problem [Balister et al., Comb. Probab. Comput. 2004] whether the same holds for chordal graphs. Similarly, we show that every connected well-partitioned chordal graph admits a (polynomial-time constructible) tree 3-spanner, while the complexity status of the TREE 3-SPANNER problem remains open on chordal graphs [Brandstädt et al., Theor. Comput. Sci. 2004]. Finally, we show that the problem of finding a minimum-size geodetic set is polynomial-time solvable on well-partitioned chordal graphs. This is the first example of a problem that is NP-hard on chordal graphs and polynomial-time solvable on well-partitioned chordal graphs. Altogether, these results reinforce the significance of this recently defined graph class as a tool to tackle problems that are hard or unsolved on chordal graphs.

Keywords: well-partitioned chordal graph · graph class · longest path transversal · tree spanner · geodetic set

1 Introduction

In this work, we deepen the structural and algorithmic understanding of the recently introduced class of well-partitioned chordal graphs [1]. This subclass of chordal graphs generalizes split graphs in two ways. Split graphs can be viewed as graphs whose vertices can be partitioned into cliques that are arranged in a star structure, the leaves of which are of size one. Well-partitioned chordal graphs are graphs whose vertex set can be partitioned into cliques that can be arranged in a tree structure, without any limitations on the size of any clique.

J. A. and O. K. are supported by the Institute for Basic Science (IBS-R029-C1). O. K. is also supported by the National Research Foundation of Korea (NRF) grant funded by the Ministry of Education (No. NRF-2018R1D1A1B07050294).

T. Calamoneri and F. Corò (Eds.): CIAC 2021, LNCS 12701, pp. 23–36, 2021.
https://doi.org/10.1007/978-3-030-75242-2_2

The star-like structure of split graphs is fairly restricted compared to the tree-like structure of chordal graphs. Questions in structural or algorithmic graph theory which are difficult to answer on chordal graphs may have an easy solution on split graphs thanks to their restricted structure. A natural path to a resolution of such questions on chordal graphs is to extend their solutions on split graphs to graph classes that are structurally closer to chordal graphs. Well-partitioned chordal graphs exhibit a tree-like structure, which makes them a natural target in such a scenario. We consider two such questions: We show that every well-partitioned chordal graph has a vertex that intersects all its longest paths (or cycles), while the corresponding question on chordal graphs has remained an open problem [4]. We also show that every well-partitioned chordal graph has a polynomial-time constructible tree 3-spanner, while the complexity of the TREE 3-SPANNER problem remains unresolved on chordal graphs [5]. We discuss these problems in more detail below.

There are several examples of algorithmic problems in the literature that are efficiently solvable in split graphs but hard on chordal graphs, see [1] and the references therein. In such cases it is worthwhile to narrow down the complexity gap between split and chordal graphs, especially due to the structural difference between the two classes. For several variants of vertex-coloring problems that are NP-hard on chordal graphs and polynomial-time solvable on split graphs, it was observed [1] that they remain NP-hard on well-partitioned chordal graphs. However, there was no example of such a problem that becomes polynomial-time solvable on well-partitioned chordal graphs. We give the first such example by showing that there is a polynomial-time algorithm that given a well-partitioned chordal graph, constructs a minimum-size geodetic set. This problem is known to be NP-hard on chordal graphs [15].

Transversals of Longest Paths and Cycles. It is well-known that in a connected graph, every two longest paths always share a common vertex. In 1966, Gallai [18] asked whether every graph contains a vertex that belongs to all of its longest paths. This question, whose answer is already known to be negative in general [33,34], was shown to have a positive answer on several well-known graph classes. It is not difficult to see that it holds for trees, and it has been shown for outerplanar graphs and 2-trees [31], which has later been generalized to series-parallel graphs, or equivalently, graphs of treewidth at most 2 [13]. (Interestingly, the couterexample for general graphs [33] has treewidth 3.) Besides that, Gallai's question has a positive answer on circular arc graphs [4,22], P_4-sparse (which includes cographs) and $(P_5, K_{1,3})$-free graphs [10], dually chordal graphs [21], and $2K_2$-free graphs [19]. As alluded to above, it has a positive answer on split graphs [24], and this result has been generalized to starlike graphs [10]. Both split graphs and starlike graphs are subclasses of well-partitioned chordal graphs [1]. It remains a challenging open problem to determine whether all chordal graphs admit a longest path transversal of size one. As a step in the direction of answering this question for chordal graphs, we prove the following theorem.

Theorem 1. *Every connected well-partitioned chordal graph contains a vertex that intersects all its longest paths.*

A closely related question is whether a 2-*connected* graph has a vertex that intersects all its longest *cycles*. This question has also been studied extensively on graph classes, and several of the above mentioned references contain positive answers to this question on the corresponding graph classes. In some cases the results are not stated explicitly, but it is not too difficult to adapt the proofs for the case of longest paths to the case of longest cycles. We answer this question positively on 2-connected well-partitioned chordal graphs as well.

Theorem 2. *Every 2-connected well-partitioned chordal graph contains a vertex that intersects all its longest cycles.*

Tree 3-Spanner. For a connected graph G and a positive integer t, a spanning tree T of G is a *tree t-spanner* of G if for every pair (v, w) of vertices in G, $\text{dist}_G(v, w) \leq t \cdot \text{dist}_T(v, w)$, where $\text{dist}_G(v, w)$ (resp., $\text{dist}_T(v, w)$) denotes the length of shortest path in G (resp., T) from v to w. The TREE t-SPANNER problem asks whether a given graph G has a tree t-spanner. Tree t-spanners are motivated from applications including network research and computational geometry [2,25]. Cai and Corneil [9] showed that TREE t-SPANNER is linear-time solvable if $t \leq 2$, and is NP-complete if $t \geq 4$. For $t = 3$, the complexity of TREE 3-SPANNER is not yet unveiled. Brandstädt et al. [5] investigated the complexity of TREE t-SPANNER on chordal graphs of small diameter. They showed that for even $t \geq 4$ (resp., odd $t \geq 5$) it is NP-complete to decide if a chordal graph of diameter at most $t + 1$ (resp., $t + 2$) has a tree t-spanner. On the other hand, for any even t (resp., odd t), every chordal graph of diameter at most $t - 1$ (resp., $t - 2$) admits a tree t-spanner which can be found in linear time. Brandstädt et al. [5] also showed that TREE 3-SPANNER is polynomial-time solvable on chordal graphs of diameter at most 2. On general chordal graphs, the complexity of TREE 3-SPANNER is still open. Several subclasses of chordal graphs, such as split [32], very strongly chordal [5], and interval [26] graphs were shown to be *tree 3-spanner admissible*, meaning that each of its members admits a tree 3-spanner. In the above mentioned cases, such tree 3-spanners can always be computed in polynomial time. We show that the same holds for well-partitioned chordal graphs, generalizing the result for split graphs [32].

Theorem 3. *Every connected well-partitioned chordal graph admits a tree 3-spanner which can be constructed in polynomial time.*

A subclass of chordal graphs that is not tree 3-spanner admissible and yet has a polynomial-time algorithm for TREE 3-SPANNER is that of 2-sep chordal graphs, as shown by Das and Panda [28]. Other (non-chordal) graph classes that are known to be tree 3-spanner admissible are bipartite ATE-free graphs [6] (which include convex graphs) and permutation graphs [26]; and there are polynomial-time algorithms for TREE 3-SPANNER on cographs and co-bipartite graphs [8], as well as planar graphs [17].

Geodetic Sets. Given a graph G and a vertex set $S \subseteq V(G)$, the *geodetic closure of S* is the set of vertices that lie on a shortest path between a pair of distinct vertices in S. Such a set S is called a *geodetic set* if the geodetic closure of S is the entire vertex set of G. The GEODETIC SET problem asks, given a graph G, for the smallest size of any geodetic set in G. The study of geodetic sets was initiated by Harary et al. [20] in 1986, and is related to convexity measures in graphs; we refer to [29] for an overview. Harary et al. [20] showed that the GEODETIC SET problem is NP-hard on general graphs, see also [3]. Dourado et al. [15] showed that GEODETIC SET remains NP-hard on chordal graphs, and that it is polynomial-time solvable on split graphs. We extend their ideas to give a polynomial-time algorithm for well-partitioned chordal graphs.

Theorem 4. *There is a polynomial-time algorithm that given a well-partitioned chordal graph G, computes a minimum-size geodetic set of G.*

The complexity of GEODETIC SET has been deeply studied on graph classes. Besides the above mentioned results, it was shown to be NP-hard on chordal bipartite [15] and bipartite [14] graphs, as well as co-bipartite [16], subcubic [7], and planar graphs [12]. Very recently, Chakraborty et al. [11] showed NP-hardness on subcubic partial grids, which unifies hardness on subcubic, planar, and bipartite graphs. Interestingly, they showed that GEODETIC SET is NP-hard even on interval graphs, while a polynomial-time algorithm for *proper* interval graphs is known due to Ekim et al. [16]. Other graph classes that are known to admit polynomial-time algorithms are cographs [15], outerplanar graphs [27], block-cactus graphs [16], and solid grid graphs [11]. Kellerhals and Koana [23] recently assessed the parameterized complexity of GEODETIC SET, and proved it to be W[1]-hard parameterized by solution size plus pathwidth and feedback vertex set, while devising FPT-algorithms for the parameter feedback edge set as well as for tree-depth.

2 Preliminaries

In this paper, all graphs are simple and finite. For graphs G and H, let $G \cup H := (V(G) \cup V(H), E(G) \cup E(H))$. For a vertex v of G, let $N_G(v)$ be the set of *neighbors* of v in G, that is, $N_G(v) := \{w \in V(G) | vw \in E(G)\}$, and $N_G[v] := N_G(v) \cup \{v\}$. For a vertex set X of G, let $N_G(X) := \bigcup_{v \in X}(N_G(v) \setminus X)$, and $N_G[X] := N_G(X) \cup X$. We may omit the subscript G if it is clear what is the base graph.

For a vertex set X of G, the *subgraph induced by X*, denoted by $G[X]$, is a graph $(X, \{vw \in E(G) | v, w \in X\})$. We let $G - X := G[V(G) \setminus X]$, and if X consists of a singleton v, then we use the shorthand $G - v$ instead of $G - \{v\}$. For vertex sets X and Y of G, we denote by $G[X, Y]$ the graph $(X \cup Y, \{xy \in E(G) \mid x \in X, y \in Y\})$. For disjoint vertex sets X and Y of G, we say that X is *complete* to Y if each vertex in X is adjacent to every vertex in Y.

A graph G is *connected* if for every nonempty proper subset $X \subsetneq V(G)$, there are vertices $x \in X$ and $y \in V(G) \setminus X$ such that $xy \in E(G)$, and *disconnected*, otherwise. A *component* of G is a maximal connected subgraph of G, that is, an induced subgraph G' of G such that for any vertex $v \in V(G) \setminus V(G')$, $G[V(G') \cup \{v\}]$ is disconnected. A graph G is *2-connected* if G is connected and has no vertex v such that $G - v$ is disconnected. A *hole* in a graph G is an induced subgraph of G isomorphic to a cycle of length at least 4. A graph is *chordal* if it has no holes.

Throughout this work, proofs of statements marked with '♣' are deferred to the full version.

Well-Partitioned Chordal Graphs

Ahn et al. [1] introduced the class of *well-partitioned chordal graphs*, which is a subclass of chordal graphs. A connected graph G is *well-partitioned chordal* if $V(G)$ admits a partition $\mathcal{P}$ and a tree $\mathcal{T}$ having $\mathcal{P}$ as a vertex set satisfying the following conditions.

(i) Each partite set $X \in \mathcal{P}$ is a clique in G.
(ii) For each edge XY of $\mathcal{T}$, there are subsets $X' \subseteq X$ and $Y' \subseteq Y$ such that

$$E(G[X, Y]) = \{xy \mid x \in X', y \in Y'\}.$$

(iii) For each pair of distinct nodes X, Y in $\mathcal{T}$ with $XY \notin E(\mathcal{T})$, $E(G[X, Y]) = \emptyset$.

We call the tree $\mathcal{T}$ a *partition tree* of G, and the elements in $\mathcal{P}$ the bags of G. A graph is well-partitioned chordal if each of its components is well-partitioned chordal. Given a partition tree $\mathcal{T}$ of a connected well-partitioned chordal graph G and distinct nodes X and Y of $\mathcal{T}$, the *boundary of X with respect to Y*, denoted by $\mathrm{bd}(X, Y)$, is the set of vertices in X having neighbors in Y. Namely, $\mathrm{bd}(X, Y) := \{x \in X | N_G(x) \cap Y \neq \emptyset\}$. Note that by the second item of the above definition, $\mathrm{bd}(X, Y)$ and $\mathrm{bd}(Y, X)$ are complete to each other.

Theorem 5 (Ahn et al. [1]). *Given a graph G, in polynomial time, one can either determine that G is not well-partitioned chordal, or find a partition tree for each component of G.*

3   Transversals of Longest Paths and Cycles

In this section, we show that well-partitioned chordal graphs admit both longest path transversals and longest cycle transversals of size one. We start with the following lemma, the proof of which exploits the Helly property of subtrees of a tree to show the existence of a bag of the partition tree that intersects all longest paths of a well-partitioned chordal graph. The same proof strategy has been used by Rautenbach and Sereni [30] to show that for any graph G, there exists a set of size $\mathbf{tw}(G) + 1$ that intersects all the longest paths of G.

Lemma 1. *Let G be a well-partitioned chordal graph with partition tree $\mathcal{T}$. Then there exists $X \in V(\mathcal{T})$ such that every longest path of G contains a vertex of X.*

Proof. Let $P_1, \ldots, P_\ell$ be the longest paths of G. Since G is connected, for each $1 \leq i \leq \ell$, the set of bags of $\mathcal{T}$ containing at least one vertex from P_i forms a subtree of $\mathcal{T}$. Let T_i be such a subtree. Since in any connected graph every two longest paths have a vertex in common, we have that $V(T_i) \cap V(T_j) \neq \emptyset$ for every $i \neq j$. By the Helly property[1] of subtrees of a tree, there exists $X \in V(\mathcal{T})$ such that $X \in V(T_i)$ for every $1 \leq i \leq \ell$. That is, X is a bag of $\mathcal{T}$ that intersects every longest path of G. □

We prove a similar lemma for longest cycles of 2-connected well-partitioned chordal graphs. The proof of this lemma follows the same lines as the one presented above, hence we omit it here.

Lemma 2. *Let G be a 2-connected well-partitioned chordal graph with partition tree $\mathcal{T}$. Then there exists $X \in V(\mathcal{T})$ such that every longest cycle of G contains a vertex of X.*

Restatement of Theorem 1. *Every connected well-partitioned chordal graph has a vertex that intersects all its longest paths.*

Proof. Let $P_1, \ldots, P_\ell$ be the longest paths of G. By Lemma 1, there exists a bag $B \in V(\mathcal{T})$ such that $V(P_i) \cap B \neq \emptyset$ for every i. Let $B_1, \ldots, B_k$ be the neighbors of B in $\mathcal{T}$. We define $\mathcal{T}_i$ to be the connected component of $\mathcal{T} - B$ containing B_i and G_i to be the subgraph of G induced by the vertices contained in the bags of $\mathcal{T}_i$. Let p_i be the length of a longest path in G_i with one endpoint in $\mathrm{bd}(B_i, B)$. We may assume without loss of generality that $p_1 \geq p_i$ for every $i > 1$. We will now show that every longest path of G contains all the vertices of $\mathrm{bd}(B, B_1)$. Let P be a longest path of G and suppose for a contradiction that there exists $v \in \mathrm{bd}(B, B_1)$ such that $v \notin V(P)$. Recall that $V(P) \cap B \neq \emptyset$. If there exists $x, y \in B$ such that $xy \in E(P)$, then we can obtain a path longer than P by inserting v between x and y in P, a contradiction with the fact that P is a longest path of G. Similarly, no endpoint of P belongs to B, otherwise we would also find a path longer than P in G. The same holds also if there exists $x \in \mathrm{bd}(B, B_1)$ and $y \in \mathrm{bd}(B_1, B)$ such that $xy \in E(P)$. Indeed, since $\mathrm{bd}(B, B_1) \cup \mathrm{bd}(B_1, B)$ is a clique, we would again find a path longer than P by inserting v between x and y in P. Therefore P contains no edge crossing from B to B_1, which implies that $V(P) \cap V(G_1) = \emptyset$. Let $P = x_1 x_2 \ldots x_t$ and let x_j be a vertex of $V(P) \cap B$ such that for every $i \geq 1$ we have $x_{j+i} \notin B$. Such a vertex exists since $x_t \notin B$. Assume without loss of generality that $x_{j+1} \in \mathrm{bd}(B_j, B)$. Note that $x_{j+1} x_{j+2} \ldots x_t$ is a path in G_j with an endpoint in $\mathrm{bd}(B_j, B)$. Hence the length of this path is at most p_1. Let $P' = x_1 x_2 \ldots x_j$ and P'' be a longest path in G_1 with an endpoint in $\mathrm{bd}(B_1, B)$. Then $P' \cdot v \cdot P''$ is a path in G that is longer than P, a contradiction. □

With a more careful argument, we can prove the analogous result for longest cycles.

---

[1] The Helly property of trees states that in every tree, every collection of pairwise intersecting subtrees has a common nonempty intersection.

Restatement of Theorem 2. *Every 2-connected well-partitioned chordal graph has a vertex that intersects all its longest cycles.*

Proof. We start as in the proof of Theorem 1. Let $C_1, \dots, C_\ell$ be the longest cycles of G. By Lemma 2, there is a bag $B \in V(\mathcal{T})$ such that $V(C_i) \cap B \neq \emptyset$ for every i. Note that we can assume B is not a leaf of $\mathcal{T}$, since if all the longest cycles intersect a bag that is a leaf, they also intersect the bag that is the neighbor of such a leaf. Let $B_1, \dots, B_k$ be the neighbors of B in $\mathcal{T}$. We define $\mathcal{T}_i$ to be a maximal subtree of $\mathcal{T}$ containing B_i and not containing B and G_i to be the subgraph of G induced by the vertices contained in the bags of $\mathcal{T}_i$.

Now, let p_i be the length of a longest path in G_i with *both* endpoints in $\mathrm{bd}(B_i, B)$. Note that this is well-defined, since $|\mathrm{bd}(B_i, B)| \geq 2$ for every i, as G is a 2-connected graph. We may assume without loss of generality that $p_1 \geq p_i$ for every $i > 1$. We will now show that every longest cycle of G contains all the vertices of $\mathrm{bd}(B, B_1)$. Let C be a longest cycle of G and suppose for a contradiction that there exists $v \in \mathrm{bd}(B, B_1)$ such that $v \notin V(C)$. We first point out the following.

Claim 1. $|V(C) \cap B| \geq 2$.

Proof. We already know that $|V(C) \cap B| \geq 1$. Suppose for a contradiction that $|V(C) \cap B| = 1$. Then there exists $x_1, x_2, x_3 \in V(C)$ such that x_1, x_2, and x_3 appear consecutively in the cycle, and $x_2 \in B$ and $x_1, x_3 \notin B$. In particular, x_2 belongs to the boundary between B and some neighboring bag B_i, and $x_1, x_3 \in \mathrm{bd}(B_i, B)$. Since G is 2-connected, there exists $u \in \mathrm{bd}(B, B_i)$, with $u \neq x_2$, such that $u \notin V(C)$. Thus, we can add u between x_2 and x_3 in C and obtain a cycle longer than C, a contradiction.

If there exists $x, y \in B$ such that $xy \in E(C)$, then we can obtain a cycle longer than C by inserting v between x and y in C, a contradiction with the fact that C is a longest cycle of G. The same holds if there exists $x \in \mathrm{bd}(B, B_1)$ and $y \in \mathrm{bd}(B_1, B)$ such that $xy \in E(C)$. Indeed, since $\mathrm{bd}(B, B_1) \cup \mathrm{bd}(B_1, B)$ is a clique, we would again find a cycle longer than C by inserting v between x and y in C. Therefore C contains no edge crossing from B to B_1, which implies that $V(C) \cap V(G_1) = \emptyset$. Consider $u \in \mathrm{bd}(B, B_1)$ such that $u \neq v$. We consider two cases.

If $u \in V(C)$, since C cannot have two consecutive vertices in B, then there exists $i \neq 1$ such that $u \in \mathrm{bd}(B, B_i)$, and there exists $u' \in \mathrm{bd}(B_i, B)$ such that $uu' \in E(C)$. Moreover, by the above claim, there exists $u'' \in V(C) \cap \mathrm{bd}(B, B_i)$ such that if P is the subpath of C starting in u, ending in u'' and containing u', then $(V(P) \setminus \{u, u''\}) \subseteq V(G_i)$. Note also that $|P| \leq p_i + 2$, since the neighbors of u and u'' in P belong to $\mathrm{bd}(B_i, B)$. Let P_1 be a longest path of G_1 with both endpoints in $\mathrm{bd}(B_1, B)$ and let $P' = u \cdot P_1 \cdot vu''$. Let C' be the cycle obtained from C by replacing P by P'. Since $|P'| = p_1 + 3$ and $p_1 \geq p_i$, we have that C' is a cycle longer than C, a contradiction.

Now we consider the case in which $u \notin V(C)$. Recall that C cannot have two consecutive vertices in B. By Claim 1, there exists $i \neq 1$ such that

$V(C) \cap V(G_i) \neq \emptyset$. Let $x, x', y, y' \in V(C)$ be such that $x, y \in \mathrm{bd}(B, B_i)$, $x', y' \in \mathrm{bd}(B_i, B)$, $xx', yy' \in E(C)$ and the subpath P of C starting in x, ending in y and containing x' and y' is such that $(V(P) \setminus \{x, y\}) \subseteq V(G_i)$. Note that it can be the case that $x' = y'$. Moreover, $|P| \leq p_i + 2$. Let P_1 be a longest path of G_1 with both endpoints in $\mathrm{bd}(B_1, B)$ and let $P' = xu \cdot P_1 \cdot vy$. Let C' be the cycle obtained from C by replacing P by P'. Since $|P'| = p_1 + 4$ and $p_1 \geq p_i$, we have that C' is a cycle longer than C, a contradiction. This concludes the proof that all the vertices of $\mathrm{bd}(B, B_1)$ are contained in all longest cycles of G. □

4   The Tree 3-Spanner Problem

In this section, we show that TREE 3-SPANNER on well-partitioned chordal graphs can be solved in polynomial time. More specifically, we show that given a connected well-partitioned chordal graph, one can always find a tree 3-spanner in polynomial time.

Restatement of Theorem 3. *Every connected well-partitioned chordal graph admits a tree 3-spanner, which one can find in polynomial time.*

Proof. Let G be a connected well-partitioned chordal graph with partition tree $\mathcal{T}$. We choose a bag R of $\mathcal{T}$ and consider it as a root bag. For each non-root bag B, let $P(B)$ denote the parent bag of B. For each non-root bag B,

- let S_B^* be a star whose center is in $\mathrm{bd}(B, P(B))$ and all leaves are exactly the vertices in $V(B) \setminus \mathrm{bd}(B, P(B))$,
- let S_B^{**} be a star whose center is in $\mathrm{bd}(P(B), B)$ and all leaves are exactly the vertices in $\mathrm{bd}(B, P(B))$, and
- let $S_B := S_B^* \cup S_B^{**}$.

Observe that the vertex set of S_B consists of all vertices of B and one vertex in $\mathrm{bd}(P(B), B)$. Moreover, S_B is a tree. For the root bag R, let S_R be a star on $V(R)$. We claim that $U := \bigcup_{B \in V(\mathcal{T})} S_B$ is a tree 3-spanner of G. It is sufficient to show that U is a spanning tree, and for every edge vw in G, $\mathrm{dist}_U(v, w) \leq 3$.

We first verify that U is a tree. Note that for each non-root bag B, S_B is a tree containing all vertices of B and at least one edge between B and $P(B)$, and furthermore, S_R is a spanning tree of R. Therefore, U is a connected subgraph containing all vertices of G. Suppose that U contains a cycle C.

Observe that for each non-root bag B of $\mathcal{T}$, the center of S_B^{**} separates $V(B)$ and $V(P(B))$ in U. Let B' be the bag containing a vertex of C such that $\mathrm{dist}_{\mathcal{T}}(R, B')$ is minimum. Since $U[V(B')]$ has no cycle, there is a child bag B'' of B' containing a vertex of C. By the above observation, $V(B') \cap V(C)$ has only one vertex that is the center of $S_{B''}^{**}$. As $|V(B') \cap V(C)| = 1$, there is no other child bag of B' containing a vertex of C.

By a repeated argument, we can see that there is no child bag of B'' containing a vertex of C. Then C contains $S_{B'}$, but by the construction, $S_{B'}$ has no cycle. We conclude that U is a spanning tree.

Now, we claim that for every edge vw in G, $\operatorname{dist}_U(v,w) \leq 3$. Choose an edge vw of G. If vw is an edge in a bag B, then $\operatorname{dist}_U(v,w) = \operatorname{dist}_{S_B}(v,w) \leq 3$. Assume that vw is an edge between a bag B and its parent $P(B)$ so that $v \in V(B)$ and $w \in V(P(B))$. If $vw \in E(S_B)$, then it is trivial. Assume that $w \notin V(S_B)$. Let z be the vertex of S_B contained in P_B. Then $\operatorname{dist}_U(v,w) = \operatorname{dist}_{S_B}(v,z) + \operatorname{dist}_{S_{P(B)}}(z,w) \leq 3$.

Our construction of a tree 3-spanner for G immediately follows the partition tree $\mathcal{T}$ of G. By Theorem 5, a partition tree of a well-partitioned chordal graph can be obtained in polynomial time, and therefore one can find a tree 3-spanner for G in polynomial time. □

5   Geodetic Sets

We now give a polynomial-time algorithm for the GEODETIC SET problem on well-partitioned chordal graphs. Recall that a geodetic set of a graph G is a subset S of its vertices such that each vertex that is not in S lies on a shortest path between some pair of vertices in S, and that the GEODETIC SET problem asks, given a graph G, for a smallest-size geodetic set of G. Throughout the following, given a vertex set $S \subseteq V(G)$, we denote by $I[S]$ the *interval* of S in G, which is the set of all vertices lying on a shortest path between a pair of vertices in S. Note that $S \subseteq I[S]$.

We first observe that any geodetic set of a graph contains all its simplicial vertices. Since the neighborhood of a simplicial vertex v is a clique, v is never an internal vertex of any shortest path: Suppose v is an internal vertex of a path P, and let u_1 and u_2 be the two neighbors of v in P. Since u_1 and u_2 are adjacent, we can obtain a shorter path P' from P by replacing u_1vu_2 with u_1u_2 such that P' has the same endpoints as P.

Observation 1. *Let G be a graph and let $v \in V(G)$ be a simplicial vertex in G. Then, every geodetic set of G contains v.*

From now on we assume that we are given a connected well-partitioned chordal graph G with partition tree $\mathcal{T}$, such that $\mathcal{T}$ has at least two nodes (otherwise, G is simply a complete graph). If G is not connected, we can apply the procedure described below to each of its connected components. As a consequence of Observation 1, we have that each leaf bag of $\mathcal{T}$ has a vertex that is contained in every geodetic set of G. Let $B \in V(\mathcal{T})$ be a leaf with neighbor C. If $\operatorname{bd}(B,C) \neq B$, then each vertex in $B \setminus \operatorname{bd}(B,C)$ is simplicial. If $\operatorname{bd}(B,C) = B$, then each vertex in B is simplicial. This also immediately implies that each non-simplicial vertex in a leaf bag is on some shortest path between two simplicial vertices: if we have a non-simplicial vertex in B, then $\operatorname{bd}(B,C) \neq B$ and the non-simplicial vertices are precisely the ones in $\operatorname{bd}(B,C)$. Since $\mathcal{T}$ has at least two nodes, there is some other leaf bag in $\mathcal{T}$ which again has some simplicial vertex, say x. Now, each shortest path from a simplicial vertex in B to x uses some vertex from $\operatorname{bd}(B,C)$, and since the vertices in $\operatorname{bd}(B,C)$ are twins in $G[B \cup C]$, each of them is on such a shortest path.

Observation 2. *Let G be a connected well-partitioned chordal graph with partition tree $\mathcal{T}$, and let S be the set of simplicial vertices of G. Each leaf bag B of $\mathcal{T}$ contains a simplicial vertex, and $B \subseteq I[S]$.*

Dourado et al. [15] showed that the geodetic number of split graphs can be computed in polynomial time. In the following, we adapt their construction to the case of internal bags of a partition tree in a well-partitioned chordal graph. First, we need a small auxiliary lemma.

Lemma 3 (♣). *Let G be a connected well-partitioned chordal graph with partition tree $\mathcal{T}$, let S denote the set of simplicial vertices of G, and let $B \in V(\mathcal{T})$ be an internal bag.*

(i) Let $u \in B$. If there are two distinct $C_1, C_2 \in N_{\mathcal{T}}(B)$ such that $u \in \mathrm{bd}(B, C_1) \cap \mathrm{bd}(B, C_2)$, then $u \in I[S]$.

(ii) For all $C_1, C_2 \in N_{\mathcal{T}}(B)$ with $\mathrm{bd}(B, C_1) \cap \mathrm{bd}(B, C_2) = \emptyset$, we have that $\mathrm{bd}(B, C_1) \cup \mathrm{bd}(B, C_2) \subseteq I[S]$.

Using the previous lemma, we can prove that any vertex in a bag that contains a simplicial vertex is on some shortest path between two simplicial vertices.

Lemma 4 (♣). *Let G be a connected well-partitioned chordal graph with partition tree $\mathcal{T}$, let S denote the set of simplicial vertices of G, and let $B \in V(\mathcal{T})$ be an internal bag. If B contains a simplicial vertex, then $B \subseteq I[S]$.*

In the remainder, we show how to deal with vertices that are not on shortest paths between simplicial vertices. We call such vertices *problematic*, and they are the ones that are contained in internal bags without simplicial vertices and do not fall under one of the cases of Lemma 3. For an illustration of a problematic vertex, see Fig. 1a.

Definition 1. *Let G be a connected well-partitioned chordal graph with partition tree $\mathcal{T}$, and let $B \in V(\mathcal{T})$ be an internal bag that does not contain any simplicial vertex. A vertex $v \in B$ is called* problematic *if*

(i) there is a unique $C \in N_{\mathcal{T}}(B)$ such that $v \in \mathrm{bd}(B, C)$, and
(ii) for each $C' \in N_{\mathcal{T}}(B) \setminus \{C\}$, $\mathrm{bd}(B, C) \cap \mathrm{bd}(B, C') \neq \emptyset$.

In this case we call C a problematic neighbor bag.

Suppose that some bag B has no simplicial vertex. Then each shortest path in G between two simplicial vertices that uses a vertex from B passes through two neighbors of B. If a vertex is problematic, then it cannot be on any such shortest path, and if it is not problematic, then it falls under one of the cases of Lemma 3, which leads to the following observation.

Observation 3. *Let G be a connected well-partitioned chordal graph with partition tree $\mathcal{T}$, let S denote the set of simplicial vertices of G, and let $B \in V(\mathcal{T})$ be an internal bag with $B \cap S = \emptyset$. Let P be the set of problematic vertices of B, then $P = B \setminus I[S]$.*

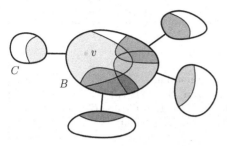
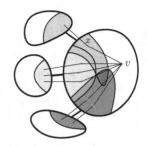

(a) Illustration of a problematic vertex v. The only boundary v is contained in is $\mathrm{bd}(B, C)$, and every other boundary in B intersects $\mathrm{bd}(B, C)$.

(b) Illustration of a problem solver v. Note that v may be in $I[S]$, and that x is a problem solver as well.

Fig. 1. Problematic vertices and problem solvers.

By similar reasoning, we observe that if a problematic vertex in B is on some shortest path, then this shortest path has to have an endpoint in B.

Observation 4. *Let G be a connected well-partitioned chordal graph with partition tree $\mathcal{T}$, and let $B \in V(\mathcal{T})$ be an internal bag. Let $v \in B$ be a problematic vertex. Any shortest path that has v as an internal vertex has one endpoint in B.*

By Observations 3 and 4, we know that if a bag B has no simplicial vertex and it has at least one problematic vertex, then we need at least one more vertex from B in any geodetic set. The following notion captures in which situation a single additional vertex suffices. We illustrate the following definition in Fig. 1b.

Definition 2. *Let G be a connected well-partitioned chordal graph with partition tree $\mathcal{T}$ and let $B \in V(\mathcal{T})$. Let $P \subseteq B$ denote the set of problematic vertices in B and $C_1, \ldots, C_\ell$ the problematic neighbor bags. A vertex $v \in B$ is called a* problem solver *if for each $i \in [\ell]$, either $v \notin \mathrm{bd}(B, C_i)$ or $\mathrm{bd}(B, C_i) \cap P = \{v\}$.*

Lemma 5 (♣). *Let G be a connected well-partitioned chordal graph with partition tree $\mathcal{T}$ and let $S \subseteq V(G)$ be the simplicial vertices of G. Let $B \in V(\mathcal{T})$ be a bag with $B \cap S = \emptyset$. For each $v \in B$, $B \subseteq I[S \cup \{v\}]$ if and only if v is a problem solver.*

If there are at least two distinct problematic neighbor bags, then two additional vertices always suffice.

Lemma 6 (♣). *Let G be a connected well-partitioned chordal graph with partition tree $\mathcal{T}$, let S denote the set of simplicial vertices of G, let $B \in V(\mathcal{T})$ be an internal bag with $B \cap S = \emptyset$. If there are two distinct problematic neighbor bags of B, then there are two vertices $v_1, v_2 \in B$ such that $B \subseteq I[S \cup \{v_1, v_2\}]$.*

Finally, in the remaining case when there is only one problematic neighbor bag and no problem solver, any geodetic set of G has to include all problematic vertices.

Input : A connected well-partitioned chordal graph G with partition tree $\mathcal{T}$.
Output: A minimum-size geodetic set of G.

1 Find the set S of simplicial vertices of G;
2 **foreach** *internal bag* $B \in V(\mathcal{T})$ **do**
3   **if** B *contains a simplicial vertex* **then** do nothing;
4   **else if** *there is a problem solver* $v \in B$ **then** $S \leftarrow S \cup \{v\}$;
5   **else if** B *has two distinct problematic neighbor bags* C_1 *and* C_2 **then**
6   Let $v_1 \in \mathrm{bd}(B, C_1)$ and $v_2 \in \mathrm{bd}(B, C_2)$ be problematic;
7          $S \leftarrow S \cup \{v_1, v_2\}$;
8   **else** $S \leftarrow S \cup P$, where P is the set of problematic vertices in B;
9 **return** S;

Algorithm 1: A polynomial-time algorithm for finding a minimum-size geodetic set of a well-partitioned chordal graph.

Lemma 7 (♣). *Let G be a connected well-partitioned chordal graph with partition tree $\mathcal{T}$, let S denote the set of simplicial vertices of G, let $B \in V(\mathcal{T})$ be an internal bag with $B \cap S = \emptyset$. Let $P \subseteq B$ be the set of problematic vertices of B, and suppose there is a neighbor $C \in N_{\mathcal{T}}(B)$ such that $P \subseteq \mathrm{bd}(B, C)$. If there is no problem solver, then every geodetic set of G contains P.*

The resulting procedure is given in Algorithm 1. We now argue its correctness. In line 1, it adds all simplicial vertices to the set it produces. This is safe by Observation 1. Moreover, by Observation 2, any vertex contained in any leaf of the partition tree is contained in the interval of the simplicial vertices. Let B be any internal bag in the partition tree. In line 3, the algorithm asserts that if B contains a simplicial vertex, then no additional vertex of B has to be added. Correctness of this decision is argued in Lemma 4. Suppose B has no simplicial vertex. By Observation 3, each vertex in B that is not in the interval of the simplicial vertices is problematic, and by Observation 4, a shortest path that has a problematic vertex as an internal vertex has one endpoint in B. Therefore, any geodetic set of G has to contain at least one vertex from B. Lemma 5 characterizes the situation in which one additional vertex (a problem solver) suffices, which is checked for next by the algorithm, in line 4. If no such vertex exists, then each geodetic set uses at least two vertices from B. If there are at least two distinct problematic neighbor bags, then two additional vertices suffice as shown in Lemma 6. The algorithm checks this next in line 5. Otherwise, there is precisely one problematic neighbor bag C, there is no problem solver, and $\mathrm{bd}(B, C)$ contains at least two problematic vertices. By Lemma 7, all these vertices are in any geodetic set of G, so the algorithm is correct in line 8.

It is easy to verify that each line in Algorithm 1 takes polynomial time, and that the main loop has a polynomial number of iterations. Since well-partitioned chordal graphs can be recognized in polynomial time by an algorithm that produces a partition tree if one exists, see Theorem 5, this proves Theorem 4.

Restatement of Theorem 4. *There is a polynomial-time algorithm that given a well-partitioned chordal graph G, computes a minimum-size geodetic set of G.*

References

1. Ahn, J., Jaffke, L., Kwon, O., Lima, P.T.: Well-partitioned chordal graphs: obstruction set and disjoint paths. In: Adler, I., Müller, H. (eds.) WG 2020. LNCS, vol. 12301, pp. 148–160. Springer, Cham (2020). https://doi.org/10.1007/978-3-030-60440-0_12

2. Althöfer, I., Das, G., Dobkin, D., Joseph, D., Soares, J.: On sparse spanners of weighted graphs. Discrete Comput. Geom. **9**(1), 81–100 (1993). https://doi.org/10.1007/BF02189308

3. Atici, M.: Computational complexity of geodetic set. Int. J. Comput. Math. **79**(5), 587–591 (2002). https://doi.org/10.1080/00207160210954

4. Balister, P.N., Györi, E., Lehel, J., Schelp, R.H.: Longest paths in circular arc graphs. Comb. Probab. Comput. **13**(3), 311–317 (2004). https://doi.org/10.1017/S0963548304006145

5. Brandstädt, A., Dragan, F.F., Le, H.O., Le, V.B.: Tree spanners on chordal graphs: complexity and algorithms. Theoret. Comput. Sci. **310**(1–3), 329–354 (2004). https://doi.org/10.1016/S0304-3975(03)00424-9

6. Brandstädt, A., Dragan, F.F., Le, H., Le, V.B., Uehara, R.: Tree spanners for bipartite graphs and probe interval graphs. Algorithmica **47**(1), 27–51 (2007). https://doi.org/10.1007/s00453-006-1209-y

7. Bueno, L.R., Penso, L.D., Protti, F., Ramos, V.R., Rautenbach, D., Souza, U.S.: On the hardness of finding the geodetic number of a subcubic graph. Inf. Process. Lett. **135**, 22–27 (2018). https://doi.org/10.1016/j.ipl.2018.02.012

8. Cai, L.: Tree spanners: spanning trees that approximate distances. Ph.D. thesis, University of Toronto (1992)

9. Cai, L., Corneil, D.G.: Tree spanners. SIAM J. Discrete Math. **8**(3), 359–387 (1995). https://doi.org/10.1137/S0895480192237403

10. Cerioli, M.R., Lima, P.T.: Intersection of longest paths in graph classes. Discrete Appl. Math. **281**, 96–105 (2020). https://doi.org/10.1016/j.dam.2019.03.022

11. Chakraborty, D., Das, S., Foucaud, F., Gahlawat, H., Lajou, D., Roy, B.: Algorithms and complexity for geodetic sets on planar and chordal graphs. arXiv:2006.16511 (2020). To appear at ISAAC 2020

12. Chakraborty, D., Foucaud, F., Gahlawat, H., Ghosh, S.K., Roy, B.: Hardness and approximation for the geodetic set problem in some graph classes. In: Changat, M., Das, S. (eds.) CALDAM 2020. LNCS, vol. 12016, pp. 102–115. Springer, Cham (2020). https://doi.org/10.1007/978-3-030-39219-2_9

13. Chen, G., et al.: Nonempty intersection of longest paths in series-parallel graphs. Discrete Math. **340**(3), 287–304 (2017). https://doi.org/10.1016/j.disc.2016.07.023

14. Dourado, M.C., Protti, F., Szwarcfiter, J.L.: On the complexity of the geodetic and convexity numbers of a graph. Lect. Notes Ramanujan Math. Soc. **7**, 497–500 (2006)

15. Dourado, M.C., Protti, F., Rautenbach, D., Szwarcfiter, J.L.: Some remarks on the geodetic number of a graph. Discrete Math. **310**(4), 832–837 (2010)

16. Ekim, T., Erey, A., Heggernes, P., van 't Hof, P., Meister, D.: Computing minimum geodetic sets of proper interval graphs. In: Fernández-Baca, D. (ed.) LATIN 2012. LNCS, vol. 7256, pp. 279–290. Springer, Heidelberg (2012). https://doi.org/10.1007/978-3-642-29344-3_24

17. Fekete, S.P., Kremer, J.: Tree spanners in planar graphs. Discrete Appl. Math. **108**(1–2), 85–103 (2001). https://doi.org/10.1016/S0166-218X(00)00226-2

18. Gallai, T.: Problem 4. In: Erdős, P., Katona, G.O.H. (eds.) Proceedings of the Colloquium on Theory of Graphs Held in Tihany, Hungary, 1966, p. 362 (1968)
19. Golan, G., Shan, S.: Nonempty intersection of longest paths in $2K_2$-free graphs. Electron. J. Comb. **25**(2), P2.37 (2018)
20. Harary, F., Loukakis, E., Tsouros, C.: The geodetic number of a graph. Math. Comput. Modell. **17**(11), 89–95 (1993)
21. Jobson, A.S., Kézdy, A.E., Lehel, J., White, S.C.: Detour trees. Discrete Appl. Math. **206**, 73–80 (2016). https://doi.org/10.1016/j.dam.2016.02.002
22. Joos, F.: A note on longest paths in circular arc graphs. Discussiones Mathematicae Graph Theory **35**(3), 419–426 (2015)
23. Kellerhals, L., Koana, T.: Parameterized complexity of geodetic set. arXiv:2001.03098 (2020). To appear at IPEC 2020
24. Klavžar, S., Petkovšek, M.: Graphs with nonempty intersection of longest paths. Ars Combinatoria **29**, 43–52 (1990)
25. Loui, M.C., Luginbuhl, D.R.: Optimal on-line simulations of tree machines by random access machines. SIAM J. Comput. **21**(5), 959–971 (1992). https://doi.org/10.1137/0221056
26. Madanlal, M.S., Venkatesan, G., Rangan, C.P.: Tree 3-spanners on interval, permutation and regular bipartite graphs. Inf. Process. Lett. **59**(2), 97–102 (1996). https://doi.org/10.1016/0020-0190(96)00078-6
27. Mezzini, M.: Polynomial time algorithm for computing a minimum geodetic set in outerplanar graphs. Theoret. Comput. Sci. **745**, 63–74 (2018). https://doi.org/10.1016/j.tcs.2018.05.032
28. Panda, B.S., Das, A.: Tree 3-spanners in 2-sep chordal graphs: characterization and algorithms. Discrete Appl. Math. **158**(17), 1913–1935 (2010). https://doi.org/10.1016/j.dam.2010.08.015
29. Pelayo, I.M.: Geodesic Convexity in Graphs. Springer, New York (2013). https://doi.org/10.1007/978-1-4614-8699-2
30. Rautenbach, D., Sereni, J.: Transversals of longest paths and cycles. SIAM J. Discrete Math. **28**(1), 335–341 (2014). https://doi.org/10.1137/130910658
31. de Rezende, S.F., Fernandes, C.G., Martin, D.M., Wakabayashi, Y.: Intersecting longest paths. Discrete Math. **313**(11), 1401–1408 (2013). https://doi.org/10.1016/j.disc.2013.02.016
32. Venkatesan, G., Rotics, U., Madanlal, M.S., Makowsky, J.A., Rangan, C.P.: Restrictions of minimum spanner problems. Inf. Comput. **136**(2), 143–164 (1997). https://doi.org/10.1006/inco.1997.2641
33. Walther, H., Voss, H.J.: Über Kreise in Graphen. Deutscher Verlag der Wissenschaften (1974)
34. Zamfirescu, T.: On longest paths and circuits in graphs. Mathematica Scandinavica **38**(2), 211–239 (1976)

Distributed Distance-r Covering Problems
on Sparse High-Girth Graphs

Saeed Akhoondian Amiri[1] and Ben Wiederhake[2(✉)]

[1] University of Cologne, Cologne, Germany
`amiri@cs.uni-koeln.de`
[2] MPII, Saarbrücken, Germany
`bwiederh@mpi-inf.mpg.de`

Abstract. We prove that the distance-r dominating set, distance-r connected dominating set, distance-r vertex cover, and distance-r connected vertex cover problems admit constant factor approximations in the CONGEST model of distributed computing in a *constant* number of rounds on classes of sparse high-girth graphs. In this paper, sparse means bounded expansion, and high-girth means girth at least $4r + 2$. Our algorithm is quite simple; however, the proof of its approximation guarantee is non-trivial. To complement the algorithmic results, we show tightness of our approximation by providing a loosely matching lower bound on rings.

Our result is the first to show the existence of constant-factor approximations in a constant number of rounds in non-trivial classes of graphs for distance-r covering problems.

1 Introduction

In the sequential setting, many APX-hard covering problems can be well approximated if they are limited to the class of sparse graphs. Hence, it is interesting to understand how sparsity enables better distributed algorithms in distributed computing models, which could mean improving the approximation factor or reducing the number of communication rounds. In the distributed setting every node is considered as a processor that can communicate with its neighbors per synchronized rounds. The aim is to reduce the total number of such rounds while providing a *good* solution.

In this work, we continue the line of study on sparse graphs and explore the algorithmic properties of a wide range of sparse graphs, namely the class of graphs of bounded expansion, with an extra combinatorial property: sparse graphs of high girth. The girth of the graph is the length of its shortest cycle and for instance girth of a tree is infinity.

Girth plays a role in understanding structural properties of graphs. Sparse graphs of high girth appear in important constructions such as spanner graphs [1]. Similarly random graphs have only a few short cycles and at the same time, depending on their parameters, they could be quite sparse. In such a graph class (we will formally specify them later) we study several central covering problems in their most generic form: distance-r covering problems.

© Springer Nature Switzerland AG 2021
T. Calamoneri and F. Corò (Eds.): CIAC 2021, LNCS 12701, pp. 37–60, 2021.
https://doi.org/10.1007/978-3-030-75242-2_3

As a result, we answer some more cases of the famous question of Naor and Stockmeyer: "What can be computed locally?" [21] More precisely, we show that the following problems on the considered graph class have constant factor approximation in a constant number of rounds in the CONGEST model of computation, if r (the distance) is constant. Whenever feasible, we also give the precise relation to r.

1. Distance-r Dominating Set   2. Distance-r Connected Dominating Set
3. Distance-r Vertex Cover and 4. Distance-r Connected Vertex Cover.

The aforementioned problems are hard in general graphs when it comes to distributed settings. For instance there is no constant-factor approximation CONGEST algorithm for the distance-2 dominating set even if we let the algorithm to run for $o(n^2)$ rounds [9], where n is the number of nodes. Observe that in order to exchange information, two nodes require at most $O(n)$ rounds of communication, however, for such a restricted case of the problem (only distance-2) the amount of data to be transferred is too big to fit in one message that respects the bandwidth of the network. Hence it needs $\Omega(n^2)$ rounds of computation to merely approximate the optimum solution. Thus, a natural approach to progress on such problems is to consider them on restricted graph families.

Throughout the paper, we assume that a graph $G = (V, E)$ is given. In the (distance 1) dominating set problem, we seek a set $D \subseteq V$ such that every other vertex of G is a neighbor of a vertex in D. In the connected version of the problem, the induced graph of G on the vertices of D should be connected.

In the vertex cover problem, we seek a set C of vertices of the graph such that every edge in the graph is incident to at least one of the vertices in C. Similarly, the connected version of the problem asks for a vertex cover $\hat{C}$ such that the induced graph of C on G is connected. In all of the above problems we would like to minimize the size of the corresponding set.

The distance-r versions of covering problems are defined similarly to the classic versions: for the dominating set problem, we consider the distance-r neighborhood. In the distance-r vertex cover problem, we say that vertex v covers edge e if and only if vertex v is within distance r of both endpoints.

We consider problems in the LOCAL and CONGEST distributed models of computation. Intuitively speaking, in both of these models, each vertex in the graph is a processor, has a unique identifier, and communicates only with its neighbors once per round. The CONGEST model restricts the bandwidth of communication links to a reasonable complexity. The aim is to solve the problem with the least number of communication rounds. A more rigorous definition follows in Sect. 2. We specifically look into the problem of finding a small distance-r dominating set, distance-r connected dominating set, distance-r vertex cover, or distance-r connected vertex cover where each vertex has to output its membership in the set.

Our main algorithmic contribution is that the distance-r dominating set problem admits a constant factor approximation in a constant number of rounds in sparse high-girth graphs (for constant r). We also extend this algorithm to the described related covering problems.

Related Work

In distributed settings, for the dominating set problem on general graphs, Kuhn et al. [14] provided a $(1 + \epsilon)(1 + \log(\Delta + 1))$-approximation in $f(n)$ rounds. In this bound, Δ is the maximum degree and $f \colon \mathbb{N} \to \mathbb{N}$ is the number of rounds that is needed to compute the network decomposition [5,6,16]. Given the recent breakthrough result of Rozhon and Ghaffari [25], the aforementioned algorithm runs in a polylogarithmic number of rounds.

For the vertex cover problem, Bar-Yehuda et al. [10] provided a constant factor approximation in a sublogarithmic number of rounds. This is complemented by the lower bound of Kuhn, Moscibroda and Wattenhofer (KMW) [17] shows that a logarithmic approximation in almost sublogarithmic time for the vertex cover problem and the minimum dominating set (and some other covering problems) is impossible in general graphs, even in the LOCAL model of computation. Their lower-bound graph for vertex cover has high girth, but it is also of unbounded arboricity (more generally unbounded average degree).

We investigate what happens in between graph classes. If we consider a graph class of very high girth and very low edge density, e.g. trees, then these problems are easy to approximate in zero rounds: take all non-leaf vertices. The above observations raises the questions: In which graph classes does the problem admit a constant approximation factor in a constant number of rounds? What about distance-r problems?

For the dominating set problem, Lenzen et al. [12,18] provided the first constant-factor approximation in a constant number of rounds in planar graphs, which was improved by Czygrinow et al. [12]. Later, Amiri et al. [3,4] provided a new analysis method to extend the result of Lenzen et al. to bounded genus graphs. This has been improved by Czygrinow et al. [13] to excluded minor graphs.

A natural generalization of excluded minor graphs is the class of bounded expansion graphs. Simply put, bounded expansion graphs only exclude minors on a localized level; there may still be large clique minors in the graph.

On graphs of bounded expansion, only a logarithmic time constant factor approximation is known for the dominating set problem; however, it seems that one can extend the algorithm of [13] to bounded expansion graphs, as they only consider "local" minors. If we go slightly beyond these graphs, to graphs of bounded arboricity (where every subgraph has a constant edge density), the situation is worse: only an $O(\log \Delta)$-approximation in $O(\log n)$ rounds is known. There is a $O(\log n)$ round $O(1)$-approximation in such graphs; however, this algorithm is randomized [19].

All these results are about the distance-1 dominating set problem. Significantly less work exists on the topic of the distance-r dominating set problem. We are only aware of the algorithm of Amiri et al. [2] for bounded expansion graphs that provides a constant factor approximation in a logarithmic number of rounds.

We are not aware of any paper tackling the distance-r vertex cover problem, except the general techniques known for bounded degree graphs and the generic algorithmic techniques that one can apply to bounded arboricity graphs.

Existing Approaches for (Distance-r) Dominating Set

There are several existing approaches one might employ to tackle the problem: 1) Take the r-th power of the graph and go back to the distance-1 dominating set, 2) Decentralize existing decomposition methods in the sequential setting and employ them, 3) Use existing fast distributed graph decomposition methods for sparse graphs. In the following, we explain how all of the above approaches, without introducing new ideas, are impractical in providing sublogarithmic round algorithms for distance-r covering problems.

For the first approach, we lose the sparsity of the graph already on stars. Hence, we cannot rely on existing algorithms for solving the domination problem in sparse graphs.

Decentralizing the existing sequential decomposition methods does not seem promising if one hopes to achieve anything better than logarithmic rounds: To the best of our knowledge, every such sequential decomposition already consumes polylogarithmically many rounds. Even assuming the decomposition is given in advance, such methods handle the clusters sequentially; however the number of clusters is usually at least logarithmic, requiring at least logarithmically many rounds.

For the third approach, these existing fast distributed graph decomposition methods are mostly inspired by existing methods in classical settings, like Baker's method [8]; this includes, e.g., the $O(\log^* n)$ round algorithm of [11]. The idea is to partition a sparse graph into connected clusters such that each cluster has a small diameter and the number of in-between cluster edges is small. Then, find the optimal solution inside each cluster efficiently, and because the number of edges between a pair of clusters is small, we can just ignore conflicts.

However, this fails already for distance-2 domination, since the number of edges between clusters in the power graph is high. As mentioned earlier, recent research has shown that there is a lowerbound of $\tilde{\Omega}(n^2)$ for distance-2 dominating set, both for solving it exactly [7], and even for constant-factor approximations [9].

Also, we cannot rely on global properties (such as tree decomposition) like in the sequential setting, since this increases the number of rounds to the diameter of the graph, which can easily be superlogarithmic.

Therefore, any distributed algorithm that solves distance-r covering problems has to exploit special properties of the underlying graph class or problem, motivating our choice of sparse high-girth graphs.

Our Approach and Results

We consider a generalization of the dominating set and vertex cover problem, and fill a gap between the lower bounds and upper bounds by analyzing the complexity of the problems on graphs of high girth, similar to the lower bound graph by Kuhn et al. Given that the lower bound graph in that work is relatively dense, we restrict the graph class further to sparse graphs, in particular, to bounded expansion graphs (similar to the work of [2]).

Let us present the algorithm for the distance-r dominating set problem: Each vertex chooses its dominator to be the neighbor that has a maximum degree in the r-th power graph. To implement such a simple algorithm in the CONGEST model without actually constructing the r-th power graph (which is basically impossible in our desired running time even for $r = 2$), we exploit the fact that the neighborhood of a vertex looks like a tree. The output of the algorithm is the set D of dominators, which solves the problem.

To prove the correctness of the algorithm, we partition the vertices of the graph into Voronoi cells where the center of each cell is one of the vertices of the optimum distance-r dominating set. Then we divide the vertices chosen by our algorithm into two subsets: those that are the dominators of vertices inside their Voronoi cell (D^I) and those that are the dominator of vertices outside their cell (D^O). We show (by non-trivial arguments) these sets are bounded in terms of the optimal distance-r dominating set, which also bounds $|D|$.

This is used to prove Theorem 1. We generalize the algorithm to handle the Distance-r Connected Dominating Set (proven by Theorem 5), Distance-r Vertex Cover (Theorem 6), and Distance-r Connected Vertex Cover (Corollary 8) problem.

Theorem 1. *Let C be a graph class of bounded expansion $f(r)$ and girth at least $4r + 3$. There is a CONGEST algorithm that runs in $O(r)$ rounds and provides an $O(r \cdot f(r))$-approximation of minimum distance-r dominating set on C.*

Given that the distance-r dominating set problem is equivalent to the dominating set problem of the r-th power of the input graph, the algorithm can also be used to provide a constant factor approximation in a non-trivial class of dense graphs for covering problems. There are very few known algorithms with a constant factor guarantee in a constant number of rounds on non-trivial dense graphs, e.g., the algorithm of Schneider et al. [26] on graphs of bounded independence number (for the independent set and the connected dominating set problem) partially falls into this category.

To show that our upper bound is reasonably tight, we provide a lower bound as well. This we obtain by a reduction from a lower bound for independent set on the ring [12,18] to the distance-r dominating set on the ring (naturally, a ring with high girth). More formally we prove Theorem 2.

Theorem 2. *Assume an arbitrary but fixed $\delta > 0$ and $r > 1$, with $r \in o(\log^* n)$. Then, there is no deterministic LOCAL algorithm that finds in $O(r)$ rounds a $(2r + 1 - \delta)$-approximation of distance-r dominating set for all $G \in C$, where C is the class of cycles of length $\gg 4r + 3$.*

We will formally introduce the notion of bounded expansion in Sect. 2. The relation between the sparse graph classes is shown in Fig. 1.

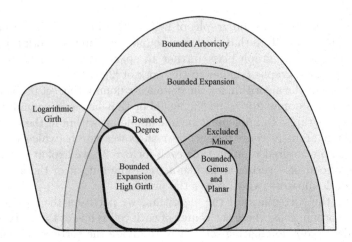

Fig. 1. Diagram of the relation of sparse graph classes. The graph class in the lower bound construction of Kuhn et al. [17] is a subclass of logarithmic girth class of unbounded arboricity. The bounded expansion class is a subclass of bounded arboricity class. Bounded expansion is also a superclass of many common sparse graph classes: planar, bounded genus, excluded minors, and bounded degree. The class of bounded expansion with high girth intersects each of the other four classes, but neither contains nor is fully contained in any of them.

2   Preliminaries and Notation

We assume basic familiarity with graph theory. In the following, we introduce basic graph notation to avoid ambiguities. We refer to the book by Diestel [15] for further reading.

Graph, Neighborhood, Distance-r: We only consider simple, connected, undirected graphs $G = (V, E)$. For $u, v \in V$, define $d(u, v)$ as the distance (in number of edges) between the two vertices. For a set $S \subseteq V$, we define $d(u, S)$ as the maximum distance between vertex u and any vertex in S. Two vertices $u, v \in V$ are neighbors in G if there is an edge $e = \{u, v\} \in E$. We extend this definition to the distance-r neighborhood $N^r[v]$ and open distance-r neighborhood $N^r(v)$ of a vertex v in the following way:

$$N^r[v] := \{u \in V \mid d(u, v) \leq r\} \qquad N^r(v) := N^r[v] \setminus \{v\}$$

Similarly for a set $S \subseteq V$ we define:

$$N^r[S] := \bigcup_{v \in S} N^r[v] \qquad N^r(S) := N^r[S] \setminus S$$

Girth, Radius: The girth g of a graph G is the length of its shortest cycle, or ∞ if acyclic. The radius R of G is the minimum integer R for which $\exists v \in V : N^R[v] = V$.

Distance-r dominating set: A set $M \subseteq V$ is a distance-r dominating set if $V = N^r[M]$. If there is no smaller such set, then M is a minimum distance-r dominating set of G.

Distance-r Connected Dominating Set: A set of vertices $D \subseteq V$ is a distance-r connected dominating set of G if D is a distance-r dominating set of G and the subgraph of G induced on vertices in D is connected.

Distance-r Vertex Cover: A set $C \subseteq V$ is a distance-r vertex cover of G if for every edge $e = \{u, v\}$, there exists a vertex $w \in C$ such that the distance of both u and v from w is at most r. The special case of $r = 1$ is the standard vertex cover problem. Note that in contrast to dominating set, there is no equivalence for vertex cover between the distance-r and power graph version.

Distance-r Connected Vertex Coverxspace: Similarly, a set $\hat{C}$ is a distance-r connected vertex cover of G if it is a distance-r vertex cover of G and the induced subgraph of G on vertices of $\hat{C}$ is connected.

Edge Density, r-Shallow Minor, Expansion: Let $G = (V, E)$ be a graph; its edge density is $|E|/|V|$. A graph H is an r-shallow minor of G if H can be obtained from G by the following operations. First, we take a subgraph S of G and then partition the vertices of S into vertex disjoint connected subgraphs $S_1, \ldots, S_t$ of S, each of radius at most r and, at the end, contract each S_i ($i \in [t]$) to a single vertex to obtain H. We denote by $\nabla_r(G)$ the maximum edge density among all r-shallow minors of the graph G.

A graph class $\mathcal{C}$ is bounded expansion if there is a function $f : \mathbb{N} \rightarrow \mathbb{N}$ such that for every graph $G \in \mathcal{C}$ and integer $r \in \mathbb{N}$ we have $\nabla_r(G) \leq f(r)$; here f is the *expansion function*. A class of graphs $\mathcal{C}$ has constant expansion if for every integer r, we have $f(r) \in O(1)$.

Every planar, bounded genus, and excluded minor graph is a constant expansion graph. Every class of bounded degree graphs is also bounded expansion, but not of constant expansion. For more information on bounded expansion graphs, we refer the reader to the book by Nešetřil and Ossona de Mendez [22].

LOCAL and CONGEST Model of Computation: The LOCAL model of computation assumes that the problem is being solved in a distributed manner: Each vertex in the graph is also a computational node, the input graph is also the communication graph, and initially, each vertex only knows its own unique identifier and its neighbors. An algorithm proceeds in synchronous rounds on each vertex in parallel. In each round, the algorithm can run an arbitrary amount of local computation, send a message of arbitrary size to its neighboring vertices, and then receive all messages from its neighbors. Each vertex can decide locally whether it halts with an output or continues. The most common metric is the number of communication rounds.

This model was first introduced by Linial [20]; later Peleg [23] named it LOCAL model.

The CONGEST model is very similar to the LOCAL model, except that identifiers can be represented in $O(\log n)$ bits, and each message can only hold $O(\log n)$ bits, where n is the number of vertices in the network.

3   Distributed Approximation Algorithm for Dominating Set

In this section we prove the following theorem.

Theorem 1. Let $\mathcal{C}$ be a graph class of bounded expansion $f(r)$ and girth at least $4r + 3$. There is a CONGEST algorithm that runs in $O(r)$ rounds and provides an $O(r \cdot f(r))$-approximation of a minimum distance-r dominating set on $\mathcal{C}$.

We prove this by providing Algorithm 1, satisfying all bounds. At its core, the algorithm is simple: Each vertex computes the size of its distance-r neighborhood, i.e., the distance-r degree. This degree is propagated so that each vertex selects in its distance-r neighborhood the vertex with the highest such degree. The output is the set of all selected vertices. We expect this to yield a good approximation because only few candidates can be selected.

Algorithm 1 defines this formally. The main technical contribution is Lemma 6, which concludes that Algorithm 1 is correct and thus proves Theorem 1.

Algorithm 1: CONGEST computation of r-MDS, on each vertex v in parallel

---

1: Compute $|N^r(v)|$, e.g. using Algorithm 2
2: // Phase 1: Select the vertex with the highest degree:
3: $(prio^v, sel^v) := (|N^r(v)|, v)$
4: **for** r rounds **do**
5:     Send $(prio^v, sel^v)$ to all neighbors
6:     Receive $(prio^u, sel^u)$ from each neighbor u
7:     $(prio^v, sel^v) := \max_{u \in N^1[v]}\{(prio^u, sel^u)\}$
8:     Remember all received messages that contained $(prio^v, sel^v)$
9: **end for**
10: // Phase 2: Propagate back to the selected vertex:
11: $D^v := \{sel^v\}$
12: **for** $r - 1$ rounds **do**
13:     **for** each neighbor $u \in N^1(v)$ **do**
14:         Determine which vertices sent by u in Phase 1 are in D^v
15:         Send these to u, encoded as a bitset of size r
16:     **end for**
17:     Receive bitsets, extend D^v accordingly
18: **end for**
19: Join the dominating set if and only if $v \in D^v$, in other words:
20: **return** $v \in D^v$

We say that Algorithm 1 computes a set D by returning $\top$ for all vertices in the set, and $\bot$ for all others. Naturally, messages and comparisons only consider the ID of vertices, and not the vertices themselves. This abuse of notation simplifies the algorithm and analysis. In line 7, we order tuples lexicographically: Tuples are ordered by the first element (the size of the distance-r neighborhood); ties are broken by the second element (the ID of the vertex).

Algorithm 2: CONGEST computation of $|N^r(v)|$, on each vertex v in parallel

---

1: $n_u := 1$ for all $u \in N^1(v)$
2: **for** $r - 1$ rounds **do**
3:     To each vertex $u \in N^1(v)$, send $1 + \sum_{w \in N^1(v) \setminus \{u\}} n_w$
4:     $n_u :=$ the number received from u, for each $u \in N^1(v)$
5: **end for**
6: **return** $\sum_{w \in N^1(v)} n_w$

Algorithm 2 implements the computation of the distance-r neighborhood. The intuition is to compute the size of a *rooted* tree, for all possible roots at once. The high girth of G and line 3 mean that each vertex is counted only once (if at all).

The remainder of this section proves Algorithm 1's correctness and approximation factor.

Correctness

First, we will show basic correctness properties. One can trivially check that all messages contain only $O(\log n)$ many bits. Specifically, the bitsets have only size $r \in o(\log^* n)$.

Lemma 1. *Algorithm 2 computes the size of $N^r(v)$ for each vertex v in parallel.*

Next, we show that Algorithm 1 selects the maximum degree neighbor:

Lemma 2. *In Algorithm 1, at the start of Phase 2 (line 10 et seq.), each vertex v has selected a vertex sel^v. This is the unique vertex $\arg\max_{u \in N^r[v]} \{(|N^r(u)|, u)\}$.*

Proof. By construction, only tuples of the form $(|N^r(w)|, w)$ with $w \in V$ are ever stored. The max operator is commutative and associative, so it is sufficient to prove that each vertex v considers precisely the tuples for $w \in N^r[v]$. This can be shown by straightforward induction: After round i, vertex v considers precisely the tuples for $w \in N^i[v]$. The base case is $i = 0$, and the induction step is straight-forward.

Hence, each vertex selects the maximum neighbor. Next, we show that this is back-propagated:

Lemma 3. *If there is a vertex u that selects v ($sel^u = v$), then $v \in D^v$.*

Proof. Consider the path along which v was forwarded during the selection phase. By straight-forward induction, one can see that after i rounds of propagation, for all vertices w on the path with $d(w, u) \leq i$, have $v \in D^w$. The path has length at most r edges, so $v \in D^v$ after r rounds.

And because no further vertices are added into any D^v, we get:

Corollary 1. *The selected vertices are precisely the computed set:* $D = \{sel^v \mid v \in V\}$

Together with Lemma 2, this shows that Algorithm 1 computes a dominating set.

Lemma 4. *The computed set D is a distance-r dominating set.*

Proof. Assume towards contradiction that a vertex v is not dominated. Lemma 2 shows that v selected a vertex sel^v in its distance-r neighborhood. Corollary 1 shows that $sel^v \in D$, which distance-r-dominates v, which is a contradiction.

The time complexity analysis is trivial.

Lemma 5. *Algorithm 1 runs in $O(r)$ rounds.*

Approximation Analysis

So far we have seen the correctness of the algorithm and the running time bound. In this subsection, we prove the approximation bound of Lemma 6. Specifically, we prove that the size of D, the set of selected vertices, is within factor $1+4{\cdot}r{\cdot}f(r)$ of $|M|$, a minimum distance-r dominating set.

Lemma 6. *If the graph class $\mathcal{C}$ has expansion $f(r)$ and girth at least $4r+3$, then the set of vertices D selected by Algorithm 1 is small: $|D|/|M| \in O(r \cdot f(r))$.*

In the remainder of this subsection, we prove Lemma 6. Note that this means that if r is constant, then the approximation factor is constant, too.

We now analyze the behavior of Algorithm 1 on a particular graph $G \in \mathcal{C}$. We begin by showing that the optimal solution implies a partition into Voronoi cells [24], which we will use throughout the analysis. First, we define what a *covering* vertex is. Note that this can be (and often is) different from the vertex selected by the algorithm.

Definition 1. *Let $c : V \to M$ be the mapping from vertices in V to corresponding dominating vertices in M, breaking ties first by distance and then by ID:*

$$c(v) := \underset{u \in N^r[v] \cap M}{\arg\min} \{(d(u,v), u)\} \tag{1}$$

We order tuples lexicographically, again. The equivalent term $\arg\min_{u \in M} \{(d(u,v), u)\}$ provides shorter notation: By construction, each vertex v has a vertex in M in its r-neighborhood, so $\arg\min$ will select from $N^r[v]$ anyway. Next, we partition V into Voronoi cells $H_m := \{v \in V \mid c(v) = m\}$ for each $m \in M$.

Corollary 2. *Each H_m is connected and has radius at most r.*

Proof. As vertex m dominates all vertices in H_m, we know that H_m has radius r.

We use the high-girth property to show that the Voronoi cells behave nicely:

Lemma 7. *The subgraph induced by H_m in G is a tree.*

Proof. Assume towards contradiction that there is a cycle C' in H_m. We construct a cycle that has length at most $2r + 1$, a contradiction.

Specifically, construct a BFS-tree of H_m rooted in m. Then, the cycle C' must contain an edge e between $u, v \in H_m$. Consider the cycle that consists of the path from u to v along the BFS-tree and the edge $\{u, v\}$. This cycle must have length at most $r + r + 1$, because the BFS-tree has depth at most r. This contradicts G having a girth of at least $4r + 3$.

Lemma 8. *For any two Voronoi cells $H_m \neq H_n$, there is at most one edge between them.*

Proof. Let $\{u, v\} \in E$ and $\{s, t\} \in E$ be two different edges between H_m and H_n. W.l.o.g. assume $c(u) = c(s) = m$ and $c(v) = c(t) = n$, and assume $v \neq t$ (but $u = s$ is possible).

By Corollary 2, we know that the subgraphs induced by H_m and H_n are each connected, so there must be a path p_m entirely in H_m between vertices u and s, possibly of length 0. Likewise, a path p_n must exist entirely in H_n between vertices v and t. The union of the paths and the assumed edges forms a cycle $C_{u,v,s,t}$, as no vertex can be repeated. We will now prove that $C_{u,v,s,t}$ is too small.

The paths p_m and p_n have length at most $2r$ each. Therefore, we have found the cycle $C_{u,v,s,t}$ to have length at most $4r + 2$, in contradiction to the minimum girth $4r + 3$.

Let $G' = (V', E')$ be the result of contracting H_m to a single vertex, for each $m \in M$.

Lemma 9. *The edge set E' is small:* $|E'| \leq f(r) \cdot |M|$

Proof. Using Corollary 2, we can apply the definition of the function $f(r)$:

$$|E'| = \frac{|E'|}{|V'|} \cdot |V'| \leq f(r) \cdot |M|$$

Now we can take a closer look at the set of vertices D selected by the algorithm. We construct two sets of bounded size whose union is D; this bounds the size of D.

Definition 2. *We consider the set D^O of vertices v that were selected by a vertex u in a different Voronoi cell (i.e. $c(v) \neq c(u)$) and the possibly overlapping set D^I of vertices v that were selected by a vertex u in the same Voronoi cell (i.e. $c(v) = c(u)$):*

$$D^O := \{d \in D \mid \exists v.\ v \text{ selected } d \text{ with } c(v) \neq c(d)\}$$
$$D^I := \{d \in D \mid \exists v.\ v \text{ selected } d \text{ with } c(v) = c(d)\}$$

Note that $D = D^I \cup D^O$, and $|D| \leq |D^I| + |D^O|$. In order to prove Lemma 6, it is therefore sufficient to show $|D^I|, |D^O| \in O(r \cdot f(r) \cdot |M|)$.

First, we consider D^O, the set of vertices selected across Voronoi cells. There are only few crossing edges, so there can only be few such selections:

Lemma 10. *The set D^O of vertices selected across Voronoi cells is small: $|D^O| \in O(r \cdot f(r) \cdot |M|)$.*

Proof. Lemmas 8 and 9 tell us that there are at most $f(r) \cdot |M|$ edges across Voronoi cells. Across each edge, at most $2r$ distinct messages are passed (i.e., r vertex proposals in each direction); therefore, there are at most $f(r)|M| \cdot 2r$ many candidates for D^O.

Next, we prove a bound on set D^I of Definition 2: the set of vertices selected from within a Voronoi cell. We see that these always fall on the spanning tree inside Voronoi cells, which are small. First, we define the candidate set:

Definition 3. *For each Voronoi cell H_m, we define the set of vertices $\mathcal{T}_m$ as the union of all shortest paths $P_{m,u}$ between vertex m and each vertex u on the Voronoi boundary:*

$$\mathcal{T}_m := \bigcup_{\substack{\{u,v\} \in E \\ c(u)=m, c(v) \neq m}} P_{m,u} \qquad \mathcal{T} := \bigcup_{m \in M} \mathcal{T}_m$$

This is well-defined due to Lemma 7. Observe that $\mathcal{T}_m$ is not necessarily equal to H_m: All leaves in $\mathcal{T}_m$ have edges in G that lead outside the Voronoi cell. A vertex in H_m that has no such edges will not be in $\mathcal{T}_m$. Next, we prove in two steps that D^I is contained in $\mathcal{T}$.

Lemma 11. *If $|M| \geq 1$, then there is always a vertex not in distance r:*

$$\forall v \in V \; \exists u \in V \quad d(u,v) = r+1 \qquad (2)$$

Proof. Assume towards contradiction that there is a vertex v for which no such vertex u exists. Then, there is also no vertex u' with $d(u',v) > r+1$, because one could pick a shortest path and construct such a u. Therefore, $D' = \{v\}$ is a dominating set, a contradiction.

With this, we can show that only vertices in $\mathcal{T}_m$ are selected.

Lemma 12. *Let v be a vertex selected by u in the same Voronoi cell. Then $v \in \mathcal{T}_{c(v)}$:*

$$(sel^u = v \wedge c(u) = c(v)) \implies v \in \mathcal{T}_{c(v)} \qquad (3)$$

Proof. Assume towards contradiction that $v \notin \mathcal{T}_{c(v)}$. For brevity, let $m := c(v)$. Observe that $m \neq v$, because $m \in \mathcal{T}_m$. Let w be the next vertex on a shortest path from v towards m; possibly m itself. We now analyze the properties of

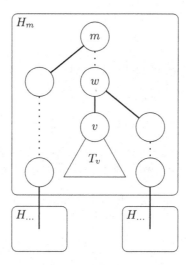

Fig. 2. Typical vertex layout in proof of Lemma 12. The identity of vertex u does not matter; hence, it is not shown.

vertex w and conclude that vertex u should not have selected v. Refer to Fig. 2 for an overview.

By Lemma 7, the subgraph induced by H_m is a tree. If we root this H_m-tree at vertex m, we can denote the subtree rooted at v as T_v. This subtree has depth at most $r - 1$, so w covers the entire subtree: $T_v \subseteq N^r(w)$. All vertices $x \in V \setminus T_v$ are closer to w than to v, as all paths from x to v must go through w. So, the neighborhood of v is included in the neighborhood of w: $N^r[v] \subseteq N^r[w]$.

Now we can use Lemma 11: There must be a vertex t that has distance $r + 1$ to vertex v, so $t \notin N^r(v)$. This means that $t \notin T_v$, and, therefore, $t \in N^r(w)$. In summary, the degree of w is strictly larger: $|N^r(w)| > |N^r(v)|$. Thus, vertex u would prefer selecting w over v.

All that is left is to show that vertex u is indeed able to select vertex w: If $u \in T_v$, then the maximum depth of $r - 1$ means the distance to w is at most r. If $u \notin T_v$, then w is on every shortest path between u and v, and therefore in reach, too.

This leads to a contradiction: Vertex u selected v, although vertex w is in reach, has a strictly larger distance-r neighborhood, and should be preferred by the algorithm.

Corollary 3. *Selections within a Voronoi cell are restricted to T: $D^I \subseteq T$*

Then we bound the size of D^I by providing a bound on the size of T:

Lemma 13. *The set T is small: $|T| \leq (1 + 2r \cdot f(r)) |M|$*

Proof. Consider an arbitrary but fixed $\{v, u\} \in E$ with $c(v) \neq c(u)$. Each path $P_{c(v),v}$ has at most r vertices not in M, because it is a shortest path, and by construction all vertices are dominated by $c(v)$. Each edge in E' corresponds to at most two such paths. With Lemma 9, this bounds the number of paths to at most $2f(r)\,|M|$.

Therefore, $\mathcal{T}$ contains at most $2r \cdot f(r)\,|M| + |M|$ vertices.

Corollary 4. *The set D^I is small:* $\left|D^I\right| \le (1 + 2r \cdot f(r))\,|M|$

Proof. Follows from Corollary 3 and Lemma 13.

As both $\left|D^I\right|$ and $\left|D^O\right|$ are in $O(r \cdot f(r) \cdot |M|)$, this proves Lemma 6, and thus Theorem 1. More specifically, we have proved the upper bound $(1 + 4 \cdot r \cdot f(r)) \cdot |M|$ on $|D|$.

It is possible to show that the above computed upper bound is tight and we postponed this proof entirely to Appendix A.

4   Lower Bound

In this section, we prove that a significantly better approximation of the problem is hard. Intuitively speaking, this is because symmetry cannot be broken in $o(\log^* n)$ rounds, and without that, it is hard to construct any non-trivial distance-r dominating set. Due to the pages limit we only present the idea of the proof and state our main theorem.

We show the hardness by a reduction from the "large" independent set problem to the distance-r dominating set problem, on the graph class of cycles. Intuitively speaking we find a distance-r dominating set D on cycle C; two consecutive vertices of D on C are of distance at most $2r + 1$ from each other, and hence, these vertices help us to break the symmetry, and as $r \in o(\log^* n)$ it yields an independent set of size $O(n)$ in $o(\log^* n)$ rounds.

Theorem 2. Assume an arbitrary but fixed $\delta > 0$ and $r > 1$, with $r \in o(\log^* n)$. Then, there is no deterministic LOCAL algorithm that finds in $O(r)$ rounds a $(2r + 1 - \delta)$-approximation of distance-r dominating set for all $G \in \mathcal{C}$, where $\mathcal{C}$ is the class of cycles of length $\gg 4r + 3$.

5   Vertex Cover, Connected Dominating Set, and Connected Vertex Cover

In this section, we extend Algorithm 1 to solve several other covering problems, namely Distance-r Vertex Cover, Distance-r Connected Vertex Cover, and Distance-r Connected Dominating Set. However, due to page limits we moved the complete proofs to Appendix C.

Proposition 3. *There is a CONGEST algorithm that computes an $O((r \cdot f(r))^2)$ approximation of Distance-r Connected Dominating Set in graphs of bounded expansion with high girth in $O(r)$ rounds.*

Proposition 4. *There is a CONGEST algorithm that computes an $O(r \cdot f(r) \cdot f(r))$ approximation of Distance-r Vertex Cover in graphs of bounded expansion with high girth in $O(r)$ rounds.*

Corollary 5. *There is a CONGEST algorithm that computes a constant factor approximation of the minimum Distance-r Connected Vertex Cover in a constant number of rounds.*

Let us explain the rough idea of proof of all of the above: for connected variations, we take the original set and connect two vertices that are within small distance via short paths. It is possible to show that due to the structure of bounded expansion graphs, this approach does not add much extra overhead and it is a constant factor approximation to the problem. For vertex cover, observe that every vertex cover is already a dominating set, to go the other way around we define boundary vertices of each Voronoi cell of the dominating set, then include them into the solution to vertex cover. It is possible to show that this approach provides a constant factor approximation for the problem (distance-r variation).

6 Conclusion

We have analyzed important distance-r covering problems in a deterministic setting on graphs of bounded expansion $f(r)$ (e.g. planar graphs) and high girth. We have provided CONGEST algorithms and proved that they achieve a constant factor approximation in constant rounds (for a constant r). We have shown that the standard $\Omega(\log^* n)$ lower bound on bounded degree graphs also holds here, even if r is super-constant. This means that Algorithm 1 can only be improved up to a factor of $f(r)$, if at all.

We believe that with our algorithmic analysis tools it is possible to prove that the lower bound of Kuhn et al. [17] extends to high-girth graphs for dominating set; we believe that this can be done without taking the line graph. On the other hand, the necessity of some kind of sparsity requirement seems clear. To what extent can this requirement be reduced?

We have covered the dominating set problem and related covering problems. The next step is to weaken either the sparsity condition or the girth requirement, with the goal of finding the most generic class of graphs with a reasonable approximation in constant rounds, and eventually discovering their connection to dense graphs.

For distance-r computation, one might also consider other interesting problems such as independent set and coloring problems. These relate to network decomposition, and it might help to find a faster network decomposition in the CONGEST model.

A Omitted Proofs of the Main Algorithm

This proves that the algorithm to count the neighborhood indeed works as intended.

Lemma 1. Algorithm 2 computes the size of $N^r(v)$ for each vertex v in parallel.

Proof. First, observe that in only $r - 1$ rounds of communication, no cycle can be detected, as the girth is at least $4r + 3$. This means that $N^i(v)$ is a tree for every $i \leq r - 1$ and $v \in V$. We define the tree $T_{u,i}^{\neg v}$ as the (set of vertices in the) tree of edge-depth i, rooted at vertex u, excluding the branch to vertex v, where v is a neighbor of vertex u.

We prove by induction: At vertex v, after the i-th round[1] (where $0 \leq i \leq r - 1$), n_u stores the size of the tree $T_{u,i}^{\neg v}$.

For the induction basis $i = 0$, we know $\forall u, v : n_u = 1 = \left|T_{u,0}^{\neg v}\right| = |\{u\}|$.

This leaves the induction step: At the beginning of the i-th round (for $1 \leq i \leq r-1$), we know that $n_u = \left|T_{u,i-1}^{\neg v}\right|$ by the induction hypothesis, for every u, v. Consider vertex v. By construction, its distance-i open neighborhood is the union of all edge-depth $i - 1$ trees of v's neighbors, so: $N^i(v) = \bigcup_{u \in N^1(v)} T_{u,i-1}^{\neg v}$. Due to the high girth requirement, we know that all sets in this union are disjoint. Vertex v can therefore compute $\left|N^i(v)\right|$ by summing up all its n_us, and can even compute $\left|T_{v,i}^{\neg u}\right|$ for an arbitrary vertex u by subtracting the corresponding n_u again. This is exactly what happens in line 3. Then v sends $\left|T_{v,i}^{\neg u}\right|$ to each neighbor u, which stores it in the corresponding variable n_v. By symmetry, this also means that vertex v now has stored $\left|T_{u,i}^{\neg v}\right|$ in n_u, thus proving the induction step.

With the meaning of n_u established, line 6 of Algorithm 2 must compute $|N^r(v)|$.

Lemma 12. Let v be a vertex selected by u in the same Voronoi cell. Then $v \in \mathcal{T}_{c(v)}$:

$$\left(sel^u = v \wedge c(u) = c(v)\right) \implies v \in \mathcal{T}_{c(v)} \tag{4}$$

Proof. Assume towards contradiction that $v \notin \mathcal{T}_{c(v)}$. For brevity, let $m := c(v)$. Observe that $m \neq v$, because $m \in \mathcal{T}_m$. Let w be the next vertex on a shortest path from v towards m; possibly m itself. We now analyze the properties of vertex w and conclude that vertex u should not have selected v. Refer to Fig. 2 for an overview.

By Lemma 7, the subgraph induced by H_m is a tree. If we root this H_m-tree at vertex m, we can denote the subtree rooted at v as T_v. This subtree has depth at most $r - 1$, so w covers the entire subtree: $T_v \subseteq N^r(w)$. All vertices $x \in V \setminus T_v$ are closer to w than to v, as all paths from x to v must go through w. So, the neighborhood of v is included in the neighborhood of w: $N^r[v] \subseteq N^r[w]$.

Now we can use Lemma 11: There must be a vertex t that has distance $r + 1$ to vertex v, so $t \notin N^r(v)$. This means that $t \notin T_v$, and, therefore, $t \in N^r(w)$. In summary, the degree of w is strictly larger: $|N^r(w)| > |N^r(v)|$. Thus, vertex u would prefer selecting w over v.

All that is left is to show that vertex u is indeed able to select vertex w: If $u \in T_v$, then the maximum depth of $r - 1$ means the distance to w is at most r.

---

[1] We interpret "after the zeroth round" as "before the first round".

If $u \notin T_v$, then w is on every shortest path between u and v, and therefore in reach, too.

This leads to a contradiction: Vertex u selected v, although vertex w is in reach, has a strictly larger distance-r neighborhood, and should be preferred by the algorithm.

Tightness of Approximation

We have seen that the algorithm is an $O(r \cdot f(r))$ approximation. Is it possible that the algorithm actually performs significantly better what the analysis guarantees? This subsection proves that there are graphs for which the algorithm yields an $\Omega(r \cdot f(r))$ approximation, meaning that the above analysis of the algorithm is asymptotically tight.

Lemma 14. *There is a computable function $f \colon \mathbb{N} \to \mathbb{N}$ and a class of graphs $\mathcal{C}$ of expansion $f(r)$ and girth at least $4r + 3$ such that Algorithm 1 takes $\Theta(r)$ rounds and provides an $\Omega(r \cdot f(r))$-approximation of minimum distance-r dominating set on $\mathcal{C}$. (Cf. Theorem 1.)*

Proof. The rest of this subsection constructs such a set $\mathcal{C}$ from graphs $G_{r,f(r)}$ for all values of $r \geq 1$ and $f(r) \geq 2$.

This does not mean that the problem is hard. It only shows that in the worst case, the presented algorithm may use up the approximation slack.

The construction is a modified version of the subdivided biclique. Let X and Y be two disjoint sets of vertices, each of size $2f(r)$. For each pair $(x, y) \in X \times Y$, connect them with a path $P_{x,y}$ of $2r$ vertices, such that $d(x, y) = 2r + 1$. This means that no vertex can simultaneously cover x and y, i.e., is within distance r of both x and y. For each $x \in X, y \in Y$, create a set $B_{x,y}$ of $k = 2r \cdot f(r)$ new vertices, and connect each vertex in $B_{x,y}$ by a single edge to the vertex closest to x of each path. Let V be the union of all these sets and E as described, then $G_{r,f(r)} = (V, E)$ is the constructed graph.

First, we prove that the graph class satisfies all requirements.

Lemma 15. *For arbitrary but fixed $r \geq 1$ and $f(r) \geq 2$, the graph class $\mathcal{C}$ has expansion $f(r)$ and girth at least $4r + 3$.*

Proof. Let $G_{r,f(r)} \in \mathcal{C}$ be a fixed graph from the constructed graph class. The girth is at least $4 \cdot (2r + 1) > 4r + 3$, as a cycle needs to pass through at least two vertices from X and two vertices from Y. The low expansion can be shown by contracting as much as possible around all vertices in $X \cup Y$, which results in the biclique $K_{2f(r),2f(r)}$, with $4f(r)$ vertices and $4f(r)^2$ edges. Therefore, the constructed graph has $f^G(r) \geq f(r)$. As this is the optimal contraction choice, this also shows $f^G(r) = f(r)$.

This graph class causes worst-case behavior. The running time is trivial:

Lemma 16. *Algorithm 1 runs in $\Omega(r)$ rounds on graphs in $\mathcal{C}$.*

Proof. Follows from the construction of Algorithm 1.

Next, we show that the algorithm computes a comparatively large dominating set:

Lemma 17. *Algorithm 1 provides an $\Omega(r \cdot f(r))$-approximation of minimum distance-r dominating set on $\mathcal{C}$.*

Proof. By construction, $X \cup Y$ is a dominating set, meaning $|M| \leq 4f(r)$. Therefore, it suffices to show that $|D| \geq 4r \cdot f(r)^2$.

We do so by simulating the algorithm on G. We only need to consider the vertices selected by vertices on the paths do. Specifically, pick a specific path $P_{x,y}$ between $x \in X$ and $y \in Y$. Vertices v_x closer to x than to y cover the attached vertices $B_{x,y}$, so $|N^r(v_x)| \geq 2r + k = 2r + 2r \cdot f(r)$. The vertices closer to vertex x cover more of the other paths ending in x, each step increases $|N^r(v_x)|$ by at least $2f(r) - 1$, and loses at most 1 vertex out of sight in the y direction. Note that we ignore the vertices in $B_{x,y'}$ with $y' \neq y$, which would only make this argument stronger. The important property is that $|N^r(v_x)|$ strictly increases towards x, among vertices v_x with $d(v_x, x) < d(v_x, y)$.

Each vertex v_y closer to y than to x does not cover the attached vertices $B_{x,y}$ close to vertex x, as distance r from them would imply distance r to x. We can compute $|N^r(v_y)| \leq r + N^r(v_r) + 1 - 1 = r + r \cdot 2f(r) < 2r + 2r \cdot f(r) \leq |N^r(v_x)|$, so vertex v_y will choose some vertex v_x. As we already established, $|N^r(v_x)|$ increases with decreasing distance to x. Therefore, each v_y will select the vertex closest to x, meaning at least half of each path will be selected, specifically the one on the v_l side.

This means the algorithm selects at least r vertices per path, and there is one such path for each $X \times Y$ combination. Hence $|D| \geq r \cdot 4 \cdot f(r)^2$. Recall that $|M| \leq 4f(r)$, so the algorithm achieves an approximation factor of at least $r \cdot f(r)$ for the constructed graph. Compared with the upper bound of $1 + (4r \cdot f(r))$ this is asymptotically tight.

This concludes the proof of Lemma 14 (tightness of approximation).

B   Lower Bound

In this section we prove the following.

Theorem 2. *Assume an arbitrary but fixed $\delta > 0$ and $r > 1$, with $r \in o(\log^* n)$. Then, there is no deterministic LOCAL algorithm that finds in $O(r)$ rounds a $(2r + 1 - \delta)$-approximation of distance-r dominating set for all $G \in \mathcal{C}$, where $\mathcal{C}$ is the class of cycles of length $\gg 4r + 3$.*

As we will see later, the trivial distance-r dominating set V (i.e., the set of all vertices), is a $(2r + 1)$-approximation in the case of cycles.

This has been proved implicitly in the work of [18]. However, we find it simpler to provide a new proof tailored for our setting, but only for n being a multiple of $2r + 1$. In essence, we show a reduction from the "large" independent set problem to the distance-r dominating set problem, on the graph class of cycles. Intuitively speaking, any algorithm that does significantly better than the trivial dominating set *anywhere* on the cycle leads to a linear sized independent set; and the bound is constructed such that the algorithm needs to do better than trivial somewhere indeed.

The idea is simple: Find a distance-r dominating set D on cycle C; we know two consecutive vertices of D on C are of distance at most $2r+1$ from each other, and hence, these vertices help us to break the symmetry, and as $r \in o(\log^* n)$ it yields an independent set of size $O(n)$ in $o(\log^* n)$ rounds. In the remainder, we formalize this argument.

Assume towards contradiction that $\mathcal{ALG}$ is such a deterministic distributed algorithm, which finds a distance-r dominating set in $G \in \mathcal{C}$ of size at most $(2r + 1 - \delta)\,|M|$, where M is a minimum distance-r dominating set.

We show that $\mathcal{ALG}$ can be used to construct an algorithm violating known lowerbounds on "large" independent set [12, 18]:

Lemma 18 (Lemma 4 of [12]). *There is no deterministic distributed algorithm that finds an independent set of size $\Omega(n/\log^* n)$ in a cycle on n vertices in $o(\log^* n)$ rounds.*

We present the reduction algorithm in Algorithm 3.

Algorithm 3: CONGEST computation of an IS on a cycle $G \in \mathcal{C}_r$, for each v in parallel

---
1: Compute a distance-r dominating set D by simulating $\mathcal{ALG}$.
2: Determine the connected components $V \setminus D$.
3: **for** each component C_i **do**
4:     Determine the two adjacent vertices to C_i, i.e. $u, v \in N(C_i)$.
5:     Let u be the vertex with the lower ID, name it representor of C_i.
6:     All vertices of odd distance to u in C_i join I.
7: **end for**
8: **return** I

We begin by showing basic correctness:

Lemma 19. *Algorithm 3 runs in $o(\log^* n)$ rounds.*

Proof. By assumption, $\mathcal{ALG}$ executes in $O(r)$ rounds. On the other hand, observe that each vertex in D only covers up to a distance of r. Because D is a dominating set, all components must have length at most $2r$. Hence, discovering the adjacent vertex of lowest ID can be done in $O(r)$ as well as propagating the distance information. By construction $r \in o(\log^* n)$, so Algorithm 3 takes $o(\log^* n)$ rounds.

Lemma 20. *Algorithm 3 computes set I, which is an independent set.*

Proof. For two distinct vertices $u, v \in I$, if they belong to different components, then there is no edge between them; otherwise, if they are in the same component, their distance is at least 2, as they are distinct vertices of odd distance from their representor.

Now we can show that this yields a large independent set:

Lemma 21. *The dominating set is not too large:* $|D| \leq (1 - \delta')n$ *for some* $\delta' > 0$.

Proof. By assumption, we know $|D| \leq (2r + 1 - \delta)|M|$, where M is the minimum distance-r dominating set. Construct M' by picking every $2r + 1$-th vertex so that $|M'| = n/(2r + 1)$. Note that M' is a distance-r dominating set, so we have $|M| \leq |M'|$. Together we get $|D| \leq (2r + 1 - \delta)n/(2r + 1) = (1 - \delta')n$, for $\delta' := 1/(2r + 1) > 0$.

Lemma 22. *The set I is large:* $|I| \in \Omega(n/\log^* n)$

Proof. Many vertices must be part of some component: $|V \setminus D| \geq \delta'n$ for some $\delta' > 0$ by Lemma 21. At least half of those vertices are taken into I, thus $|I| \geq \delta'n/2 \in \Omega(n/\log^* n)$.

Proof (Proof of Theorem 2). Lemmas 18 and 22 imply that algorithm $\mathcal{ALG}$ cannot exist.

Note that this does not preclude randomized algorithms. This is because randomized algorithms can indeed achieve a better approximation quality, at least on cycles, by randomly joining the dominating set with sufficiently small probability if necessary, for several rounds, and finally all uncovered vertices join.

C   Omitted Proofs of the Extensions

The basic version of these problems are well studied in the literature; our extension to distance-r can be found in Sect. 2. We explain the extension of Algorithm 1 to each of these problems and show their correctness. Again, we assume the input graph is $G = (V, E)$ so that we can directly refer to its edge and vertex set.

We begin with the Distance-r Connected Dominating Set problem:

Proposition 5. *There is a CONGEST algorithm that computes an $O((r \cdot f(r))^2)$ approximation of Distance-r Connected Dominating Set in graphs of bounded expansion with high girth in $O(r)$ rounds.*

Proof. We prove this by constructing Algorithm 4 as a simple extension of Algorithm 1, or any other appropriate CONGEST distance-r dominating set algorithm.

Algorithm 4 is a CONGEST algorithm, as the distance-r dominating set algorithm is CONGEST, and all other messages only contain a constant amount of identifiers.

Algorithm 4: CONGEST computation of connected r-MDS, on each vertex v in parallel

---

1: Compute a Distance-r Dominating Set D of the graph
2: Determine the closest dominating vertex sel^v
3: Determine the path P^v from v to sel^v
4: If any neighbor u has a $sel^u \neq sel^v$, call vertex v a border
5: **return** $\hat{D}$ as the union of D, all border vertices, and all paths P^v

---

Algorithm 4 takes $O(r)$ rounds, because the distance-r dominating set algorithm does so, too, and all other steps also only take $O(r)$ rounds.

$\hat{D}$ is a dominating set because $D \subseteq \hat{D}$ is a dominating set.

Define Voronoi cells H_m according to sel^v for $v \in V$. Note that Corollary 2 and Lemma 7 apply analogously.

We show that $\hat{D}$ is connected by construction, if G is connected: Vertices v within a Voronoi cell $H_m \cap \hat{D}$ are connected by construction, as they are all connected to $m \in \hat{D}$. Furthermore, for every path P_G in the input graph G, one can construct a corresponding walk W_H in the Voronoi graph by mapping each vertex to its Voronoi cell (i.e. sel^v). Thus, the Voronoi cells are connected.

Finally, we show the approximation quality: Consider the minimal Distance-r Connected Dominating Set $\hat{M}$. One can easily see that the minimum distance-r dominating set M is not larger: $|M| \leq \left|\hat{M}\right|$. An argument similar to Lemma 9 shows that the number of border vertices is bounded in $|D|$; and by construction of D the Voronoi cells have radius at most r (and therefore so do the paths). By Theorem 1, we can now deduce: $\left|\hat{D}\right| \in O(r \cdot f(r) \cdot |D|) \subseteq O((r \cdot f(r))^2 \cdot |M|) \subseteq O((r \cdot f(r))^2 \cdot \left|\hat{M}\right|)$

For constant r, the terms simplify to:

Corollary 6. *There is a CONGEST algorithm that computes a constant factor approximation of Distance-r Connected Dominating Set for constant r in graphs of bounded expansion with high girth in constant number of rounds.*

Likewise, we can solve the related vertex cover problem. Intuitively speaking, we can define Voronoi cells according to the computed dominating set, determine borders between cells, and include all borders into the vertex cover. More formally:

Proposition 6. *There is a CONGEST algorithm that computes an $O(r \cdot f(r) \cdot f(r))$ approximation of Distance-r Vertex Cover in graphs of bounded expansion with high girth in $O(r)$ rounds.*

Proof. We prove this by constructing Algorithm 5 as a simple extension of Algorithm 1, or any other appropriate CONGEST distance-r dominating set algorithm.

Algorithm 5: CONGEST computation of distance-r vertex cover, on each vertex v in parallel

---

1: Compute a Distance-r Dominating Set D of the graph
2: Determine the closest dominating vertex sel^v
3: If any neighbor u of vertex v has a $sel^u \neq sel^v$, call v a border
4: **return** C as the union of D and all border vertices

Algorithm 5 is a CONGEST algorithm as the Distance-r Dominating Set algorithm is CONGEST, and all other messages only contain a constant amount of identifiers. Algorithm 5 takes $O(r)$ rounds, because the Distance-r Dominating Set algorithm does so, too, and all other steps also only take $O(r)$ rounds.

Define Voronoi cells according to sel^v for $v \in V$. The set C is a distance-r vertex cover by simple case distinction: All edges $e = u, v$ that are fully inside a Voronoi cell, i.e., there is a vertex $w \in C$ with $w = sel^u = sel^v$, is covered by vertex w. All edges $e = u, v$ with $sel^u \neq sel^v$ are covered by vertices u and v, as both vertices were detected as borders.

Finally, we show the approximation quality: Consider the minimal Distance-r Vertex Cover M^{VC}. One can easily see that the minimum distance-r dominating set M is smaller: $|M| \leq |M^{VC}|$. An argument similar to Lemma 9 shows that the number of border vertices is bounded in $|D|$. By Theorem 1, we can now deduce: $|C| \in O(f(r) \cdot |D|) \subseteq O(r \cdot f(r) \cdot f(r) \cdot |M|) \subseteq O(r \cdot f(r) \cdot f(r) \cdot |M^{VC}|)$

Again, for constant r, the terms simplify to:

Corollary 7. *There is a CONGEST algorithm that computes a constant factor approximation of the minimum Distance-r Vertex Cover in a constant number of rounds.*

From Theorem 5 and Corollary 7, the following is straight-forward to see:

Corollary 8. *There is a CONGEST algorithm that computes a constant factor approximation of the minimum Distance-r Connected Vertex Cover in a constant number of rounds.*

The proof is omitted for brevity and follows exactly the same scheme.

This proves that Algorithm 1 can be extended to compute a Distance-r Vertex Cover instead, as mentioned in Sect. 5.

References

1. Althöfer, I., Das, G., Dobkin, D., Joseph, D., Soares, J.: On sparse spanners of weighted graphs. Discrete Comput. Geometry **9**(1), 81–100 (1993). https://doi.org/10.1007/BF02189308
2. Amiri, S.A., Ossona de Mendez, P., Rabinovich, R., Siebertz, S.: Distributed domination on graph classes of bounded expansion. In: Proceedings of the 30th on Symposium on Parallelism in Algorithms and Architectures, pp. 143–151. ACM (2018)

3. Amiri, S.A., Schmid, S.: Brief announcement: a log-time local MDS approximation scheme for bounded genus graphs. In: Proceedings of 30th International Symposium on Distributed Computing (DISC), pp. 480–483. Springer, Heidelberg (2016). https://doi.org/10.1007/978-3-662-53426-7
4. Amiri, S.A., Schmid, S., Siebertz, S.: A local constant factor MDS approximation for bounded genus graphs. In: Proceedings. ACM Symposium on Principles of Distributed Computing (PODC) (2016)
5. Awerbuch, B., Berger, B., Cowen, L., Peleg, D.: Fast network decomposition. In: Proceedings of the Eleventh Annual ACM Symposium on Principles of Distributed Computing, PODC 1992, pp. 169–177. ACM (1992)
6. Awerbuch, B., Goldberg, A.V., Luby, M., Plotkin, S.A.: Network decomposition and locality in distributed computation. In: 30th Annual Symposium on Foundations of Computer Science, 1989, pp. 364–369 (1989)
7. Bacrach, N., Censor-Hillel, K., Dory, M., Efron, Y., Leitersdorf, D., Paz, A.: Hardness of distributed optimization. In: Proceedings of the 2019 ACM Symposium on Principles of Distributed Computing, PODC 2019, pp. 238–247. ACM (2019)
8. Baker, B.S.: Approximation algorithms for NP-complete problems on planar graphs. J. ACM (JACM) **41**(1), 153–180 (1994)
9. Bar-Yehuda, R., Censor-Hillel, K., Maus, Y., Pai, S., Pemmaraju, S.V.: Distributed approximation on power graphs. In: Proceedings of the 2020 ACM Symposium on Principles of Distributed Computing, PODC 2020, pp. 501–510. ACM (2020)
10. Bar-Yehuda, R., Censor-Hillel, K., Schwartzman, G.: A distributed $(2+\epsilon)$-approximation for vertex cover in $o(\log\delta/\epsilon \log\log \delta)$ rounds. In: Proceedings of the 2016 ACM Symposium on Principles of Distributed Computing, PODC 2016, Chicago, IL, USA, pp. 3–8 (2016)
11. Czygrinow, A., Hańćkowiak, M., Szymańska, E.: Distributed approximation algorithms for planar graphs. In: Calamoneri, T., Finocchi, I., Italiano, G.F. (eds.) CIAC 2006. LNCS, vol. 3998, pp. 296–307. Springer, Heidelberg (2006). https://doi.org/10.1007/11758471_29
12. Czygrinow, A., Hańćkowiak, M., Wawrzyniak, W.: Fast distributed approximations in planar graphs. In: Taubenfeld, G. (ed.) DISC 2008. LNCS, vol. 5218, pp. 78–92. Springer, Heidelberg (2008). https://doi.org/10.1007/978-3-540-87779-0_6
13. Czygrinow, A., Hańćkowiak, M., Wawrzyniak, W., Witkowski, M.: Distributed approximation algorithms for the minimum dominating set in K_h-minor-free graphs. In: 29th International Symposium on Algorithms and Computation, ISAAC, pp. 22:1–22:12 (2018)
14. Deurer, J., Kuhn, F., Maus., Y.: Deterministic distributed dominating set approximation in the CONGEST model. In: Proceedings of the 2019 ACM Symposium on Principles of Distributed Computing, PODC 2019, pp. 94–103 (2019)
15. Diestel, R.: Graph Theory. GTM, vol. 173. Springer, Heidelberg (2017). https://doi.org/10.1007/978-3-662-53622-3
16. Ghaffari, M.: Distributed maximal independent set using small messages. In: Proceedings of the Thirtieth Annual ACM-SIAM Symposium on Discrete Algorithms, SODA 2019, pp. 805–820 (2019)
17. Kuhn, F., Moscibroda, T., Wattenhofer, R.: Local computation: lower and upper bounds. J. ACM **63**(2), 17:1–17:44 (2016)
18. Lenzen, C., Pignolet, Y.A., Wattenhofer, R.: Distributed minimum dominating set approximations in restricted families of graphs. Distrib. Comput. **26**(2), 119–137 (2013)

19. Lenzen, C., Wattenhofer, R.: Minimum dominating set approximation in graphs of bounded arboricity. In: Lynch, N.A., Shvartsman, A.A. (eds.) DISC 2010. LNCS, vol. 6343, pp. 510–524. Springer, Heidelberg (2010). https://doi.org/10.1007/978-3-642-15763-9_48

20. Linial, N.: Locality in distributed graph algorithms. SIAM J. Comput. **21**(1), 193–201 (1992)

21. Naor, M., Stockmeyer, L.: What can be computed locally? In: Proceedings of ACM 25th Annual ACM Symposium on Theory of Computing (STOC), pp. 184–193 (1993)

22. Nešetřil, J., Ossona de Mendez, P.: Sparsity. AC, vol. 28. Springer, Heidelberg (2012). https://doi.org/10.1007/978-3-642-27875-4

23. Peleg, D.: Distributed Computing: A Locality-sensitive Approach. Society for Industrial and Applied Mathematics (2000)

24. Preparata, F.P., Shamos, M.I.: Computational Geometry - An Introduction. Texts and Monographs in Computer Science. Springer, New York (1985). https://doi.org/10.1007/978-1-4612-1098-6

25. Rozhon, V., Ghaffari, M.: Polylogarithmic-time deterministic network decomposition and distributed derandomization. In: Proceedings of ACM 52nd Annual ACM SIGACT Symposium on Theory of Computing (STOC), pp. 350–363. ACM (2020)

26. Schneider, J., Wattenhofer, R.: A log-star distributed maximal independent set algorithm for growth-bounded graphs. In: Proceedings of the Twenty-Seventh Annual ACM Symposium on Principles of Distributed Computing, PODC 2008, pp. 35–44 (2008)

Reconfiguration of Connected Graph Partitions via Recombination

Hugo A. Akitaya[1]([✉])(iD), Matias Korman[2], Oliver Korten[3], Diane L. Souvaine[4], and Csaba D. Tóth[4,5](iD)

[1] Department of Computer Science, University of Massachusetts Lowell, Lowell, MA, USA
hugo_akitaya@uml.edu
[2] Siemens Electronic Design Automation, Wilsonville, OR, USA
[3] Department of Computer Science, Columbia University, New York, NY, USA
[4] Department of Computer Science, Tufts University, Medford, MA, USA
[5] Department of Mathematics, Cal State Northridge, Los Angeles, CA, USA

Abstract. Motivated by applications in gerrymandering detection, we study a reconfiguration problem on connected partitions of a connected graph G. A partition of $V(G)$ is **connected** if every part induces a connected subgraph. In many applications, it is desirable to obtain parts of roughly the same size, possibly with some slack s. A **Balanced Connected k-Partition with slack** s, denoted (k,s)-**BCP**, is a partition of $V(G)$ into k nonempty subsets, of sizes $n_1, \ldots, n_k$ with $|n_i - n/k| \leq s$, each of which induces a connected subgraph (when $s = 0$, the k parts are perfectly balanced, and we call it k-**BCP** for short).

A **recombination** is an operation that takes a (k,s)-BCP of a graph G and produces another by merging two adjacent subgraphs and repartitioning them. Given two k-BCPs, A and B, of G and a slack $s \geq 0$, we wish to determine whether there exists a sequence of recombinations that transform A into B via (k,s)-BCPs. We obtain four results related to this problem: (1) When s is unbounded, the transformation is always possible using at most $6(k-1)$ recombinations. (2) If G is Hamiltonian, the transformation is possible using $O(kn)$ recombinations for any $s \geq n/k$, and (3) we provide negative instances for $s \leq n/(3k)$. (4) We show that the problem is PSPACE-complete when $k \in O(n^\varepsilon)$ and $s \in O(n^{1-\varepsilon})$, for any constant $0 < \varepsilon \leq 1$, even for restricted settings such as when G is an edge-maximal planar graph or when $k \geq 3$ and G is planar.

1 Introduction

Partitioning the vertex set of a graph $G = (V, E)$ into k nonempty subsets $V = \bigcup_{i=1}^{k} V_i$ that each induces a connected graph $G[V_i]$ is a classical problem, known as the **Connected Graph Partition** problem [9,16]. Motivated by fault-tolerant network design and facility location problems, it is part of a broader

Supported by NSF CCF-1422311, CCF-1423615, DMS-1800734, OIA-1937095, and NSERC.

© Springer Nature Switzerland AG 2021
T. Calamoneri and F. Corò (Eds.): CIAC 2021, LNCS 12701, pp. 61–74, 2021.
https://doi.org/10.1007/978-3-030-75242-2_4

family of problems where each induced graph $G[V_i]$ must have a certain graph property (e.g., ℓ-connected or H-minor-free). In some instances, it is desirable that the parts $V_1, \ldots, V_k$ have the approximately the same size (depending on some pre-established threshold). A **Balanced Connected k-Partition** (for short, k-**BCP**) is a connected partition requiring that $|V_i| = n/k$, for $i \in \{1, \ldots, k\}$ where $n = |V(G)|$ is the total number of vertices. Dyer and Frieze [7] proved that finding a k-BCP is NP-hard for all $2 \leq k \leq n/3$. For $k = 2, 3$ the problem can be solved efficiently when G is bi- or triconnected, respectively [20,22], and is equivalent to the perfect matching problem for $k = n/2$. Later Chlebíková [4] and Chataigner et al. [3] obtained approximation and inapproximability results for maximizing the "balance" ratio $\max_i |V_i| / \min_j |V_j|$ over all connected k-partitions. See also [12,14,19,23] for variants under various other optimization criteria.

In this paper, our basic element is a connected k-partition of a graph $G = (V, E)$ that is balanced up to some additive threshold that we call a **slack** $s \geq 0$, denoted (k, s)-**BCP**. We explore the space of all (k, s)-**BCPs** of the graph $G = (V, E)$. Note that the total number of (k, s)-**BCPs** for all $s \geq 0$, is bounded above by the number k-partitions of V, which is the Stirling number of the second kind $S(n, k)$, and asymptotically equals $(1 + o(1))k^n / k!$ for constant k. This bound is the best possible for the complete graph $G = K_n$.

In a recent application [1,6,18], $G = (V, E)$ represents the adjacency graph of precincts in an electoral map, which should be partitioned into k districts $V_1, \ldots, V_k$ where each district will elect one representative. Motivated by the design and evaluation of electoral maps under optimization criteria designed to promote electoral fairness, practitioners developed empirical methods to sample the configuration space of potentialdistrict maps by a random walk on the graph where each step corresponds to some elementary **reconfiguration move** [8]. From a theoretical perspective, the stochastic process converges to uniform sampling [13,15]. However, the move should be **local**, i.e., it must affect a constant number of districts, to allow efficient computation of each move, and it should support rapid mixing (i.e., the random walk should converge, in total variation distance, to its steady state distribution in time polynomial in n). Crucially, the space of (approximately balanced) k-partitions of G must be **connected** under the proposed move. Previous research considered the **single switch** move, in which a single vertex $v \in V$ switches from one set V_i to another set V_j (assuming that both $G[V_i]$ and $G[V_j]$ remain connected). Akitaya et al. [2] proved that the configuration space is connected under single switch moves if G is biconnected, but in general it is NP-hard both to decide whether the space is connected and to find a shortest path between two valid k-partitions. While the single switch is local, both worst-case constructions and empirical evidence [5,18] indicate that it does not support rapid mixing.

In this paper we consider a different move. Specifically, we consider the configuration space of k-partitions under the **recombination** move, proposed by DeFord et al. [5], in which the vertices in $V_i \cup V_j$, for some $i, j \in \{1, \ldots, k\}$, are re-partitioned into $V_i' \cup V_j'$ such that both $G[V_i']$ and $G[V_j']$ are connected. We also study variants restricted to balanced or near-balanced partitions, that is, when $|V_i| = n/k$ for all $i \in \{1, \ldots, k\}$, or when $\big||V_i| - n/k\big| \leq s$ for a given slack $s \geq 0$. In application domains mentioned above, the underlying graph G is often planar or near-planar, and in some cases it is a triangulation (i.e., an edge-maximal planar graph). Results pertaining to these special cases are of particular interest. Our results lay down theoretical foundations for this model in graph theory and computational tractability. Although our results imply lower bounds in the mixing time of worst-case instances, they have no direct implication for the average case analysis.

Definitions. Let $G = (V, E)$ be a graph with $n = |V(G)|$. For a positive integer k, a **connected k-partition** Π of G is a partition of $V(G)$ into disjoint nonempty subsets $\{V_1, \ldots, V_k\}$ such that the induced subgraph $G[V_i]$ is connected for all $i \in \{1, \ldots, k\}$. Each subgraph induced by V_i is called a **district**. We write $\Pi(v)$ for the subset in Π that contains vertex v.

Denote by $\text{Part}(G, k)$ the set of connected k-partitions on G. We also consider subsets of $\text{Part}(G, k)$ in which all districts have the same or almost the same number of vertices. A connected k-partition of G is **balanced** (k-BCP) if every district has precisely n/k vertices (which implies that n is a multiple of k); and it is **balanced with slack** $s \geq 0$ ((k, s)-BCP), if $\big||U| - n/k\big| \leq s$ for every district $U \subset V$. Let $\text{Bal}_s(G, k)$ denote the set of connected k-partitions on G that are balanced with slack s, i.e., the set of all (k, s)-BCPs. The set of balanced k-partitions is denoted $\text{Bal}(G, k) = \text{Bal}_0(G, k)$; and $\text{Part}(G, k) = \text{Bal}_\infty(G, k)$.

We now formally define a **recombination move** as a binary relation on $\text{Bal}_s(G, k)$. Two non-identical (k, s)-BCPs, $\Pi_1 = \{V_1, \ldots, V_k\}$ and $\Pi_2 = \{W_1, \ldots, W_k\}$, are related by a recombination move if there exist $i, j \in \{1, \ldots, k\}$, and a permutation π on $\{1, \ldots, k\}$ such that $V_i \cup V_j = W_{\pi(i)} \cup W_{\pi(j)}$ and $V_\ell = W_{\pi(\ell)}$ for all $\ell \in \{1, \ldots, k\} \backslash \{i, j\}$. We say that Π_1 and Π_2 are a recombination of each other. This binary relation is symmetric and defines a graph on $\text{Bal}_s(G, k)$ for all $s \geq 0$. This graph is the **configuration space** of $\text{Bal}_s(G, k)$ under recombination, denoted by $\mathcal{R}_s(G, k)$.

Balanced Recombination Problem $BR(G, k, s)$: Given a graph $G = (V, E)$ with $|V| = n$ vertices and two (k, s)-BCPs A and B, decide whether there exists a path between A and B in $\mathcal{R}_s(G, k)$, i.e. whether there is a sequence of recombination moves that carries A to B such that every intermediate partition is a (k, s)-BCP.

Our Results. We prove, in Sect. 2, that the configuration space $\mathcal{R}_\infty(G, k)$ is connected whenever the underlying graph G is connected and the size of the districts is unrestricted. It is easy, however, to construct a graph G where $\mathcal{R}_0(G, k)$ is disconnected. We study what is the minimum slack s, as a function of n and k, that guarantees that $\mathcal{R}_s(G, k)$ is connected for all connected (or possibly biconnected) graphs G with n vertices. We prove that $\mathcal{R}_s(G, k)$ is connected and its diameter is $O(nk)$ for $s = n/k$ when G is a Hamiltonian graph (Sect. 3). As a counterpart, we construct a family of Hamiltonian planar graphs G such that $\mathcal{R}_s(G, k)$ is disconnected for $s < n/(3k)$ (Sect. 4).

We prove in Sect. 5 that $BR(G, k, s)$ is PSPACE-complete even for the special case when G is a triangulation (i.e., an edge-maximal planar graph), k is $O(n^\varepsilon)$

and s is $O(n^{1-\varepsilon})$ for constant $0 < \varepsilon \le 1$. As a consequence we show that finding a (k,s)-BCP of G is NP-hard in the same setting. Note that the previously known hardness proofs for finding k-BCPs either require that G is weighted and nonplanar [3] or G contain cut vertices [7]. In contrast, if G is planar and 4-connected, then G admits a Hamilton cycle [21] and, therefore, a (k,s)-BCP is easily obtained by partitioning a Hamilton cycle into the desired pieces. Finally, we modify our construction to also show that $BR(G,k,s)$ is PSPACE-complete even for the special case when G is planar, $k \ge 3$, and s is bounded above by $O(n^{1-\varepsilon})$ for constant $0 < \varepsilon \le 1$.

2   Recombination with Unbounded Slack

In this section, we show that the configuration space $\mathcal{R}_\infty(G,k)$ is connected under recombination moves if G is connected (cf. Theorem 1). The proof proceeds by induction on k, where the induction step depends on Lemma 2 below.

We briefly review some standard graph terminology. A **block** of a graph G is a maximal biconnected component of G. A vertex $v \in V(G)$ is a **cut vertex** if it lies in two or more blocks of G, otherwise it is a **block vertex**. In particular, if v is a block vertex, then $G - v$ is connected. If G is a connected graph with two or more vertices, then every block has at least two vertices. A block is a **leaf-block** if it contains precisely one cut vertex of G. Every connected graph either is biconnected or has at least two leaf blocks. The **arboricity** of a graph G is the minimum number of forests that cover all edges in $E = (G)$. The **degeneracy** of G is the largest minimum vertex degree over all induced subgraphs of G. It is well known that if the arboricity of a graph is a, then its degeneracy is between a and $2a - 1$.

Lemma 1. *If the arboricity of a graph is a, then it contains a block vertex of degree at most $2a - 1$.*

The proof of Lemma 1 can be found in [17]. The heart of the induction step of our main result hinges on the following lemma.

Lemma 2. *Let G be a connected graph, $k \ge 2$ an integer, and $\Pi_1, \Pi_2 \in Part(G,k)$ be two k-partitions of G. Then there exists a block vertex $v \in V(G)$ such that up to three recombination moves can transform Π_1 and Π_2 each to two new k-partitions in which $\{v\}$ is a singleton distinct.*

Proof. Let $\Pi_1 = \{V_1, \ldots, V_k\}$ and $\Pi_2 = \{W_1, \ldots, W_k\}$. We construct two spanning trees, T_1 and T_2, for G that each contain $k-1$ edges between the districts of Π_1 and Π_2, respectively. Specifically, for $i \in \{1, \ldots, k\}$, let $T(V_i)$ be a spanning tree of $G[V_i]$, $T(W_i)$ a spanning tree of $G[W_i]$. As G is connected, we can augment the forest $\bigcup_{i=1}^{k} T(V_i)$ to a spanning tree T_1 of G, using $k-1$ new edges, which connect vertices in distinct districts. Similarly, we can augment $\bigcup_{i=1}^{k} T(W_i)$ to a spanning tree T_2 of G. Now, let $G' = T_1 \cup T_2$. By definition, the arboricity of G' is at most 2. By Lemma 1, G' contains a block vertex v with $\deg_{G'}(v) \le 3$.

We show that we can modify Π_1 (resp., Π_2) to create a singleton district $\{v\}$ in at most three moves. Assume without loss of generality that $v \in V_1$ and $v \in W_1$. Since $\deg_{G'}(v) \leq 3$, we have $\deg_{T(V_1)}(v) \leq 3$ and $\deg_{T(W_1)}(v) \leq 3$. Consequently, $T(V_1) - v$ (resp., $T(W_1) - v$) has at most three components, each of which is adjacent to some other district, since $G' - v$ is connected. Up to three successive recombinations can decrease the district V_1 with the components of $T(V_1) - v$, and reduce V_1 to $\{v\}$. Similarly, at most three successive recombinations can reduce W_1 to $\{v\}$.   □

Theorem 1. *Let G be a connected graph and $k \geq 1$ a positive integer. For all $\Pi_1, \Pi_2 \in Part(G, k)$, there exists a sequence of at most $6(k-1)$ recombination moves that transforms Π_1 to Π_2.*

Proof. We proceed by induction on k. In the base case, $k = 1$, and $\Pi_1 = \Pi_2$. Assume that $k > 1$ and claim holds for $k - 1$. By Lemma 2, we can find a block vertex $v \in V(G)$ and up to six recombination moves transform Π_1 and Π_2 into Π_1' and Π_2' such that both contain $\{v\}$ as a singleton district. Since v is a block vertex, $G - v$ is connected; and since $\{v\}$ is a singleton district in both Π_1' and Π_2', we have $\Pi_1 - \{v\}, \Pi_2 - \{v\} \in Part(G-v, k-1)$. By induction, a sequence of up to $6(k-2)$ recombination moves in $G-v$ can transform $\Pi_1 - \{v\}$ into $\Pi_2 - \{v\}$. These moves remain valid recombination moves in G if we add singleton district $\{v\}$. Overall, the combination of these sequences yields a sequence of up to $6 + 6(k-2) = 6(k-1)$ recombination moves that transforms Π_1 to Π_2. This completes the induction step.   □

3   Recombination with Slack

In this section, we prove that the configuration space $\mathcal{R}_s(G, k)$ is connected if the slack is greater or equal to the average district size, that is, $s \geq n/k$, and the underlying graph G is Hamiltonian (Theorem 2).

Let G be a graph with n vertices that contains a Hamilton cycle C. Assume that n is a multiple of k. A k-partition in $\mathsf{Bal}_s(G, k)$ is **canonical** if each district consists of consecutive vertices along C. Using a slack of $s \geq n/k$, we can transform any canonical k-partition to any other using $O(k^2)$ reconfigurations.

Lemma 3. *Let G be a graph with n vertices and a Hamilton cycle C, $k \geq 1$ is a divisor of n, and $s \geq n/k$. Then the subgraph of $\mathcal{R}_s(G, k)$ induced by canonical k-partitions is connected and its diameter is at most $k^2 + 1$.*

Proof Sketch. We proceed by induction: We assume that the first $\ell \in \{0, \ldots, k\}$ districts each have size $\frac{n}{k}$, and we change the size of the $(\ell + 1)$st district to $\frac{n}{k}$ using at most $k - \ell - 1$ recombinations. Since the average size of the remaining $k - \ell$ districts is n/k, there are two consecutive districts of size at most $\frac{n}{k}$ and at least $\frac{n}{k}$, respectively. We recombine the first such pair of districts, and propagate the changes to the $(\ell + 1)$st district, completing the induction step.   □

In the remainder of this section, we show that every k-partition in $\mathsf{Bal}_s(G, k)$ can be brought into canonical form by a sequence of $O(nk)$ recombinations.

Preliminaries. We introduce some terminology. Let $\Pi = \{V_1, \ldots, V_k\} \in$ $\mathsf{Bal}_s(G, k)$ with a slack of $s \geq n/k$. For every $i \in \{1, \ldots, k\}$, a **fragment** of $G[V_i]$ is a maximum set $F \subset V_i$ of vertices that are contiguous along C. Every set V_i is the disjoint union of one or more fragments. The k-partition Π is canonical if and only if every district has precisely one fragment. Our strategy is to "defragment" Π if it is not canonical, that is, we reduce the number of fragments using recombination moves.

We distinguish between two types of districts in Π: A district V_i is **small** if $|V_i| \leq n/k$, otherwise it is **large**. Every edge in $E(G)$ is either an edge or a **chord** of the cycle C. For every $i \in \{1, \ldots, k\}$, let f_i be the number of fragments of V_i. Let T_i be a spanning tree of $G[V_i]$ that contains the minimum number of chords. The edges of $G[V_i]$ along C form a forest of f_i paths; we can construct T_i by augmenting this forest to a spanning tree of $G[V_i]$ using $f_i - 1$ chords.

The **center** of a tree T is a vertex $v \in V(T)$ such that each component of $T - v$ has up to $|V(T)|/2$ vertices. It is well known that every tree has a center. For $i \in \{1, \ldots, k\}$, let c_i be a center of the spanning tree T_i of $G[V_i]$. Let the fragment of V_i be **heavy** if it contains c_i; and **light** otherwise. We also define a parent-child relation between the fragments of V_i. Fragments A and B are in a parent-child relation if they are adjacent in T_i and if c_i is closer to A than to B in T_i. Note that a light fragment and its descendants jointly contain less than $|V_i|/2 \leq (n/k + s)/2$ vertices; see Fig. 1.

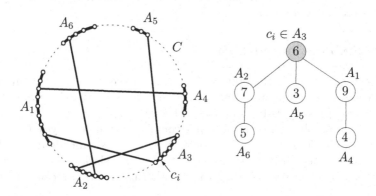

Fig. 1. Left: A distinct V_i with 26 vertices (hollow dots) in six fragments (bold arcs) along C. The spanning tree T_i of $G[V_i]$ contains five edges, with a center at c_i. Right: Parent-child relationship between fragments is defined by the tree rooted at the fragment containing c_i.

The following four lemmas show that we can decrease the number of fragments under some conditions. In all four lemmas, we assume that G is a graph with a Hamiltonian cycle C, and Π is a noncanonical (k, s)-BCP with $s \geq n/k$.

Lemma 4. *If a light fragment of a large district is adjacent to a small district along C, then a recombination move can decrease the number of fragments.*

Proof. Assume without loss of generality that v_1v_2 is an edge of C, where $v_1 \in F_1 \subset V_1$, $v_2 \in F_2 \subseteq V_2$, F_1 is a light fragment of a large district V_1, and F_2 is some fragment of a small district V_2. Let $\overline{F}_1$ be the union of fragment F_1 and all its descendants. By the definition of the center c_1, we have $|\overline{F}_1| < |V_1|/2$. Apply a recombination replacing V_1 and V_2 with $W_1 = V_1 \backslash \overline{F}_1$ and $W_2 = V_2 \cup \overline{F}_1$.

We show that the resulting partition is a (k, s)-BCP. Note also that both $G[\overline{F}_1]$ and $G[V_1 \backslash \overline{F}_1] = G[W_1]$ are connected. Since $v_1v_2 \in E(G)$, then $G[V_2 \cup \overline{F}_1] = G[W_2]$ is also connected. As W_1 contains the center of V_1, we have $|W_1| \geq 1$ and $|W_1| < |V_1| \leq n/k+s$. As V_2 is small, have $|W_2| = |V_2|+|\overline{F}_1| < n/k+n/k \leq 2n/k \leq n/k + s$. Finally, note that $F_1 \cup F_2$ is a single fragment in the resulting k-partition, hence the number of fragments decreased by at least 1. $\qquad\square$

Lemma 5. *If no light fragment of a large district is adjacent to any small district along C, then there exists two adjacent districts along C whose combined size is at most $2n/k$.*

Proof. Suppose, to the contrary, that every small district is adjacent only to heavy fragments along C, and the combined size of every pair of adjacent districts along C is greater than $\frac{2n}{k}$, meaning that at least one district is large. We assign every small district to an adjacent large district as follows. For every small district V_i, let F_i be one of its arbitrary fragments. We assign V_i to the large district whose heavy fragment is adjacent to F_i in the clockwise direction along C. Since every large district has a unique heavy fragment, and at most one district precedes it in clockwise order along C, the assignment is a matching of the small districts to large districts. Denote this matching by M. Every district that is not part of a pair in M must be large. By assumption, every pair in M has combined size greater than $\frac{2n}{k}$, so the average district size over the districts in M is greater than $\frac{n}{k}$. The districts not in M are large so their average size also exceeds $\frac{n}{k}$. Overall the average district size exceeds $\frac{n}{k}$. But Π is a k-partition of n vertices, hence the average district size is exactly $\frac{n}{k}$, a contradiction. $\qquad\square$

Lemma 6. *If districts V_1 and V_2 are adjacent along C and $|V_1 \cup V_2| \leq n/k + s$, then there is a recombination move that either decreases the number of fragments, or maintains the same number of fragments and creates a singleton district.*

Proof. Assume, w.l.o.g., that $v_1 \in F_1 \subseteq V_1$, $v_2 \in F_2 \subseteq V_2$, where v_1v_2 is an edge of C, and F_1 and F_2 are fragments of V_1 and V_2, respectively. The induced graph $G[V_1 \cup V_2]$ is connected, and $T_1 \cup T_2 \cup v_1v_2$ is one of its spanning trees. If T_1 or T_2 contains a chord, say e, then $(T_1 \cup T_2 \cup v_1v_2) - e$ has two components, T_3 and T_4, each of size at most $n/k + s - 1$. A recombination move can replace V_1 and V_2 with $V(T_3)$ and $V(T_4)$. Since fragments F_1 and F_2 merge into one fragment, the number of fragments decreases by at least one. Otherwise, neither T_1 nor T_2 contains a chord. Then V_1 and V_2 each has a single fragment, so $V_1 \cup V_2$ is a chain of vertices along C. Let v be the first vertex in this chain. A recombination move can replace V_1 and V_2 with $W_1 = \{v\}$ and $W_2 = (V_1 \cup V_2) \backslash \{v\}$. By construction both $G[W_1]$ and $G[W_2]$ are connected, $|W_1| = 1$, $|W_2| = |V_1 \cup V_2| - 1 \leq n/k + s - 1$, and the number of fragments does not change. $\qquad\square$

Lemma 7. *If there exists a singleton district, then there exists a sequence of at most $k - 1$ recombination moves that decreases the number of fragments.*

Proof. Let $C = (v_1, \ldots, v_n)$. Assume without loss of generality that $V_1 = \{v_1\}$ is a singleton district, and $v_2 \in F_2 \subseteq V_2$, where F_2 is a fragment of district V_2. Since not all districts are singletons, we may further assume that $|V_2| \geq 2$. We distinguish between two cases.

Case 1: $F_2 \neq V_2$ (i.e., V_2 has two or more fragments). Let e be an arbitrary chord in T_2, and denote the two subtrees of $T_2 - e$ by T_2^- and T_2^+ such that v_2 is T_2^-. Since $|V_2| \leq n/k + s$, the subtrees T_2^- and T_2^+ each have at most $n/k + s - 1$ vertices. We can recombine V_1 and V_2 into $W_1 = V_1 \cup V(T_2^-)$ and $W_2 = V(T_2^+)$. Then $|W_1| \leq 1 + (n/k + s - 1) = n/k + s$ and $|W_2| \leq n/k + s - 1$; they both induce a connected subgraph of G. As the singleton fragment V_1 and F_2 merge into one fragment of W_1, the number of fragments decreases by at least one.

Case 2: $F_2 = V_2$ (i.e., district V_2 has only one fragment). Let $t > 2$ be the smallest index such that v_t is in a district that has two or more fragments (such district exists since Π is not canonical). Then the chain $(v_1, \ldots, v_{t-1})$ is covered by single-fragment districts that we denote by $V_1, \ldots, V_\ell$ along C. By recombining V_i and V_{i+1} for $i = 1, \ldots, \ell - 1$, we create new single-fragment districts $W_1, \ldots, W_\ell$ such that $|W_i| = |V_{i+1}|$ for $i = 1, \ldots, \ell + 1$ and $|W_\ell| = |V_1| = 1$. Now we can apply Case 1 for the singleton district W_ℓ. □

We are now ready to prove the main result of this section.

Theorem 2. *If G is a Hamiltonian graph on n vertices and $s \geq n/k$, then $\mathcal{R}_s(G, k)$ is connected and its diameter is $O(nk)$.*

Proof. Based on Lemmas 4–7, the following algorithm successively reduces the number of fragments to k, thereby transforming any balanced k-partition to a canonical partition. While the number of fragments is more than k, do:

1. If a fragment of a small district is adjacent to a light fragment of a large district along C, then apply the recombination move in Lemma 4, which decreases the number of fragments.
2. Else, by Lemma 5, there are two adjacent districts along C whose combined size is at most $2n/k$. Apply a recombination move in Lemma 6. If this move does not decrease the number of fragments, it creates a singleton district, and then up to $k - 1$ recombination moves in Lemma 7 decrease the number of fragments by at least one.

There can be at most n different fragments in a k-partition of a set of n vertices. We can reduce the number of fragments using up to k recombination moves. Overall, $O(nk)$ recombination moves can bring any two (k, s)-BCPs to canonical form, which are within $k^2 + 1$ moves apart by Lemma 3. □

4   Disconnected Configuration Space

In this section we show that the configuration space is not always connected, even in Hamiltonian graphs. Specifically, we show the following result:

Theorem 3. *For any $k \geq 4$ and $s > 0$ there exists a Hamiltonian planar graph G of $n = k(3s + 2)$ vertices such that $\mathcal{R}_s(G, k)$ is disconnected.*

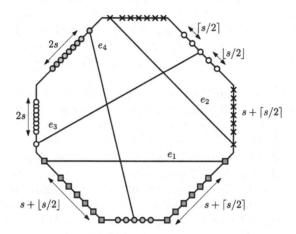

Fig. 2. Problem instance showing that $\mathcal{R}_s(G, k)$ is not always connected (for $k = 4$, $n = 56$ and $s = 4 = \frac{n}{3k} - O(1)$). (Color figure online)

Proof Sketch. The proof is constructive and can be found in [17]. We construct an instance that consists of a cycle and 4 chords (shown in Fig. 2). Each district consists of two contiguous arcs along the cycle, which are connected by a unique chord. We prove that no sequence of recombinations can change this fact. Indeed, the chords are sufficiently limiting that a district can only gain/lose vertices in a very restricted fashion (e.g., the district of represented by orange squares can gain up to s vertices from the district of blue circles). □

5   Hardness Results

This section presents our hardness results. Only a sketch of our reductions are included in this extended abstract; see [17] for full details. Our reductions are from Nondeterministic Constrained Logic (NCL) reconfiguration which is PSPACE-complete [10,11]. An instance of NCL is given by a planar cubic undirected graph G_{NCL} where each edge is colored either red or blue. Each vertex is either incident to three blue edges or incident to two red and one blue edges. We respectively call such vertices **OR** and **AND** vertices. An orientation of G_{NCL} must satisfy the constraint that at every vertex $v \in V(G_{NCL})$, at least one blue edge or at least two red edges are oriented towards v. A **move** is an operation that transforms a satisfying orientation to another by reversing the orientation of a single edge. The problem gives two satisfying orientations A and B of G_{NCL} and asks for a sequence of moves to transform A into B. As in [2], we subdivide each edge in G_{NCL} obtaining a bipartite graph G'_{NCL} with one part formed by original vertices in $V(G_{NCL})$ and another part formed by degree-2 vertices. We require that an orientation must additionally satisfy the constraint that each degree-2 vertex v must have an edge oriented towards v. The question of whether there exists a sequence of moves transforming orientation A' into B' of G'_{NCL} remains PSPACE-complete. We follow the framework in [2] with a few crucial differences. The main technical challenge is dealing with the slack constraints while maintaining the desired behavior for the gadgets. We first describe the reduction to instances with slack equals zero, and We then generalize the proofs.

Zero Slack. In the following reduction, we are given a bipartite instance of NCL given by (G'_{NCL}, A', B'), and we produce an instance of $BR(G, k, s)$ of the balanced recombination problem consisting of two (k, s)-BCP of a planar graph G, Π_A and Π_B, with $k = O(|V(G_{NCL})|)$ districts, and slack $s = 0$. Here we give a brief overview of the reduction. Details can be found in [17].

The AND, OR and degree-2 gadgets are shown in Figs. 3(a), (b) and (c) respectively. The green (black) dots are called **heavy** (**light**) vertices and are considered to be weighed with integer weight more than one (equal to one). We can implement weights by attaching an appropriate number of degree-1 vertices to a heavy vertex so that, in order for a k-BCP to be connected, whichever district contains the heavy vertex must also contain all degree-1 vertices attached to it. Every edge $e \in E(G'_{NCL})$ is represented by two light vertices of G, e^+ and e^-, that belong to two neighboring gadgets as shown in Figs. 3(d).

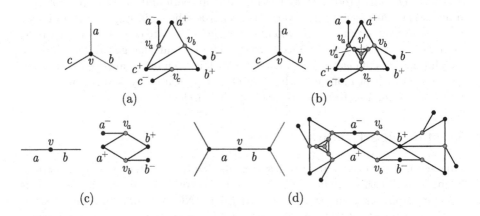

Fig. 3. Gadgets for the reduction from NCL to $BR(G, k, 0)$. (Color figure online)

The weights of heavy vertices are set up so that, for each AND or OR gadget, a district must contain its heavy vertices v_a, v_b and v_c and no heavy vertices of neighbor degree-2 gadgets. Then, for all degree-2 gadget, there is a district that contains v_a and v_b and no other heavy vertex. Additionally, for all OR gadgets, a district must contain v' and exactly one vertex in $\{v'_a, v'_b, v'_c\}$. We encode whether an edge e points toward a vertex by whether a district of the corresponding gadget contains e^+. Then, the connectivity constrains of the districts simulate the NCL constraints. See Fig. 4.

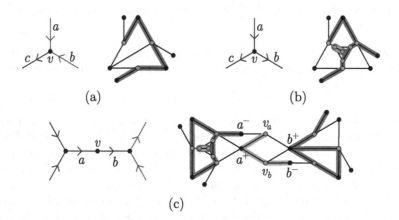

Fig. 4. Equivalence between a satisfying orientation of G'_{NCL} and a k-BCP of G.

Lemma 8. *$BR(G, k, 0)$ is PSPACE-complete even for a planar graph G with constant maximum degree.*

Generalizations. We generalize the reduction of Lemma 8.

Bounded-Degree Triangulation G. The main new technical tool presented in this section is the **filler gadget** shown in Fig. 5(b). Each face marked with a dot is called a **heavy face** associated with an integer weight, and whose recursive construction is shown in Fig. 5(a). Figure 5(c) shows how to use copies of the filler gadget to transform G in a triangulation. The main property of the filler gadget is that we set the weights of heavy faces so that each red vertex must belong to a different district and the gadget only intersects 5 districts. Then, such districts are "trapped" in the filler gadget and don't interfere with the other gadgets.

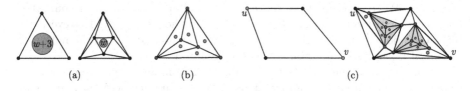

Fig. 5. Construction of the filler gadget. (Color figure online)

Theorem 4. *$BR(G, k, s)$ is PSPACE-complete even if G is maximal planar of constant maximum degree, $k \in O(n^{\varepsilon})$, and $s \in O(n^{1-\varepsilon})$ for $0 < \varepsilon \leq 1$.*

Finding Balanced Connected Partitions. The **NCL orientation problem** is defined by an input undirected graph G_{NCL} edge colored as before, and asks whether there exist an orientation of G_{NCL} that satisfies the NCL constraints. This problem is NP-complete [11]. We remark that our construction implies the following theorem.

Theorem 5. *It is NP-complete to decide whether there exist a (k,s)-BCP of a graph G, even if G is maximal planar of constant maximum degree, $k \in O(n^\varepsilon)$, and $s \in O(n^{1-\varepsilon})$ for $0 < \varepsilon \leq 1$.*

Constant Number of Districts. The drawback of the previous construction is that it requires 5 new districts for each filler gadget. We obtain PSPACE-hardness with $k = 3$, but we lose the restriction that G is a triangulation, and instead we only require that G is a bounded-degree planar graph. The main technical difficulty is to guarantee that the same subset of heavy vertices is always contained in the same district. This property is obtained by a careful setting of the weights so that there is a unique partition of weights of the heavy vertices that allow for the 3 districts to be balanced within s slack. This allow us to label the districts according to the heavy vertices that is contains. One of the districts then locally acts like the districts that previously contained v_a, v_b and v_3 in each AND and OR gadgets, maintaining the equivalency between the connectedness of this district and the NCL constraints. Our proof can be adapted for any $k \geq 3$.

Theorem 6. *$BR(G, 3, s)$ is PSPACE-complete even if G is planar with constant maximum degree, and $s \in O(n^{1-\varepsilon})$ for $0 < \varepsilon \leq 1$.*

6   Conclusion and Open Problems

We have shown that the configuration space $\mathcal{R}_s(G, k)$ of (k, s)-BCPs is connected when G is connected and $s = \infty$, or when G is Hamiltonian and $s \geq n/k$. We hope that our results inform future research on the properties of G, k, and s that are sufficient to obtain an efficient sampling of $\mathsf{Bal}_s(G, k)$. We also leave it as an open problem whether our results in Sect. 3 generalize to other classes of graphs. We conjecture that the configuration space $\mathcal{R}_s(n, k)$ is connected for every **biconnected** graph G on n vertices when $s \geq n/k$. However, our techniques do not directly generalize; it is unclear how to extend the notion of canonical k-partitions in the absence of a Hamilton cycle.

We have shown that $BR(G, k, s)$ is PSPACE-complete even in specific settings that are of interest in applications such as sampling electoral maps. Our results imply that the configuration space $\mathcal{R}_s(G, k)$ has diameter exponential in n, establishing as well an exponential lower bound on the mixing time of a Markov chain on $\mathcal{R}_s(G, k)$ for these settings. We note that Theorems 4 and 6 do not include other settings of interest such as when G is maximal planar (or even 3-connected) and k is a constant. We leave these as open problems.

References

1. Abrishami, T., et al.: Geometry of graph partitions via optimal transport. SIAM J. Sci. Comput. **42**(5), A3340–A3366 (2020). https://doi.org/10.1137/19M1295258
2. Akitaya, H.A., et al.: Reconfiguration of connected graph partitions. Preprint (2019). http://arxiv.org/abs/1902.10765
3. Chataigner, F., Salgado, L.B., Wakabayashi, Y.: Approximation and inapproximability results on balanced connected partitions of graphs. Discret. Math. Theor. Comput. Sci. **9**(1) (2007). http://dmtcs.episciences.org/384
4. Chlebíková, J.: Approximating the maximally balanced connected partition problem in graphs. Inf. Process. Lett. **60**(5), 223–230 (1996). https://doi.org/10.1016/S0020-0190(96)00175-5
5. DeFord, D.R., Duchin, M., Solomon, J.: Recombination: a family of Markov chains for redistricting. Preprint (2019). http://arxiv.org/abs/1911.05725
6. Duchin, M.: Gerrymandering metrics: how to measure? What's the baseline? Preprint (2018). https://arxiv.org/abs/1801.02064
7. Dyer, M.E., Frieze, A.M.: On the complexity of partitioning graphs into connected subgraphs. Discrete Appl. Math. **10**(2), 139–153 (1985). https://doi.org/10.1016/0166-218X(85)90008-3
8. Fifield, B., Higgins, M., Imai, K., Tarr, A.: Automated redistricting simulation using Markov chain Monte Carlo. J. Comput. Graph. Stat. 1–14 (2020). https://doi.org/10.1080/10618600.2020.1739532
9. Győri, E.: On division of graphs to connected subgraphs. In: Combinatorics (Proceedings of the Fifth Hungarian Combinatorial Colloquia, 1976, Keszthely), pp. 485–494. Bolyai (1978)
10. Hearn, R.A., Demaine, E.D.: PSPACE-completeness of sliding-block puzzles and other problems through the nondeterministic constraint logic model of computation. Theor. Comput. Sci. **343**(1), 72–96 (2005). https://doi.org/10.1016/j.tcs.2005.05.008
11. Hearn, R.A., Demaine, E.D.: Games, Puzzles, and Computation. AK Peters/CRC Press, Wellesley (2009)
12. Ito, T., Zhou, X., Nishizeki, T.: Partitioning a graph of bounded tree-width to connected subgraphs of almost uniform size. J. Discrete Algorithms **4**(1), 142–154 (2006). https://doi.org/10.1016/j.jda.2005.01.005
13. Jerrum, M., Valiant, L.G., Vazirani, V.V.: Random generation of combinatorial structures from a uniform distribution. Theor. Comput. Sci. **43**, 169–188 (1986). https://doi.org/10.1016/0304-3975(86)90174-X
14. Lari, I., Ricca, F., Puerto, J., Scozzari, A.: Partitioning a graph into connected components with fixed centers and optimizing cost-based objective functions or equipartition criteria. Networks **67**(1), 69–81 (2016). https://doi.org/10.1002/net.21661
15. Levin, D.A., Peres, Y.: Markov Chains and Mixing Times, 2nd edn. AMS, Providence (2017)
16. Lovász, L.: A homology theory for spanning tress of a graph. Acta Mathematica Academiae Scientiarum Hungarica **30**(3), 241–251 (1977). https://doi.org/10.1007/BF01896190
17. Korman, M., Akitaya, H.A., Korten, O., Souvaine, D.L., Tóth, C.D.: Reconfiguration of connected graph partitions via recombination. Preprint (2020). https://arxiv.org/abs/2011.07378

18. Najt, L., DeFord, D.R., Solomon, J.: Complexity and geometry of sampling connected graph partitions. Preprint (2019)
19. Soltan, S., Yannakakis, M., Zussman, G.: Doubly balanced connected graph partitioning. ACM Trans. Algorithms **16**(2), 20:1–20:24 (2020). https://doi.org/10.1145/3381419
20. Suzuki, H., Takahashi, N., Nishizeki, T.: A linear algorithm for bipartition of biconnected graphs. Inf. Process. Lett. **33**(5), 227–231 (1990). https://doi.org/10.1016/0020-0190(90)90189-5
21. Tutte, W.T.: A theorem on planar graphs. Trans. Am. Math. Soc. **82**(1), 99–116 (1956)
22. Wada, K., Kawaguchi, K.: Efficient algorithms for tripartitioning triconnected graphs and 3-edge-connected graphs. In: van Leeuwen, J. (ed.) WG 1993. LNCS, vol. 790, pp. 132–143. Springer, Heidelberg (1994). https://doi.org/10.1007/3-540-57899-4_47
23. Wu, D., Zhang, Z., Wu, W.: Approximation algorithm for the balanced 2-connected k-partition problem. Theor. Comput. Sci. **609**, 627–638 (2016). https://doi.org/10.1016/j.tcs.2015.02.001

Algorithms for Energy Conservation in Heterogeneous Data Centers

Susanne Albers and Jens Quedenfeld[✉]

Technical University of Munich, 85748 Garching, Germany
{albers,jens.quedenfeld}@in.tum.de

Abstract. Power consumption is the major cost factor in data centers. It can be reduced by dynamically right-sizing the data center according to the currently arriving jobs. If there is a long period with low load, servers can be powered down to save energy. For identical machines, the problem has already been solved optimally by [25] and [1].

In this paper, we study how a data-center with heterogeneous servers can dynamically be right-sized to minimize the energy consumption. There are d different server types with various operating and switching costs. We present a deterministic online algorithm that achieves a competitive ratio of $2d$ as well as a randomized version that is $1.58d$-competitive. Furthermore, we show that there is no deterministic online algorithm that attains a competitive ratio smaller than $2d$. Hence our deterministic algorithm is optimal. In contrast to related problems like convex body chasing and convex function chasing [17,30], we investigate the discrete setting where the number of active servers must be an integral, so we gain truly feasible solutions.

1 Introduction

Energy management is an important issue in data centers. A huge amount of a data center's financial budget is spent on electricity that is needed to operate the servers as well as to cool them [12,20]. However, server utilization is typically low. In fact there are data centers where the average server utilization is as low as 12% [16]; only for a few days a year is full processing power needed. Unfortunately, idle servers still consume about half of their peak power [29]. Therefore, right-sizing a data center by powering down idle servers can save a significant amount of energy. However, shutting down a server and powering it up immediately afterwards incurs much more cost than holding the server in the active state during this time period. The cost for powering up and down does not only contain the increased energy consumption but also, for example, wear-and-tear costs or the risk that the server does not work properly after restarting [26]. Consequently, algorithms are needed that manage the number of active servers to minimize the total cost, without knowing when new jobs will arrive in the future. Since about 3% of the global electricity production

Work supported by the European Research Council, Grant Agreement No. 691672.

T. Calamoneri and F. Corò (Eds.): CIAC 2021, LNCS 12701, pp. 75–89, 2021.
https://doi.org/10.1007/978-3-030-75242-2_5

is consumed by data centers [11], a reduction of their energy consumption can also decrease greenhouse emissions. Thus, right-sizing data centers is not only important for economical but also for ecological reasons.

Modern data centers usually contain heterogeneous servers. If the capacity of a data center is no longer sufficient, it is extended by including new servers. The old servers are still used however. Hence, there are different server types with various operating and switching costs in a data center. Heterogeneous data centers may also include different processing architectures. There can be servers that use GPUs to perform massive parallel calculations. However, GPUs are not suitable for all jobs. For example, tasks with many branches can be computed much faster on common CPUs than on GPUs [31].

Problem Formulation. We consider a data center with d different server types. There are m_j servers of type j. Each server has an active state where it is able to process jobs, and an inactive state where no energy is consumed. Powering up a server of type j (i.e., switching from the inactive into the active state) incurs a cost of β_j (called *switching* cost); powering down does not cost anything. We consider a finite time horizon consisting of the time slots $\{1, \ldots, T\}$. For each time slot $t \in \{1, \ldots, T\}$, jobs of total volume $\lambda_t \in \mathbb{N}_0$ arrive and have to be processed during the time slot. There must be at least λ_t active servers to process the arriving jobs. We consider a basic setting where the operating cost of a server of type j is load and time independent and denoted by $l_j \in \mathbb{R}_{\geq 0}$. Hence, an active server incurs a constant but type-dependent operating cost per time slot.

A schedule X is a sequence $x_1, \ldots, x_T$ with $x_t = (x_{t,1}, \ldots, x_{t,d})$ where each $x_{t,j}$ indicates the number of active servers of type j during time slot t. At the beginning and the end of the considered time horizon all servers are shut down, i.e., $x_0 = x_{T+1} = (0, \ldots, 0)$. A schedule is called *feasible* if there are enough active servers to process the arriving jobs and if there are not more active servers than available, i.e., $\sum_{j=1}^{d} x_{t,j} \geq \lambda_t$ and $x_{t,j} \in \{0, 1, \ldots, m_j\}$ for all $t \in \{1, \ldots, T\}$ and $j \in \{1, \ldots, d\}$. The cost of a feasible schedule is defined by

$$C(X) := \sum_{t=1}^{T} \left(\sum_{j=1}^{d} l_j x_{t,j} + \sum_{j=1}^{d} \beta_j (x_{t,j} - x_{t-1,j})^+ \right) \tag{1}$$

where $(x)^+ := \max(x, 0)$. The switching cost is only paid for powering up. However, this is not a restriction, since all servers are inactive at the beginning and end of the workload. Thus the cost of powering down can be folded into the cost of powering up. A problem instance is specified by the tuple $\mathcal{I} = (T, d, m, \beta, l, \Lambda)$ where $m = (m_1, \ldots, m_d)$, $\beta = (\beta_1, \ldots, \beta_d)$, $l = (l_1, \ldots, l_d)$ and $\Lambda = (\lambda_1, \ldots, \lambda_T)$. The task is to find a schedule with minimum cost.

We focus on the central case without *inefficient* server types. A server type j is called *inefficient* if there is another server type $j' \neq j$ with both smaller (or equal) operating and switching costs, i.e., $l_j \geq l_{j'}$ and $\beta_j \geq \beta_{j'}$. This assumption is natural because a better server type with a lower operating cost usually has a higher switching cost. An inefficient server of type j is only powered up, if all servers of all types j' with $\beta_{j'} \leq \beta_j$ and $l_{j'} \leq l_j$ are already running. Therefore, excluding inefficient servers is not a relevant restriction in practice. In related work, Augustine et al. [6] exclude inefficient states when operating a single server.

Our Contribution. We analyze the online setting of this problem where the job volumes λ_t arrive one-by-one. The vector of the active servers x_t has to be determined without knowledge of future jobs $\lambda_{t'}$ with $t' > t$. A main contribution of our work, compared to previous results, is that we investigate heterogeneous data centers and examine the online setting when truly feasible (integral) solutions are sought.

In Sect. 2, we present a $2d$-competitive deterministic online algorithm, i.e., the total cost of the schedule calculated by our algorithm is at most $2d$ times larger than the cost of an optimal offline solution. Roughly, our algorithm works as follows. It calculates an optimal schedule for the jobs received so far and ensures that the operating cost of the active servers is at most as large as the operating cost of the active servers in the optimal schedule. If this is not the case, servers with high operating cost are replaced by servers with low operating cost. If a server is not used for a specific duration depending on its switching and operating costs, it is shut down.

In Sect. 3, we devise a randomized version of our algorithm achieving a competitive ratio of $\frac{e}{e-1}d \approx 1.582d$ against an oblivious adversary.

In Sect. 4, we show that there is no deterministic online algorithm that achieves a competitive ratio smaller than $2d$. Therefore, our algorithm is optimal. Additionally, for a data center that contains m unique servers (that is $m_j = 1$ for all $j \in \{1, \ldots, d\}$), we show that the best achievable competitive ratio is $2m$.

Related Work. The design of energy-efficient algorithms has received quite some research interest over the last years, see e.g. [3,10,21] and references therein. Specifically, data center right-sizing has attracted considerable attention lately. Lin and Wierman [25,26] analyzed the data-center right-sizing problem for data centers with identical servers ($d = 1$). The operating cost is load dependent and modeled by a convex function. In contrast to our setting, continuous solutions are allowed, i.e., the number of active server x_t can be fractional. This allows for other techniques in the design and analysis of an algorithm, but the created schedules cannot be used directly in practice. They gave a 3-competitive deterministic online algorithm for this problem. Bansal et al. [9] improved this result by randomization and developed a 2-competitive online algorithm. In our previous paper [1] we showed that 2 is a lower bound for randomized algorithms in the continuous setting; this result was independently shown by [4]. Furthermore, we analyzed the discrete setting of the problem where the number of active servers is integral ($x_t \in \mathbb{N}_0$). We presented a 3-competitive deterministic and a 2-competitive randomized online algorithm. Moreover, we proved that these competitive ratios are optimal.

Data-center right-sizing of heterogeneous data centers is related to convex function chasing, which is also known as smoothed online convex optimization [15]. At each time slot t, a convex function f_t arrives. The algorithm then has to choose a point x_t and pay the cost $f_t(x_t)$ as well as the movement cost $\|x_t - x_{t-1}\|$ where $\|\cdot\|$ is any metric. The problem described by Eq. (1) is a special case of convex function chasing if fractional schedules are allowed, i.e., $x_{t,j} \in [0, m_j]$ instead of $x_{t,j} \in \{0, \ldots, m_j\}$. The operating cost $\sum_{j=1}^{d} l_j x_{t,j}$ in Eq. (1) together with the feasibility requirements can be modeled as a convex function that is infinite for $\sum_{j=1}^{d} x_{t,j} < \lambda_t$ and $x_{t,j} \notin [0, m_j]$. The switching cost equals the Manhattan metric if the number of servers is scaled appropriately. Sellke [30] gave a $(d+1)$-competitive algorithm for convex function chasing. A similar result was found by Argue et al. [5].

In the discrete setting, convex function chasing has at least an exponential competitive ratio, as the following setting shows. Let $m_j = 1$ and $\beta_j = 1$ for all $j \in \{1, \ldots, d\}$, so the possible server configurations are $\{0, 1\}^d$. The arriving convex functions f_t are infinite for the current position x_{t-1} of the online algorithm and 0 for all other positions $\{0, 1\}^d \backslash \{x_{t-1}\}$. After $T := 2^d - 1$ functions arrived, the switching cost paid by the algorithm is at least $2^d - 1$ (otherwise it has to pay infinite operating costs), whereas the offline schedule can go directly to a position without any operating cost and only pays a switching cost of at most d.

Already for the 1-dimensional case (i.e. identical machines), it is not trivial to round a fractional schedule without increasing the competitive ratio (see [26] and [2]). In d-dimensional space, it is completely unclear, if continuous solutions can be rounded without arbitrarily increasing the total cost. Simply rounding up can lead to arbitrarily large switching costs, for example if the fractional solution rapidly switches between 1 and $1 + \epsilon$. Using a randomized rounding scheme like in [2] (that was used for homogeneous data centers) independently for each dimension can result in an infeasible schedule (for example, if $\lambda_t = 1$ and $x_t = (1/d, \ldots, 1/d)$ is rounded down to $(0, \ldots, 0)$). Therefore, Sellke's result does not help us for analyzing the discrete setting. Other publications handling convex function chasing or convex body chasing are [8,13,17].

Goel and Wierman [19] developed a $(3 + \mathcal{O}(1/\mu))$-competitive algorithm called Online Balanced Descent (OBD) for convex function chasing, where the arriving functions were required to be μ-strongly convex. We remark that the operating cost defined by Eq. (1) is not strongly convex, i.e., $\mu = 0$. Hence their result cannot be used for our problem. A similar result is given by Chen et al. [15] who showed that OBD is $(3 + \mathcal{O}(1/\alpha))$-competitive if the arriving functions are locally α-polyhedral. In our case, $\alpha = \min_{j \in \{1, \ldots, d\}} l_j / \beta_j$, so α can be arbitrarily small depending on the problem instance.

Another similar problem is the Parking Permit Problem by Meyerson [28]. There are d different permits which can be purchased for β_j dollars and have a duration of D_j days. Certain days are driving days where at least one parking permit is needed ($\lambda_t \in \{0, 1\}$). The permit cost corresponds to our switching cost. However, the duration of the permit is fixed to D_j, whereas in our problem the online algorithm can choose for each time slot if it wants to power down a server. Furthermore, there is no operating cost. Even if each server type is replaced by an infinite number of permits with the duration t and the cost $\beta_j + l_j \cdot t$, it is still a different problem, because the algorithm has to choose the time slot for powering down in advance (when the server is powered up).

Data-center right-sizing of heterogeneous data centers is related to geographical load balancing analyzed in [24] and [27]. Other applications are shown in [7,14,18,22,23,32,33].

2   Deterministic Online Algorithm

In this section we present a deterministic $2d$-competitive online algorithm for the problem described in the preceding section. The basic idea of our algorithm is to calculate an optimal schedule for the problem instance that ends at the current time slot. Based on this schedule, we decide when a server is powered up. If a server is idle for a specific time, it is powered down.

Formally, given the original problem instance $\mathcal{I} = (T, d, \boldsymbol{m}, \boldsymbol{\beta}, \boldsymbol{l}, \Lambda)$, the shortened problem instance $\mathcal{I}^t$ is defined by $\mathcal{I}^t := (t, d, \boldsymbol{m}, \boldsymbol{\beta}, \boldsymbol{l}, \Lambda^t)$ with $\Lambda^t = (\lambda_1, \dots, \lambda_t)$. Let $\hat{X}^t$ denote an optimal schedule for $\mathcal{I}^t$ and let $X^{\mathcal{A}}$ be the schedule calculated by our algorithm $\mathcal{A}$.

W.l.o.g. there are no server types with the same operating and switching costs, i.e., $\beta_j = \beta_{j'}$ and $l_j = l_{j'}$ implies $j = j'$. Furthermore, let $l_1 > \cdots > l_d$, i.e., the server types are sorted by their operating costs. Since inefficient server types are excluded, this implies that $\beta_1 < \cdots < \beta_d$.

Let $[n] := \{1, \dots, n\}$ where $n \in \mathbb{N}$. We separate a problem instance into $m := \sum_{j=1}^{d} m_j$ lanes. At time slot t, there is a single job in lane $k \in [m]$, if and only if $k \leq \lambda_t$. We can assume that $\lambda_t \leq m$ holds for all $t \in [T]$, because otherwise there is no feasible schedule for the problem instance. Let X be an arbitrary feasible schedule with $\boldsymbol{x}_t = (x_{t,1}, \dots, x_{t,d})$. We define

$$y_{t,k} := \begin{cases} \max\{j \in [d] \mid \sum_{j'=j}^{d} x_{t,j'} \geq k\} & \text{if } k \in \left[\sum_{j=1}^{d} x_{t,j}\right] \\ 0 & \text{else} \end{cases} \qquad (2)$$

to be the server type that handles the k-th lane during time slot t. If $y_{t,k} = 0$, then there is no active server in lane k at time slot t. By definition, the values $y_{t,1}, \dots, y_{t,m}$ are sorted in descending order, i.e., $y_{t,k} \geq y_{t,k'}$ for $k < k'$. Note that $y_{t,k} = 0$ implies $\lambda_t < k$, because otherwise there are not enough active servers to handle the jobs at time t. For the schedule $\hat{X}^t$, the server type used in lane k at time slot t' is denoted by $\hat{y}^t_{t',k}$. Our algorithm calculates $y^{\mathcal{A}}_{t,k}$ directly, the corresponding variables $x^{\mathcal{A}}_{t,j}$ can be determined by $x^{\mathcal{A}}_{t,j} = |\{k \in [m] \mid y^{\mathcal{A}}_{t,k} = j\}|$.

Our algorithm works as follows: First, an optimal solution $\hat{X}^t$ is calculated. If there are several optimal schedules, we choose a schedule that fulfills the inequality $\hat{y}^t_{t',k} \geq \hat{y}^{t-1}_{t',k}$ for all time slots $t' \in [t]$ and lanes $k \in [m]$, so $\hat{X}^t$ never uses smaller server types than the previous schedule $\hat{X}^{t-1}$. We will see in Lemma 2 that such a schedule exists and how to construct it.

If there is a server type j with $l_j = 0$, then in an optimal schedule such a server can be powered up before it is needed, although $\lambda_t = 0$ holds for this time slot. Similarly, such a server can run for more time slots than necessary. W.l.o.g. let $\hat{X}^t$ be a schedule where servers are powered up as late as possible and powered down as early as possible.

Beginning from the lowest lane ($k = 1$), it is ensured that $\mathcal{A}$ uses a server type that is not smaller than the server type used by $\hat{X}^t$, i.e., $y^{\mathcal{A}}_{t,k} \geq \hat{y}^t_{t,k}$ must be fulfilled. If the server type $y^{\mathcal{A}}_{t-1,k}$ used in the previous time slot is smaller than $\hat{y}^t_{t,k}$, it is powered down and server type $\hat{y}^t_{t,k}$ is powered up. A server of type j that is not replaced by a greater server type stays active for $\bar{t}_j := \lfloor \beta_j / l_j \rfloor$ time slots. If $\hat{X}^t$ uses a smaller server type $j' \leq j$ in the meantime, then server type j will run for at least $\bar{t}_{j'}$ further time slots (including time slot t). Formally, a server of type j in lane k is powered down at time slot t, if $\hat{y}^{t'}_{t',k} \neq j'$ holds for all server types $j' \leq j$ and time slots $t' \in [t - \bar{t}_{j'} + 1 : t]$ with $[a : b] := \{a, a+1, \ldots, b\}$.

The pseudocode below clarifies how algorithm $\mathcal{A}$ works. The variables e_k for $k \in [m]$ store the time slot when the server in the corresponding lane will be powered down.

Algorithm 1. Algorithm $\mathcal{A}$

1: **for** $t := 1$ **to** T **do**
2:     Calculate $\hat{X}^t$ such that $\hat{y}^t_{t',k} \geq \hat{y}^{t-1}_{t',k}$ for all $t' \in [t]$ and $k \in [m]$
3:     **for** $k := 1$ **to** m **do**
4:         **if** $y^{\mathcal{A}}_{t-1,k} < \hat{y}^t_{t,k}$ **or** $t \geq e_k$ **then**
5:             $y^{\mathcal{A}}_{t,k} := \hat{y}^t_{t,k}$
6:             $e_k := t + \bar{t}_{y^{\mathcal{A}}_{t,k}}$
7:         **else**
8:             $y^{\mathcal{A}}_{t,k} := y^{\mathcal{A}}_{t-1,k}$
9:             $e_k := \max\{e_k, t + \bar{t}_{\hat{y}^t_{t,k}}\}$ where $\bar{t}_0 := 0$

Structure of Optimal Schedules. Before we can analyze the competitiveness of algorithm $\mathcal{A}$, we have to show that an optimal schedule with the desired properties required by line 2 actually exists. First, we will investigate basic properties of optimal schedules. In an optimal schedule $\hat{X}$, a server of type j that runs in lane k does not change the lane while running. Formally, if $\hat{y}_{t-1,k} = j$ and $\hat{y}_{t,k} \neq j$, then there exists no other lane $k' \neq k$ with $\hat{y}_{t-1,k'} \neq j$ and $\hat{y}_{t,k'} = j$. Furthermore, a server is only powered up or powered down if the number of jobs is increased or decreased, respectively. Finally, in a given lane k, the server type does not change immediately, i.e., there must be at least one time slot, where no server is running in lane k. These properties are proven in the full version of this paper.

Given the optimal schedules $\hat{X}^u$ and $\hat{X}^v$ with $u < v$, we construct a *minimum* schedule $X^{\min(u,v)}$ with $y^{\min(u,v)}_{t,k} := \min\{\hat{y}^u_{t,k}, \hat{y}^v_{t,k}\}$. Furthermore, we construct a *maximum* schedule $X^{\max(u,v)}$ as follows. Let $z_l(t,k)$ be the last time slot $t' < t$ with $\hat{y}^u_{t',k} = \hat{y}^v_{t',k} = 0$ (no active servers in both schedules) and let $z_r(t,k)$ be the first time slot $t' > t$ with $\hat{y}^u_{t',k} = \hat{y}^v_{t',k} = 0$. The schedule $X^{\max(u,v)}$ is defined by

$$y^{\max(u,v)}_{t,k} := \max_{t' \in [z_l(t,k)+1 : z_r(t,k)-1]} \{\hat{y}^u_{t',k}, \hat{y}^v_{t',k}\}. \tag{3}$$

Another way to construct $X^{\max(u,v)}$ is as follows. First, we take the maximum of both schedules (analogously to $X^{\min(u,v)}$). However, this can lead to situations where the server type changes immediately, so the necessary condition for optimal schedules would not be fulfilled. Therefore, we replace the lower server type by the greater one until there are no more immediate server changes. This construction is equivalent to Eq. (3).

We will see in Lemma 2 that the *maximum* schedule is an optimal schedule for $\mathcal{I}^v$ and fulfills the property required by algorithm $\mathcal{A}$ in line 2, which says that the server type used in lane k at time t never decreases when the considered problem instance is expanded. To prove this property, first we have to show that $X^{\min(u,v)}$ and $X^{\max(u,v)}$ are feasible schedules for the problem instances $\mathcal{I}^u$ and $\mathcal{I}^v$, respectively.

Lemma 1. $X^{min(u,v)}$ *and* $X^{max(u,v)}$ *are feasible for* $\mathcal{I}^u$ *and* $\mathcal{I}^v$, *respectively.*

The proof can be found in the full version of this paper. Now, we are able to show that the maximum schedule is optimal for the problem instance $\mathcal{I}^v$.

Lemma 2. *Let* $u, v \in [T]$ *with* $u < v$. $X^{max(u,v)}$ *is optimal for* $\mathcal{I}^v$.

The works roughly as follows (the complete proof can be found in the full paper). First, we prove that the sum of the operating costs of $\hat{X}^u$ and $\hat{X}^v$ is greater than or equal to the sum of the operating cost of $X^{\min(u,v)}$ and $X^{\max(u,v)}$. Each server activation in $X^{\min(u,v)}$ and $X^{\max(u,v)}$ can be mapped to exactly one server activation in $\hat{X}^u$ and $\hat{X}^v$ with the same or a greater server type. Therefore, $C(X^{\min(u,v)}) + C(X^{\max(u,v)}) \le C(\hat{X}^u) + C(\hat{X}^v)$ holds and by using Lemma 1, it is shown that $X^{\max(u,v)}$ is optimal for $\mathcal{I}^v$.

Feasibility. In the following, let $\{\hat{X}^1, \ldots, \hat{X}^T\}$ be optimal schedules that fulfill the inequality $\hat{y}_{t',k}^t \ge \hat{y}_{t',k}^{t-1}$ for all $t, t' \in [T]$ and $k \in [m]$ as required by algorithm $\mathcal{A}$. Lemma 2 ensures that such a schedule sequence exists (and also shows how to construct it). Before we can prove that algorithm $\mathcal{A}$ is $2d$-competitive, we have to show that the computed schedule $X^{\mathcal{A}}$ is feasible. In an optimal schedule $\hat{X}^t$, the values $\hat{y}_{t',1}^t, \ldots, \hat{y}_{t',m}^t$ are sorted in descending order by definition. This also holds for schedule calculated by our algorithm.

Lemma 3. *For all time slots* $t \in [T]$, *the values* $y_{t,1}^{\mathcal{A}}, \ldots, y_{t,m}^{\mathcal{A}}$ *are sorted in descending order, i.e.,* $y_{t,k}^{\mathcal{A}} \ge y_{t,k'}^{\mathcal{A}}$ *for* $k < k'$.

The proof uses the fact that the running times $\bar{t}_j$ are sorted in ascending order, i.e., $\bar{t}_1 \le \cdots \le \bar{t}_d$, because $l_1 > \cdots > l_d$ and $\beta_1 < \cdots < \beta_d$. In other words, the higher the server type is, the longer it stays in the active state. See the full paper for more details. By means of Lemma 3, we are able to prove the feasibility of $X^{\mathcal{A}}$.

Lemma 4. *The schedule* $X^{\mathcal{A}}$ *is feasible.*

Proof Idea. A schedule is feasible, if (1) there are enough active servers to handle the incoming jobs (i.e., $\sum_{j=1}^{d} x_{t,j}^{\mathcal{A}} \ge \lambda_t$) and (2) there are not more active servers than available (i.e., $x_{t,j}^{\mathcal{A}} \leq m_j$). The first property directly follows from the definition of algorithm $\mathcal{A}$, since $\sum_{j=1}^{d} x_{t,j}^{\mathcal{A}} \geq \sum_{j=1}^{d} \hat{x}_{t,j}^{t} \geq \lambda_t$. Lemma 3 is used to prove that $x_{t,j}^{\mathcal{A}} \leq m_j$ is always fulfilled after setting $y_{t,k}^{\mathcal{A}}$ in line 5 or 8. The complete proof is presented in the full paper. □

Competitiveness. To show the competitiveness of $\mathcal{A}$, we divide the schedule $X^{\mathcal{A}}$ into blocks $A_{t,k}$ with $t \in [T]$ and $k \in [m]$. Each block $A_{t,k}$ is described by its creation time t, its start time $s_{t,k}$, its end time $e_{t,k}$, the used server type $j_{t,k}$ and the corresponding lane k. The start time is the time slot when $j_{t,k}$ is powered up and the end time is the first time slot, when $j_{t,k}$ is inactive, i.e., during the time interval $[s_{t,k} : e_{t,k} - 1]$ the server of type $j_{t,k}$ is in the active state.

There are two types of blocks: *new* blocks and *extended* blocks. A *new* block starts when a new server is powered up, i.e., lines 5 and 6 of algorithm $\mathcal{A}$ are executed because $y_{t-1,k}^{\mathcal{A}} < \hat{y}_{t,k}^{t}$ or $t \geq e_k \wedge y_{t-1,k}^{\mathcal{A}} > \hat{y}_{t,k}^{t} \wedge \hat{y}_{t,k}^{t} > 0$ (in words: the previous block ends and $\hat{X}^t$ has an active server in lane k, but the server type is smaller than the server type used by $\mathcal{A}$ in the previous time slot). It ends after $\bar{t}_{y_{t,k}^{\mathcal{A}}}$ time slots. Thus $s_{t,k} := t$ and $e_{t,k} := t + \bar{t}_{y_{t,k}^{\mathcal{A}}}$ (i.e., $e_{t,k}$ equals e_k after executing line 6).

An *extended* block is created when the running time of a server is extended, i.e., the value of e_k is updated, but the server type remains the same (that is $y_{t-1,k}^{\mathcal{A}} = y_{t,k}^{\mathcal{A}}$). We have $e_{t,k} := t + \bar{t}_{\hat{y}_{t,k}^{t}}$ (i.e., the value of e_k after executing line 9 or 6) and $s_{t,k} := e_{t',k}$, where $A_{t',k}$ is the previous block in the same lane. Note that an *extended* block can be created not only in line 9, but also in line 6, if $t = e_k$ and $y_{t-1,k}^{\mathcal{A}} = \hat{y}_{t,k}^{t}$. If line 8 and 9 are executed, but the value of e_k does not change (because $t + \bar{t}_{\hat{y}_{t,k}^{t}}$ is smaller than or equal to the previous value of e_k), then the block $A_{t,k}$ does not exist.

Let $d_{t,k} := e_{t,k} - s_{t,k}$ be the duration of the block $A_{t,k}$ and let $C(A_{t,k})$ be the cost caused by $A_{t,k}$ if the block $A_{t,k}$ exists or 0 otherwise. The next lemma describes how the cost of a block can be estimated.

Lemma 5. *The cost of the block $A_{t,k}$ is upper bounded by*

$$C(A_{t,k}) \leq \begin{cases} 2\beta_{j_{t,k}} & \text{if } A_{t,k} \text{ is a new block} \\ l_{j_{t,k}} d_{t,k} & \text{if } A_{t,k} \text{ is an extended block} \end{cases} \tag{4}$$

The lemma follows from the definition of $\bar{t}_j$ (see the full paper for more details). To show the competitiveness of algorithm $\mathcal{A}$, we introduce another variable that will be used in Lemmas 7 and 8. Let

$$\tilde{y}_{t,k}^{u} := \max_{t' \in [t:u]} \hat{y}_{t',k}^{t'}$$

be the largest server type used in lane k by the schedule $\hat{X}^{t'}$ at time slot t' for $t' \in [t : u]$. The next lemma shows that $\tilde{y}_{t,k}^{u}$ is monotonically decreasing with respect to t as well as k and increasing with respect to u.

Lemma 6. *Let $u' \geq u$, $t' \leq t$ and $k' \leq k$. It is $\tilde{y}_{t,k}^{u} \leq \tilde{y}_{t',k'}^{u'}$.*

This lemma follows from the definition of $\tilde{y}^u_{t,k}$. A proof can be found in the full paper. The cost of schedule X in lane k during time slot t is denoted by

$$C_{t,k}(X) := \begin{cases} l_{y_{t,k}} + \beta_{y_{t,k}} & \text{if } y_{t-1,k} \neq y_{t,k} > 0 \\ l_{y_{t,k}} & \text{if } y_{t-1,k} = y_{t,k} > 0 \\ 0 & \text{otherwise.} \end{cases} \tag{5}$$

The total cost of X can be written as $C(X) = \sum_{t=1}^{T} \sum_{k=1}^{m} C_{t,k}(X)$. The technical lemma below will be needed for our induction proof in Theorem 1. Given the optimal schedules $\hat{X}^u$ and $\hat{X}^v$ with $u < v$, the inequality $\sum_{k=1}^{m} \sum_{t=1}^{u} C_{t,k}(\hat{X}^u) \leq \sum_{k=1}^{m} \sum_{t=1}^{u} C_{t,k}(\hat{X}^v)$ is obviously fulfilled (because $\hat{X}^u$ is an optimal schedule for $\mathcal{I}^u$, so $\hat{X}^v$ cannot be better). The lemma below shows that this inequality still holds if the cost $C_{t,k}(\cdot)$ is scaled by $\tilde{y}^u_{t,k}$.

Lemma 7. *Let $u, v \in [T]$ with $u < v$. It holds that*

$$\sum_{k=1}^{m} \sum_{t=1}^{u} \tilde{y}^u_{t,k} C_{t,k}(\hat{X}^u) \leq \sum_{k=1}^{m} \sum_{t=1}^{u} \tilde{y}^u_{t,k} C_{t,k}(\hat{X}^v). \tag{6}$$

The proof is shown in the full paper. The next lemma shows how the cost of a single block $A_{v,k}$ can be folded into the term $2 \sum_{t=1}^{v-1} \tilde{y}^{v-1}_{t,k} C_{t,k}(\hat{X}^v)$ which is the right hand side of Eq. (6) given in the previous lemma with $u = v - 1$.

Lemma 8. *For all lanes $k \in [m]$ and time slots $v \in [T]$, it is*

$$2 \sum_{t=1}^{v-1} \tilde{y}^{v-1}_{t,k} C_{t,k}(\hat{X}^v) + C(A_{v,k}) \leq 2 \sum_{t=1}^{v} \tilde{y}^v_{t,k} C_{t,k}(\hat{X}^v). \tag{7}$$

Proof. If the block $A_{v,k}$ does not exists, Eq. (7) holds by Lemma 6 and $C(A_{v,k}) = 0$.

If $A_{v,k}$ is a *new* block, then $C(A_{v,k}) \leq 2\beta_j$ with $j := j_{v,k} = \hat{y}^v_{v,k}$ by Lemma 5. Since $A_{v,k}$ is a *new* block, server type j was not used in the last time slot of the last $\bar{t}_j$ schedules, i.e., $\hat{y}^t_{t,k} \leq j - 1$ for $t \in [v - \bar{t}_j : v - 1]$. If $\hat{y}^{v-\bar{t}_j}_{v-\bar{t}_j,k} = j$ would hold, then $y^A_{v-1,k} = j$ and there would be an *extended* block at time slot v. By using the facts above and the definition of $\tilde{t}^v_{t,k}$, for $t \in [v - \bar{t}_j : v - 1]$, we get

$$\tilde{y}^{v-1}_{t,k} = \max_{t' \in [t:v-1]} \hat{y}^{t'}_{t',k} \leq j - 1 = \hat{y}^v_{v,k} - 1 \leq \max_{t' \in [t:v]} \hat{y}^{t'}_{t',k} - 1 = \tilde{y}^v_{t,k} - 1. \tag{8}$$

By using Lemma 6 and Eq. (8), we can estimate the first sum in (7):

$$\sum_{t=1}^{v-1} \tilde{y}^{v-1}_{t,k} C_{t,k}(\hat{X}^v) \overset{L6,(8)}{\leq} \sum_{t=1}^{v-\bar{t}_j-1} \tilde{y}^v_{t,k} C_{t,k}(\hat{X}^v) + \sum_{t=v-\bar{t}_j}^{v-1} (\tilde{y}^v_{t,k} - 1) C_{t,k}(\hat{X}^v)$$

$$\leq \sum_{t=1}^{v} \tilde{y}^v_{t,k} C_{t,k}(\hat{X}^v) - \beta_j. \tag{9}$$

For the second inequality, we add $(\tilde{y}_{v,k}^v - 1) \cdot C_{v,k}(\hat{X}^v) \geq 0$ and use $\sum_{t=v-\bar{t}_j}^{v} C_{t,k}(\hat{X}^v) \geq \beta_j$ which holds because either j was powered up in $\hat{X}^v$ during $[v - \bar{t}_j : v]$ (then there is the switching cost of β_j) or j runs for $\bar{t}_j + 1$ time slots resulting in an operating cost of $l_j \cdot (\bar{t}_j + 1) = l_j \cdot (\lfloor \beta_j/l_j \rfloor + 1) \geq \beta_j$. Altogether, we get (beginning from the left hand side of Eq. (7) that has to be shown)

$$2 \sum_{t=1}^{v-1} \tilde{y}_{t,k}^{v-1} C_{t,k}(\hat{X}^v) \overset{(9),L5}{\leq} 2 \sum_{t=1}^{v} \tilde{y}_{t,k}^v C_{t,k}(\hat{X}^v) - 2\beta_j + 2\beta_j$$

$$\leq 2 \sum_{t=1}^{v} \tilde{y}_{t,k}^v C_{t,k}(\hat{X}^v).$$

If $A_{v,k}$ is an *extended* block, the proof of Eq. (7) is quite similar (see the full version of this paper for more details). □

Theorem 1. *Algorithm $\mathcal{A}$ is 2d-competitive.*

Proof. The feasibility of $X^{\mathcal{A}}$ was already proven in Lemma 4, so we have to show that $C(X^{\mathcal{A}}) \leq 2d \cdot C(\hat{X}^T)$. Let $C_v(X^{\mathcal{A}}) := \sum_{t=1}^{v} \sum_{k=1}^{m} C(A_{t,k})$ denote the cost of algorithm $\mathcal{A}$ up to time slot v. We will show by induction that

$$C_v(X^{\mathcal{A}}) \leq 2 \sum_{k=1}^{m} \sum_{t=1}^{v} \tilde{y}_{t,k}^v C_{t,k}(\hat{X}^v) \tag{10}$$

holds for all $v \in [T]_0$.

For $v = 0$, we have no costs for both $X^{\mathcal{A}}$ and $\hat{X}^v$, so inequality (10) is fulfilled. Assume that inequality (10) holds for $v - 1$. By using the induction hypothesis as well as Lemmas 7 and 8, we get

$$C_v(X^{\mathcal{A}}) = C_{v-1}(X^{\mathcal{A}}) + \sum_{k=1}^{m} C(A_{v,k})$$

$$\overset{\text{I.H.}}{\leq} 2 \sum_{k=1}^{m} \sum_{t=1}^{v-1} \tilde{y}_{t,k}^{v-1} C_{t,k}(\hat{X}^{v-1}) + \sum_{k=1}^{m} C(A_{v,k})$$

$$\overset{L7,L8}{\leq} 2 \sum_{k=1}^{m} \sum_{t=1}^{v} \tilde{y}_{t,k}^v C_{t,k}(\hat{X}^v). \tag{11}$$

Since $\tilde{y}_{t,k}^v \leq d$, we get

$$C_T(X^{\mathcal{A}}) \overset{(11)}{\leq} 2 \sum_{k=1}^{m} \sum_{t=1}^{T} \tilde{y}_{t,k}^T C_{t,k}(\hat{X}^T) \leq 2d \sum_{k=1}^{m} \sum_{t=1}^{T} C_{t,k}(\hat{X}^T) \leq 2d \cdot C(\hat{X}^T).$$

The schedule $\hat{X}^T$ is optimal for the problem instance $\mathcal{I}$, so algorithm $\mathcal{A}$ is 2d-competitive. □

3   Randomized Online Algorithm

The $2d$-competitive algorithm can be randomized to achieve a competitive ratio of $\frac{e}{e-1}d \approx 1.582d$ against an oblivious adversary. The randomized algorithm $\mathcal{B}$ chooses $\gamma \in [0,1]$ according to the probability density function $f_\gamma(x) = e^x/(e-1)$ for $x \in [0,1]$. The variables $\bar{t}_j$ are set to $\lfloor \gamma \cdot \beta_j/l_j \rfloor$, so the running time of a server is randomized. Then, algorithm $\mathcal{A}$ is executed. Note that γ is determined at the beginning of the algorithm and not for each block.

Theorem 2. *Algorithm $\mathcal{B}$ is $\frac{e}{e-1}d$-competitive against an oblivious adversary.*

The complete proof of this theorem is shown in the full paper. Most lemmas introduced in the previous section still hold, because they do not depend on the exact value of $\bar{t}_j$, only Lemmas 5 and 8 have to be adapted. For the proof of Theorem 2, we first give an upper bound for the expected cost of block $A_{t,k}$ (replacing Lemma 5). This bound is used to show that

$$\frac{e}{e-1} \cdot \sum_{t=1}^{v-1} \tilde{y}_{t,k}^{v-1} C_{t,k}(\hat{X}^v) + \mathbb{E}[C(A_{v,k})] \leq \frac{e}{e-1} \cdot \sum_{t=1}^{v} \tilde{y}_{t,k}^{v} C_{t,k}(\hat{X}^v)$$

holds for all lanes $k \in [m]$ and time slots $v \in [T]$ (similar to Lemma 8). Finally, Theorem 2 is proven by induction.

4   Lower Bound

In this section, we show that there is no deterministic online algorithm that achieves a competitive ratio that is better than $2d$.

We consider the following problem instance: Let $\beta_j := N^{2j}$ and $l_j := 1/N^{2j}$ where N is a sufficiently large number that depends on the number of servers types d. The value of N will be determined later. The adversary will send a job for the current time slot if and only if the online algorithm has no active server during the previous time slot. This implies that the online algorithm has to power up a server immediately after powering down any server. Note that $\lambda_t \in \{0,1\}$, i.e., it is never necessary to power up more than one server. The optimal schedule is denoted by X^*. Let $\mathcal{A}$ be an arbitrary deterministic online algorithm and let $X^\mathcal{A}$ be the schedule computed by $\mathcal{A}$.

W.l.o.g. in $X^\mathcal{A}$ there is no time slot with more than one active server. If this were not the case, we could easily convert the schedule into one where the assumption holds without increasing the cost. Assume that at time slot t a new server of type k is powered up such that there are (at least) two active servers at time t. If we power up the server at $t+1$, the schedule is still feasible, but the total costs are reduced by l_k. We can repeat this procedure until there is at most one active server for each time slot.

Lemma 9. *Let $k \in [d]$. If $X^\mathcal{A}$ only uses servers of type lower than or equal to k and if the cost of $\mathcal{A}$ is at least $C(X^\mathcal{A}) \geq N\beta_k$, then the cost of $\mathcal{A}$ is at least*

$$C(X^{\mathcal{A}}) \geq (2k - \epsilon_k) \cdot C(X^*) \tag{12}$$

with $\epsilon_k = 9k^2/N$ and $N \geq 6k$.

Proof Idea. We will prove the lemma by induction. The base case $k = 1$ is shown in the full version of this paper, so we assume that Lemma 9 holds for $k - 1$.

We divide the schedule $X^{\mathcal{A}}$ into phases $L_0, K_1, L_1, K_2, \ldots, L_n$ such that in the phases $K_1, \ldots, K_n$ server type k is used exactly once, while in the intermediate phases $L_0, \ldots, L_n$ the other server types $1, \ldots, k-1$ are used. A phase K_i begins when a server of type k is powered up and ends when it is powered down. The phases L_i can have zero length (if the server type k is powered up immediately after it is powered down, so between K_i and K_{i+1} an empty phase L_i is inserted).

The operating cost during phase K_i is denoted by $\delta_i \beta_k$. The operating and switching costs during phase L_i are denoted by $p_i \beta_k$. We divide the intermediate phases L_i into long phases where $p_i > 1/N$ holds and short phases where $p_i \leq 1/N$. Note that we can use the induction hypothesis only for long phases. The index sets of the long and short phases are denoted by $\mathcal{L}$ and $\mathcal{S}$, respectively.

To estimate the cost of an optimal schedule we consider two strategies: In the first strategy, a server of type k is powered up at the first time slot and runs for the whole time except for phases K_i with $\delta_i > 1$, then powering down and powering up are cheaper than keeping the server in the active state (β_k vs. $\delta_i \beta_k$). The operating cost for the phases K_i is $\delta_i^* \beta_k$ with $\delta_i^* := \min\{1, \delta_i\}$ and the operating cost for the phases L_i is at most $\frac{1}{N^2} p_i \beta_k$, because algorithm $\mathcal{A}$ uses servers whose types are lower than k and therefore the operating cost of $\mathcal{A}$ is at least N^2 times larger. Thus, the total cost of this strategy is at most

$$\beta_k \left(1 + \sum_{i=1}^{n} \delta_i^* + \sum_{i \in \mathcal{L} \cup \mathcal{S}} \frac{1}{N^2} p_i \right) \geq C(X^*).$$

In the second strategy, for the long phases L we use the strategy given by our induction hypothesis, while for the short phases S we behave like algorithm $\mathcal{A}$ and in the phases K_i we run the server type 1 for exactly one time slot (note that in K_i we only have $\lambda_t = 1$ in the first time slot of the phase). Therefore the total cost is upper bounded by

$$\beta_k \left(\sum_{i \in \mathcal{L}} \frac{1}{\alpha} p_i + \sum_{i \in \mathcal{S}} p_i + 2n\beta_1/\beta_k \right) \geq C(X^*)$$

with $\alpha := 2k - 2 - \epsilon_{k-1}$.

The total cost of $\mathcal{A}$ is equal to $\beta_k \left(\sum_{i=1}^{n}(1 + \delta_i) + \sum_{i \in \mathcal{L} \cup \mathcal{S}} p_i \right)$, so the competitive ratio is given by

$$\frac{C(X^{\mathcal{A}})}{C(X^*)} \geq \frac{\sum_{i=1}^{n}(1 + \delta_i) + \sum_{i \in \mathcal{L} \cup \mathcal{S}} p_i}{C(X^*)/\beta_k}.$$

By cleverly separating the nominator into two terms and by estimating $C(X^*)$ with strategy 1 and 2, respectively, it can be shown that $\frac{C(X^A)}{C(X^*)} \geq 2 + \alpha - \frac{16k}{N} \geq 2k - \epsilon_k$. The complete calculation including all intermediate steps is shown in the full paper. □

Theorem 3. *There is no deterministic online algorithm for the data-center optimization problem with heterogeneous servers and time and load independent operating costs whose competitive ratio is smaller than 2d.*

Proof Idea. Assume that there is an $(2d - \epsilon)$-competitive deterministic online algorithm $\mathcal{A}$. We construct a workload as described at the beginning of this section until the cost of $\mathcal{A}$ is greater than $N\beta_d$ (note that $l_j > 0$ for all $j \in [d]$, so the cost of $\mathcal{A}$ can be arbitrarily large). By using Lemma 9 with $k = d$ and $N := \max\{6d, \lceil 9k^2/\epsilon + 1 \rceil\}$, we get $C(X^A) > (2d - \epsilon) \cdot C(X^*)$ which is a contradiction to our assumption. See the full paper for more details. □

The schedule constructed for the lower bound only uses at most one job in each time slot, so there is no reason for an online algorithm to utilize more than one server of a specific type. Thus, for a data center with m unique servers (i.e. $m_j = 1$ for all $j \in [d]$), the best achievable competitive ratio is $2d = 2m$.

References

1. Albers, S., Quedenfeld, J.: Optimal algorithms for right-sizing data centers. In: Proceedings of the 30th on Symposium on Parallelism in Algorithms and Architectures, pp. 363–372. ACM (2018)
2. Albers, S., Quedenfeld, J.: Optimal algorithms for right-sizing data centers–extended version. arXiv preprint arXiv:1807.05112 (2018)
3. Antoniadis, A., Garg, N., Kumar, G., Kumar, N.: Parallel machine scheduling to minimize energy consumption. In: Proceedings of the Fourteenth Annual ACM-SIAM Symposium on Discrete Algorithms, pp. 2758–2769. SIAM (2020)
4. Antoniadis, A., Schewior, K.: A tight lower bound for online convex optimization with switching costs. In: Solis-Oba, R., Fleischer, R. (eds.) WAOA 2017. LNCS, vol. 10787, pp. 164–175. Springer, Cham (2018). https://doi.org/10.1007/978-3-319-89441-6_13
5. Argue, C., Gupta, A., Guruganesh, G., Tang, Z.: Chasing convex bodies with linear competitive ratio. In: Proceedings of the Fourteenth Annual ACM-SIAM Symposium on Discrete Algorithms, pp. 1519–1524. SIAM (2020)
6. Augustine, J., Irani, S., Swamy, C.: Optimal power-down strategies. SIAM J. Comput. **37**(5), 1499–1516 (2008)
7. Badiei, M., Li, N., Wierman, A.: Online convex optimization with ramp constraints. In: 2015 54th IEEE Conference on Decision and Control (CDC), pp. 6730–6736. IEEE (2015)
8. Bansal, N., Böhm, M., Eliáš, M., Koumoutsos, G., Umboh, S.W.: Nested convex bodies are chaseable. In: Proceedings of the Twenty-Ninth Annual ACM-SIAM Symposium on Discrete Algorithms, pp. 1253–1260. SIAM (2018)

9. Bansal, N., Gupta, A., Krishnaswamy, R., Pruhs, K., Schewior, K., Stein, C.: A 2-competitive algorithm for online convex optimization with switching costs. In: LIPIcs-Leibniz International Proceedings in Informatics, vol. 40. Schloss Dagstuhl-Leibniz-Zentrum fuer Informatik (2015)
10. Bansal, N., Kimbrel, T., Pruhs, K.: Speed scaling to manage energy and temperature. J. ACM (JACM) **54**(1), 1–39 (2007)
11. Bawden, T.: Global warming: data centres to consume three times as much energy in next decade, experts warn (2016). http://www.independent.co.uk/environment/global-warming-data-centres-to-consume-three-times-as-much-energy-in-next-decade-experts-warn-a6830086.html
12. Brill, K.G.: The invisible crisis in the data center: the economic meltdown of Moore's law. White paper, Uptime Institute, pp. 2–5 (2007)
13. Bubeck, S., Klartag, B., Lee, Y.T., Li, Y., Sellke, M.: Chasing nested convex bodies nearly optimally. In: Proceedings of the Fourteenth Annual ACM-SIAM Symposium on Discrete Algorithms, pp. 1496–1508. SIAM (2020)
14. Chen, N., Agarwal, A., Wierman, A., Barman, S., Andrew, L.L.: Online convex optimization using predictions. In: ACM SIGMETRICS Performance Evaluation Review, vol. 43, pp. 191–204. ACM (2015)
15. Chen, N., Goel, G., Wierman, A.: Smoothed online convex optimization in high dimensions via online balanced descent. Proc. Mach. Learn. Res. **75**, 1574–1594 (2018)
16. Delforge, P., et al.: Data center efficiency assessment (2014). https://www.nrdc.org/sites/default/files/data-center-efficiency-assessment-IP.pdf
17. Friedman, J., Linial, N.: On convex body chasing. Discrete Comput. Geom. **9**(1), 293–321 (1993). https://doi.org/10.1007/BF02189324
18. Goel, G., Chen, N., Wierman, A.: Thinking fast and slow: optimization decomposition across timescales. In: 2017 IEEE 56th Annual Conference on Decision and Control (CDC), pp. 1291–1298. IEEE (2017)
19. Goel, G., Wierman, A.: An online algorithm for smoothed regression and LQR control. Proc. Mach. Learn. Res. **89**, 2504–2513 (2019)
20. Hamilton, J.: Cost of power in large-scale data centers (2008). http://perspectives.mvdirona.com/2008/11/cost-of-power-in-large-scale-data-centers/
21. Irani, S., Pruhs, K.R.: Algorithmic problems in power management. ACM SIGACT News **36**(2), 63–76 (2005)
22. Kim, S.J., Giannakis, G.B.: Real-time electricity pricing for demand response using online convex optimization. In: ISGT 2014, pp. 1–5. IEEE (2014)
23. Kim, T., Yue, Y., Taylor, S., Matthews, I.: A decision tree framework for spatiotemporal sequence prediction. In: Proceedings of the 21th ACM SIGKDD International Conference on Knowledge Discovery and Data Mining, pp. 577–586. ACM (2015)
24. Lin, M., Liu, Z., Wierman, A., Andrew, L.L.: Online algorithms for geographical load balancing. In: Green Computing Conference (IGCC), pp. 1–10. IEEE (2012)
25. Lin, M., Wierman, A., Andrew, L.L., Thereska, E.: Dynamic right-sizing for power-proportional data centers. IEEE/ACM Trans. Netw. (TON) **21**(5), 1378–1391 (2013)
26. Lin, M., Wierman, A., Andrew, L.L., Thereska, E.: Dynamic right-sizing for power-proportional data centers – extended version (2013)
27. Liu, Z., Lin, M., Wierman, A., Low, S.H., Andrew, L.L.: Greening geographical load balancing. In: Proceedings of the ACM SIGMETRICS Joint International Conference on Measurement and Modeling of Computer Systems, pp. 233–244. ACM (2011)

28. Meyerson, A.: The parking permit problem. In: 46th Annual IEEE Symposium on Foundations of Computer Science (FOCS 2005), pp. 274–282. IEEE (2005)
29. Schmid, P., Roos, A.: Overclocking core i7: power versus performance (2009). www.tomshardware.com/reviews/overclock-core-i7,2268-10.html
30. Sellke, M.: Chasing convex bodies optimally. In: Proceedings of the Fourteenth Annual ACM-SIAM Symposium on Discrete Algorithms, pp. 1509–1518. SIAM (2020)
31. Shan, A.: Heterogeneous processing: a strategy for augmenting Moore's law. Linux J. **2006**(142), 7 (2006)
32. Wang, H., Huang, J., Lin, X., Mohsenian-Rad, H.: Exploring smart grid and data center interactions for electric power load balancing. ACM SIGMETRICS Perform. Eval. Rev. **41**(3), 89–94 (2014)
33. Zhang, M., Zheng, Z., Shroff, N.B.: An online algorithm for power-proportional data centers with switching cost. In: 2018 IEEE Conference on Decision and Control (CDC), pp. 6025–6032. IEEE (2018)

On Vertex-Weighted Graph Realizations

Amotz Bar-Noy[1], Toni Böhnlein[3(✉)], David Peleg[3], and Dror Rawitz[2]

[1] City University of New York (CUNY), New York, USA
amotz@sci.brooklyn.cuny.edu
[2] Bar Ilan University, Ramat-Gan, Israel
dror.rawitz@biu.ac.il
[3] Weizmann Institute of Science, Rehovot, Israel
{toni.bohnlein,david.peleg}@weizmann.ac.il

Abstract. Given a degree sequence d of length n, the DEGREE REALIZATION problem is to decide if there exists a graph whose degree sequence is d, and if so, to construct one such graph. Consider the following natural variant of the problem. Let $G = (V, E)$ be a simple undirected graph of order n. Let $f \in \mathbb{R}_{\geq 0}^n$ be a vector of vertex *requirements*, and let $w \in \mathbb{R}_{\geq 0}^n$ be a vector of *provided services* at the vertices. Then w *satisfies* f on G if the constraints $\sum_{j \in N(i)} w_j = f_i$ are satisfied for all $i \in V$, where $N(i)$ denotes the neighborhood of i. Given a requirements vector f, the WEIGHTED GRAPH REALIZATION problem asks for a suitable graph G and a vector w of provided services that satisfy f on G. In the original degree realization problem, all the provided services must be equal to one.

In this paper, we consider two avenues. We initiate a study that focuses on weighted realizations where the graph is required to be of a specific class by providing a full characterization of realizable requirement vectors for paths and acyclic graphs. However, checking the respective criteria is shown to be NP-hard.

In the second part, we advance the study in general graphs. In [7] it was observed that any requirements vector f where n is even can be realized. For odd n, the question of whether f is realizable is framed as whether f_n (largest requirement) lies within certain intervals whose boundaries depend on the requirements $f_1, \ldots, f_{n-1}$. Intervals were identified where f can be realized but for their complements the question is left open. We describe several new, realizable intervals and show the existence of an interval that cannot be realized. The complete classification for general graphs is an open problem.

1 Introduction

Background. Given a degree sequence d of length n, the degree realization problem is to decide if d has a realization, that is, an n-vertex graph whose

---

This work was supported by US-Israel BSF grant 2018043 and ARL Cooperative Grant ARL Network Science CTA W911NF-09-2-0053.

T. Calamoneri and F. Corò (Eds.): CIAC 2021, LNCS 12701, pp. 90–102, 2021.
https://doi.org/10.1007/978-3-030-75242-2_6

degree sequence is d, and if so, to construct such a realization (see [1,10,12–14,16–19]). The problem was well researched over the recent decades and plays an important role in the field of Social Networks (cf. [8,11,15]). For additional graph realization problems see [2–4,6] including a survey [5].

Bar-Noy, Peleg, and Rawitz [7] introduced a natural variant of the problem: Let $G = (V, E)$ be a simple undirected graph on $V = \{1, 2, \ldots, n\}$. Let $f \in \mathbb{R}^n_{\geq 0}$ be a vector of *vertex requirements*, and let $w \in \mathbb{R}^n_{\geq 0}$ be a vector of *vertex weights*. Vector w *satisfies* the requirement vector f on G if the constraints $\sum_{j \in N(i)} w_j = f_i$ are satisfied for all $i \in V$, where $N(i)$ denotes the (open) neighborhood of vertex i. The vertex-weighted realization problem is now as follows: Given a requirement vector f, find a suitable graph G and a weight vector w that satisfy f on G (if exist). This yields a conceptual generalization of the original degree realization problem, which corresponds to the case where it is required that all vertex weights are equal to one.

As noted in [7], any vector f of even length can be realized by a graph composed of $\frac{n}{2}$ independent edges (v_i, u_i), using the weights $w_{u_i} = f_{v_i}$ and $w_{v_i} = f_{u_i}$ for every i.

Theorem 1 ([7]). *Any requirements vector of even length can be realized.*

The problem becomes significantly harder for odd n. A preliminary observation shows that for odd n, f can be realized if either $\min\{f_i\} = 0$, or $f_i = f_j$ for two distinct indices $i, j \in [1, n]$, hence we focus w.l.o.g. on the domain

$$\mathcal{F}_n \triangleq \left\{ f \in \mathbb{R}^n_{\geq 0} : 0 < f_1 < f_2 < \cdots < f_n \right\}.$$

As a simple example, consider the domain $\mathcal{F}_3$. Note that any graph that potentially realizes some $f \in \mathcal{F}_3$ must be connected since $f_1 > 0$. For $n = 3$, the only two connected graphs are the path P_3 (Fig. 1a) and the complete graph K_3 (Fig. 1c). The graph layout implies an equation system that the requirements and weights must satisfy; see Figs. 1b and 1d for P_3 and K_3, respectively.

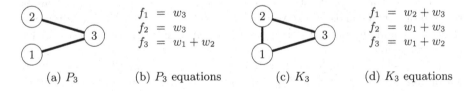

(a) P_3 (b) P_3 equations   (c) K_3 (d) K_3 equations

Fig. 1. Graphs that realize vectors of $\mathcal{F}_3$ and their equation systems.

The system in Fig. 1b implies that $f_1 = f_2$ must hold. In general, P_3 implies that f must satisfy $f_i = f_j$ where i, j are the labels of the two vertices of degree one. As a consequence, P_3 cannot realize a vector $f \in \mathcal{F}_3$. For K_3, the labeling is immaterial due to the graph's symmetry. Solving this system for the weights yields:

$$w_1 = (f_2 + f_3 - f_1)/2, \quad w_2 = (f_1 + f_3 - f_2)/2, \quad w_3 = (f_1 + f_2 - f_3)/2.$$

The problem requires the weights to be non-negative, so each equation implies a constraint, yielding

$$0 \leq f_2 + f_3 - f_1, \qquad 0 \leq f_1 + f_3 - f_2, \qquad 0 \leq f_1 + f_2 - f_3.$$

The first two equations are satisfied by any $f \in \mathcal{F}_3$; the third one yields the constraint that $f \in \mathcal{F}_3$ can be realized if and only if $f_3 \leq f_1 + f_2$. The example demonstrates an approach used in this study: Given a graph G, we deduce constraints, and use them to define the domain realizable by G.

A full characterization of the vertex weighted problem up to $n = 5$ is presented in [7]. For odd n, it was shown that a vector $f \in \mathcal{F}_n$ cannot be realized if it belongs to the *exponential growth* domain

$$\mathcal{D}_n^{\mathrm{exp}} = \left\{ f : \forall i \in [1, n], f_i > \sum_{j < i} f_j \right\}.$$

On the other hand, f *can* be realized if it falls in the *sub-exponential growth* domain

$$\mathcal{D}_n^{\mathrm{sub}} = \left\{ f : \exists i \in [1, n-1], f_i \leq \sum_{j < i} f_j \right\}.$$

Theorem 2 ([7]). *Let $n \geq 3$ be an odd integer. Then,*

1. *a requirements vector $f \in \mathcal{D}_n^{exp}$ cannot be realized.*
2. *a requirements vector $f \in \mathcal{D}_n^{sub}$ can be realized.*

Note that in the definition of $\mathcal{D}_n^{\mathrm{sub}}$, there is no inequality for bounding f_n. The "unknown domain" at this point, for which the realizability problem is still unsettled, is the "almost exponential" domain

$$\mathcal{D}_n^{\mathrm{exp\text{-}}} = \left\{ f : \forall i \in [1, n-1], f_i > \sum_{j < i} f_j \text{ and } f_n \leq \sum_{j < n} f_j \right\}.$$

Hence, subsequent analysis should concentrate on the domain $\mathcal{D}_n^{\mathrm{exp\text{-}}}$, resolve the status of some of its subdomains and thus narrow down the unknown regions.

Based on these results, the question whether a vector $f \in \mathcal{D}_n^{\mathrm{exp\text{-}}}$ can be realized or not, depends on the value of f_n in relation to the other requirements. Hence, subdomains of $\mathcal{D}_n^{\mathrm{exp\text{-}}}$ are typically defined in terms of *intervals* in the range of possible values for f_n. The situation at the two extremes of this range is clear. If f_n is larger than $\sum_{i < n} f_i$, then a vector $f \in \mathcal{F}_n$ cannot be realized due to Theorem 2. At the other end, if $f_n \leq f_{n-1} + f_{n-2}$, then there exists a realization for f that uses K_3 and a matching graph as described in [7]. Consequently, our analysis concentrates on vectors $f \in \mathcal{D}_n^{\mathrm{exp\text{-}}}$ where f_n is in the intermediate range, $f_n \in [f_{n-1} + f_{n-2}, \sum_{i < n} f_i]$.

It is shown in [7] that parts of this interval can be realized by two types of domains called the *windmill* and the *kite* domains that are both defined by a lower and upper bound on f_n (see Sect. 3 for more details). Combining these domains, the following collection of ranges are identified as realizable: for even $\ell \in [2, n-5]$,

$$\mathcal{D}_{n,\ell}^{\mathrm{W \cup K}} = \left\{ f \in \mathcal{D}_n^{\mathrm{exp\text{-}}} : \sum_{j=\ell+1}^{n-1} (f_j - f_{\ell+1}) \leq f_n \leq \sum_{j=\ell-1}^{n-1} f_j - f_\ell \right\}.$$

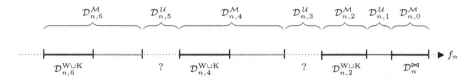

Fig. 2. Coverage of f_n's line plot by the domains $\mathcal{D}_{n,\ell}^{\mathcal{M}}$ and $\mathcal{D}_{n,k}^{\mathcal{U}}$. The blue intervals are the known, realizable intervals due to [7]. The green intervals are shown to be realizable in Sect. 3, and the red interval $\mathcal{D}_{n,1}^{\mathcal{U}}$ is shown to be unrealizable in Sect. 4.

Our Results. We consider the task of classifying requirement vectors that can be realized by specific graph classes. More specifically, we focus on paths and acyclic graphs. For any even sequence, Theorem 1 provides a realization with the caveat that the graph is disconnected. We redeem this shortcoming by showing how to realize any even sequence with a path. Additionally, we classify odd sequences that can be realized by a path. Alas, deciding whether a given odd sequence can be realized using a path is shown to be NP-hard.

We then turn to acyclic (not necessarily connected) graphs. For even sequences, the results mentioned above yield realizations. For odd sequences, we provide a full characterization: f can be realized by a forest if and only if there exist two disjoint nonempty index sets I and J such that $\sum_{i \in I} f_i = \sum_{j \in J} f_j$ and $|I| + |J| \leq \lceil n/2 \rceil$. Determining whether this condition holds is shown to be NP-hard as well.

For general graphs, we present *extended windmill* and *kite* domains, which are based on graphs that were used in [7] to define the windmill and kite domains. The extensions result from using different vertex-labelings. Moreover, we show that certain collections of these domains have pairwise overlapping intervals, i.e., they form a single, larger interval. We use the larger intervals to define the meta-domains $\mathcal{D}_{n,\ell}^{\mathcal{M}}$ for every even integer $0 \leq \ell \leq n - 5$. Given this result, we define the complements of the meta-domains as the (so far) *unknown domains* $\mathcal{D}_{n,k}^{\mathcal{U}}$, for odd integers $1 \leq k \leq n - 4$. Figure 2 illustrates the placements of the domains' intervals on a line plot of f_n.

The second result for general graphs focuses on the first unknown domain of [7], namely on $\mathcal{D}_{n,1}^{\mathcal{U}}$ (see Fig. 2). We analyze it using an opposite approach to the earlier one (of generating a set of constraints from a graph). Assume that a vector $f \in \mathcal{D}_{n,1}^{\mathcal{U}}$ is realized by a graph G and weights w. Then f is subject to a set of constraints, namely, upper and lower bounds on f_n, and exponential growth constraints for $f_1, \ldots, f_{n-1}$, implied by the definition of $\mathcal{D}_n^{\exp-}$. From these, we deduce structural properties of G. For example, we show that vertex n must be adjacent to at least $n - 2$ vertices. Based on the exponential growth of $f_1, \ldots, f_{n-1}$, we show that each vertex must have a neighbor with a *dedicated weight* which ensures that its requirement is met. We show that this dependency is pairwise by deducing a one-to-one correspondence between weights and requirements, revealed by decomposing the graph G. The decomposition process can be viewed as removing pairs of vertices in $\frac{n-1}{2}$ many steps. Each pair of

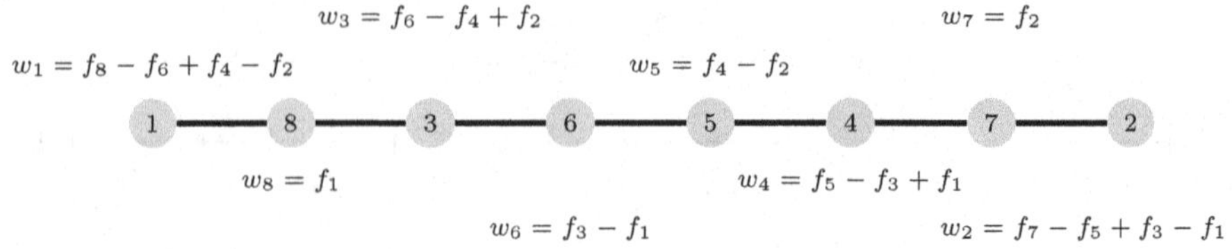

$$w_3 = f_6 - f_4 + f_2 \qquad\qquad w_7 = f_2$$
$$w_1 = f_8 - f_6 + f_4 - f_2 \qquad\qquad w_5 = f_4 - f_2$$

①——⑧——③——⑥——⑤——④——⑦——②

$$w_8 = f_1 \qquad\qquad w_4 = f_5 - f_3 + f_1$$
$$w_6 = f_3 - f_1 \qquad\qquad w_2 = f_7 - f_5 + f_3 - f_1$$

Fig. 3. The alternating path realization for the domain $0 < f_1 \le f_2 \cdots \le f_n$ with $n = 8$. The alternating permutation is $\pi(i) = i$, if i is odd, and $\pi(i) = n + 2 - i$, if i is even.

vertices is connected by an edge and all these edges form a matching in G, such that the matching partner of a vertex carries its dedicated weight. Based on the knowledge of $G's$ structure, we deduce constraints on f_n which contradict some of the constraints implied by $\mathcal{D}_{n,1}^{\mathcal{U}}$. Consequently, a requirement vector $f \in \mathcal{D}_{n,1}^{\mathcal{U}}$ cannot be realized. As mentioned above, Bar-Noy et al. [7] give a full characterization for $n = 5$ and show that $\mathcal{D}_{5,1}^{\mathcal{U}}$ cannot be realized. We generalize some of their ideas.

Organization. In Sect. 2, we characterize requirement vectors realizable by paths and acyclic graphs. Section 3 presents the extended windmill and kite domains. In Sect. 4, we show that $\mathcal{D}_{n,1}^{\mathcal{U}}$ is an un-realizable domain.

2   Realizations with Acyclic Graphs

In this section, we consider vertex weighted realizations where the graph must be acyclic. For an even n, we show that any sequence can be realized by a path graph, and therefore by an acyclic graph. We provide a necessary and sufficient condition for acyclic and path realizations when n is odd.

Path Realizations for Even n. We first show that every requirement vector in $\mathcal{F}_n$ can be realized using a path when n is even. Denote the vertices of the path graph $G = (V, E)$ by $V = \{1, 2, \ldots, n\}$. Let the edge set of G be $E = \{(\pi(i), \pi(i+1)) : 1 \le i < n\}$, where π is a permutation of V. A permutation is called *alternating* if the following two conditions are satisfied:

$$\pi(1) \le \pi(3) \le \cdots \le \pi(n-3) \le \pi(n-1) \tag{C1}$$
$$\pi(2) \ge \pi(4) \ge \cdots \ge \pi(n-2) \ge \pi(n). \tag{C2}$$

That is, odd indexed vertices generate an increasing sequence while even indexed vertices generate a decreasing sequence. An example of such a permutation is given in Fig. 3.

Theorem 3. *Let $f \in \mathbb{R}_{\ge 0}^n$, where n is even. Also, let π be an alternating permutation. Then, there exists a nonnegative weight vector w that realizes f with the path graph whose labeling is π.*

Proof. On this graph, the requirement constraints take the form: $f_{\pi(1)} = w_{\pi(2)}$, $f_{\pi(n)} = w_{\pi(n-1)}$, and $f_{\pi(i)} = w_{\pi(i-1)} + w_{\pi(i+1)}$, for $i \in \{2, \ldots, n-1\}$. It is not hard to verify that these requirements are satisfied by the weight assignment $w_{\pi(2i)} = \sum_{j=1}^{i}(-1)^{i-j} f_{\pi(2j-1)}$ and $w_{\pi(n+1-2i)} = \sum_{j=1}^{i}(-1)^{i-j} f_{\pi(n+2-2j)}$, and that all of these weights are nonnegative. □

Path Realizations of Odd n. Next, we identify requirement vectors in $\mathcal{F}_n$ that can be realized using a path, when n is odd. Let $G = (V, E)$ be a path graph, $V = \{1, 2, \ldots, n\}$, and $E = \{(\pi(i), \pi(i+1)) : 1 \leq i < n\}$, where π is a permutation of V. A permutation is called *sound* if the following conditions hold:

1. $\pi(i) \leq \pi(i+2)$, for every even $i \in \{2, 4, \ldots, n-3\}$.
2. $\sum_{i=1}^{k/2}(-1)^{i-1} f_{\pi(k-2i+1)} \geq 0$, for all even $k < n$.
3. $\sum_{i=1}^{(n+1)/2}(-1)^{i} f_{\pi(2i-1)} = 0$.

Theorem 4. *A vector $f \in \mathbb{R}_{\geq 0}^{n}$ can be realized by a path if and only if there exists a sound permutation of $\overline{V}$.*

Proof. First assume that there exists a sound permutation π of V. Define the following weights: For the vertex $\pi(k)$ in vertex k on the path,

$$w_{\pi(k)} = \begin{cases} \sum_{i=1}^{k/2}(-1)^{i-1} f_{\pi(k-2i+1)}, & k \text{ is even}, \\ \sum_{j=1}^{(k-1)/2}(-1)^{(k-1)/2-j} f_{\pi(2j)}, & k \text{ is odd} \end{cases}$$

Since π is sound, the weights are nonnegative. It is not hard to verify that G and w satisfy f.

For the other direction, assume that f can be realized by a path P of length n, and let π be a corresponding permutation. Observe that weights on even nodes of P (namely, those placed in the even locations on P) satisfy the requirements of the odd nodes. To prove Conditions 2 and 3, observe that for an even k we have

$$f_{\pi(k-1)} - w_{\pi(k)} = w_{\pi(k-2)} = f_{\pi(k-3)} - w_{\pi(k-4)}$$
$$= f_{\pi(k-3)} - f_{\pi(k-5)} + w_{\pi(k-6)}$$
$$= \ldots$$
$$= \sum_{i=1}^{k/2-1}(-1)^{i-1} f_{\pi(k-2i-1)}.$$

Condition 2 holds since, for $k < n$, we have

$$f_{\pi(k-1)} - \sum_{i=1}^{k/2-1}(-1)^{i-1} f_{\pi(k-2i-1)} = \sum_{i=1}^{k/2}(-1)^{i-1} f_{\pi(k-2i+1)} = w_{\pi(k)} \geq 0.$$

Moreover, Condition 3 is implied, since

$$f_{\pi(n)} = w_{\pi(n-1)} = \sum_{i=1}^{(n-1)/2}(-1)^{i-1} f_{\pi(n-2i)}.$$

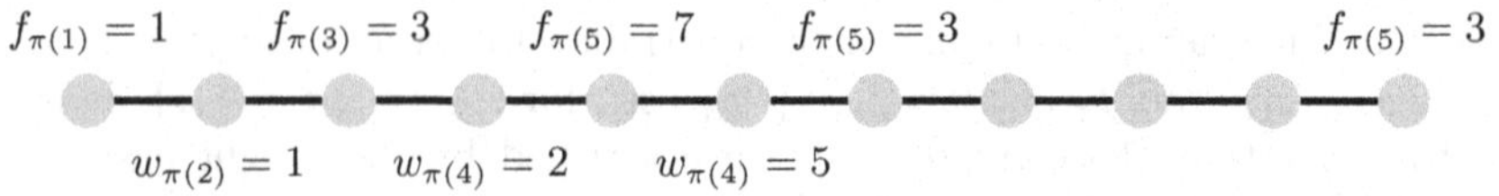

$$f_{\pi(1)} = 1 \qquad f_{\pi(3)} = 3 \qquad f_{\pi(5)} = 7 \qquad f_{\pi(5)} = 3 \qquad\qquad f_{\pi(5)} = 3$$

$$w_{\pi(2)} = 1 \qquad w_{\pi(4)} = 2 \qquad w_{\pi(4)} = 5$$

Fig. 4. An example that shows that Condition 2 is needed.

Since the weights on the odd nodes of P are only used for satisfying the requirements of the even nodes, we can change the order of the requirements in the even locations on the path so that their requirements appear in nonincreasing order, thus getting a new permutation π' that satisfies Condition 1. (The *weights* assigned to the even locations in the given realization remain in place, though, and are not moved along with the requirements.) Consequently, we need to modify the weights of the odd nodes by setting $w_{\pi(k)} = \sum_{j=1}^{(k-1)/2}(-1)^{(k-1)/2-j}f_{\pi(2j)}$ for every odd node k, to obtain a sound permutation. □

The following example shows that Condition 2 is indeed necessary. Consider the vector $f = (1,1,3,3,3,7,25,50,100,200,400)$. Assume towards a contradiction that f can be realized by a path of length $n = 11$, and let π be a corresponding permutation. It must be that

$$f_{\pi(1)} + f_{\pi(5)} + f_{\pi(9)} = \sum_{i=1}^{5} w_{\pi(2i)} = f_{\pi(3)} + f_{\pi(7)} + f_{\pi(11)},$$

Hence, without loss of generality we have that

$$\{\pi(1), \pi(2), \pi(6)\} = \{1,5,9\} \qquad \{\pi(3), \pi(4), \pi(5)\} = \{3,7,11\}.$$

Since $f_{\pi(3)} = 3$, it must be that $f_{\pi(1)} = 1$. Hence $w_{\pi(2)} = 1$ and $w_{\pi(4)} = 2$. It follows that $f_{\pi(5)} = 7$. This implies that $w_{\pi(6)} = 5$ and we get a contradiction. See Fig. 4.

Alas, deciding whether a given requirement vector can be realized by a weighted path is NP-hard. The proof is given in the full version of the paper.

Theorem 5. *Deciding if a vector $f \in \mathbb{R}^n_{\geq 0}$ admits a sound permutation is NP-hard.*

Acyclic Realizations of Odd n. Finally, we turn to realizations for arbitrary acyclic graphs. For an even n, the realizations from Theorems 1 & 3 are done with acyclic graphs. For an odd n, we classify all the requirement vectors that can be realized by a forest. The proofs are presented in the full version.

Theorem 6. *A vector $f \in \mathbb{R}^n_{\geq 0}$ can be realized by a forest if and only if there exist two disjoint index sets I and J such that $\sum_{i \in I} f_i = \sum_{j \in J} f_j$ and $|I|+|J| \leq \lceil n/2 \rceil$.*

Theorem 7. *Deciding whether a vector f of order n_f satisfies the condition in Theorem 6, namely if there exist two disjoint index sets I and J such that $\sum_{i \in I} f_i = \sum_{j \in J} f_j$ and $|I| + |J| \leq \lceil n_f/2 \rceil$, is NP-hard.*

The proof uses a reduction from the EQUAL SUM SUBSETS problem, which was shown to be NP-hard in [20].

3 Extended Windmill and Kite

In this section, we provide extended versions of the kite and windmill domains introduced in [7]. The details are presented in the full paper where we also combining several of these new domains to define meta domains. Let $n \geq 5$ and let $\ell \in [2, n-1]$ be an even number. Let

$$\mathcal{D}_{n,\ell}^{\mathcal{M}} = \left\{ f \in \mathcal{D}_n^{\text{exp-}} : \sum_{j=\ell+1}^{n-1}(f_j - f_{\ell+1}) \leq f_n \leq \sum_{j=1}^{n-1} f_j - f_\ell \right\}, \text{ and}$$

$$\mathcal{D}_{n,0}^{\mathcal{M}} = \left\{ f \in \mathcal{D}_n^{\text{exp-}} : \sum_{j=2}^{n-1}(f_j - f_1) \leq f_n \leq \sum_{j=1}^{n-1} f_j \right\}.$$

We refer to this interval as ℓ-th *meta domain* (see Fig. 5).

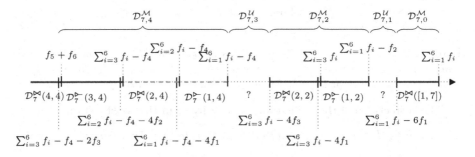

Fig. 5. Coverage of the interval U for $n = 7$ by the windmill and kite domains. Values for f_7 are plotted on the line. The two gaps marked by a green, dotted line are analyzed in this section. The (up to this point) unknown domains are marked by question marks.

Theorem 8. *Let $n \geq 5$ and let $\ell < n$ be an even number. The vector f in the meta domain $\mathcal{D}_{n,\ell}^{\mathcal{M}}$ can be realized.*

We now describe the *unknown* domains located between the meta domains. Let k be an odd index such that $k \leq n - 4$. The k-th unknown domain is:

$$\mathcal{D}_{n,k}^{\mathcal{U}} = \left\{ f \in \mathcal{D}_n^{\text{exp-}} : \sum_{j=1}^{n-1} f_j - f_{k+1} < f_n < \sum_{j=k}^{n-1}(f_j - f_k) \right\}.$$

Note that such a domain only exists if $\sum_{j=1}^{k-1} f_j + (n-k)f_k < f_{k+1}$.

4 An Unrealizable Domain

In this section, we show that a requirement vector $f \in \mathcal{D}_{n,1}^{\mathcal{U}}$ cannot be realized.

We start with two observations that we need to prove the main result, Theorem 9. Given a vertex-weight function $w : V \rightarrow \mathbb{R}$, let π be a permutation of V for which $w_{\pi(i)} \leq w_{\pi(i+1)}$, for every $i \in [1, n-1]$.

Lemma 1. *Let n be an integer, and let f be a requirement vector such that $\sum_{j<i} f_j < f_i$, for all $i \in [1, \ell]$, where $\ell \in [1, n - \deg(1) + 1]$. If (G, w) realize the vector f, then (i) $\sum_{j=1}^{\deg(1)} w_{\pi(j)} \le f_1$, and (ii) $w_{\pi(j+\deg(1)-1)} \le f_j$, for every $j \in [2, \ell]$.*

Proof. First, observe that $\deg(1) \ge 1$. Also, $\sum_{j=1}^{\deg(1)} w_{\pi(j)} \le \sum_{j \in N(1)} w_j = f_1$ showing the first claim. We prove the second claim by induction on i. For the base case, consider $\sum_{j=1}^{\deg(1)} w_{\pi(j)} \le f_1 < f_2$. Since G realizes f, there is a vertex $k \in N(2)$ whose weight is larger than $w_{\pi(\deg(1))}$. Hence, $w_{\pi(\deg(1)+1)} \le w_k \le f_2$.

For the inductive step, let $w_{\pi(j+\deg(1)-1)} \le f_j$ for every $j < i$. Hence,

$$\sum_{j=1}^{\deg(1)+i-2} w_{\pi(j)} = \sum_{j=1}^{\deg(1)} w_{\pi(j)} + \sum_{j=2}^{i-1} w_{\pi(\deg(1)+j-1)} \le \sum_{j=1}^{i-1} f_j < f_i$$

where the final, strict inequality holds by assumption. Similar to the base case, there is a vertex $k \in N(i)$ whose weight is larger than $w_{\pi(\deg(1)+i-2)}$. Hence, $w_{\pi(\deg(1)+i-1)} \le w_k \le f_i$.

Lemma 2. *Let $f \in \mathcal{F}_n$ be a vector that is realized by (G, w). Also, let $V' \subseteq V$ and let f' be a new requirement vector which is defined on V' as follows: $f_i' \triangleq f_i - \sum_{j \in N(i) \setminus V'} w_j$, for $i \in V'$. Then, $G[V']$ and $w|_{V'}$ realize f'.*

The correctness of the lemma follows readily from the construction.

Theorem 9. *Let $n \ge 5$ be an odd integer. A vector $f \in \mathcal{D}_{n,1}^{\mathcal{U}}$ cannot be realized.*

Proof. Towards a contradiction, suppose that $f \in \mathcal{D}_{n,1}^{\mathcal{U}}$ is be realized by a graph $G = (V, E)$ and a weight function w. Since $f_1 > 0$, the graph G cannot have an isolated vertex. By definition of $\mathcal{D}_{n,1}^{\mathcal{U}}$, the following equations hold:

$$\sum_{j<i} f_j < f_i, \quad \text{for } i \in [1, n-1] \tag{1}$$

$$\sum_{i=1}^{n-1} f_i - f_2 < f_n \tag{2}$$

$$f_n < \sum_{i=1}^{n-1} f_i - (n-1) f_1 \tag{3}$$

To show the theorem, we consider three cases depending on the size of neighborhood of vertex 1: (1) $\deg(1) \ge 3$, (2) $\deg(1) = 1$, and (3) $\deg(1) = 2$.

Case 1. If $\deg(1) \ge 3$, then $\sum_{j=1}^{\deg(1)} w_{\pi(j)} \le \sum_{j \in N(1)} w_j = f_1$. Lemma 1 implies that

$$\sum_{j=1}^{n} w_{\pi(j)} \le f_1 + \sum_{j=2}^{n-\deg(1)+1} f_j < f_{n-\deg(1)+2}$$

where the last inequality is due to Eq. (1) which can be used since $n - \deg(1) + 2 \le n - 1$. Hence, $f_{n-\deg(1)+2}$ cannot be realized and we reach the desired contradiction.

Case 2. Let vertex $N(1) = \{k\}$. It follows that $w_k = f_1$. We use Lemma 2 with $V' = V \setminus \{1, k\}$. Hence, the new requirements vector

$$f_i' = \begin{cases} f_i - f_1 & i \in N(k), \\ f_i & i \notin N(k), \end{cases}$$

is realized by $G' = G[V']$ and $w|_{V'}$. Consider $i \in V'$. If $i < n$, we have that

$$\sum_{j<i, j\in V'} f'_j \leq \sum_{j<i, j\in V'} f_j \leq \sum_{j<i} f_j - f_1 < f_i - f_1 \leq f'_i,$$

where the third inequality follows from Eq. (1). If $i = n$, we have that

$$\sum_{j<n, j\in V'} f'_j \leq \sum_{j<n, j\in V'} f_j = \sum_{j=1}^{n-1} f_j - f_k - f_1 \leq \sum_{j=1}^{n-1} f_j - f_2 - f_1$$
$$< f_n - f_1 \leq f'_i,$$

where the third inequality is due to Eq. (2). We reach a contradiction since $f' \in \mathcal{D}_{n-2}^{\exp}$.

Case 3. In this case, $N(1) = \{k, k'\}$. Hence, $w_{\pi(1)} + w_{\pi(2)} \leq w_k + w_{k'} = f_1$. By Eq. (2) and Lemma 1, we have that

$$w_{\pi(1)} + w_{\pi(2)} + \sum_{j=4}^{n} w_{\pi(j)} \leq f_1 + \sum_{j=3}^{n-1} f_j < f_n.$$

It follows that $f_n \geq \sum_{j=3}^{n} w_{\pi(j)}$ and that $\deg(n) \geq n - 2$. Hence, vertex n has a central role like in the windmill and kite graphs of Sect. 3. There are two cases: If $\deg(n) = n - 2$, then it must be that $N(n) = \{\pi(3), \ldots, \pi(n)\}$. Otherwise, $N(n) = \{\pi(1/2), \pi(3), \ldots, \pi(n)\}$, where $\pi(1/2)$ stands for either $\pi(1)$ or $\pi(2)$. It follows that $w_n = w_{\pi(1/2)}$.

Claim. $n \in N(1)$

Proof. Towards a contradiction, suppose that $n \notin N(1)$. Consequently, $\deg(n) = n - 2$. By Lemma 1 and Eq. (2), we have that

$$\sum_{j=4}^{n} w_{\pi(j)} \leq \sum_{j=3}^{n-1} f_j < f_n - f_1 = \sum_{j=3}^{n} w_{\pi(j)} - (w_k + w_{k'}) \leq \sum_{j=5}^{n} w_{\pi(j)}$$

since k and k' are adjacent to n. Hence, we reach a contradiction. □

With $N(1) = \{k, n\}$ we have $f_1 = w_k + w_n$. Due to Lemma 1 and Eq. (2) we have

$$w_n + w_k + \sum_{j=4}^{n} w_{\pi(j)} \leq f_1 + \sum_{j=3}^{n-1} f_j < f_n.$$

If $\deg(n) = n-2$, then $N(n) = \{\pi(3), \ldots, \pi(n)\}$ and $f_n = \sum_{j=3}^{n} w_{\pi(j)}$. Therefore it must be that $w_k = w_{\pi(2/1)}$. It follows that $k \notin N(n)$.

The main part of the remaining proof is to show that vertices $2, \ldots, n-1$, except k, have degree 2 in G. For that, define $V' = V \setminus \{1, n, k\}$ and consider $G' = G[V']$. Notice that in both cases (i.e., $\deg(n) = n - 1$ or $\deg(n) = n - 2$) we have that $V' \subseteq N(n)$.

Let $i \in V'$. Since $w_n + w_k = f_1 < f_i$, it follows that i must have at least one neighbor, apart from n, in G. Hence, $\deg'(i) \geq 1$, for every $i \in V'$. The next step in our proof is to show that $\deg'(i) \leq 1$, for every $i \in V'$. We do so by constructing a perfect matching M' in G' and then proving that $E' = M'$.

Claim. Graph G' contains a perfect matching M'.

Proof. Using Lemma 2 we get that the following requirement vector

$$f_i' = \begin{cases} f_i - w_n - w_k, & i \in N(k), \\ f_i - w_n & i \notin N(k). \end{cases}$$

which is defined on V' and is realized by G' and $w|_{V'}$. With $f_1 = w_n + w_k$ and Eq. (1) we get

$$\sum_{j<i, j\neq 1, k} f_j' \leq \sum_{j<i, j\neq 1, k} f_j < f_i - f_1 \leq f_i'$$

for $i \in V'$. Thus, $f' \in \mathcal{D}^{\exp}$. Let H be a component of G' and let $h = \{f_{i_1}, \ldots, f_{i_m}\}$ be the subsequence of f' induced by the vertices of H. where $m = |V(H)|$. Since $f' \in \mathcal{D}^{\exp}$, it must be that $h \in \mathcal{D}^{\exp}$. In addition, since H is realized by h, it follows by Theorem 2 that H contains an even number of vertices. If $\deg'(i_1) \geq 2$, then $w_{\pi(i_1)} + w_{\pi(i_2)} \leq f_{i_1}$, and thus by Lemma 1 we have that $\sum_{j=1}^m w_{\pi(i_j)} \leq \sum_{j=1}^{m-1} f_{i_j} < f_{i_m}$, which means that f_{h_m} cannot be realized. Hence, $\deg'(i_1) = 1$. The edge connecting i_1 and its only neighbor i_1' is added to the matching M. Next remove i_1 and i_1' from H. Let $H' = H \backslash \{i_1, i_1'\}$. Since $w_{i_1'} = h_{i_1}$, by Lemma 2 the requirement vector

$$h_j' = \begin{cases} h_j - h_{i_1} & i \in N(h_1'), \\ h_j & i \notin N(h_1'). \end{cases}$$

is realized by $G[H']$ and $w|_{H'}$. Also, observe that $h' \in \mathcal{D}^{\exp}$. Therefore we can again argue that the vertex with the smallest requirement in H' has exactly one neighbor. The edge connecting this vertex and its neighbor are part of our matching M. We continue until the remaining subgraph is empty. By performing this process on all components of G' we obtain a perfect matching M' in G'. $\square$

Let $M = M' \cup \{1, k\}$ and define a function $\varphi : V \backslash \{n\} \to V \backslash \{n\}$ by $\varphi(i) = j$ if $(i, j) \in M$. Observe that $\varphi(\varphi(i)) = i$, $w_{\varphi(i)} \leq f_i$ and $w_i \leq f_{\varphi(i)}$.

Claim. For all $i \in V'$, $\deg'(i) = 1$, i.e., $E' = M'$.

Proof. Towards a contradiction, suppose that there is a vertex $i \in V'$ such that $\deg'(i) \geq 2$. Let $p, q \in N'(i)$, where $\varphi(i) = p \neq q$. It follows that $w_p + w_q \leq f_i' \leq f_i$. By the Eq. (2) we have that

$$f_1 + \sum_{j=3}^{n-1} f_j < f_n \leq \sum_{j=1}^{n-1} w_j \leq \sum_{j\neq p,q,n} w_j + f_i \leq \sum_{j\neq p,q,n} f_{\varphi(j)} + f_i$$
$$= \sum_{j\neq q,n} f_{\varphi(j)} < \sum_{j\neq\varphi(q),n} f_j.$$

Since $q \in V'$, we know that $\varphi(q) \neq 1$, and we get a contradiction. It follows that $\deg'(i) = 1$ for every $i \in V'$, namely that $E' = M'$. Similarly, we show that $N(k) \subseteq \{1, n\}$. Suppose k has a neighbor $p \in V'$. In this case,

$$f_1 + \sum_{j=3}^{n-1} f_j < f_n \leq \sum_{j=1}^{n-1} w_j \leq \sum_{j\neq 1,p,n} w_j + f_k \leq \sum_{j\neq 1,q,n} f_{\varphi(j)} + f_k$$
$$= \sum_{j\neq q,n} f_{\varphi(j)} < \sum_{j\neq\varphi(q),n} f_j,$$

and we reach a contradiction. $\square$

Since each vertex in V' has exactly one neighbor in G' and k has no neighbors in V', we get that $f_i = w_{\varphi(i)} + w_n$, for all $i \in V'$. Also, if $k \in N(n)$, we have that $f_k = w_1 + w_n$, and otherwise $f_k = w_1$. If $\deg(n) = n - 1$,

$$\sum_{j=1}^{n-1} f_j = \sum_{j=1}^{n-1}(w_{\varphi(j)} + w_n) = \sum_{j=1}^{n-1}(w_j + w_n) = f_n + (n-1)w_n$$
$$\leq f_n + (n-1)f_1,$$

and we have a contradiction with Eq. (3). On the other hand, if $\deg(n) = n - 2$ and $k \notin N(n)$ we have that

$$\sum_{j=1}^{n-1} f_j = \sum_{j \neq k, n}(w_{\varphi(j)} + w_n) + w_1 = \sum_{j \neq n} w_j + (n-2)w_n$$
$$= f_n + w_k + (n-2)w_n \leq f_n + (n-1)f_1,$$

yielding a contradiction to Eq. (3). Hence, f cannot be realized and the theorem follows. $\qquad\square$

5   Conclusion

Based on the work of Bar-Noy et al. [7], we advanced the understanding of the WEIGHTED GRAPH REALIZATION problem. With the extended kite and windmill we found new, realizable domains. Qualitatively more important, we showed the existence of an un-realizable domain, namely $\mathcal{D}^{\mathcal{U}}_{n,1}$ for any odd n (see Fig. 2). Up to this point we are left with several, *unknown* domains ($\mathcal{D}^{\mathcal{U}}_{n,k}$, for $k > 1$) where it is not clear whether a requirement vector can be realized or not. Initial results exists showing that there are realizable and un-realizable sub-domains, but these sub-domains do not cover the unknown domains entirely. A full classification may result in a polynomial time algorithm to decide whether a sequence can be realized since our domains are typically defined using a linear function of the given sequence.

We classified the sequences realizable by paths and acyclic graphs. Even sequences can always be realized by a path. We classified the odd sequences that can be realized by a path and by a forest. As an intermediate result, it would be interesting to classify odd sequences realizable by a tree. Note that the DEGREE REALIZATION problem was studied in trees (see, e.g., [9]). The graphs that we use in Sect. 3 can have many odd cycles. Analyzing bipartite graphs, which generalize forests, may help to understand how even cycles can be used to realize additional sequences. Moreover, the question for connected graphs is wide open. To show that $\mathcal{D}^{\mathrm{sub}}_n$ is realizable, Bar-Noy et al. [7] use graphs that are disconnected as are some of the graphs that we use in Sect. 3.

It might also be intriguing to study weighted graph realizations where the neighborhoods are closed, i.e., vertices see also their own weight. There is a simple realization for any sequence that uses a graph without edges and the weight of a vertex is equal to its requirement. To make the problem interesting, one would have to forbid isolated vertices or study the problem on specific classes like connected graphs.

References

1. Aigner, M., Triesch, E.: Realizability and uniqueness in graphs. Discrete Math. **136**, 3–20 (1994)
2. Bar-Noy, A., Böhnlein, T., Lotker, Z., Peleg, D., Rawitz, D.: The generalized microscopic image reconstruction problem. In: 30th ISAAC, pp. 42:1–42:15 (2019)
3. Bar-Noy, A., Böhnlein, T., Lotker, Z., Peleg, D., Rawitz, D.: Weighted microscopic image reconstruction. In: Bureš, T., et al. (eds.) SOFSEM 2021. LNCS, vol. 12607, pp. 373–386. Springer, Cham (2021). https://doi.org/10.1007/978-3-030-67731-2_27
4. Bar-Noy, A., Choudhary, K., Cohen, A., Peleg, D., Rawitz, D.: Minimum neighboring degree realizations in graphs and trees. In: 28th ESA (2020)
5. Bar-Noy, A., Choudhary, K., Peleg, D., Rawitz, D.: Realizability of graph specifications: characterizations and algorithms. In: Lotker, Z., Patt-Shamir, B. (eds.) SIROCCO 2018. LNCS, vol. 11085, pp. 3–13. Springer, Cham (2018). https://doi.org/10.1007/978-3-030-01325-7_1
6. Bar-Noy, A., Choudhary, K., Peleg, D., Rawitz, D.: Graph realizations: maximum degree in vertex neighborhoods. In: 17th SWAT, vol. 162 (2020)
7. Bar-Noy, A., Peleg, D., Rawitz, D.: Vertex-weighted realizations of graphs. TCS **807**, 56–72 (2020)
8. Blitzstein, J.K., Diaconis, P.: A sequential importance sampling algorithm for generating random graphs with prescribed degrees. Internet Math. **6**, 489–522 (2010)
9. Buckley, F., Lewinter, M.: Introductory Graph Theory with Applications. Waveland Press, Long Grove (2013)
10. Choudum, S.A.: A simple proof of the Erdös-Gallai theorem on graph sequences. Bull. Aust. Math. Soc. **33**, 67–70 (1986)
11. Cloteaux, B.: Fast sequential creation of random realizations of degree sequences. Internet Math. **12**, 205–219 (2016)
12. Erdös, P., Gallai, T.: Graphs with prescribed degrees of vertices. Matemati. Lapok **11**, 264–274 (1960)
13. Hakimi, S.L.: On realizability of a set of integers as degrees of the vertices of a linear graph-I. SIAM J. Appl. Math. **10**(3), 496–506 (1962)
14. Havel, V.: A remark on the existence of finite graphs. Casopis Pest. Mat. **80**, 477–480 (1955). (in Czech)
15. Mihail, M., Vishnoi, N.: On generating graphs with prescribed degree sequences for complex network modeling applications. 3rd ARACNE (2002)
16. Sierksma, G., Hoogeveen, H.: Seven criteria for integer sequences being graphic. J. Graph Theory **15**(2), 223–231 (1991)
17. Tripathi, A., Tyagi, H.: A simple criterion on degree sequences of graphs. DAM **156**(18), 3513–3517 (2008)
18. Tripathi, A., Venugopalan, S., West, D.B.: A short constructive proof of the Erdös-Gallai characterization of graphic lists. Discrete Math. **310**(4), 843–844 (2010)
19. Tripathi, A., Vijay, S.: A note on a theorem of Erdös & Gallai. Discrete Math. **265**(1–3), 417–420 (2003)
20. Woeginger, G.J., Yu, Z.: On the equal-subset-sum problem. Inf. Process. Lett. **42**(6), 299–302 (1992)

On the Role of 3's for the 1-2-3 Conjecture

Julien Bensmail[1], Foivos Fioravantes[1(✉)], and Fionn Mc Inerney[2]

[1] Université Côte d'Azur, Inria, CNRS, I3S, Nice, France
{julien.bensmail,fioravantes.foivos}@inria.fr
[2] Laboratoire d'Informatique et Systèmes, Aix-Marseille Université, CNRS, and Université de Toulon Faculté des Sciences de Luminy, Marseille, France

Abstract. The 1-2-3 Conjecture states that every connected graph different from K_2 admits a proper 3-labelling, i.e., can have its edges labelled with $1, 2, 3$ so that no two adjacent vertices are incident to the same sum of labels. In connection with recent optimisation variants of this conjecture, we study the role of label 3 in proper 3-labellings of graphs. Previous studies suggest that, in general, it should always be possible to produce proper 3-labellings assigning label 3 to a only few edges. We prove that, for every $p \geq 0$, there are various graphs needing exactly p 3's in their proper 3-labellings. Actually, deciding whether a given graph can be labelled with p 3's is NP-complete for every $p \geq 0$. We also focus on particular classes of 3-chromatic graphs (cacti, triangle-free planar graphs, etc.), for which we prove there is no $p \geq 1$ such that they all admit proper 3-labellings assigning label 3 to at most p edges. In such cases, we give lower and upper bounds on the number of needed 3's.

Keywords: Proper labellings · 3-chromatic graphs · 1-2-3 Conjecture

1 Introduction

This work is mainly motivated by the so-called **1-2-3 Conjecture**, which can be defined through the following terminology and notation. Let G be a graph and consider a k-labelling $\ell : E(G) \to \{1, \ldots, k\}$, i.e., an assignment of labels $1, \ldots, k$ to the edges of G. To every vertex $v \in V(G)$, we associate, as its *colour* $c_\ell(v)$, the sum of labels assigned by ℓ to its incident edges. That is, $c_\ell(v) = \sum_{u \in N(v)} \ell(vu)$. We say that ℓ is *proper* if we have $c_\ell(u) \neq c_\ell(v)$ for every $uv \in E(G)$, that is, if no two adjacent vertices of G get incident to the same sum of labels by ℓ.

The complete graph on two vertices, K_2, is the only connected graph admitting no proper labellings. Thus, when studying the 1-2-3 Conjecture, we focus on *nice graphs*, which are those graphs with no connected component isomorphic to K_2, i.e., admitting proper labellings. If a graph G is nice, then we can investigate

This work was supported by the ANR project DISTANCIA (ANR-14-CE25-0006). For a full version of the paper go to: https://hal.archives-ouvertes.fr/hal-02975031.

T. Calamoneri and F. Corò (Eds.): CIAC 2021, LNCS 12701, pp. 103–115, 2021.
https://doi.org/10.1007/978-3-030-75242-2_7

the smallest $k \geq 1$ such that proper k-labellings of G exist. This parameter is denoted by $\chi_\Sigma(G)$. A natural question to ask, is whether this parameter $\chi_\Sigma(G)$ can be large for a given graph G. This question is precisely at the heart of the 1-2-3 Conjecture [11], which states that if G is a nice graph, then $\chi_\Sigma(G) \leq 3$.

To date, most of the progress towards the 1-2-3 Conjecture can be found in [13]. Let us highlight that the conjecture was verified mainly for 3-colourable graphs [11] and complete graphs [4]. Regarding the tightness of the conjecture, it was proved that deciding if a given graph G verifies $\chi_\Sigma(G) \leq 2$ (denoted as the 2-LABELLING problem) is NP-complete in general [8], and remains so even in the case of cubic graphs [6]. Hence, there is no nice characterisation of graphs admitting proper 2-labellings (or, the other way round, of graphs needing 3's in their proper 3-labellings), unless P=NP. Lastly, to date, the best result towards the 1-2-3 Conjecture, from [10], is that $\chi_\Sigma(G) \leq 5$ holds for every nice graph G.

This work takes place in a recent line of research studying optimisation problems related to the 1-2-3 Conjecture which arise when considering proper labellings fulfilling additional constraints. In a way, one of the main sources of motivation here is further understanding the very mechanisms that lie behind proper labellings. In particular, towards better comprehending the connection between proper labellings and proper vertex-colourings, the authors of [1,3] studied proper labellings ℓ for which the resulting vertex-colouring c_ℓ is required to be close to an optimal proper vertex-colouring (i.e., with the number of distinct resulting vertex colours being close to the chromatic number). Due to one of the core motivations behind the 1-2-3 Conjecture, the authors of [2] also investigated proper labellings minimising the sum of labels assigned to the edges.

Each of these previous studies led to presumptions of independent interest. In particular, it is believed in [3], that every nice graph G admits a proper labelling where the maximum vertex colour is at most $2\Delta(G)$ (recall that $\Delta(G)$ and $\delta(G)$ are used to denote the maximum and the minimum, resp., degree of any vertex of G), while, from [2], it is believed that every G should admit a proper labelling where the sum of assigned labels is at most $2|E(G)|$. One of the main reasons why these presumptions are supposed to hold is that, in general, it seems that nice graphs admit 2-labellings that are almost proper, in the sense that they need only a few 3's to design proper 3-labellings. This belief on the number of 3's is long-standing, as, in a way, it lies behind the 1-2 Conjecture of Przybyło and Woźniak [12], which states that we should be able to build a proper 2-labelling of every graph if we can also locally alter each vertex colour a bit.

Our goal in this work is to study and formally establish the intuition that, in general, graphs should admit proper 3-labellings assigning only a few 3's. First, we study whether, given a (possibly infinite) class $\mathcal{F}$ of graphs, the members of $\mathcal{F}$ admit proper 3-labellings assigning only a constant number of 3's. Note that this holds, for instance, for all nice trees since they admit proper 2-labellings [4]. In case $\mathcal{F}$ admits no such constant $c_\mathcal{F}$, i.e., the number of 3's the members of $\mathcal{F}$ need in their proper 3-labellings is a function of their number of edges, the second question we consider is whether the number of 3's needed can be "large" for a given member of $\mathcal{F}$, with respect to the number of its edges.

In this work, we investigate these two questions in general and for restricted classes of graphs. We begin in Sect. 2 by formally introducing the terminology that we employ throughout this work. In Sect. 3, we introduce proof techniques for establishing lower and upper bounds on the number of 3's needed in proper 3-labellings for some graph classes. In Sect. 4, we use these tools to establish that, for several classes of graphs, the number of required 3's in their proper 3-labellings is not bounded by an absolute constant. In such cases, we exhibit bounds (functions depending on the size of said graphs) on this number.

2 Terminology and a Conjecture

For any notation on graph theory not defined in the paper, we refer the reader to [7]. Let G be a (nice) graph, and ℓ be a k-labelling of G. For any $i \in \{1, \ldots, k\}$, we denote by $\mathrm{nb}_\ell(i)$ the number of edges assigned label i by ℓ. Focusing now on proper 3-labellings, we denote by $\mathrm{mT}(G)$ the minimum number of edges assigned label 3 by a proper 3-labelling of G. That is, $\mathrm{mT}(G) = \min\{\mathrm{nb}_\ell(3) : \ell$ is a proper 3-labelling of $G\}$. We extend this parameter mT to classes $\mathcal{F}$ of graphs by defining $\mathrm{mT}(\mathcal{F})$ as the maximum value of $\mathrm{mT}(G)$ over the members G of $\mathcal{F}$. Clearly, $\mathrm{mT}(\mathcal{F}) = 0$ for every class $\mathcal{F}$ of graphs admitting proper 2-labellings (i.e., $\chi_\Sigma(G) \leq 2$ for every $G \in \mathcal{F}$). Given a graph class $\mathcal{F}$, we are interested in determining whether $\mathrm{mT}(\mathcal{F}) \leq p$ for some $p \geq 0$. From that perspective, for every $p \geq 0$, we denote by $\mathcal{G}_p$ the class of graphs G with $\mathrm{mT}(G) = p$. For convenience, we also define $\mathcal{G}_{\leq p} := \mathcal{G}_0 \cup \cdots \cup \mathcal{G}_p$.

Since nice trees admit proper 2-labellings [4], if $\mathcal{T}$ is the class of all nice trees, then the notation above allows us to state that $\mathcal{T} \subset \mathcal{G}_0$. More generally speaking, bipartite graphs form perhaps the most investigated class of graphs in the context of the 1-2-3 Conjecture. A notable result, due to Thomassen, Wu, and Zhan [14], is that bipartite graphs verify the 1-2-3 Conjecture. These graphs were further studied in several works, such as [2], in which it was proved that:

Theorem 1 ([2]). *If G is a nice bipartite graph, then $G \in \mathcal{G}_{\leq 2}$.*

Theorem 1 is worrisome since, even without additional constraints, we do not know much about how proper 3-labellings behave beyond the scope of bipartite graphs. Our take in this work is to focus on the next natural case to consider, that of 3-chromatic graphs, which fulfil the 1-2-3 Conjecture [11]. Unfortunately, as will be seen later on, a result equivalent to Theorem 1 for 3-chromatic graphs does not exist, even for very restricted classes of 3-chromatic graphs.

As mentioned earlier, we will see throughout this work that, for several graph classes $\mathcal{F}$, there is no $p \geq 0$ such that $\mathcal{F} \subset \mathcal{G}_{\leq p}$. For such a class, we want to know whether the proper 3-labellings of their members require assigning label 3 many times, with respect to their number of edges. We study this aspect through the following terminology. For a nice graph G, we define $\rho_3(G) := \mathrm{mT}(G)/|E(G)|$. We extend this ratio to a class $\mathcal{F}$ by setting $\rho_3(\mathcal{F}) = \max\{\rho_3(G) : G \in \mathcal{F}\}$.

In this work, we are interested in determining bounds on $\rho_3(\mathcal{F})$ for graph classes $\mathcal{F}$ of 3-chromatic graphs, and, generally speaking, in how large this ratio

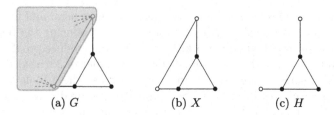

(a) G (b) X (c) H

Fig. 1. A graph G containing a graph H as a weakly induced subgraph X. In G, the white vertices can have arbitrarily many neighbours in the red part, while the full neighbourhood of the black vertices is as displayed. In H, the white vertices are the border vertices, while the black vertices are the core vertices. (Color figure online)

can be. Among the sample of small connected graphs (e.g., of order at most 6), the maximum ratio ρ_3 is exactly $1/3$, and is attained by C_3 and C_6. These are the worst graphs we know of, which leads us to raising the following conjecture.

Conjecture 1. If G is a nice connected graph, then $\rho_3(G) \leq \frac{1}{3}$.

3   Tools for Establishing Bounds on mT and ρ_3

3.1   Weakly Induced Subgraphs – A Tool for Lower Bounds

Our lower bounds on mT and ρ_3 exhibited in Sect. 4 are through a graph construction requiring the following terminology. For two graphs G and H, we say that G *contains H as a weakly induced subgraph X* (see Fig. 1) if there exists an induced subgraph X of G such that H is a spanning subgraph of X, and, for every vertex $v \in V(H)$, either $d_H(v) = 1$ or $d_H(v) = d_G(v)$ (note that $d_G(v) = |\{v \in V(G) : uv \in E(G)\}|$ denotes the degree of v in a (sub)graph G). In other words, if we add to H the edges of G that connect the vertices of degree 1 in H, we get X. That is, for every edge $uv \in E(G)$, if $u \in V(X)$ and $v \in V(G) \backslash V(X)$, then $d_H(u) = 1$; we call these the *border vertices* of H. Also, we call the other vertices of H (those that are not border vertices) its *core vertices*. By the definitions, if G contains H as a weakly induced subgraph and $\delta(H) \geq 2$, then G is isomorphic to H, and thus, this notion makes more sense when $\delta(H) = 1$.

Two weakly induced subgraphs of a graph G, X_1 and X_2, are *disjoint* (in G) if they share no core vertices. It follows from the definition that, for every $v \in V(G)$, if $v \in V(X_1) \cap V(X_2)$, then v is a border vertex of both X_1 and X_2.

Let ℓ be a labelling of G. For a subgraph H of G, we denote by $\ell|_H$ the *restriction* of ℓ to the edges of H, i.e., we have $\ell|_H(e) = \ell(e)$ for every edge $e \in E(H)$. Assume now that G contains H as a weakly induced subgraph X. Abusing the notations, we will sometimes write $\ell|_H$, which refers to the labelling of H inferred from $\ell|_X$, i.e., where $\ell|_H(e) = \ell|_X(e)$ for every $e \in E(H)$.

The key result is that, if a graph G contains other graphs $H_1, \ldots, H_n$ as pairwise disjoint weakly induced subgraphs, then $\mathrm{mT}(G) \geq \sum_{i=1}^{n} \mathrm{mT}(H_i)$.

Lemma 1. *Let G be a graph containing nice graphs $H_1, \ldots, H_n$ as pairwise disjoint weakly induced subgraphs $X_1, \ldots, X_n$. If ℓ is a proper 3-labelling of G, then $\ell|_{H_i}$ is a proper 3-labelling of H_i for every $i \in \{1, \ldots, n\}$. Consequently, $\mathrm{mT}(G) \geq \sum_{i=1}^{n} \mathrm{mT}(H_i)$.*

Proof. Consider H_j for some $1 \leq j \leq n$. Since, by any labelling of a nice graph, a vertex of degree 1 cannot get the same colour as its unique neighbour, then it cannot be involved in a conflict. This implies that $\ell|_{H_j}$ is proper if and only if any two adjacent core vertices of H_j get distinct colours by $\ell|_{H_j}$. By the definition of a weakly induced subgraph, recall that we have $d_{H_j}(v) = d_{X_j}(v) = d_G(v)$ for every core vertex v of H_j, which implies that $c_{\ell|_{H_j}}(v) = c_{\ell|_{X_j}}(v) = c_\ell(v)$. Thus, for every edge $uv \in E(H_j)$ joining core vertices, we have $c_\ell(u) = c_{\ell|_{H_j}}(u) = c_{\ell|_{X_j}}(u) \neq c_{\ell|_{X_j}}(v) = c_{\ell|_{H_j}}(v) = c_\ell(v)$ since ℓ is proper, meaning that $\ell|_{H_j}$ is also proper. Now, since G contains nice graphs $H_1, \ldots, H_n$ as pairwise disjoint weakly induced subgraphs $X_1, \ldots, X_n$, then $\mathrm{mT}(G) \geq \sum_{i=1}^{n} \mathrm{mT}(H_i)$. $\square$

The next lemma points out that, in some contexts, we can add some structure to a given graph without altering its value of mT. This will be useful for applying inductive arguments or simplifying the structure of a considered graph later on.

Lemma 2. *Let G be a nice graph with minimum degree 1 and $v \in V(G)$ be such that $d(v) = 1$. If G' is the graph obtained from G by adding $x > 0$ vertices of degree 1 adjacent to v, then $\mathrm{mT}(G') = \mathrm{mT}(G)$.*

Next, we prove each graph class $\mathcal{G}_p$ ($p \geq 1$) contains infinitely many graphs.

Theorem 2. *Given a graph G and any (fixed) integer $p > 1$, deciding if $G \in \mathcal{G}_{\leq p}$ is NP-complete.*

Sketch of Proof. We do a reduction from the 2-LABELLING problem, which is NP-hard even when the graph has minimum degree 1 [8]. Given an instance H of 2-LABELLING such that $\delta(H) = 1$, we construct a graph G such that $\mathrm{mT}(G) = p$ if and only if H admits a proper 2-labelling. The graph G is constructed by identifying (all to one vertex w) a vertex of degree 1 from each of p copies of a nice graph H' and a vertex of degree 1 of H, where $\delta(H') = 1$ and $\mathrm{mT}(H') = 1$ (H' exists, see Sect. 4), and adding many leaves adjacent to w. The result follows from Lemma 1 since G contains p copies of H' and one copy of H as pairwise disjoint weakly induced subgraphs (and since it is easy to deduce a proper 3-labelling of G using $p + mT(H)$ 3's). $\diamond$

3.2 Switching Closed Walks – A Tool for Upper Bounds

Due to Theorem 1, investigating the parameters mT and ρ_3 only makes sense for graphs with chromatic number at least 3, i.e., that are not bipartite.

Theorem 3. *If G is a connected 3-chromatic graph, then $\mathrm{mT}(G) \leq |V(G)|$, and thus $\rho_3(G) \leq \frac{|V(G)|}{|E(G)|}$.*

Proof. Since G is not bipartite, there exists an odd-length cycle C in G. Let H be a subgraph of G constructed as follows. Start from $C = H$. Then, until $V(H) = V(G)$, repeatedly choose a vertex $v \in V(G) \backslash V(H)$ such that there exists a vertex $u \in V(H)$ with $uv \in E(G)$, and add the edge uv to H. In the end, H is a connected spanning subgraph of G containing only one cycle, C, which is of odd length. Then, we have $|E(H)| = |V(G)|$.

Let $\phi : V(G) \rightarrow \{0, 1, 2\}$ be a proper 3-vertex-colouring of G. In what follows, we construct a 3-labelling ℓ of G such that $c_\ell(v) \equiv \phi(v) \bmod 3$ for every vertex $v \in V(G)$, thus making ℓ proper. To prove the full statement, we also want ℓ to satisfy $\mathrm{nb}_\ell(3) \leq |V(G)|/|E(G)|$. Aiming at vertex colours modulo 3, we can instead assume that ℓ assigns labels $0, 1, 2$, and require $\mathrm{nb}_\ell(0) \leq |V(G)|/|E(G)|$. To obtain such a labelling, we start from ℓ assigning label 2 to all edges of G. We then modify ℓ iteratively until all vertex colours are as desired modulo 3.

As long as G has a vertex v with $c_\ell(v) \not\equiv \phi(v) \bmod 3$, we apply the following procedure. Choose $W = (v, v_1, \ldots, v_n, v)$, a closed walk[1] of odd length in G starting and ending at v, and going through edges of H only. This walk exists. Indeed, consider, in H, a (possibly empty) path P from v to the closest vertex u of C (if v lies on C, then note that $u = v$ and P has no edge). Then, the closed walk $vPuCuPv$ is a possible W. We then follow the consecutive edges of W, starting from v and ending at v, and, going along, we apply $+2, -2, +2, -2, \ldots, +2$ (modulo 3) to the labels assigned by ℓ to the traversed edges. As a result, $c_\ell(x)$ is not altered modulo 3 for every vertex $x \neq v$, while $c_\ell(v)$ is incremented by 1 modulo 3. If $c_\ell(v) \equiv \phi(v) \bmod 3$, then we are done with v. Otherwise, we repeat this switching procedure once again, so that v fulfils that property.

Eventually, $c_\ell(v) \equiv \phi(v) \bmod 3$ for every $v \in V(G)$, meaning that ℓ is proper. Recall that we have $\ell(e) = 2$ for every $e \in E(G) \backslash E(H)$. Thus, only the edges of H can be assigned label 0 by ℓ. Since there are exactly $|V(G)|$ such edges, and we can replace all assigned 0's with 3's without breaking the modulo 3 property, we have $\mathrm{mT}(G) \leq |V(G)|$, which implies that $\rho_3(G) \leq |V(G)|/|E(G)|$.   □

In the next lemma, we show a way to play with ϕ in order to reduce the number of 3's assigned by ℓ to certain sets of edges.

Lemma 3. *Let G be a graph and ℓ be a proper $\{0, 1, 2\}$-labelling of G such that $c_\ell(u) \not\equiv c_\ell(v) \bmod 3$ for every edge $uv \in E(G)$. If H is a (not necessarily connected) spanning d-regular subgraph of G for some $d \geq 1$, then there exists a proper $\{0, 1, 2\}$-labelling ℓ' of G such that $c_{\ell'}(u) \not\equiv c_{\ell'}(v) \bmod 3$ for every edge $uv \in E(G)$ and that assigns label 0 to at most a third of the edges of $E(H)$. Moreover, for every edge $e \in E(G) \backslash E(H)$, $\ell'(e) = \ell(e)$.*

Proof. We construct the following labelling: starting from ℓ, add 1 (modulo 3) to all the labels assigned by ℓ to the edges of H. The resulting labelling ℓ_1 is a proper $\{0, 1, 2\}$-labelling of G such that $c_{\ell_1}(u) \not\equiv c_{\ell_1}(v) \bmod 3$ for every edge $uv \in E(G)$. Indeed, for every $v \in V(G)$, we have $c_{\ell_1}(v) \equiv c_\ell(v) + d \bmod 3$. Thus, if there exist two vertices $u, v \in V(G)$ such that $c_{\ell_1}(u) \equiv c_{\ell_1}(v) \bmod 3$, then

---

[1] Recall that a *walk* in a graph is a path in which vertices and edges can be repeated.

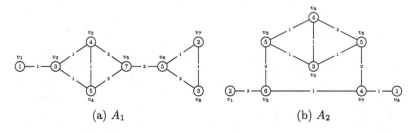

Fig. 2. Proper 3-labellings ℓ of A_1 and A_2 with $\mathrm{nb}_\ell(3) = 1$. The colours by c_ℓ are indicated by integers within the vertices.

$c_\ell(u) \equiv c_\ell(v) \bmod 3$, a contradiction. We define ℓ_2 in a similar way, by adding 1 (modulo 3) to all the labels assigned by ℓ_1 to the edges of H. Similarly, ℓ_2 is proper. Since, for every edge $e \in E(H)$, we have $\{\ell(e), \ell_1(e), \ell_2(e)\} = \{0, 1, 2\}$, then at least one of ℓ, ℓ_1, ℓ_2 assigns label 0 to at most a third of the edges of $E(H)$. Since none of the labels of the edges of $E(G)\backslash E(H)$ were changed to obtain ℓ_1 from ℓ and to get ℓ_2 from ℓ_1, the last statement of the lemma holds. □

In Lemma 3, if $d = 2$, then H forms a cycle cover of G. Thus, when H is also a unicyclic spanning connected subgraph of G, an application of Lemma 3 in conjunction with the proof of Theorem 3 gives the following:

Corollary 1. *If G is Hamiltonian, of odd order, and $\chi(G) = 3$, then $\rho_3(G) \leq \frac{1}{3}$.*

4   Results for mT and ρ_3 for Some Graph Classes

We now use the tools from Sect. 3 to exhibit results on the parameters mT and ρ_3 for some classes of 3-chromatic graphs. In particular, we prove that, for many classes $\mathcal{F}$ of 3-chromatic graphs, there is no $p \geq 1$ such that $\mathcal{F} \subset \mathcal{G}_{\leq p}$. In most cases, we provide upper bounds for $\rho_3(\mathcal{F})$.

4.1   Connected Graphs Needing Lots of 3's

As mentioned before, we are aware of only two connected graphs for which ρ_3 is exactly $1/3$, and these are C_3 and $C_6{}^2$. One question to ask, is if the bound in Conjecture 1 is accurate in general, i.e., whether it can be attained by arbitrarily large graphs. In light of these thoughts, our goal in this section is to provide a class of arbitrarily large connected graphs achieving the largest possible ratio ρ_3.

We ran computer programs to find graphs H with $\delta(H) = 1$, $\mathrm{mT}(H) \geq 1$, and with the fewest edges possible. It turns out that the smallest such graphs have 10 edges. Two such graphs, which we call A_1 and A_2, are depicted in Fig. 2. The following observation, proven by simple case analysis, allows us to use these two graphs to build arbitrarily large connected graphs with large ρ_3.

---

[2] Conjecture 1 focuses on connected graphs since any disjoint union of C_3's and C_6's reaches that value.

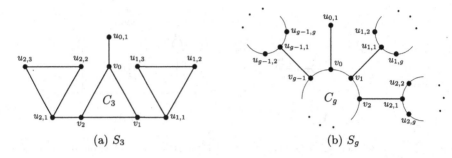

Fig. 3. The planar graphs S_3 (left) and S_g (right) of girths 3 and g, respectively.

Observation 4. $mT(A_1) = mT(A_2) = 1$.

Theorem 5. *There are arbitrarily large connected graphs G with $\rho_3(G) \geq \frac{1}{10}$.*

Proof. Let $p \geq 1$ be fixed. We construct a connected graph G with $10p$ edges, such that $nb_\ell(3) \geq p$ for any proper 3-labelling ℓ of G, which implies that $\rho_3(G) \geq 1/10$. Start, as G, with p disjoint copies of A_1 (or A_2), and identify a vertex of degree 1 from each of these p copies to a single vertex. The labelling property follows from Lemma 1 and Observation 4, since G contains p copies of A_1 or A_2 as pairwise disjoint weakly induced subgraphs.   □

4.2 Bounds for Connected Planar Graphs with Large Girth

The *girth* $g(G)$ of a graph G is the length of its shortest cycle. For any $g \geq 3$, we denote by $\mathcal{P}_g$ the class of planar graphs with girth at least g. Note that $\mathcal{P}_3$ is the class of all planar graphs, and that $\mathcal{P}_4$ is the class of all triangle-free planar graphs. Recall that the girth of a tree is set to ∞, since it has no cycle.

To date, it is still unknown whether planar graphs verify the 1-2-3 Conjecture, which makes the study of the parameters mT and ρ_3 adventurous for this class of graphs. However, there is no $p \geq 1$ such that planar graphs lie in $\mathcal{G}_{\leq p}$ by the construction in the proof of Theorem 5, since the graphs A_1 and A_2 are planar. Thus, there exist arbitrarily large connected planar graphs G with $\rho_3(G) \geq 1/10$.

To go further, we consider planar graphs with large girth. By Grötzsch's Theorem [9], triangle-free planar graphs are 3-colourable, which means they verify the 1-2-3 Conjecture (see [11]). In what follows, first, we prove that, for every $g \geq 3$, there is no $p \geq 1$ such that $\mathcal{P}_g \subset \mathcal{G}_{\leq p}$. Second, we prove that, as the girth $g(G)$ of a planar graph G grows, the ratio $\rho_3(G)$ decreases. As a side result, we prove Conjecture 1 for planar graphs with girth at least 36. To prove the first result above, we cannot use the graphs A_1 and A_2 introduced previously, as they contain triangles. Instead, we use the graph S_g illustrated in Fig. 3.

Lemma 4. *For every $g \geq 3$ with $g \equiv 3 \mod 4$, we have $mT(S_g) = 1$.*

Sketch of Proof. Let ℓ be a proper 2-labelling of S_g. Due to the length of each outer cycle, it follows that $c_\ell(u_{i,1}) = \ell(u_{i,1}v_i) + 3$. Moreover, due to C_g being of odd length, we can deduce that there must exist a vertex $v_x \in V(C_g)$ such that $v_x \neq v_0$ and $c_\ell(v_x) = \ell(v_x u_{x,1}) + 3 = c_\ell(u_{x,1})$, a contradiction. The rest of the result follows from identifying a proper 3-labelling ℓ' of S_g with $\text{nb}_{\ell'}(3) = 1$. $\diamond$

By taking arbitrarily many copies of the S_g graph and identifying their respective roots (the vertex $u_{0,1}$ in Fig. 3), we can prove that:

Theorem 6. *For every $g' \geq 3$, there exist arbitrarily large connected planar graphs G with $g(G) \geq g'$ and $\rho_3(G) \geq \frac{1}{g^2+g}$, where g is the smallest natural number such that $g \geq g'$ and $g \equiv 3 \bmod 4$.*

We now proceed to prove that $\rho_3(G) \leq \frac{2}{k-1}$ for any planar graph G of girth $g \geq 5k + 1$, when $k \geq 7$. The next theorem from [5] is one of the tools we use to prove this result. Note that, for any $k \geq 1$, a *k-thread* in a graph G is a path $(u_1, \ldots, u_{k+2})$, where the k inner vertices $u_2, \ldots, u_{k+1}$ all have degree 2 in G.

Theorem 7 ([5]). *For any integer $k \geq 1$, every planar graph with minimum degree at least 2 and girth at least $5k + 1$ contains a k-thread.*

We can now proceed with the main theorem.

Theorem 8. *Let $k \geq 7$. If G is a nice planar graph with $g(G) \geq 5k + 1$, then $\rho_3(G) \leq \frac{2}{k-1}$.*

Proof. Throughout this proof, we set $g = g(G)$. The proof is by induction on the order of G. The base case is when $|V(G)| = 3$. In that case, G must be a path of length 2 (due to the girth assumption), and the claim is clearly true. So let us focus on proving the general case.

If G is a tree, then $\chi_\Sigma(G) \leq 2$ and we have $\rho_3(G) = 0$. So, from now on, we may assume that G is not a tree. We first deal with the case of planar graphs G of girth $g \geq 5k+1$ for which there exists at least one cut vertex $v \in V(G)$ (meaning that $G-\{v\}$ has more connected components than G) such that $G-\{v\}$ contains a connected component T' that is a tree such that $|E(T')| \geq 1$ and the induced subgraph of G formed by the vertices of T' and v is also a tree. Let $u \in V(T')$ be the neighbour of v. By the inductive hypothesis, $\rho_3(G - V(T')) \leq \frac{2}{k-1}$ since removing a pendant tree from G can neither decrease its girth nor result in a tree, and thus, by recursively applying Lemma 2, there is a proper 3-labelling of G such that $\rho_3(G) \leq \frac{2}{k-1}$. The same arguments can be applied for all such connected components of $G-\{v\}$. Hence, we can assume that G does not contain any such cut vertex v. Another way to state this, is that if G contains a vertex v to which a pending tree T' is attached, then T' is a star with center v.

Let G^- be the graph obtained from G by removing all vertices of degree 1. Note that removing vertices of degree 1 from G can neither decrease its girth nor result in a tree. Since G has girth $g \geq 5k + 1$ and does not contain any cut vertex $v \in V(G)$ as described above, the graph G^- has minimum degree

2. By Theorem 7, G^- contains a k-thread P. Let $u_1, \ldots, u_{k+2}$ be the vertices of P, where $d_H(u_i) = 2$ for all $2 \leq i \leq k+1$. Thus, the vertices of P exist in G except that each of the vertices u_i (for $2 \leq i \leq k+1$) may be adjacent to some vertices of degree 1 in addition to their adjacencies in G^-. Let G' be the graph obtained from G by removing the vertices $u_3, \ldots, u_k$ and all of their neighbours that have degree 1 in G. Note that G' might contain up to two connected components. In case G' has exactly two connected components, then, due to a previous assumption, none of these can be a tree, which implies that G' is nice. If G' is connected, then, because it has at least two edges ($u_1 u_2$ and $u_{k+1} u_{k+2}$), it must be nice. Furthermore, in both cases, the girth of G' is at least that of G. Then, by combining the inductive hypothesis and the fact that every nice tree T verifies $\rho_3(T) = 0$, we deduce that $\rho_3(G') \leq \frac{2}{k-1}$.

To obtain a proper 3-labelling ℓ of G such that $\rho_3(G) \leq \frac{2}{k-1}$, we extend a proper 3-labelling ℓ' of G' corresponding to $\rho_3(G') \leq \frac{2}{k-1}$, as follows. First, label all of the edges incident to the vertices of degree 1 and the vertices $u_3, \ldots, u_k$ with 1's. Note that none of these vertices of degree 1 can, later on, be in conflict with their neighbour since they have degree 1. Now, for each $2 \leq j \leq k-2$, in increasing order of j, label the edge $u_j u_{j+1}$ with 1 or 2, so that the resulting colour of u_j does not conflict with the colour of u_{j-1}. Finally, label the edges $u_{k-1} u_k$ and $u_k u_{k+1}$ with 1, 2 or 3, so that the resulting colour of u_{k-1} does not conflict with that of u_{k-2}, the resulting colour of u_k does not conflict with that of u_{k-1} nor with that of u_{k+1}, and the resulting colour of u_{k+1} does not conflict with that of u_{k+2}. Indeed, this is possible since there exist at least two distinct labels $\{\alpha, \beta\}$ ($\{\alpha', \beta'\}$, respectively) in $\{1, 2, 3\}$ for $u_{k-1} u_k$ ($u_k u_{k+1}$, respectively) such that the colour of u_{k-1} (u_{k+1}, respectively) is not in conflict with that of u_{k-2} (u_{k+2}, respectively). Thus, w.l.o.g., choose α and α' for the labels of $u_{k-1} u_k$ and $u_k u_{k+1}$, respectively. If the colour of u_k does not conflict with that of u_{k-1} nor with that of u_{k+1}, then we are done. If the colour of u_k conflicts with both that of u_{k-1} and that of u_{k+1}, then it suffices to change both the labels of $u_{k-1} u_k$ and $u_k u_{k+1}$ to β and β', respectively. Lastly, w.l.o.g., if the colour of u_k only conflicts with that of u_{k-1}, then it suffices to change the label of $u_k u_{k+1}$ to β'. The resulting labelling ℓ of G is thus proper. Moreover, $|E(G) \backslash E(G')| \geq k-1$ and ℓ uses label 3 at most twice more than ℓ', and so, the result follows.   □

4.3   Bounds for Connected Cacti

A *cactus* is a graph in which any two cycles have at most one vertex in common. The graphs S_g introduced in Sect. 4.2, and those constructed in order to prove Theorem 6, are all cacti. Since the smallest graph S_g is S_3, which has 12 edges, that theorem directly implies that there exist arbitrarily large connected cacti G with $\rho_3(G) \geq 1/12$. We now prove Conjecture 1 for cacti.

Theorem 9. *If G is a nice cactus, then $\rho_3(G) \leq \frac{1}{3}$.*

Sketch of Proof. The proof is by induction on $|V(G)|$. The general case is proven by focusing on *end-cycles* $C_1, \ldots, C_q$, to which pending trees might be attached, and sharing a root vertex r that separates these cycles from the rest of the graph. By analysing the C_i's, it can be proved that the induction hypothesis can be invoked to get a desired labelling of G, as soon as one of their inner vertices has a pending tree attached, or the pending tree attached to r is not a star with center r. So the C_i's can be assumed to be mostly cycles, in which case we can remove their inner vertices, invoke the induction hypothesis, and extend a labelling of the remaining graph to a desired one of G, by labelling the edges of the C_i's so that only a few 3's are assigned.   ◇

4.4   An Upper Bound for Halin Graphs

A *Halin graph* is a planar graph with minimum degree 3 obtained as follows. Start from a tree T with no vertex of degree 2, and consider a planar embedding of T. Then, add edges to form a cycle going through all the leaves of T in the clockwise ordering in this embedding. A Halin graph is called a *wheel* if it is constructed from a tree T with diameter at most 2. Halin graphs have triangles and Hamiltonian cycles going through any given edge [15]. Also, Halin graphs are 3-degenerate, so, due to the presence of triangles, each of them has chromatic number 3 or 4. The dichotomy is well understood, as a Halin graph has chromatic number 4 if and only if it is a wheel of even order [16]. It is easy to see that these wheels admit proper 2-labellings, and so, we focus on 3-chromatic Halin graphs in the proof of the next theorem. In particular, we use our tools from Sect. 3 to establish an upper bound on ρ_3 for the 3-chromatic Halin graphs.

Theorem 10. *If G is a Halin graph, then $\rho_3(G) \leq \frac{1}{3}$.*

Proof. As said above, we can assume that G is not a wheel of even order. Then $\chi(G) = 3$. If $|V(G)|$ is odd, then the result follows from Corollary 1. Thus, we can assume that $|V(G)|$ is even.

By considering any non-leaf vertex r of T in G, and defining a usual root-to-leaf (virtual) orientation, since no vertex has degree 2 in T, then G has a triangle (u, v, w, u), where v, w are leaves in T with parent u. Furthermore, $d_G(v) = d_G(w) = 3$, while $d_G(u) \geq 3$. Due to these degree properties, if we consider C a Hamiltonian cycle traversing uv, then C must also include either wu or vw. Precisely, if we orient the edges of C, resulting in a spanning oriented cycle $\mathbf{C}$, then, at some point, $\mathbf{C}$ enters (u, v, w, u) through one of its vertices, goes through another vertex of the triangle and then through the third of its vertices, before leaving the triangle. In other words, $\mathbf{C}$ traverses all vertices of (u, v, w, u) at once.

Up to relabelling the vertices of (u, v, w, u), we can assume that $\mathbf{C}$ enters the triangle through u, then goes to v, before going to w and leaving the triangle. Let us consider H, the subgraph of G containing the three edges of (u, v, w, u), and all successive edges traversed by C after leaving the triangle except for the edge going back to u. Note that H is a unicyclic spanning connected subgraph of G, in which the only cycle is the triangle (u, v, w, u) to which is attached a hanging path $(w, x_1, \ldots, x_{n-3})$ containing all other vertices of G (i.e., $n = |V(G)|$). Furthermore, in $E(G) \backslash E(H)$, if we set $x = x_{n-3}$, then the edge xu exists. Since H is spanning, connected, and unicyclic, $|E(H)| = |V(G)|$, which is at most $2|E(G)|/3$, since $\delta(G) \geq 3$.

All conditions are now met to invoke the arguments in the proof of Theorem 3, from which we can deduce a proper $\{0, 1, 2\}$-labelling of G where adjacent vertices get distinct colours modulo 3 and in which only the edges of the chosen H are possibly assigned label 0. Let us consider the subgraph H' of G obtained from H by adding the edge xu, which is present in G. Recall that $\ell(xu) = 2$ by default. Note that H' contains at least two disjoint perfect matchings M_1, M_2. Indeed, since $|V(G)|$ is even, then, in H, the hanging path attached at w has odd length. A first perfect matching M_1 of H' contains $x_{n-3}x_{n-4}, x_{n-5}x_{n-6}, \ldots, wx_1$ and uv. A second perfect matching M_2 of H' contains $x_{n-4}x_{n-5}, x_{n-6}x_{n-7}, \ldots, x_2x_1$, and wv and xu. By Lemma 3, we can assume that at most a third of the edges in $M_1 \cup M_2$ are assigned label 0 by ℓ. Since $|M_1| + |M_2| = |E(H')| - 1 = |E(H)|$, this gives $\mathrm{nb}_\ell(0) \leq \frac{E(H)}{3} + 1$, which is less than $|E(G)|/3$ since $|E(G)| \geq 3|V(G)|/2$. By turning 0's by ℓ into 3's, we get a proper 3-labelling of G with the same upper bound on the number of assigned 3's. $\square$

4.5   Bounds for Outerplanar Graphs

First off, we note that the construction described in the proof of Theorem 5, when performed with copies of A_1 only, provides graphs that are outerplanar[3], since A_1 is itself outerplanar. Recall as well that outerplanar graphs form a subclass of series-parallel graphs. Thus, there exist arbitrarily large connected outerplanar (series-parallel, resp.) graphs G (H, resp.) with $\rho_3(G) \geq 1/10$ ($\rho_3(H) \geq 1/10$, resp.). Note however that the outerplanar graphs constructed above have cut vertices. So the question remains, whether or not this lower bound still holds when considering 2-connected outerplanar graphs (recall that outerplanar graphs are 2-degenerate, and thus, each of them is 3-chromatic and either separable or 2-connected). As for an upper bound, we can provide the following:

Theorem 11. *If G is a 2-connected outerplanar graph such that $|E(G)| \geq |V(G)| + 3$, then $\rho_3(G) \leq \frac{1}{3}$.*

Sketch of Proof. If $|V(G)|$ is odd, the result follows from Corollary 1. If $|V(G)|$ is even, then there is an odd-length cycle of G that consists of consecutive vertices of the outer face of G, and thus, since G is Hamiltonian, there is a unicyclic spanning connected subgraph H containing an odd cycle. Theorem 3 can be applied using this H, and Lemma 3 can be applied to two disjoint perfect matchings containing all the edges of H but one, as in the proof of Theorem 10. $\diamond$

Theorem 11 covers all 2-connected outerplanar graphs with at least three chords but it can also be shown to hold when there are at most two chords.

---

[3] An *outerplanar graph* admits a planar embedding with all vertices on the outer face.

5   Further Work

A first direction for further research is to prove Conjecture 1 for more classes of graphs such as for other classes of 3-chromatic graphs like separable outerplanar graphs and, more generally, series-parallel graphs. Another one is to investigate whether the bound of 1/3 in that conjecture is close to being tight or not, in particular for large graphs. Also, we were not able to come up with examples of arbitrarily large Halin graphs needing many 3's in their proper 3-labellings.

References

1. Baudon, O., Bensmail, J., Hocquard, H., Senhaji, M., Sopena, E.: Edge weights and vertex colours: minimizing sum count. Discrete Appl. Math. **270**, 13–24 (2019)
2. Bensmail, J., Fioravantes, F., Nisse, N.: On proper labellings of graphs with minimum label sum. In: Gąsieniec, L., Klasing, R., Radzik, T. (eds.) IWOCA 2020. LNCS, vol. 12126, pp. 56–68. Springer, Cham (2020). https://doi.org/10.1007/978-3-030-48966-3_5
3. Bensmail, J., Li, B., Li, B., Nisse, N.: On minimizing the maximum color for the 1-2-3 conjecture. Discrete Appl. Math. **289**, 32–51 (2021)
4. Chang, G., Lu, C., Wu, J., Yu, Q.: Vertex-coloring edge-weightings of graphs. Taiwan. J. Math. **15**, 1807–1813 (2011)
5. Chang, G.J., Duh, G.H.: On the precise value of the strong chromatic index of a planar graph with a large girth. Discrete Appl. Math. **247**, 389–397 (2018)
6. Dehghan, A., Sadeghi, M.R., Ahadi, A.: Algorithmic complexity of proper labeling problems. Theor. Comput. Sci. **495**, 25–36 (2013)
7. Diestel, R.: Graph Theory: Graduate Texts in Mathematics, vol. 173, 4th edn. Springer, Heidelberg (2017). https://doi.org/10.1007/978-3-662-53622-3
8. Dudek, A., Wajc, D.: On the complexity of vertex-coloring edge-weightings. Discrete Math. Theor. Comput. Sci. **13**, 45–50 (2011)
9. Grötzsch, H.: Zur theorie der diskreten gebilde. VII. Ein Dreifarbensatz für dreikreisfreie Netze auf der Kugel. (German) Wiss Z, Martin-Luther-Univ Halle-Wittenb, Math-Natwiss Reihe **8**, 109–120 (1958)
10. Kalkowski, M., Karoński, M., Pfender, F.: Vertex-coloring edge-weightings: towards the 1-2-3-conjecture. J. Comb. Theory Ser. B **100**(3), 347–349 (2010)
11. Karoński, M., Łuczak, T., Thomason, A.: Edge weights and vertex colours. J. Comb. Theory Ser. B **91**(1), 151–157 (2004)
12. Przybylo, J., Woźniak, M.: On a 1, 2 conjecture. Discrete Math. Theor. Comput. Sci. **12**(1), 101–108 (2010)
13. Seamone, B.: The 1-2-3 conjecture and related problems: a survey (2012)
14. Thomassen, C., Wu, Y., Zhang, C.Q.: The 3-flow conjecture, factors modulo k, and the 1-2-3-conjecture. J. Comb. Theory Ser. B **121**, 308–325 (2016)
15. Wang, W., Bu, Y., Montassier, M., Raspaud, A.: On backbone coloring of graphs. J. Comb. Optim. **23**(1), 79–93 (2012). https://doi.org/10.1007/s10878-010-9342-6
16. Wang, W., Lih, K.: List coloring halin graphs. Ars Combinatoria Waterloo then Winnipeg **77**(10), 53–63 (2005)

Upper Tail Analysis of Bucket Sort and Random Tries

Ioana O. Bercea$^{(\boxtimes)}$ and Guy Even

Tel Aviv University, Tel Aviv, Israel
ioana@cs.umd.edu, guy@eng.tau.ac.il

Abstract. Bucket Sort is known to run in expected linear time when the
input keys are distributed independently and uniformly at random in the
interval $[0, 1)$. The analysis holds even when a quadratic time algorithm
is used to sort the keys in each bucket. We show how to obtain linear
time guarantees on the running time of Bucket Sort that hold with *very
high probability*. Specifically, we investigate the asymptotic behavior of
the exponent in the upper tail probability of the running time of Bucket
Sort. We consider large additive deviations from the expectation, of the
form cn for large enough (constant) c, where n is the number of keys
that are sorted.

Our analysis shows a profound difference between variants of Bucket
Sort that use a quadratic time algorithm within each bucket and variants
that use a $\Theta(b \log b)$ time algorithm for sorting b keys in a bucket. When
a quadratic time algorithm is used to sort the keys in a bucket, the probability that Bucket Sort takes cn more time than expected is exponential
in $\Theta(\sqrt{n} \log n)$. When a $\Theta(b \log b)$ algorithm is used to sort the keys in
a bucket, the exponent becomes $\Theta(n)$. We prove this latter theorem by
showing an upper bound on the tail of a random variable defined on
tries, a result which we believe is of independent interest. This result
also enables us to analyze the upper tail probability of a well-studied
trie parameter, the external path length, and show that the probability
that it deviates from its expected value by an additive factor of cn is
exponential in $\Theta(n)$.

Keywords: Bucket Sort · Upper tail analysis · Running time

1 Introduction

The Bucket Sort algorithm sorts n keys in the interval $[0, 1)$ as follows: (i) Distribute the keys among n buckets, where the jth bucket consists of all the keys
in the interval $[j/n, (j + 1)/n)$. (ii) Sort the keys in each bucket. (iii) Scan the

This research was supported by a grant from the United States-Israel Binational Sci-
ence Foundation (BSF), Jerusalem, Israel, and the United States National Science
Foundation (NSF).
A full version of this paper can be found at [1].

© Springer Nature Switzerland AG 2021
T. Calamoneri and F. Corò (Eds.): CIAC 2021, LNCS 12701, pp. 116–129, 2021.
https://doi.org/10.1007/978-3-030-75242-2_8

buckets and output the keys in each bucket in their sorted order. We consider two natural classes of Bucket Sort algorithms that differ in how the keys inside each bucket are sorted. The first class of BucketSort algorithms that we consider sorts the keys inside a bucket using a quadratic time algorithm (such as Insertion Sort). We refer to algorithms in this class as b^2-Bucket Sort. The second class of algorithms sorts the keys in a bucket using a $\Theta(b \log b)$ algorithm for sorting b keys (such as Merge Sort). We refer to this variant as $b \log b$-Bucket Sort.

When the n keys are distributed independently and uniformly at random, the expected running time of Bucket Sort is $\Theta(n)$, even when a quadratic time algorithm is used to sort the keys in each bucket [3,17,19]. A natural question is whether such linear time guarantees hold with high probability. For Quick Sort, analyses of this sort have a long and rich history [7,10,16].

In this paper, we focus on analyzing the running time of Bucket Sort with respect to large deviations, e.g., running times that exceed the expectation by $10n$ comparisons. In particular, we study the asymptotic behavior of the exponent in the upper tail of the running time.

Rate of the Upper Tail. We analyze the upper tail probability of a random variable using the notion of rate, defined as follows[1].

Definition 1. *Given a random variable Y with expected value μ, we define the rate of the upper tail of Y to be the function defined on $t > 0$ as follows:*

$$R_Y(t) \triangleq -\ln\left(\Pr\left[Y \geq \mu + t\right]\right).$$

Note that we consider an additive deviation from the expectation, i.e., we bound the probability that the random variable deviates from its expected value by an additive term of t, for sufficiently large values of t^2. In particular, we consider values of $t = cn$, where n is the size of the input and c is a constant greater than some threshold. Finally, we abbreviate and refer to $R_Y(t)$ as the *rate* of Y.

We study the rates of the running times of deterministic Bucket Sort algorithms in which the input is sampled from a uniform probability distribution. We also consider parameters of tries induced by infinite prefix-free binary strings chosen independently and uniformly at random.

1.1 Our Contributions

Our first two results derive the rates of the two classes of Bucket Sort algorithms and show that they are different. Specifically, we prove the following:

Theorem 1. *There exists a constant $C > 0$ such that, for all $c > C$, the rate $R_{b^2}(\cdot)$ of the b^2-Bucket Sort algorithm on n keys chosen independently and uniformly at random in $[0, 1)$ satisfies $R_{b^2}(cn) = \Theta(\sqrt{n} \log n)$.*

---

[1] Throughout the paper, $\ln x$ denotes the natural logarithm of x and $\log x$ denotes the logarithm of base 2 of x.

[2] One should not confuse this analysis with concentration bounds that address small deviations from the expectation.

Since the expected running time of b^2-Bucket Sort is $\Theta(n)$, Theorem 1 states that the probability that b^2-Bucket Sort on random keys takes more than dn time is $e^{-\Theta(\sqrt{n}\log n)}$ (for a sufficiently large constant d)[3]. Theorem 1 proves both a lower bound and an upper bound on the asymptotic rate $R_{b^2}(cn)$. In particular, Theorem 1 rules out the possibility that the probability that the running time of b^2-Bucket Sort is greater than $100n$ is bounded by $e^{-\Theta(n)}$.

We prove the lower bound on $R_{b^2}(cn)$ by applying multiplicative Chernoff bounds in different regimes of large (superconstant, in fact) deviations from the mean. In such settings, the dependency of the exponent of the Chernoff bound on the deviation from the mean can have a significant impact on the quality of the bounds we obtain. Indeed, we employ a rarely used form of the Chernoff bound that exhibits a $\delta \log \delta$ dependency in the exponent when the deviation from the mean is δ (see Chapter 10.1.1 in [5]). Although the proof of this bound is straightforward, the proof of Theorem 1 crucially relies on this additional (superconstant) $\log \delta$ factor (see Claim 4).

For $b \log b$-Bucket Sort on random keys, we show that the rate is linear in the size of the input:

Theorem 2. *There exists a constant $C > 0$ such that, for all $c > C$, the rate $R_{b \log b}(\cdot)$ of the $b \log b$-Bucket Sort algorithm on n keys chosen independently and uniformly at random in $[0, 1)$ satisfies $R_{b \log b}(cn) = \Theta(n)$.*

We prove the lower bound on $R_{b \log b}(cn)$ by analyzing a random variable arising in random tries. Specifically, we consider tries on infinite binary strings in which each bit is chosen independently and uniformly at random. The parameter we study is called the *excess path length* and is defined formally in Sect. 2. We show that the time it takes to sort the buckets in $b \log b$-Bucket Sort can be upper bounded by the excess path length in a random trie (Lemma 1). We then bound the upper tail of the excess path length (Theorem 4) and use it to lower bound $R_{b \log b}(cn)$.

We also use the upper tail of the excess path length to derive the rate of a well-studied trie parameter, the sum of root to leaf paths in a minimal trie, called the *nonvoid external path length* [13,20]. It is known that the expected value of the nonvoid external path length in a random trie is $n \log n + \Theta(n)$ [13,20,22]. We show the following:

Theorem 3. *There exists a constant $C > 0$ such that, for all $c > C$, the rate $R_0(\cdot)$ of the nonvoid external path length of a minimal trie on n infinite binary strings chosen independently and uniformly at random satisfies $R_0(cn) = \Theta(n)$.*

Note that Theorem 3 implies that the probability that the nonvoid external path length is more than $n \log n + dn$ is $e^{-\Theta(n)}$ (for a sufficiently large constant d).

---

[3] The threshold C depends on: (1) the constant that appears in the sorting algorithm used within each bucket, and (2) the constant that appears in the expected running time of b^2-Bucket Sort.

1.2 Related Work

Showing that Bucket Sort runs in linear expected time when the keys are distributed independently and uniformly at random in $[0,1)$ is a classic textbook result [3,17,19]. Bounds on the expectation as well as limiting distributions for the running time have also been studied for different versions of Bucket Sort [4,14]. We are not aware of any work that directly addresses the rate of the running time of Bucket Sort. The upper and lower tails of the running time of Quick Sort have been studied in depth [7,10,18], including in the regime of large deviations [16].

The expected value of the nonvoid external path length of a trie is a classic result in applying the methods of analytic combinatorics to the analysis of algorithms [2,13,15,20,22]. We consider the case in which the binary strings are independent and random (i.e., the bits are independent and unbiased). In [13,15,20,22] it is shown that for random strings, the expected value of the nonvoid external path length is $n \log n + \Theta(n)$. The variance of the nonvoid external path length and limiting distributions for it have also been studied extensively for different string distributions [9,12,23].

In Knuth [13, Section 5.2.2], the nonvoid external path length is shown to be proportional to the number of bit comparisons of radix exchange sort. The bound in Theorem 3 therefore applies to the rate of the number of bit comparisons of radix exchange sort when the strings are distributed independently and uniformly at random.

The connection between the running time of sorting algorithms and various trie parameters (including external path length) has also been studied by Seidel [21], albeit in a significantly different model than ours. Specifically, [21] analyzes the expected number of bit comparisons of Quick Sort and Merge Sort when the input is a randomly permuted set of strings sampled from a given distribution. In Seidel's model, the cost of comparing two strings is proportional to the length of their longest common prefix. Seidel shows that the running time of these algorithms can be naturally expressed in terms of parameters of the trie induced by the input strings. We emphasize that our analysis connects the running time of Bucket Sort to the excess path length in the comparison model (in which the cost of comparing two keys does not depend on their binary representation).

1.3 Paper Organization

Preliminaries and definitions are in Sect. 2. In Sect. 3, we present reductions from the running time of $b \log b$-Bucket Sort and the nonvoid external path length to the excess path length. The bound on the upper tail of the excess path length is proved in Sect. 4. Section 5 proves a lower bound on the rate of b^2-Bucket Sort. Upper bounds on the rates are proved in the full version of the paper [1]. Theorems 1, 2 and 3 are completed in Sect. 6. Finally, in the full version of the paper, we also include a discussion on the difference between the rate of Bucket Sort and that of Quick Sort.

2 Preliminaries and Definitions

Bucket Sort. The input to Bucket Sort consists of n keys $X \triangleq \{x_1, \ldots, x_n\}$ in the interval $[0, 1)$. We define bucket j to be the set of keys in the interval $[j/n, (j+1)/n)$. Let $\mathbf{b}(X) \triangleq (B_0, \ldots B_{n-1})$ be the *occupancy vector* for input X, where B_j denotes the number of keys in X that fall in bucket j.

The buckets are separately sorted and the final output is computed by scanning the sorted buckets in increasing order. The initial assignment of keys to buckets and the final scanning of the sorted buckets takes $\Theta(n)$ time. We henceforth focus only on the time spent on sorting the keys in each bucket, i.e., the number of integer comparisons performed.

We consider the two natural options for sorting buckets: (i) Sort b keys in time $\Theta(b^2)$, using a sorting algorithm such as Insertion Sort or Bubble Sort. We refer to this option as b^2-Bucket Sort. (ii) Sort b keys in time $\Theta(b \log b)$ using a sorting algorithm such as Merge Sort or Heap Sort. We refer to this option as $b \log b$-Bucket Sort. Let $[n]$ denote the set $\{0, \ldots, n-1\}$ and let $\mathbf{b} = (B_0, \ldots, B_{n-1})$ denote an arbitrary occupancy vector. We define the functions

$$f(\mathbf{b}) \triangleq \sum_{j \in [n]} B_j^2 \qquad g(\mathbf{b}) \triangleq \sum_{j \in [n], B_j > 0} B_j \log B_j.$$

We let $T_{b^2}(X)$ and $T_{b \log b}(X)$ denote the running time on input X of b^2-Bucket Sort and $b \log b$-Bucket Sort, respectively. Then, $T_{b^2}(X) = \Theta(n + f(\mathbf{b}(X)))$ and $T_{b \log b}(X) = \Theta(n + g(\mathbf{b}(X)))^4$.

Excess Path Length and Tries. We let $|\alpha|$ denote the length of a binary string $\alpha \in \{0, 1\}^*$. For a set L, let $|L|$ denote the cardinality of L.

Definition 2. *A set of strings $\{\alpha_1, \ldots, \alpha_s\}$ is* prefix-free *if, for every $i \neq j$, the string α_i is not a prefix of α_j.*

A *trie* is a rooted binary tree with edges labeled $\{0, 1\}$ such that the two edges emanating from the same trie node are labeled differently. Every binary string can be mapped to a node in a trie. Futhermore, any set L of prefix-free binary strings can be represented by a unique trie $T(L)$ in which the strings in L correspond to leaves of the trie. Conversely, any trie corresponds to a unique set of prefix-free strings corresponding to all root-to-leaf paths.

Given a set L of prefix-free binary strings, we are interested in the smallest subtree of $T(L)$ that can still separate the strings in L, i.e., we want to trim $T(L)$ of trailing node-to-leaf paths that only consist of nodes of degree one. Namely, let $\varphi_0(L)$ denote the set of minimal prefixes of strings in L such that every string in L has a distinct prefix in $\varphi_0(L)$ and $\varphi_0(L)$ is prefix-free. The trie $T(\varphi_0(L))$ is called the *minimal* trie on L. Note that the structure of $\varphi_0(L)$ (or of $T(\varphi_0(L))$) does not change if we append more bits to the strings in L

The following definition extends the definition of $\varphi_0(L)$ by requiring that the prefixes have length at least k, i.e., each leaf must be at a depth of at least k.

---

[4] Interestingly, the sum of squares of bin occupancies, i.e., $f(\mathbf{b})$, also appears in the FKS perfect hashing construction [8].

Definition 3 (minimal k-prefixes). *Let* $L = \{\alpha_0, \ldots, \alpha_{n-1}\}$ *denote a set of* n *distinct infinite binary strings. Given a parameter* $k \geq 0$, *the set of* minimal *k-prefixes of* L, *denoted by* $\varphi_k(L) \triangleq \{\beta_0, \ldots, \beta_{n-1}\}$, *is the set that satisfies the following properties:*

1. *for all* $i \in [n]$, *the string* β_i *is a prefix of* α_i,
2. *for all* $i \in [n]$, $|\beta_i| \geq k$,
3. *the set* $\varphi_k(L)$ *is prefix-free,*
4. *for all* $i \in [n]$, β_i *is minimal with respect to all other prefixes of* α_i *that satisfy the first* 3 *conditions.*

The definition of $\varphi_k(L)$ can be modified to handle strings of length less than k by appending an arbitrary infinite string (say, zeros) to them.

In this paper, we are interested in the following trie parameter defined on $\varphi_k(L)$:

Definition 4. *The* k-excess path length $p_k(L)$ *of a set* L *of distinct infinite binary strings is defined as:*

$$p_k(L) \triangleq \sum_{\beta \in \varphi_k(L)} (|\beta| - k).$$

In [20], $p_0(L)$ is called the *nonvoid external path length* of the minimal trie $T(\varphi_0(L))$ on L. When $k = \lceil \log |L| \rceil$, we simply refer to $p_k(L)$ as the *excess path length* of L.

Distributions. Let $\mathcal{X}_n$ denote the uniform distribution over $[0, 1)^n$. Note that if the set $X = \{x_0, \ldots, x_{n-1}\}$ is chosen according to $\mathcal{X}_n$, then $x_0, \ldots, x_{n-1}$ are chosen independently and uniformly at random from the interval $[0, 1)$. Let μ_{b^2} (res. $\mu_{b \log b}$) denote the expected running time $T_{b^2}(X)$ (resp., $T_{b \log b}(X)$) when $X \sim \mathcal{X}_n$. Similarly, let μ_f (res. μ_g) denote the expected values of $f(\mathbf{b}(X))$ (resp., $g(\mathbf{b}(X))$) when $X \sim \mathcal{X}_n$. It is known that $\mu_f = 2n - 1$ (see [3,17]), and consequently, we have that $\mu_{b^2} = \Theta(n)$. Since $g \leq f$, we also have $\mu_g = \Theta(n)$ as well as $\mu_{b \log b} = \Theta(n)$.

Let $\mathcal{L}_n$ denote the uniform distribution over all sets of n infinite-length binary strings. Note that if $L = \{\alpha_0, \ldots, \alpha_{n-1}\}$ is chosen according to $\mathcal{L}_n$, then all the bits of the strings are independent and unbiased. We let μ_0 denote the expected value of the external nonvoid path $p_0(L)$ when $L \sim \mathcal{L}_n$. It is know that $\mu_0 = n \log n + \Theta(n)$ (see [13,20,22]).

Rates. Let $R_{b^2}(\cdot)$ (resp., $R_{b \log b}(\cdot)$) denote the rate of $T_{b^2}(X)$ (resp., $T_{b \log b}(X)$) when $X \sim \mathcal{X}_n$. Similarly, let $R_f(\cdot)$ (resp., $R_g(\cdot)$) denote the rate of $f(\mathbf{b}(X))$ (resp., $g(\mathbf{b}(X))$) when $X \sim \mathcal{X}_n$.

We first note that, to study the asymptotic behavior of R_{b^2} (for sufficiently large deviations) it suffices to study the asymptotic behavior of R_f. The proof of the following observation appears in the full version [1].

Observation 1. *For every* $c > 0$, *there exist constants* $\delta_1 = \Theta(c)$ *and* $\delta_2 = \Theta(c)$ *such that* $R_f(\delta_1 \cdot n) \leq R_{b^2}(c \cdot n) \leq R_f(\delta_2 \cdot n)$.

An analogous statement holds for the rates $R_{b \log b}$ and R_g. The rate of the nonvoid external path length $p_0(L)$ is denoted by $R_0(\cdot)$.

3   Reductions

3.1   Balls-into-Bins Abstraction

We interpret the assignment of keys to buckets using a balls-into-bins abstraction. The keys correspond to balls, and the buckets correspond to bins. The assumption that $X \sim \mathcal{X}_n$ implies that the balls choose the bins independently and uniformly at random. The value B_j then equals the occupancy of bin j.

A similar balls-into-bins abstraction holds for the embedding of the minimal $(\log n)$-prefixes of $L \sim \mathcal{L}_n$ in a trie (assuming n is a power of 2). Indeed, let $\{v_0, \ldots, v_{n-1}\}$ denote the n nodes of the trie $T(L)$ at depth $\log n$. For a node v_j, we say that a string α *chooses* v_j, if the path labeled α contains v_j. Since the strings are random, each string chooses a node of depth $\log n$ independently and uniformly at random. Let C_j denote the number of strings in L who choose node v_j.[5] We refer to C_j as the *occupancy* of v_j with respect to L and define the vector $\mathbf{c}(L) \triangleq (C_0, \ldots, C_{n-1})$.

Observation 2. *When $X \sim \mathcal{X}_n$ and $L \sim \mathcal{L}_n$, the occupancy vector $\mathbf{b}(X)$ has the same joint probability distribution as $\mathbf{c}(L)$.*

3.2   Lower Bounding the Rate of $b \log b$-Bucket Sort

By Observation 1, to prove a lower bound in $R_{b \log b}$ it suffices to prove a lower bound on R_g. In this section we show how to lower bound R_g by bounding the upper tail probability of the excess path length $p_{\log n}(L)$. We begin with the following observation about the nonvoid external path length $p_0(L)$:

Observation 3. *For every set L of n infinite prefix-free binary strings, $p_0(L) \geq n \log n$.*

Now consider an arbitrary vector $\mathbf{c}(L)$ and apply Observation 3 to each node of depth $\log n$ in $T(\varphi_{\log n}(L))$ separately. We obtain the following corollary:

Corollary 1. *For every set L of n infinite prefix-free binary strings,*

$$p_{\log n}(L) \geq \sum\nolimits_{j=0}^{n-1} C_j \log C_j = g(\mathbf{c}(L)).$$

In other words, the contribution to the excess path length $p_{\log n}(L)$ of the strings that choose node v_j is lower bounded by the number of comparisons that $b \log b$-Bucket Sort performs to sort a bin of occupancy C_j (up to a constant multiplicative factor). Since the distribution on occupancies of $\{v_j\}_{j=0}^{n-1}$ is the same as the distribution of bin occupancies in Bucket Sort (Observation 2), we can lower bound the rate of $g(\mathbf{b}(X))$ by bounding the upper tail of the excess path length as follows:

---

[5] Formally, $T(L)$ may contain a subset of these n nodes. If a node v_j at depth $\log n$ is not chosen by any string, then define $C_j = 0$.

Lemma 1. *For every $c > 0$,*

$$\Pr_{X \leftarrow \mathcal{X}_n} [g(\mathbf{b}(X)) \geq \mu_g + cn] \leq \Pr_{L \leftarrow \mathcal{L}_n} [p_{\log n}(L) \geq cn] . \tag{1}$$

Proof. Recall that μ_g denotes the expected value of $g(\mathbf{b}(X))$. Since $\mu_g > 0$, we have that $\Pr[g(\mathbf{b}(X)) \geq \mu_g + cn] \leq \Pr[g(\mathbf{b}(X)) \geq cn]$. Observation 2 implies that $\Pr_{X \leftarrow \mathcal{X}_n} [g(\mathbf{b}(X)) \geq cn] = \Pr_{L \leftarrow \mathcal{L}_n} [g(\mathbf{c}(L)) \geq cn]$. The claim then follows by Corollary 1. □

3.3 Lower Bounding the Rate of the Nonvoid External Path Length

In this section, we show how to use the upper tail of $p_{\log n}(L)$ to lower bound the rate of the nonvoid external path length $p_0(L)$.

Observation 4. *For every set L of n infinite prefix-free binary strings, we have that $p_0(L) \leq n \log n + p_{\log n}(L)$.*

Proof. The strings in $\varphi_0(L)$ are themselves prefixes of strings in $\varphi_{\log n}(L)$. We therefore get that $\sum_{\alpha \in \varphi_0(L)} |\alpha| \leq \sum_{\beta \in \varphi_{\log n}(L)} |\beta|$, and the claim follows. □

Observation 3 implies that $\mu_0 \geq n \log n$. Together with Observation 4 this implies that:

Corollary 2. *For every $L \sim \mathcal{L}_n$ and every $c > 0$.*

$$\Pr[p_0(L) \geq \mu_0 + cn] \leq \Pr[p_{\log n}(L) \geq cn].$$

4 The Upper Tail of the Excess Path Length

We bound the upper tail of $p_{\log n}(L)$ as follows:

Theorem 4. *Let $L \sim \mathcal{L}_n$. For every $c > 0$:*

$$\Pr[p_{\log n}(L) \geq cn] \leq \exp\left(-\frac{c - 1 - \ln c}{4} \cdot n\right) .$$

Proof. Let $L = \{\alpha_1, \ldots, \alpha_n\}$ be a set of infinite random binary strings. We consider the evolution of the set $\varphi_{\log n}(L)$ of minimal $\log n$-prefixes as we process the strings α_i one by one. Specifically, let $L^{(i)} \triangleq \{\alpha_1, \ldots, \alpha_i\}$, for $1 \leq i \leq n$, and $L_0 = \emptyset$.

Let $\varphi(L^{(i)}) \triangleq \left\{ s_j \circ \delta_j^{(i)} \mid |s_j| = \lceil \log n \rceil \text{ and } s_j \circ \delta_j^{(i)} \text{ is a prefix of } \alpha_j, \text{ for } 1 \leq j \leq i \right\}$. Note that

$$p_{\log n}(L^{(i)}) = \sum_{j \in [i]} \left| \delta_j^{(i)} \right|.$$

We bound $p_{\log n}(L)$ by considering the increase $\Delta_i \triangleq p_{\log n}(L^{(i)}) - p_{\log n}(L^{(i-1)})$. Since $p_{\log n}(L^{(0)}) = 0$ and $p_{\log n}(L^{(n)}) = p_{\log n}(L)$, then $p_{\log n}(L) = \sum_{i=1}^{n} \Delta_i$.

The addition of the string α_i has two types of contributions to Δ_i. The first contribution is $\delta_i^{(i)}$. The second contribution is due to the need to extend colliding strings. Indeed, since the set $L^{(i-1)}$ is prefix-free, there exists at most one $j < i$ such that $s_j \circ \delta_j^{(i-1)}$ is a prefix of α_i. If $s_j \circ \delta_j^{(i-1)}$ is a prefix of α_i, then $\Delta_i = \left|\delta_j^{(i)}\right| - \left|\delta_j^{(i-1)}\right| + \left|\delta_i^{(i)}\right|$. Because $\delta_j^{(i)}$ and $\delta_i^{(i)}$ are minimal subject to being prefix-free, we also have that $\left|\delta_j^{(i)}\right| = \left|\delta_i^{(i)}\right|$. Hence, $\Delta_i \leq 2 \cdot \left|\delta_i^{(i)}\right|$. This implies that, for every τ:

$$\Pr[\Delta_i \geq 2\tau] \leq \Pr\left[\left|\delta_i^{(i)}\right| \geq \tau\right] .$$

We now proceed to bound $\Pr\left[\left|\delta_i^{(i)}\right| \geq \tau\right]$. Fix $i \geq 1$ and let $\delta_i(\ell)$ denote the prefix of length ℓ of $\delta_i^{(i)}$. We denote by n_ℓ the number of leaves in the subtree rooted at $s_i \circ \delta_i(\ell)$ in the trie $T(L^{(i-1)})$ (i.e., right before the string α_i is processed). Formally,

$$n_\ell \triangleq \left|\left\{j < i \mid s_i \circ \delta_i(\ell) \text{ is a prefix of } s_j \circ \delta_j^{(i-1)}\right\}\right| .$$

Clearly, $n_0 = \left|\{j < i \mid s_i = s_j\}\right|$ and $n_{\left|\delta_i^{(i)}\right|} = 0$. We bound $\left|\delta_i^{(i)}\right|$ by bounding the minimum ℓ for which n_ℓ becomes zero as follows:

$$\mathbb{E}[n_\ell] = \sum_{j<i} \Pr\left[s_i \circ \delta_i(\ell) \text{ is a prefix of } s_j \circ \delta_j^{(i-1)}\right]$$

$$= \sum_{j<i} \frac{1}{2^{\log n + \ell}} \leq \frac{n}{2^{\log n + \ell}} = \frac{1}{2^\ell} .$$

In fact $\mathbb{E}\left[n_\ell \mid \bigwedge_{j<i} \Delta_j = \xi_j\right] \leq 1/2^\ell$ for every realization $\{\xi_j\}_{j<i}$ of $\{\Delta_j\}_{j<i}$. By Markov's inequality:

$$\Pr\left[|\delta_i^{(i)}| \geq \tau\right] = \Pr[n_{\tau-1} \geq 1] \leq \mathbb{E}[n_{\tau-1}] \leq \frac{1}{2^{\tau-1}} .$$

Therefore,

$$\Pr[\Delta_i \geq 2\tau] \leq 2^{-\tau+1} . \tag{2}$$

Note that Eq. 2 also holds under every conditioning on the realizations of $\{\Delta_j\}_{j<i}$.

Let $\Delta_i' \triangleq \frac{1}{2} \cdot \Delta_i$ and note that $\Pr[\Delta_i' \geq \tau] \leq 2^{-\tau+1}$. Let $\{G_i\}_i$ denote independent geometric random variables, where $G_i \sim Ge(1/2)$. Since $\Pr[G_i \geq \tau] =$

$2^{-(\tau-1)}$, we conclude that Δ_i' is stochastically dominated by G_i. In fact, the random variables $\{\Delta_i'\}_{i \in [f]}$ are unconditionally sequentially dominated by $\{G_i\}_{i \in [n]}$. By [5, Lemma 8.8], it follows that $\sum_{i \in [n]} \Delta_i'$ is stochastically dominated by $\sum_{i \in [n]} G_i$[6]. The sum of independent geometric random variables is concentrated [11] and so we get:

$$\Pr\left[\sum_{i \in [n]} \Delta_i' \geq c \cdot n/2\right] \leq \Pr\left[\sum_{i \in [n]} G_i \geq c \cdot n/2\right] \leq \exp\left(-\frac{c - 1 - \ln c}{4} \cdot n\right)$$

as required.   □

5   Lower Bound for b^2-Bucket Sort

This section deals with proving the following lower bound on the rate R_f. By Observation 1, this also implies a lower bound on the rate R_{b^2}.

Lemma 2. *There exists a constant $C > 0$ such that, for all $c > C$, we have that $R_f(cn) = \Omega(\sqrt{n}\log n)$, for all sufficiently large n.*

5.1   Preliminaries

Given an input X of n keys and its associated occupancy vector $\mathbf{b}(X) = (B_0, B_1, \ldots, B_{n-1})$, define $\mathcal{S}_i \triangleq \{j \in [n] \mid B_j \geq i\}$ to be the set of buckets with at least i keys assigned to them. Note that the random variables $\{|\mathcal{S}_i|\}_i$ are negatively associated because they are monotone functions of bin occupancies, which are a classical example of negatively associated RVs [6].

Claim 1. *For every occupancy vector $(B_0, B_1, \ldots, B_{n-1})$, the following holds:*

$$\sum_{j \in [n]} \binom{|B_j| + 1}{2} = \sum_{i \in [n+1]} i \cdot |\mathcal{S}_i| . \tag{3}$$

Proof. Consider an $n \times n$ matrix A in which $A_{i,j} = i$ if $B_j \geq i$ and 0 otherwise. Let $S \triangleq \sum_{i,j} A_{i,j}$. The sum of entries in column j equals $\binom{|B_j|+1}{2}$. The sum of entries in row i equals $i \cdot |\mathcal{S}_i|$. Hence both sides of Eq. 3 equal S.   □

The following lemma states that, in order to prove Lemma 2, it suffices to prove a lower bound on the upper tail probability of the random variable $\sum_{i \in [n+1]} i \cdot |\mathcal{S}_i|$.

Lemma 3. *For every c, we have that*

$$\Pr\left[f(\mathbf{b}(X)) \geq \mu_f + cn\right] = \Pr\left[\sum_{i \in [n+1]} i |\mathcal{S}_i| \geq \frac{(3+c)n - 1}{2}\right] .$$

---

[6] Note that RVs $\{\Delta\}_i$ are not independent and probably not even negatively associated. Hence, standard concentration bounds do not apply to $\sum \Delta_i$.

Proof. By Claim 1, $f(\mathbf{b}(X)) = 2 \cdot \sum_{i \in [n+1]} i\,|\mathcal{S}_i| - n$. The lemma follows from the fact that $\mu_f = 2n - 1$ [3,17].   □

Next, we upper bound $\mathbb{E}\left[|\mathcal{S}_i|\right]$. Let $E_i \triangleq \left(\frac{e}{i}\right)^i$ and note the following:

Claim 2. *For every* $i \in \{1, \ldots, n\}$, *we have that* $\mathbb{E}\left[|\mathcal{S}_i|\right] \leq n \cdot E_i$.

Proof. Fix i and let $X_{i,j}$ be the indicator random variable that is 1 if $B_j \geq i$ and 0 otherwise. We get that $|\mathcal{S}_i| = \sum_j X_{i,j}$. Because each key chooses a bucket independently and uniformly at random, we have that:

$$\Pr\left[B_j \geq i\right] \leq \binom{n}{i} \cdot \left(\frac{1}{n}\right)^i \leq \left(\frac{en}{i}\right)^i \cdot \left(\frac{1}{n}\right)^i = \left(\frac{e}{i}\right)^i = E_i .$$

The claim follows by linearity of expectation.   □

One can analytically show that:

Observation 5. $\sum_{i=1}^{\infty} i \cdot E_i \leq 10.$

5.2   Applying Chernoff Bounds in Different Regimes

In order to prove a lower bound on the upper tail of $\sum_{i \in [n+1]} i \cdot |\mathcal{S}_i|$, we partition the sum according to three thresholds on bin occupancies $\tau_1 \leq \tau_2 \leq \tau_3$, defined as follows:

$$\tau_1 \triangleq \max\left\{i \,\Big|\, E_i \geq \frac{c \log n}{\sqrt{n}}\right\}, \qquad \tau_2 \triangleq \frac{n^{1/4}}{\sqrt{\log n}}, \qquad \tau_3 \triangleq \sqrt{n}.$$

We invoke different versions of the multiplicative Chernoff bound depending on the specific partition we are considering. We include here the two most interesting cases, including the proof that invokes the rarely used form of the Chernoff bound mentioned in the introduction (Claim 4). We defer the rest to the full version of the paper [1].

Claim 3. *For every* $c > 0$, *there exists a* $\gamma = \gamma(c) > 0$ *such that for* n *sufficiently large:*

$$\Pr\left[\sum_{i=\lceil \tau_1 \rceil}^{\lfloor \tau_2 \rfloor} i\,|\mathcal{S}_i| \geq cn + \sum_{i=\lceil \tau_1 \rceil}^{\lfloor \tau_2 \rfloor} iE_i \cdot n\right] \leq \exp\left(-\gamma\sqrt{n}\log n\right).$$

Proof. For every $\tau_1(c) < i \leq \tau_2$, define $\delta_i \triangleq (c \log n)/(E_i \sqrt{n})$ so that

$$\sum_{i \leq \tau_2} \delta_i \cdot iE_i = \sum_{i \leq \tau_2} i \cdot \frac{c \log n}{\sqrt{n}} \leq (\tau_2)^2 \cdot \frac{c \log n}{\sqrt{n}} = c . \tag{4}$$

Since $\delta_i > 1$ for every $i > \tau_1$, by the standard Chernoff bound:

$$\Pr\left[|\mathcal{S}_i| > (1 + \delta_i) \cdot E_i \cdot n\right] \leq \exp\left(-\delta_i \cdot n \cdot E_i/3\right) = \exp\left(-c/3 \cdot \sqrt{n}\log n\right) .$$

By applying a union bound over all $\tau_1 \leq i \leq \tau_2$, it follows that there exists a constant $\gamma > 0$ such that:

$$\Pr\left[\sum_{i=\lceil\tau_1\rceil}^{\lfloor\tau_2\rfloor} i\,|\mathcal{S}_i| \geq \sum_{i=\lceil\tau_1\rceil}^{\lfloor\tau_2\rfloor} (1 + \delta_i) \cdot iE_i \cdot n\right] \leq \exp\left(-\delta\sqrt{n}\log n\right) . \tag{5}$$

The claim follows by Eq. 4 and 5. $\square$

Claim 4. *For every $c > 0$, there exists a $\gamma = \gamma(c) > 0$ such that for sufficiently large n, we have that:*

$$\Pr\left[\sum_{i=\lceil\tau_2\rceil}^{\lfloor\tau_3\rfloor} i\,|\mathcal{S}_i| > cn + \sum_{i=\lceil\tau_2\rceil}^{\lfloor\tau_3\rfloor} iE_i \cdot n\right] \leq \exp\left(-\gamma\sqrt{n}\log(n)\right) .$$

Proof. For every $\tau_2 \leq i \leq \tau_3$, define $\delta_i \triangleq \frac{c}{5} \cdot \frac{\log n}{i\log i \cdot E_i \cdot \sqrt{n}}$ so that the following holds for sufficiently large n:

$$\sum_{i=\lceil\tau_2\rceil}^{\lfloor\tau_3\rfloor} \delta_i \cdot iE_i = \frac{c}{5} \cdot \left(\sum_{i=\lceil\tau_2\rceil}^{\lfloor\tau_3\rfloor} \frac{1}{\log i}\right) \cdot \frac{\log n}{\sqrt{n}} \tag{6}$$

$$\leq \frac{c}{5} \cdot \frac{\tau_3}{\log \tau_2} \cdot \frac{\log n}{\sqrt{n}} \leq \frac{c}{5} \cdot \frac{\log n}{0.25\log n - 0.5\log\log n} \leq c . \tag{7}$$

For a sufficiently large n, it holds that $\delta_i > 1$; moreover $\log \delta_i \geq \Omega(i\log i)$ for every $\tau_2 \leq i \leq \tau_3$. By the Chernoff bound with a $\delta_i \ln(\delta_i)$ dependency in the exponent 9 (see Chapter 10.1.1 in [5]):

$$\Pr\left[|\mathcal{S}_i| > (1 + \delta_i) \cdot n \cdot E_i\right] \leq \exp\left(-\delta_i \ln(\delta_i) \cdot n \cdot E_i/2\right) \leq \exp\left(-\Omega(\sqrt{n}\log n)\right) .$$

By applying a union bound over all $\tau_2 \leq i \leq \tau_3$, it follows that there exists a $\delta(c) > 0$ such that:

$$\Pr\left[\sum_{i=\lceil\tau_2\rceil}^{\lfloor\tau_3\rfloor} i\,|\mathcal{S}_i| > \sum_{i=\lceil\tau_2\rceil}^{\lfloor\tau_3\rfloor} (1 + \delta_i) \cdot i \cdot E_i \cdot n\right] \leq \exp\left(-\delta\sqrt{n}\log n\right) . \tag{8}$$

The claim follows by Eq. 7 and 8. $\square$

6   Proofs of Theorems 1, 2 and 3

To prove Theorems 1 and 2, we employ Observation 1 that shows a reduction from R_f (and R_g, respectively) to R_{b^2} (and $R_{b\log b}$ respectively). The lower bound for R_f is discussed in Lemma 2. The lower bound for R_g follows from Lemma 1 and Theorem 4. The lower bound for R_0 follows from Corollary 2 and Theorem 4. Matching upper bounds on R_f, R_g and R_0 are discussed in the full version of the paper [1].

References

1. Bercea, I.O., Even, G.: Upper tail analysis of bucket sort and random tries. CoRR abs/2002.10499 (2020). https://arxiv.org/abs/2002.10499
2. Clément, J., Flajolet, P., Vallée, B.: Dynamical sources in information theory: a general analysis of trie structures. Algorithmica $29(1–2)$, 307–369 (2001)
3. Cormen, T.H., Leiserson, C.E., Rivest, R.L., Stein, C.: Introduction to Algorithms. MIT Press, Cambridge (2009)
4. Devroye, L.: Lecture Notes on Bucket Algorithms, vol. 12. Birkhäuser Boston (1986)
5. Doerr, B.: Probabilistic tools for the analysis of randomized optimization heuristics. CoRR abs/1801.06733 (2018). http://arxiv.org/abs/1801.06733
6. Dubhashi, D.P., Panconesi, A.: Concentration of Measure for the Analysis of Randomized Algorithms. Cambridge University Press, Cambridge (2009)
7. Fill, J.A., Janson, S.: Quicksort asymptotics. J. Algorithms $44(1)$, 4–28 (2002)
8. Fredman, M.L., Komlós, J., Szemerédi, E.: Storing a sparse table with $o(1)$ worst case access time. In: 23rd Annual Symposium on Foundations of Computer Science, pp. 165–169. IEEE (1982)
9. Jacquet, P., Regnier, M.: Normal limiting distribution for the size and the external path length of tries (1988)
10. Janson, S.: On the tails of the limiting quicksort distribution. Electron. Commun. Prob. **20** (2015)
11. Janson, S.: Tail bounds for sums of geometric and exponential variables. Stat. Prob. Lett. **135**, 1–6 (2018)
12. Kirschenhofer, P., Prodinger, H., Szpankowski, W.: On the variance of the external path length in a symmetric digital trie. Discrete Appl. Math. $25(1–2)$, 129–143 (1989)
13. Knuth, D.E.: The Art of Computer Programming, vol. III, 2nd edn. Addison-Wesley, Boston (1998)
14. Mahmoud, H., Flajolet, P., Jacquet, P., Régnier, M.: Analytic variations on bucket selection and sorting. Acta Informatica $36(9–10)$, 735–760 (2000)
15. Mahmoud, H.M., Lueker, G.S.: Evolution of Random Search Trees, vol. 200. Wiley, New York (1992)
16. McDiarmid, C., Hayward, R.: Large deviations for quicksort. J. Algorithms $21(3)$, 476–507 (1996). https://doi.org/10.1006/jagm.1996.0055
17. Mitzenmacher, M., Upfal, E.: Probability and Computing: Randomization and Probabilistic Techniques in Algorithms and Data Analysis. Cambridge University Press, Cambridge (2017)
18. Régnier, M.: A limiting distribution for quicksort. RAIRO-Theoretical Inform. Appl.-Informatique Théorique et Appl. **23**(3), 335–343 (1989)
19. Sanders, P., Mehlhorn, K., Dietzfelbinger, M., Dementiev, R.: Sorting and selection. In: Sequential and Parallel Algorithms and Data Structures, pp. 153–210. Springer, Cham (2019). https://doi.org/10.1007/978-3-030-25209-0_5
20. Sedgewick, R., Flajolet, P.: An Introduction to the Analysis of Algorithms. Pearson Education India, Chennai (2013)
21. Seidel, R.: Data-specific analysis of string sorting. In: Proceedings of the Twenty-First Annual ACM-SIAM Symposium on Discrete Algorithms, pp. 1278–1286. Society for Industrial and Applied Mathematics (2010)

22. Szpankowski, W.: Average Case Analysis of Algorithms on Sequences, vol. 50. John Wiley & Sons, New York (2011)
23. Vitter, J.S., Flajolet, P.: Average-case analysis of algorithms and data structures. In: Handbook of Theoretical Computer Science, Volume A: Algorithms and Complexity, pp. 431–524 (1990)

Throughput Scheduling with Equal Additive Laxity

Martin Böhm[1], Nicole Megow[2], and Jens Schlöter[2]

[1] Institute of Computer Science, University of Wrocław, Wrocław, Poland
boehm@cs.uni.wroc.pl
[2] Department of Mathematics and Computer Science, University of Bremen, Bremen, Germany
{nmegow,jschloet}@uni-bremen.de

Abstract. We study a special case of a classical throughput maximization problem. There is given a set of jobs, each job j having a processing time p_j, a release time r_j, a deadline d_j, and possibly a weight. The jobs have to be scheduled non-preemptively on m identical parallel machines. The goal is to find a schedule for a subset of jobs of maximum cardinality (or maximum total weight) that start and finish within their feasible time window $[r_j, d_j]$. In our special case, the additive laxity of every job is equal, i.e., $d_j - p_j - r_j = \delta$ with a common δ for all jobs. Throughput scheduling has been studied extensively over decades. Understanding important special cases is of major interest. From a practical point of view, our special case was raised as important in the context of last-mile meal deliveries. As a main result we show that single-machine throughput scheduling with equal additive laxity can be solved optimally in polynomial time. This contrasts the strong NP-hardness of the problem variant with arbitrary (and even equal multiplicative) laxity. Further, we give a fully polynomial-time approximation scheme for the weakly NP-hard weighted problem. Our single-machine algorithm can be used repeatedly to schedule jobs on multiple machines, such as in the greedy framework by Bar-Noy et al. [STOC '99], with an approximation ratio of $\frac{(m)^m}{(m)^m - (m-1)^m} < \frac{e}{e-1}$. Finally, we present a pseudo-polynomial time algorithm for our weighted problem on a constant number of machines.

1 Introduction

We study one of the classical models for job scheduling that has been studied since the Seventies [13,15]. There is given a set of n jobs to be assigned to m parallel identical machines. Each job j is associated with a processing time $p_j \in \mathbb{N}$, a release time $r_j \in \mathbb{N}$, a deadline $d_j \in \mathbb{N}$ and a weight $w_j \in \mathbb{N}$ with $w_j \geq 1$. The goal is to find a *feasible schedule* $\mathcal{S}$ for a subset of jobs U that maximizes $\sum_{j \in U} w_j$. A feasible schedule $\mathcal{S}$ for a set of jobs U assigns a starting time $S_j \geq r_j$ to each job $j \in U$ such that the completion time for each job $C_j = S_j + p_j$ is not later than the job's deadline, i.e., $C_j \leq d_j$. Each machine can run at most one job at each point in time, without the possibility of preemption. When jobs have unit

© Springer Nature Switzerland AG 2021
T. Calamoneri and F. Corò (Eds.): CIAC 2021, LNCS 12701, pp. 130–143, 2021.
https://doi.org/10.1007/978-3-030-75242-2_9

(resp. arbitrary) weights, the problem is called *(weighted) throughput scheduling* or, in the standard scheduling notation, $P|r_j| \sum(1-U_j)$ resp. $P|r_j| \sum w_j(1-U_j)$. When considering the single machine setting, we refer to $1|r_j| \sum(1-U_j)$ and $1|r_j| \sum w_j(1-U_j)$. The problems are all known to be strongly NP-complete [14].

Over decades, research efforts [3,4,6,16,17] have aimed for approximation algorithms, that is, polynomial-time algorithms with a bounded worst-case ratio of the objective values achieved by the algorithm and the optimal solution. The best known approximation algorithm for weighted throughput scheduling, $P|r_j| \sum w_j(1-U_j)$, yields a worst-case ratio of $\frac{(m+1)^m}{(m+1)^m-(m+\epsilon)^m}$, which is $2/(1-\epsilon)$ for $m = 1$ and any $0 < \epsilon$ and monotonously decreases towards $e/(e-1) \approx 1.582$ for increasing m [4]. For the unweighted problem on any constant number of machines, $Pm|r_j| \sum(1-U_j)$, there exists an algorithm with approximation ratio 1.55 and running time exponentially dependent on m. This algorithm is complemented by an algorithm with running time polynomial in n and m, whose approximation ratio is converging to 1 as the number of machines m tends to infinity [17]. The second part of the result also holds for the weighted problem.

Concurrently to the approximation efforts, progress has been made on identifying subinstances of the problem which are still tractable in polynomial time. For example, in the case of all equal processing times, maximizing the weighted throughput is polynomial-time solvable [2]. However, if the processing times have at least two non-divisible options (such as $\{2,3\}$), the decision problem of scheduling all jobs becomes NP-complete [12].

In this paper, we study a special case regarding the scheduling flexibility. Any job with processing time p_j that is selected to be scheduled must be started in the time interval $[r_j, d_j - p_j)$. We call the length of this time interval the *laxity* of a job and denote it by $l_j := d_j - p_j - r_j$. We investigate the special case in which all jobs have a normalized additive laxity, that is, $l_j = \delta$ for some $\delta \geq 0$ common for all jobs. We call this problem *(weighted) throughput scheduling with equal additive laxity*.

This special case is motivated also from a practical point of view. The single machine case has been raised as an open problem in the context of the scheduling of last-mile meal delivery processes by Cosmi et al. [8–10]. Consider a restaurant that schedules food deliveries using a single courier. Often restaurants guarantee that orders are delivered within an interval of a specific length, e.g., 30 min, after the delivery time requested by the customer. The time needed to travel from the restaurant to a customer and back can be encoded by the processing time p_j of a corresponding job j, and the fixed length of the guaranteed time interval can by modeled by a common additive laxity δ. To deliver within the requested interval, the courier cannot start the delivery more than $p_j/2$ time units before the requested delivery time (modeled by r_j) and, assuming the courier returns immediately after the delivery, cannot return to the restaurant more than $p_j/2+\delta$ time units after the requested delivery time (modeled by d_j).

Our Contribution. We resolve the complexity status of throughput scheduling with equal additive laxity on single and parallel machines. As a main result, we show how to solve the single-machine problem optimally in polynomial time.

Theorem 1. *The problem $1|r_j|\sum(1-U_j)$ with equal additive laxity can be solved optimally in polynomial time.*

This result is in stark contrast to the NP-hardness of the problem with arbitrary laxity. Further, it contrasts the complexity for the alternative model with normalized *multiplicative* laxity, in which $l_j = p_j \cdot \delta$ for a multiplier δ common for all jobs. In this setting, the problem continues to be strongly NP-hard, even on a single machine, as was shown by reduction from 3-Partition [11].

The key to our main result is a characterization of job inversions deviating from the release date order in an optimal schedule. While scheduling in release date order (or using similar greedy strategies based, e.g., on deadlines or processing times) does not solve the problem optimally, we show that there always exists an optimal schedule that deviates from the release date order in a very limited and particular way. By properly restricting the type of required inversions, we define canonical schedules and find an optimal one by dynamic programming.

We transfer the techniques from the unweighted problem to the weighted version. In that setting, even the single-machine problem is weakly NP-hard, as it contains the knapsack problem. We give a *fully polynomial-time approximation scheme (FPTAS)*, which is the best we can hope for in this case, unless P=NP. An FPTAS is a family of algorithms computing a $(1+\epsilon)$-approximation for every $\epsilon > 0$ with a running time polynomial in the input size and $\frac{1}{\epsilon}$.

Theorem 2. *There is an FPTAS for $1|r_j|\sum w_j(1-U_j)$ with equal additive laxity.*

On identical parallel machines, throughput scheduling with equal additive laxity is strongly NP-hard in general and weakly NP-hard if the number of machines is constant. This follows directly by a standard reduction from Partition. A natural greedy algorithm for the multi-machine problem iteratively executes a single-machine algorithm on each machine using only the jobs that have not yet been scheduled on previous machines. The analysis of Bar-Noy et al. [3, Theorem 3.3] and our single machine results imply the following result.

Corollary 1. *Consider $P|r_j|\sum(1-U_j)$ with equal additive laxity. The greedy algorithm yields an approximation factor $\frac{m^m}{m^m-(m-1)^m}$. For the weighted problem variant, the approximation factor is $\frac{(m+\epsilon)^m}{(m+\epsilon)^m-(m+\epsilon-1)^m}$ with $\epsilon > 0$ and the running time is polynomial in the input and $\frac{1}{\epsilon}$.*

Our algorithm is optimal resp. $(1 + \epsilon)$-approximate for $m = 1$, and has an approximation ratio monotonously increasing in m with its limit at $e/(e - 1) \approx 1.582$ (plus $\epsilon > 0$).

In the unweighted setting, our result improves upon the best-known 1.55-approximation ratio by Im et al. [17] if the number of machines is small, $m \leq 14$. In contrast to the 1.55-approximation, our algorithm is deterministic and purely combinatorial, i.e., it does not involve solving any linear program, which often is an advantage for the efficiency of an implementation. In the weighted case, we improve, for any m, upon the best-known approximation ratio of $\frac{(m+1)^m}{(m+1)^m-(m+\epsilon)^m}$

for arbitrary laxity by Berman and DasGupta [4]. For $m = 1, 2, 3, \ldots$ and assuming an infinitesimally small ϵ, our result compares with the ratio in [4] as 1 (our result) vs. 2 ([4]), 1.33 vs. 1.8, 1.42 vs. 1.73, 1.46 vs. 1.69, $\ldots$ with the gap decreasing to 0 for m towards infinity.

As a final result, we give a pseudo-polynomial time algorithm for our problem when $m = \mathcal{O}(1)$. More precisely, we show the following theorem.

Theorem 3. *There is an optimal algorithm for $Pm|r_j| \sum w_j(1 - U_j)$ with equal additive laxity with its running time polynomial in $\min\{\delta, \max_{j \in J} p_j\}$, $W = \sum_{j \in J} w_j$ and the input size. The running time dependency on W can be replaced by a dependency on $1/\epsilon$, for any $\epsilon > 0$, at the cost of a $(1 + \epsilon)$-factor in the objective.*

In the case of arbitrary laxity, the best-known pseudo-polynomial time algorithm for a constant number of machines and uniform weights is the 1.55-approximation by [17]. For the problem with normalized additive laxity, our result rules out strong NP-hardness on a constant number of machines.

Related Work. Throughput scheduling has been studied extensively under several names and closely related variants, including job interval selection or real-time scheduling. Bar-Noy et al. [3] initiated a line of research on approximation algorithms for the weighted and unweighted setting.

The currently best known algorithm for $P|r_j| \sum w_j(1 - U_j)$ by Berman and DasGupta [4] has an approximation ratio of $\frac{(m+1)^m}{(m+1)^m - (m+\epsilon)^m}$. When the weights are uniform, Im, Li and Moseley [17] provide an approximation algorithm with ratio 1.55 for $m \in \mathcal{O}(1)$. When m tends to infinity the ratio improves to 1. This latter result extends also to the weighted case. Recently, Hyatt-Denesik, Rahgoshay and Salavatipour [16] gave a $(1 + \epsilon)$-approximation with a running time exponential in $1/\epsilon$, m and c where c is the number of different processing times. That is, if $m, c \in \mathcal{O}(1)$, the algorithm yields a PTAS.

Regarding polynomially tractable subcases, an important positive result is due to Baptiste [2], who provides an algorithm for the problem $1|r_j| \sum w_j(1 - U_j)$ when all processing times are equal with an asymptotic running time of $O(n^7)$. For unweighted throughput, the fastest known, optimal algorithm by Chrobak et al. [5] runs in time $O(n^5)$. Another polynomially tractable subcase is the scheduling of *agreeable instances*, i.e., $r_j < r_{j'}$ implies $d_j < d_{j'}$ for all jobs $j, j' \in J$, on a single machine [18]. This result exploits that for agreeable instances on a single machine there is always an optimal schedule which schedules all feasible jobs in release date order.

For the case when all processing times p_j are either 1 or p, Sgall [21] provides a polynomial-time algorithm to check if all jobs on input can be scheduled. The complexity of the unweighted and weighted maximization in this setting remains open. Additionally, the existence of a polynomial-time algorithm for the case when all jobs are equal and m is a part of the input is also open [21].

Special cases with assumptions on the laxity have been considered in different flavors. Chuzhoy, Ostrovski and Rabani [6] assume an upper bound on the *multiplicative* laxity of all jobs, $l_{\mathrm{mult}} := \max_{j \in J}(d_j - r_j)/p_j$. They present an exact algorithm for $1|r_j|\sum w_j(1 - U_j)$ that is polynomial in n and T, where T is the time horizon of the instance, and exponential in l_{mult}. For the same problem, Berman and DasGupta [4] also give a pseudo-polynomial time $2/(1 + 1/(2^{\lfloor l_{\text{mult}} \rfloor + 1} - 2 - \lfloor l_{\text{mult}} \rfloor))$-approximation.

Recently, the problem of throughput scheduling with bounded *additive* laxity, $l_{\text{add}} := \max_{j \in J} l_j$, received quite some attention in the context of meal deliveries [9,10]. Cosmi et al. [10] consider throughput scheduling parameterized by l_{add}. Here, a dependence of the running time on l_{add} is necessary, because a large value of l_{add} will contain many instances of the general problem of throughput scheduling. The main result of [10] is a dynamic algorithm for unweighted throughput scheduling on a single machine with running time $O(n \cdot (l_{\text{add}} + 2)^{2l_{\text{add}}+1} + n \log n)$. For the case of unweighted throughput on m identical machines, they adapt the dynamic algorithm to find an optimal solution in time $O(n(l_{\text{mult}} + 2)^{2m(l_{\text{mult}}+1)} + n \log n)$. Comparing these results to our contributions, we observe that the algorithm of [10] can solve a more general set of instances than our algorithms if l_{add} is bounded by a constant small enough such that the running time is still reasonable. On the other hand, the running time of our algorithms is independent of δ, meaning that they perform well for every choice of δ. For further research around meal deliveries we refer to [1,19,20,23,24,26].

Van Bevern, Niedermeier and Suchý [25] also investigate scheduling instances parametrized by l_{add}, l_{mult}, or both, but they focus on designing algorithms that solve the decision variant of the problem, where the goal is to decide whether all jobs can be scheduled or not. Aside from showing hardness for several variants, they contribute optimal algorithms with running times $(l_{\text{add}})^{O(l_{\text{mult}} \cdot m)} \cdot n + n \log n$ and $\mathcal{O}((l_{\text{add}} + 1)^{(2l_{\text{add}}+1) \cdot m} \cdot n \cdot l_{\text{add}} \cdot m \cdot \log(l_{\text{add}} \cdot m) + n \log n)$, respectively.

Further lines of research in non-preemptive throughput scheduling address non-continuous scheduling intervals [6], minimizing the number of machines to schedule all jobs, also with bounded laxity [7] and unrelated machines [3,4,22].

Overview: In Sect. 2 we prove crucial structural properties and characterize canonical single-machine solutions. In Sect. 3 we derive a dynamic program (DP) that optimally solves $1|r_j|\sum w_j(1 - U_j)$ when jobs have equal laxity. We turn the DP into an FPTAS for arbitrary weights and give a pseudo-polynomial time algorithm for $m \in \mathcal{O}(1)$.

2   Canonical Solutions

In this section, we consider throughput scheduling with equal additive laxity on a single machine. Firstly, we define the classical earliest release date first order.

Definition 1. *Given a set of jobs J, the earliest release date first (ERF) ordering of J is an ordering $\prec_E$ where the jobs are ordered the same way as their release dates. We extend the ordering to a linear ordering by giving priority to jobs with smaller processing times, i.e. if $r_i = r_j$ and $p_i < p_j$, then we set $i \prec_E j$. Finally, all remaining ties are broken arbitrarily.*

Scheduling jobs in ERF order is not optimal as the example in Fig. 1 shows. ERF schedules the long job before the two short jobs which leads to a solution with two jobs, whereas scheduling the short jobs before the long job leads to three feasibly scheduled jobs. There are similar examples showing that other greedy strategies, such as *earliest deadline first* (EDF) and *longest processing time first* (LPT), are not optimal in our setting, either. However, we show that there always exists an optimal schedule which deviates from the ERF order in a very limited and particular way. To characterize this, we study inversions.

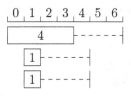

Fig. 1. An instance with uniform weights for which scheduling in ERF order is not optimal.

2.1 Inversions

We give a formal definition of a schedule deviation from ERF order.

Definition 2. *A job i inverts a job j (short i inv j) in a schedule $\mathcal{S}$ if i is before j in the ERF order, but j is started earlier than i in $\mathcal{S}$. Symbolically, $i \prec_E j$ but $S_j < S_i$. We say that there is an* inversion *between i and j if i inverts j.*

Our goal is to show that each feasible schedule can be transformed into a feasible schedule for the same set of jobs with limited inversions. We define a *safe exchange* operation, which can be used to remove certain inversions from a feasible schedule while maintaining feasibility. The operation consists of two steps.

Suppose that there is an inversion i inv j. We partition the jobs scheduled between i and j into a set A of jobs that are inverted by i and set a B of jobs which are not inverted by i. In the first step of the operation, we alter the schedule by scheduling the set B first, then j followed by i, and finally the set A. In the second step of the operation, we also swap job j with i. See Fig. 2 for an illustration. This exchange operation is not universally applicable, but it leads to a decrease in the number of inversions in an important case.

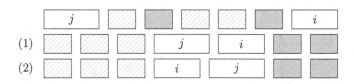

Fig. 2. The safe exchange operation applied to an inversion pair i inv j. The jobs scheduled between i and j are partitioned into the set A (solid, red) and B (hatched, blue). In Step (1), the jobs in B and A are moved so that the jobs i and j are adjacent. In Step (2), the jobs i and j are swapped. (Color figure online)

Lemma 1. *A feasible schedule S with an inversion i inv j, which satisfies that $d_{j'} \geq S_i + p_i$ holds for all jobs j' with $S_j \leq S_{j'} < S_i$, can be transformed into a feasible schedule S' for the same job set without inversion i inv j.*

Proof. Let S be a feasible schedule as described in the Lemma. Consider the schedule S' after applying the safe exchange operation to inversion i inv j. In S', all jobs in B are scheduled in the same order as in S, afterwards appears i followed by j, and then all jobs in A in the same order as in S. All other jobs are scheduled the same way as in S. Notice that S' contains the same jobs as S.

We now show that S' is feasible. Denote the starting time of a job l in S' by S'_l. The job i in S' respects its release date, since $S'_i \geq S_j$ holds by construction of S', $r_i \leq r_j$ holds by assumption and S is feasible. As i is scheduled earlier in S' than in S, it also does not miss its deadline. By construction of S', each $j' \in A \cup \{j\}$ starts in S' (at least by p_i) later than in S and completes at the latest by $S_i + p_i \leq d_{j'}$. Thus, each $j' \in A \cup \{j\}$ is feasibly scheduled in S'. For each $l \in B$ holds $r_l \leq r_i$ as l started before i in S but is not inverted by i. By construction of S', $S_l \geq S'_l \geq S_j \geq r_j \geq r_i \geq r_l$ holds and, thus, l is feasibly scheduled in S'. Since jobs in $J \setminus (A \cup B \cup \{i,j\})$ are scheduled identically in S as in S', it follows that S' is feasible. □

Applying Lemma 1 repeatedly, we restrict ourselves to schedules where for each inversion pair i inv j there exists an obstruction job j' with $d_{j'} < S_i + p_i$.

Definition 3. *We call an inversion pair i inv j a critical inversion if $d_j < S_i + p_i$ and j is the job with this property minimizing $S_j - S_i$. We also call a triple of jobs i, j, k a nested inversion if i inv j and j inv k.*

An example of a nested inversion is shown in Fig. 3a). The key that allows us to formulate a polynomial-time algorithm is the fact that there is always an optimal schedule without nested inversions.

Lemma 2. *Let S be a feasible schedule such that no inversion can be safely exchanged. Then there are no three jobs i, j, k with i inv j and j inv k in S.*

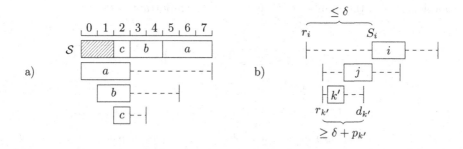

Fig. 3. a) An instance of throughput scheduling (with non-equal additive laxity) where a nested inversion a inv b, b inv c occurs for a feasible schedule S. The additional dashed job in the instance is tight. b) A visual proof of Lemma 2. The top interval in a brace is of length at most δ, but its sub-interval on the bottom has length at least $\delta + p_{k'}$.

Proof. By contradiction, assume that such a triple of jobs i, j and k exists in a feasible schedule $\mathcal{S}$. See Fig. 3b) for an illustration. In our equal additive laxity setting, any job j has its time window set to $d_j - r_j = p_j + \delta$. By definition, any feasible schedule satisfies $S_i - r_i \leq \delta$ for every job i. Applying Lemma 1 on j inv k gives us a job k' for which $d_{k'} < S_j + p_j$. Since i is scheduled after both j and k' in our triple inversion, we have that $d_{k'} \leq S_j + p_j \leq S_i$. However, this implies a contradiction, as $p_{k'} + \delta = d_{k'} - r_{k'} \leq S_i - r_{k'} \leq S_i - r_i \leq \delta$. The above equation is false for job k' with non-zero processing time. $\qquad\square$

In addition to the absence of inversion triples, we can also strengthen the properties of a critical inversion pair. We have defined the critical inversion pair i inv j' to be as close in the schedule $\mathcal{S}$ as possible, but we can actually assume that they are scheduled next to each other:

Lemma 3. *Any feasible schedule $\mathcal{S}$ with a critical inversion i inv j' can be transformed into a feasible schedule $\mathcal{S}'$ for the same set of jobs that schedules j' immediately before i, without affecting any other critical inversion pair.*

Proof. Consider the critical inversion pair i inv j'. By definition, any job k scheduled between j' and i is either not inverted with i or the deadline of k satisfies $d_k \leq S_i + p_i$. Therefore, we can apply the first step of the safe exchange operation on the pair j' and i. After the exchange, the job j' is scheduled immediately before i. By Lemma 2, this does not affect any other critical inversion pair. $\qquad\square$

2.2 Canonical Schedules

With Lemmas 2 and 3 setting the structure of inversions in feasible schedules, we wish to formally limit the set of feasible schedules that we consider (and that our algorithm can create). Our goal is to characterize feasible schedules entirely based on their critical inversion pairs. Let $\mathcal{S}$ be a feasible schedule that schedules the subset of jobs $U \subseteq J$, for each critical inversion pair i inv j' in $\mathcal{S}$, we define $J_{ij'} = \{j \in U \setminus \{i, j'\} \mid i \prec_E j \prec_E j'\}$.

Definition 4. *A feasible schedule is* canonical *iff the following constraints hold:*

1. *For each critical inversion pair i inv j', the set $J_{ij'}$ is scheduled in ERF order before j' and i as a block. That is, no other jobs are scheduled in between jobs of $J_{ij'} \cup \{i, j'\}$.*
2. *For two critical inversion pairs i_1 inv j_1 and i_2 inv j_2 with $(i_1, j_1) \neq (i_2, j_2)$ it holds that $\{i_1, j_1\} \cap \{i_2, j_2\} = \emptyset$ and $J_{i_1 j_1} \cap J_{i_2 j_2} = \emptyset$.*
3. *Let i_1 inv j_1 and i_2 inv j_2 be two* neighboring *critical inversion pairs, i.e., no critical inversions are scheduled between the two blocks $J_{i_1 j_1} \cup \{i_1, j_1\}$ and $J_{i_2 j_2} \cup \{i_2, j_2\}$. Then all jobs l with $j_1 \prec_E l \prec_E i_2$ are scheduled in ERF order between the two blocks. If i inv j is the first (resp. last) critical inversion pair in the schedule, then all jobs that are before i (resp. after j) in the ERF order are scheduled in ERF order before (resp. after) job block $J_{ij} \cup \{ij\}$.*

A canonical feasible schedule is defined entirely by the set of jobs that are scheduled and by the set of critical inversion pairs. Given the set of critical inversion pairs, Definition 4 defines a total order over all scheduled jobs, and scheduling all jobs according to that order gives us the canonical schedule. Our main result in this section is the following lemma:

Lemma 4. *Every feasible schedule can be transformed into a canonical feasible schedule that schedules the same set of jobs.*

Proof. By Lemma 3 we can restrict ourselves to schedules where for each inversion i inv j the job j', that is scheduled directly before i, forms a critical inversion pair with i. Further, by Lemma 2 there are no three jobs i, j and k with i inv j and j inv k. It remains to argue about the constraints of Definition 4.

Constraint 1. We first show that all jobs in $J_{ij'}$ are scheduled in ERF order before j'. Assume a $k \in J_{ij'}$ was scheduled after $C_i = S_i + p_i$. As j' is the critical inversion of i, we know that $d_{j'} < S_i + p_i$. Since $k \in J_{ij'}$ and $r_k \leq r_{j'}$, but k is scheduled after d'_j, more than δ time units pass between r_k and S_k. This contradicts the feasibility of $\mathcal{S}$, and so any $k \in J_{ij'}$ must be started before $S_i + p_i$. If two jobs $k, k' \in J_{ij'}$ are not scheduled in ERF order, then k inv k' follows, which contradicts Lemma 2.

It remains to show that no other jobs are scheduled in between jobs of $J_{ij'}$. Thus, we consider jobs k with either $j' \prec_E k$ or $k \prec_E i$. If $i \prec_E j' \prec_E k$, then the absence of nested inversions (Lemma 2) implies that k is scheduled after i.

For case $k \prec_E i$, we know that k must be scheduled before i since i and j' being a critical inversion implies that more than δ time units pass between r_k and $C_i = S_i + p_i$. We show that no job $l \in J_{ij'} \cup \{j'\}$ is scheduled before k. Assume that there exists such a l with $k \prec_E i \prec_E l$ but $S_l < S_k < S_i$. The pair l, k forms an inversion but the deadline d_l is (by feasibility of the schedule) only after the starting time of S_i. This property of the deadline is true for any job k' with $S_l \leq S_{k'} < S_k$, which follows from Lemma 2 and we can apply Lemma 1 to remove the inversion k inv l. Note that this transformation does not affect the fulfillment of the other criteria of the definition.

Constraint 2. Lemmas 2 and 3 directly imply $\{i_1, j_1\} \cap \{i_2, j_2\} = \emptyset$. Assume w.l.o.g. that i_1 and j_1 are scheduled before i_2 and j_2. Since $\{i_1, j_1\} \cap \{i_2, j_2\} = \emptyset$, the schedule is feasible, and both i_1 inv j_1 and i_2 inv j_2 are critical inversions, it follows $r_{j_1} < r_{i_2}$. Thus, there cannot be any $l \in J_{i_1 j_1} \cap J_{i_2 j_2}$.

Constraint 3. Let i_1 inv j_1 and i_2 inv j_2 be two neighboring critical inversion pairs. Each job l with $j_1 \prec_E l \prec_E i_2$ must be scheduled between the two blocks $J_{i_1 j_1} \cup \{i_1, j_1\}$ and $J_{i_2, j_2} \cup \{i_2, j_2\}$. Otherwise, it would form an inversion with either j_1 or i_2 and would therefore contradict Lemma 2. If two jobs scheduled between the two blocks are not in ERF order, they are inverted and Lemma 3 implies that there must also be a critical inversion. This contradicts i_1 inv j_1 and i_2 inv j_2 being neighbors. The case for jobs scheduled before the first (resp. after the last) critical inversion pair can be shown analogous.   □

3   Algorithmic Results

In this section, we first show that the structure of canonical schedules admits a dynamic program (DP) that optimally solves $1|r_j|\sum w_j(1-U_j)$ when all jobs have equal laxity with running time polynomial in n and $W = \sum_{j\in J} w_j$. Afterwards, we show how the DP can be turned into an FPTAS and finally derive a pseudo-polynomial time algorithm for the multi-machine setting with $m \in \mathcal{O}(1)$.

3.1   Optimal Algorithm for the Single Machine Case

The main result of this section is the following lemma, which implies a polynomial time algorithm for the unweighted case and thus Theorem 1.

Lemma 5. *The problem $1|r_j|\sum w_j(1-U_j)$ with equal laxity can be solved optimally with a running time polynomial in n and $W = \sum_{j\in J} w_j$.*

To prove the lemma, we introduce a DP that heavily exploits the structure of canonical schedules. An algorithm that iteratively constructs a canonical schedule starting at time $t = 0$ either schedules the first available job i in ERF order, discards job i, or decides that i forms a critical inversion with some job j that is released after i. If i is chosen to form a critical inversion pair with j, then the algorithm next schedules a subset of $\bar{J}_{ij} = \{l \in J \setminus \{i,j\} \mid i \prec_E l \prec_E j\}$ in ERF order followed by j and i. Whenever the algorithm schedules some job j, all jobs that are before j in the ERF order must have been already scheduled or discarded, except for (possibly) a single job i that inverts j.

The DP is represented by table $T[j,i]$ where j denotes the current job in the ERF order and i denotes a job that was selected to invert j. According to Definition 4, each job in a canonical schedule can only be inverted by a single other job. If no job was selected to invert j, then $i = 0$.

The cell $T[j,i]$ stores the set of all tuples (t,w) for which a feasible canonical (sub-) schedule of jobs $U \subseteq \{j' \in J \setminus \{i\} \mid j' \prec_E j\}$ exists with a makespan of t and $\sum_{j\in U} w_j = w$. In case of $i \neq 0$, i was selected to invert j and $i \prec_E j$ must hold. In those cases, the (sub-)schedule corresponding to $(t,w) \in T[j,i]$ must be of a structure that ensures the schedule to be canonical once the inversion job i is finally scheduled, i.e., prevent triple inversions.

The cell $T[n,0]$ then contains (t,w), where w is the maximum throughput for the instance. The corresponding schedule can be determined via backtracking.

To achieve the running time polynomial in n and W we, w.l.o.g., restrict the tuples that are stored by a cell $T[i,j]$ using a standard technique. We say that (t,w) is *dominated* by (t',w') if $t \geq t'$ and $w \leq w'$. We define the cells of our DP to only contain pairs that are not dominated by other contained pairs. After computing the values of a DP cell including dominated pairs, we define our algorithm to remove all dominated pairs. After removing all dominated pairs, the pairs $(t_1,w_1),\ldots,(t_l,w_l)$ of a DP cell can be ordered by $t_1 < t_2 < \ldots < t_l$ and $w_1 < w_2 < \ldots < w_l$. As start times, processing times and weights are integers, a DP cell can contain at most W elements.

We now recursively define the DP table T and start with cells $T[j,i]$ with $i = 0$. In order to define the cells, we use the following auxiliary function:

$$h(t,w,j) = \begin{cases} \{(\max\{t + p_j, r_j + p_j\}, w + w_j)\} & | \max\{t + p_j, r_j + p_j\} \le d_j \\ \emptyset & | \text{ otherwise} \end{cases}$$

In cells $T[j,0]$, no job is selected to invert j. In the first case, j is discarded or scheduled after all jobs that are in front of j in the ERF order have already been scheduled or discarded (first two parts of the $T[j,0]$ definition below). In the second case, j has already been scheduled or discarded but some job i was chosen to invert j and is now scheduled (last part of the definition below).

$$T[1,0] := \{(0,0)\} \cup h(0,0,1)$$

$$T[j,0] := T[j-1,0] \cup \bigcup_{(t,w)\in T[j-1,0]} h(t,w,j) \bigcup_{(t,w)\in T[j,i]:i\in J\wedge i\prec_E j} h(t,w,i)$$

We continue by defining cells $T[j,i]$ with $i > 0$, i.e., cells where a job i is selected to invert j and not scheduled yet. The following definition distinguishes between the case where the inversion job i is the current job in the ERF order ($j = i$), and the case where the current job j is placed after i in the ERF order ($j > i$).

$$T[i,i] := T[i-1,0]$$

$$T[j,i] := T[j-1,i] \cup \bigcup_{(t,w)\in T[j-1,i]} h(t,w,j) \qquad | j > i$$

In the first case, job i is the current job and selected as an inversion. By Definition 4, all jobs released before i must have been either scheduled or discarded and there cannot be a different ongoing inversion. In the latter case, job j is either scheduled or discarded after all jobs before j in the ERF order except i have already been scheduled or discarded and i was already selected as an inversion job. Since the DP only considers canonical schedules, j cannot be chosen to invert another job. Otherwise, there would be a contradiction to Definition 4.

As all previous cells have been pre-computed and only contain $\mathcal{O}(W)$ tuples, a new cell can be computed in time $\mathcal{O}(nW \cdot \log nW)$, leading to a total running time of $\mathcal{O}(n^3 W \cdot \log nW)$. Once $T[n,0]$ is computed, the schedule with the highest objective value can be constructed via backtracking. A straightforward induction on the DP definition implies the optimality of the algorithm and thus Lemma 5.

3.2   FPTAS for the Weighted Single Machine Case

While the DP of the previous section implies a polynomial-time algorithm for uniform weights, in general the running time is pseudo-polynomial in $W = \sum_{j\in J} w_j$. We turn the DP into an FPTAS by, for each $j \in J$, rounding the weights to $\hat{w}_j = \lfloor \frac{w_j}{\mu} \rfloor$ with $\mu = \epsilon w_{\max}/n$ and $w_{\max} = \max_{j\in J} w_j$ for an $\epsilon > 0$ and executing the DP on the rounded instance.

Theorem 2. *There is an FPTAS for* $1|r_j| \sum w_j(1 - U_j)$ *with equal additive laxity.*

Proof. Since the sum of rounded weights $\hat{W} = \sum_{j \in J} w_j \leq n^2 \cdot \frac{1}{\epsilon}$ is polynomial in n and $\frac{1}{\epsilon}$, the running time of the algorithm is polynomial in n and $\frac{1}{\epsilon}$ as well. Let I be an arbitrary instance, let $OPT(I)$ be the optimal solution value for I and let $A_\epsilon(I)$ be the solution value of the algorithm for an $\epsilon > 0$. We show the approximation factor by using a standard analysis where U and U^* are the sets of feasible jobs scheduled by A_ϵ resp. OPT:

$$A_\epsilon(I) = \sum_{j \in U} w_j \geq \mu \cdot \sum_{j \in U} \hat{w}_j \geq \mu \sum_{j \in U^*} \hat{w}_j \geq \sum_{j \in U^*} w_j - \epsilon w_{\max} \geq (1 - \epsilon) \cdot OPT(I).$$

□

3.3   Optimal Pseudo-polynomial Time Algorithm for $m \in \mathcal{O}(1)$

To extend the DP to the multiple-machine setting with $m \in \mathcal{O}(1)$, we can assume that the schedule of each individual machine is canonical since Lemma 4 can be applied to each individual machine. Similar to the single-machine case, we can index the DP cells according to the ERF order. In the single-machine case we, given an ERF index j, represent a state by the weight w that has already been scheduled, the makespan t of the corresponding subschedule and the index i of the unique job which is earlier in the ERF order than j but not yet scheduled.

We can extend the state representation to multiple machines and represent a state $\mathcal{S}$, corresponding to scheduling a subset of some prefix of the ERF order with a total weight of w, as an m-tuple of pairs (t, i). The total number of states is bounded by $O(T^m \cdot n^{m+1} \cdot W)$ with $T = \max_{j \in J} d_j$.

A possible dynamic programming algorithm updates for each job j in the ERF order the list of all feasible states after considering this job – either discarding j, placing j in the ERF order on one machine, or marking j as an inversion on a machine where no inversion is yet planned – and stores the total scheduled weight with each state.

Instead of storing the last completion time t for a machine M, an alternative DP can instead store the pair (k, σ_k), where k is the currently last scheduled job on M and $\sigma_k \in [0, \delta]$ is the relative start time of k, with $\sigma_k := r_k - S_k$. This observation is also present in [10]. Choosing the better of the two representations and using the rounding scheme of Sect. 3.2, we obtain the following result.

Theorem 3. *There is an optimal algorithm for* $Pm|r_j| \sum w_j(1 - U_j)$ *with equal additive laxity with its running time polynomial in* $\min\{\delta, \max_{j \in J} p_j\}$, $W = \sum_{j \in J} w_j$ *and the input size. The running time dependency on W can be replaced by a dependency on $1/\epsilon$, for any $\epsilon > 0$, at the cost of a $(1 + \epsilon)$-factor in the objective.*

4   Final Remarks

We answer open questions on a special case of throughput scheduling in which all jobs have the same laxity. It remains open whether there is an FPTAS for $Pm|r_j|\sum w_j(1 - U_j)$ with equal additive laxity for any constant number of machines $m \geq 2$. In Theorem 3 we rule out strong NP-hardness.

In the meal delivery context, Cosmi et al. [9] consider the possibility of aggregating orders into a single delivery. It would be interesting to incorporate this feature into the scheduling model and investigate whether our algorithmic results can be extended.

In general, it is a major open problem to find an approximation algorithm for weighted throughput scheduling with arbitrary laxity on a small number of machines improving upon the known approximation factor of $\frac{(m+1)^m}{(m+1)^m - (m+\epsilon)^m}$ [4].

Acknowledgements. We thank Ulrich Pferschy for bringing this problem to our attention. For further initial discussions we also thank Franziska Eberle, Ruben Hoeksma, Jannik Matuschke, Lukas Nölke and Bertrand Simon.

References

1. Auad, R., Erera, A., Savelsbergh, M.: Using simple integer programs to assess capacity requirements and demand management strategies in meal delivery. Preprint, Optimization Online (2020)
2. Baptiste, P.: Polynomial time algorithms for minimizing the weighted number of late jobs on a single machine with equal processing times. J. Sched. **2**(6), 245–252 (1999)
3. Bar-Noy, A., Guha, S., Naor, J., Schieber, B.: Approximating the throughput of multiple machines in real-time scheduling. SIAM J. Comput. **31**(2), 331–352 (2001)
4. Berman, P., DasGupta, B.: Improvements in throughout maximization for real-time scheduling. In: STOC, pp. 680–687. ACM (2000)
5. Chrobak, M., Dürr, C., Jawor, W., Kowalik, L., Kurowski, M.: A note on scheduling equal-length jobs to maximize throughput. J. Sched. **9**(1), 71–73 (2006)
6. Chuzhoy, J., Ostrovsky, R., Rabani, Y.: Approximation algorithms for the job interval selection problem and related scheduling problems. Math. Oper. Res. **31**(4), 730–738 (2006)
7. Cieliebak, M., Erlebach, T., Hennecke, F., Weber, B., Widmayer, P.: Scheduling with release times and deadlines on a minimum number of machines. In: Levy, J.-J., Mayr, E.W., Mitchell, J.C. (eds.) TCS 2004. IIFIP, vol. 155, pp. 209–222. Springer, Boston, MA (2004). https://doi.org/10.1007/1-4020-8141-3_18
8. Cosmi, M., Nicosia, G., Pacifici, A.: Lower bounds for a meal pickup-and-delivery scheduling problem. In: 17th Cologne-Twente Workshop on Graphs and Combinatorial Optimization (CTW), pp. 33–36 (2019)
9. Cosmi, M., Nicosia, G., Pacifici, A.: Scheduling for last-mile meal-delivery processes. IFAC-PapersOnLine **52**(13), 511–516 (2019)
10. Cosmi, M., Oriolo, G., Piccialli, V., Ventura, P.: Single courier single restaurant meal delivery (without routing). Oper. Res. Lett. **47**(6), 537–541 (2019)
11. Eberle, F., Hoeksma, R., Nölke, L., Simon, B.: Personal communication

12. Elffers, J., de Weerdt, M.: Scheduling with two non-unit task lengths is NP-complete. arXiv preprint arXiv:1412.3095 (2014)
13. Garey, M., Johnson, D.S., Simons, B.B., Tarjan, R.E.: Scheduling unit-time tasks with arbitrary release times and deadlines. SIAM J. Comput. **10**(2), 256–269 (1981)
14. Garey, M.R., Johnson, D.S.: Two-processor scheduling with start-times and deadlines. SIAM J. Comput. **6**(3), 416–426 (1977)
15. Garey, M.R., Johnson, D.S.: Computers and Intractability: A Guide to the Theory of NP-Completeness. W.H. Freeman and Company, New York (1979)
16. Hyatt-Denesik, D., Rahgoshay, M., Salavatipour, M.R.: Approximations for throughput maximization. CoRR abs/2001.10037 (2020)
17. Im, S., Li, S., Moseley, B.: Breaking $1 - 1/e$ barrier for non-preemptive throughput maximization. In: Eisenbrand, F., Koenemann, J. (eds.) IPCO 2017. LNCS, vol. 10328, pp. 292–304. Springer, Cham (2017). https://doi.org/10.1007/978-3-319-59250-3_24
18. Kise, H., Ibaraki, T., Mine, H.: A solvable case of the one-machine scheduling problem with ready and due times. Oper. Res. **26**(1), 121–126 (1978)
19. Ozbaygin, G., Karasan, O.E., Savelsbergh, M., Yaman, H.: A branch-and-price algorithm for the vehicle routing problem with roaming delivery locations. Transp. Res. Part B Methodol. **100**, 115–137 (2017)
20. Reyes, D., Erera, A., Savelsbergh, M., Sahasrabudhe, S., O'Neil, R.: The meal delivery routing problem. Optimization Online (2018)
21. Sgall, J.: Open problems in throughput scheduling. In: Epstein, L., Ferragina, P. (eds.) ESA 2012. LNCS, vol. 7501, pp. 2–11. Springer, Heidelberg (2012). https://doi.org/10.1007/978-3-642-33090-2_2
22. Spieksma, F.C.: On the approximability of an interval scheduling problem. J. Sched. **2**(5), 215–227 (1999)
23. Steever, Z., Karwan, M., Murray, C.: Dynamic courier routing for a food delivery service. Comput. Oper. Res. **107**, 173–188 (2019)
24. Ulmer, M.W., Thomas, B.W., Campbell, A.M., Woyak, N.: The restaurant meal delivery problem: Dynamic pickup and delivery with deadlines and random ready times. Transp. Sci. (2020)
25. van Bevern, R., Niedermeier, R., Suchý, O.: A parameterized complexity view on non-preemptively scheduling interval-constrained jobs: few machines, small looseness, and small slack. J. Sched. **20**(3), 255–265 (2016). https://doi.org/10.1007/s10951-016-0478-9
26. Yildiz, B., Savelsbergh, M.W.P.: Provably high-quality solutions for the meal delivery routing problem. Transp. Sci. **53**(5), 1372–1388 (2019)

Fragile Complexity of Adaptive Algorithms

Prosenjit Bose[1] , Pilar Cano[2(✉)] , Rolf Fagerberg[3] , John Iacono[2,4] ,
Riko Jacob[5] , and Stefan Langerman[2]

[1] School of Computer Science, Carleton University, Ottawa, Canada
`jit@scs.carleton.ca`
[2] Université libre de Bruxelles, Brussels, Belgium
{`pilar.cano,jiacono,stefan.langerman`}`@ulb.ac.be`
[3] University of Southern Denmark, Odense, Denmark
`rolf@imada.sdu.dk`
[4] New York University, New York, USA
[5] IT University of Copenhagen, Copenhagen, Denmark
`rikj@itu.dk`

Abstract. The fragile complexity of a comparison-based algorithm is
$f(n)$ if each input element participates in $O(f(n))$ comparisons. In this
paper, we explore the fragile complexity of algorithms adaptive to various restrictions on the input, i.e., algorithms with a fragile complexity
parameterized by a quantity other than the input size n. We show that
searching for the predecessor in a sorted array has fragile complexity
$\Theta(\log k)$, where k is the rank of the query element, both in a randomized and a deterministic setting. For predecessor searches, we also show
how to optimally reduce the amortized fragile complexity of the elements in the array. We also prove the following results: Selecting the kth
smallest element has expected fragile complexity $O(\log \log k)$ for the element selected. Deterministically finding the minimum element has fragile
complexity $\Theta(\log(\mathrm{Inv}))$ and $\Theta(\log(\mathrm{Runs}))$, where Inv is the number of
inversions in a sequence and Runs is the number of increasing runs in
a sequence. Deterministically finding the median has fragile complexity
$O(\log(\mathrm{Runs}) + \log \log n)$ and $\Theta(\log(\mathrm{Inv}))$. Deterministic sorting has fragile complexity $\Theta(\log(\mathrm{Inv}))$ but it has fragile complexity $\Theta(\log n)$ regardless of the number of runs.

Keywords: Algorithms · Comparison based algorithms · Fragile
complexity

1 Introduction

Comparison-based algorithms have been thoroughly studied in computer science.
This includes algorithms for problems such as MINIMUM, MEDIAN, SORTING,
SEARCHING, DICTIONARIES, PRIORITY QUEUES, and many others. The cost
measure analyzed is almost always the total number of comparisons performed
by the algorithm, either in the worst case or the expected case. Recently, another

© Springer Nature Switzerland AG 2021
T. Calamoneri and F. Corò (Eds.): CIAC 2021, LNCS 12701, pp. 144–157, 2021.
https://doi.org/10.1007/978-3-030-75242-2_10

type of cost measure has been introduced [1] which instead considers how many comparisons each individual element is subjected during the course of the algorithm. In [1], a comparison-based algorithm is defined to have *fragile complexity* $f(n)$ if each individual input element participates in at most $f(n)$ comparisons. The fragile complexity of a computational problem is the best possible fragile complexity of any comparison-based algorithm solving the problem.

This cost measure has both theoretical and practical motivations. On the theoretical side, it raises the question of to what extent the comparisons necessary to solve a given problem can be spread evenly across the input elements. On the practical side, this question is relevant in any real world situation where comparisons involve some amount of destructive impact on the elements being compared (hence the name of the cost measure). As argued in [1], one example of such a situation is ranking of any type of consumable objects (wine, beer, food, produce), where each comparison reduces the available amount of the objects compared. Here, an algorithm like QUICKSORT, which takes a single object and partitions the whole set with it, may use up this pivot element long before the algorithm completes. Another example is sports, where each comparison constitutes a match and takes a physical toll on the athletes involved. If a comparison scheme subjects one contestant to many more matches than others, both fairness to contestants and quality of result are impacted—finding a winner may not be very useful if this winner has a high risk of being injured in the process. The negative impact of comparisons may also be of non-physical nature, for instance when there is a privacy risk for the elements compared, or when bias grows if few elements are used extensively in comparisons.

1.1 Previous Work

In [1], the study of algorithms' fragile complexity was initiated and a number of upper and lower bounds on the fragile complexity for fundamental problems was given. The problems studied included MINIMUM, the SELECTION, SORTING, and HEAP CONSTRUCTION, and both deterministic and randomized settings were considered. In the deterministic setting, MINIMUM was shown to have fragile complexity $\Omega(\log n)$ and SORTING to have fragile complexity $O(\log n)$. Since SORTING can solve SELECTION, which can solve MINIMUM, the fragile complexity of all three problems is $\Theta(\log n)$. The authors then consider randomized algorithms, as well as a more fine-grained notion of fragile complexity, where the objective is to protect selected elements such as the minimum or median (i.e., the element to be returned by the algorithm), possibly at the expense of the remaining elements. Among other results, it is shown in [1] that MINIMUM can be solved incurring expected $O(1)$ comparisons on the minimum element itself, at a price of incurring expected $O(n^\varepsilon)$ on each of the rest. Also a more general trade-off between the two costs is shown, as well as a close to matching lower bound. For SELECTION, similar results are given, including an algorithm incurring expected $O(\log \log n)$ comparisons on the returned element itself, at a price of incurring expected $O(\sqrt{n})$ on each of the rest.

An earlier body of work relevant for the concept of fragile complexity is the study of sorting networks, started in 1968 by Batcher [5]. In sorting networks, and more generally comparator networks, the notion of depth (the number of layers, where each layer consists of non-overlapping comparators) and size (the total number of comparators) correspond to fragile complexity and standard worst case complexity, respectively, in the sense that a network with depth $f(n)$ and size $s(n)$ can be converted into a comparison-based algorithm with fragile complexity $f(n)$ and standard complexity $s(n)$ by simply simulating the network.

Batcher, as well as a number of later authors [10,17,18,21], gave sorting networks with $\mathcal{O}(\log^2 n)$ depth and $\mathcal{O}(n \log^2 n)$ size. For a long time it was an open question whether better results were possible. In 1983, Ajtai, Komlós, and Szemerédi [2,3] answered this in the affirmative by constructing a sorting network of $\mathcal{O}(\log n)$ depth and $\mathcal{O}(n \log n)$ size. This construction is quite complex and involves expander graphs [23,24]. It was later modified by others [9,13,19, 22], but finding a simple, optimal sorting network, in particular one not based on expander graphs, remains an open problem. Comparator networks for other problems, such as selection and heap construction have also been studied [4,7, 16,20,27].

While comparator networks are related to fragile complexity in the sense that results for comparator networks can be transferred to the fragile complexity setting by simple simulation, it is demonstrated in [1] that the two models are not equivalent: there are problems where one can construct fragile algorithms with the same fragile complexity, but with strictly lower standard complexity (i.e., total number of comparisons) than what is possible by simulation of comparison networks. These problems include SELECTION and HEAP CONSTRUCTION.

1.2   Our Contribution

In many settings, the classical worst case complexity of comparison-based algo-rithms can be lowered if additional information on the input is known. For instance, sorting becomes easier than $\Theta(n \log n)$ if the input is known to be close to sorted. Another example is searching in a sorted set of elements, which becomes easier than $O(\log n)$ if we know an element of rank close to the element searched for. Such algorithms may be described as *adaptive* to input restrictions (using the terminology from the sorting setting [11]). Given that the total num-ber of comparisons can be lowered in such situations, the question arises whether also reductions in the fragile complexity are possible under these types of input restrictions.

In this paper, we expand the study of the fragile complexity of comparison-based algorithms to consider the impact of a number of classic input restrictions. We show that searching for the predecessor in a sorted array has fragile com-plexity $\Theta(\log k)$, where k is the rank of the query element, both in a randomized and a deterministic setting. For predecessor searches, we also show how to opti-mally reduce the amortized fragile complexity of the elements in the array. We also prove the following results: Selecting the kth smallest element has expected fragile complexity $O(\log \log k)$ for the element selected. Deterministically finding the minimum element has fragile complexity $\Theta(\log(\mathrm{Inv}))$ and $\Theta(\log(\mathrm{Runs}))$, where Inv is the number of inversions in a sequence and Runs is the number of increasing runs in a sequence. Deterministically finding the median has fragile complexity $O(\log(\mathrm{Runs}) + \log \log n)$ and $\Theta(\log(\mathrm{Inv}))$. Deterministic sorting has fragile complexity $\Theta(\log(\mathrm{Inv}))$ but it has fragile complexity $\Theta(\log n)$ regardless of the number of runs.

2   Searching

The problem of predecessor searching is, given a sorted array A with n elements, $A[0]..A[n-1]$, answer queries of the form "What is the index of the largest element in A smaller than x?" Binary search is the classic solution to the predecessor search problem. It achieves $\log n$ fragile complexity for x, and fragile complexity at most one for each element of A. We can improve on this in two ways. The first is where we try to keep the fragile complexity of x small, which is possible if we know something about the rank of x. We show that the optimal dependency on the rank of x is $\Theta(\log k)$ where k is its rank, both for deterministic and randomized algorithms.[1] The second setting is where we are concerned with the fragile complexity of the other elements. While there is no way to improve a single search, classical deterministic binary search will always do the first comparison with the same element (typically the median). Hence we consider deterministic algorithms that improve the amortized fragile complexity of any element of the array A over a sequence of searches.

2.1   Single Search

Theorem 1. *Let A be a sorted array. Determining the predecessor of an element x within A has fragile complexity $\Theta(\log k)$ for deterministic and randomized algorithms, where k is the rank of x in A.*

Proof. The upper bound follows from standard exponential search [12]: We compare x to $A[2], A[4], A[8], \ldots$ until we find the smallest i such that $x < A[2^i]$. We perform a binary search with the initial interval $[2^{i-1}, 2^i]$. If x has the predecessor $A[k]$, this requires $O(\log k)$ comparisons.

   For the lower bound assume we have a deterministic algorithm to determine the rank of an element x. If the answer of the algorithm is k, let B_k be the bit-string resulting from concatenating the sequence of the outcomes of the comparisons performed by the algorithm, the i-th bit $B_k[i] = 0$ for $x < A[k]$, otherwise it is 1. Because the algorithm is deterministic and correct, all these bit-strings are different and they are a code for the numbers $1, \ldots, n$. Now, for any k, consider the uniform distribution on the numbers $0, \ldots, k-1$, a distribution with entropy $\log k$. By Shannon's source coding theorem, the average code length must be at least $\log k$, i.e., $\sum_{i=0}^{k-1} |B_i| \geq k \log k$.

---

[1] For simplicity of exposition, we assume the rank is close to one, but the result clearly holds for rank distance to other positions in A.

For a contradiction, assume there would be an algorithm with $|B_i| \leq \log i$ (the binary logarithm itself). Then for $k > 1$, $\sum_{i=0}^{k-1} |B_i| < k \log k$, in contrast to Shannon's theorem.

The bound $\sum_{i=0}^{k-1} |B_i| \geq k \log k$ also holds for randomized algorithms if the queries are drawn uniformly from $[1, \ldots, k]$, following Yao's principle: Any randomized algorithm can be understood as a collection of deterministic algorithms from which the 'real' algorithm is drawn according to some distribution. Now each deterministic algorithm has the lower bound, and the average number of comparisons of the randomized algorithm is a weighted average of these. Hence the lower bound also holds for randomized algorithms. □

2.2   Sequence of Searches

As mentioned, in binary search, the median element of the array will be compared with every query element. Our goal here is to develop a search strategy so as to ensure that data far away from the query will only infrequently be involved in a comparison. Data close to the query must be queried more frequently. While we prove this formally in Theorem 3, it is easy to see that predecessor and successor of a query must be involved in comparisons with the query in order to answer the query correctly.

Theorem 2. *There is a search algorithm that for any sequence of predecessor searches $x_1, x_2, \ldots, x_m$ in a sorted array A of size n the number of comparisons with any $y \in A$ is $O\left(\log n + \sum_{i=1}^{m} \frac{1}{d(x_i, y)}\right)$ where $d(x, y)$ is the number of elements between x and y in A, inclusive. The runtime is $O(\log n)$ per search and the structure uses $O(n)$ bits of additional space.*

Proof. We use the word interval to refer to a contiguous range of A; when we index an interval, we are indexing A relative to the start of the interval. Call an aligned interval I of A of rank i to be $(A[k \cdot 2^i] \ldots A[(k+1) \cdot 2^i])$ for some integer k, i.e., the aligned intervals of A are the dyadic intervals of A. There are $O(n)$ aligned intervals of A, and for each aligned interval I of rank i we store an offset $I.offset$ which is in the range $[0, 2^i)$, and it is initialized to 0.

The predecessor search algorithm with query x is a variant of recursive binary search, where at each step an interval I_q of A is under consideration, and the initial recursive call considers the whole array A. Each recursive call proceeds as follows: Find the largest i such that there are at least three rank-i aligned intervals in I_q, use I_m to denote the middle such interval (or an arbitrary non-extreme one if there are more than three), and we henceforth refer to this recursive call as a rank-i recursion. Compare $I_m[I_m.offset]$ with x, and then increment $I_m.offset$ modulo 2^i. Based on the result of the comparison, proceed recursively as in binary search. The intuition is by moving the offset with every comparison, this prevents a single element far from the search from being accessed too frequently. We note that the total space used by the offsets is $O(n)$ words, which can be reduced to $O(n)$ bits if the offsets are stored in a compact representation. First, several observations:

1. In a rank-i recursion, I_q has size at least $3 \cdot 2^i$ (since there must be at least three rank-i, size 2^i aligned intervals in I_q) and at most $8 \cdot 2^i$, the latter being true as if it was this size there would be three rank-$i+1$ intervals in I_q, which would contradict I_m having rank i.

2. If I_q has size k then if there is a recursive call, it is called with an interval of size at most $\frac{7}{8}k$. This is true by virtue of I_m being rank-i aligned with at least one rank-i aligned interval on either side of I_m in i. Since I_q has size at most $8 \cdot 2^i$, this guarantees an eighth of the elements of I_q will be removed from consideration as a result of the comparison in any recursive call.

3. From the previous two points, one can conclude that for a given rank i, during any search there are at most 7 recursions with of rank i. This is because after eight recursions any rank-i search will be reduced below the minimum for rank i: $8 \cdot 2^i \cdot \left(\frac{7}{8}\right)^8 < 3 \cdot 2^i$.

For the analysis, we fix an arbitrary element y in A and use the potential method to analyse the comparisons involving y. Let $\mathcal{I}_y = \{I_y^1, I_y^2 \ldots\}$ be the $O(\log n)$ aligned intervals that contain y, numbered such that I_y^i has rank i. Element y will be assigned a potential relative to each aligned interval $I_y^i \in \mathcal{I}_y$ which we will denote as $\varphi_y(I_y^i)$. Let $t_y(I_y^i)$ be number of times $I_y^i.offset$ needs to be incremented before $I_y^i[I_y^i.offset] = y$, which is in the range $[0, 2^i)$. The potential relative to I_y^i is then defined as $\varphi_y(I_y^i) := \frac{2^i - t_y(I_y^i)}{2^i}$, and the potential relative to y is defined to be the sum of the potentials relative to the intervals in $\mathcal{I}_y$: $\varphi_y := \sum_{I_y^i \in \mathcal{I}_y} \varphi_y(I_y^i)$.

How does $\varphi_y(I_y^i)$ change during a search? First, if there is no rank-i recursive call during the search to an interval containing y, it does not change as $I_y^i.offset$ is unchanged. Second, observe from point 3 that a search can increase $\varphi_y(I_y^i)$ by only $\frac{7}{2^i}$. Furthermore if y was involved in a comparison during a rank-i recursion, there will be a loss of $1 - \frac{1}{2^i}$ units of potential in $\varphi_y(I_y^i)$ as the offset of I_y^i changes from 0 to $2^i - 1$.

Following standard potential-based amortized analysis, the amortized number of comparisons involving y during a search is the actual number of comparisons (zero or one) plus the change in the potential φ_y. Let $i_{\min}$ be the smallest value of i for which there was a rank-i recursion that included y. As the maximum gain telescopes, the potential gain is at most $\frac{14}{2^{i_{\min}}}$, minus 1 if y was involved in a comparison. Thus the amortized number of comparisons with y in the search is at most $\frac{14}{2^{i_{\min}}}$.

Observe that if there was a rank-i recursion that included y, that $d(x, y)$ is at most $8 \cdot 2^i$ by point 1. This gives $d(x, y) \leq 8 \cdot 2^i \leq 8 \cdot 2^{i_{\min}}$. Thus the amortized cost can be restated as being at most $\frac{14}{2^{i_{\min}}} \leq \frac{112}{d(x,y)}$.

To complete the proof, the total number of comparisons involving y over a sequence of searches is the sum of the amortized costs plus any potential loss. As the potential φ_y is always nonnegative and at most $\lceil \log n \rceil$ (1 for each $\varphi_y(I_y^i)$), this gives the total cost as $O\left(\log n + \sum_{i=1}^{m} \frac{1}{d(x_i,y)}\right)$. □

Note that the above proof was designed for easy presentation and not an optimal constant. Also note that this theorem implies that if the sequence of searches is uniformly random, the expected fragility of all elements is $O(\frac{\log n}{n})$, which is asymptotically the best possible since random searches require $\Omega(\log n)$ comparisons in expectation.

2.3   Lower Bounds

It is well-known that comparison-based searching requires $\Omega(\log n)$ comparisons per search. In our method, taking a single search x_i summing over the upper bound on amortized cost of the number of comparisons with y, $\frac{42}{d(x_i,y)}$, for all y yields a harmonic series which sums to $O(\log n)$. But we can prove something stronger:

Theorem 3. *There is a constant c such that if a predecessor search algorithm has an amortized number of comparisons of $f(d(x_i, y))$ for an arbitrary y for every sequence of predecessor searches $x_1, x_2, \ldots x_m$, then $\sum_{k=1}^{p} f(k) \geq c \log p$ for all $p \leq n$.*

Proof. This can be seen by looking at a random sequence of predecessor searches for which the answers are uniform among $A[0] \ldots A[p-1]$, if the theorem was false, similarly to the proof of Theorem 1, this would imply the ability to execute such a sequence in $o(\log p)$ amortized time per operation.   □

This shows that a flatter asymptotic tradeoff between $d(x_i, y)$ and the amortized comparison cost is impossible; more comparisons are needed in the vicinity of the search than farther away. For example, a flat amortized number of comparisons of $\frac{\log n}{n}$ for all elements would sum up to $O(\log n)$ amortized comparisons over all elements, but yet would violate this theorem.

2.4   Extensions

Here we discuss extensions to the search method above. We omit the proofs as they are simply more tedious variants of the above.

One can save the additional space used by the offsets of the intervals through the use of randomization. The offsets force each item in the interval to take its turn as the one to be compared with, instead one can pick an item at random from the interval. This can be further simplified into a binary search where at each step one simply picks a random element for the comparison amongst those (in the middle half) of the part of the array under consideration.

To allow for insertions and deletions, two approaches are possible. The first is to keep the same array-centric view and simply use the packed-memory array [15,25,26] to maintain the items in sorted order in the array. This will give rise to a cost of $O(\log^2 n)$ time which is inherent in maintaining a dynamic collection of items ordered in an array [8] (but no additional fragility beyond searching for the item to insert or delete as these are structural changes). The second approach would be to use a balanced search tree such as a red-black tree [14]. This will reduce the insertion/deletion cost to $O(\log n)$ but will cause the search cost to increase to $O(\log^2 n)$ as it will take $O(\log n)$ time to move to the item in each interval indicated by the offset, or to randomly choose an item in an interval. The intervals themselves would need to allow insertions and deletions, and would, in effect be defined by the subtrees of the red-back tree. It remains open whether there is a dynamic structure with the fragility results of Theorem 2 where insertions and deletions can be done in $O(\log n)$ time.

3 Selection

In this section we consider the problem of finding the k-th smallest element of an unsorted array. There is a randomized algorithm that selects the k-th smallest element with expected fragile complexity of $O(\log \log n)$ for the selected element [1]. We consider the question if this complexity can be improved for small k. In this section we define a sampling method that, combined with the algorithm given in [1], selects the k-th smallest element with expected $O(\log \log k)$ comparisons.

Next, we define the filtering method RESET in a tail-recursive fashion.

1: **procedure** RESET(X, k) ▷ Returns a small subset $\mathcal{C}$ of X that contains the k-th element
2:    Let $n = |X|$ and $\mathcal{C} = \emptyset$
3:    **if** $k \geq \frac{n}{2} - 1$ ▷ The set has size $O(k+1)$
4:       Let $A' = X$
5:    **else**   ▷ Recursively construct a sample of expected size $O(k+1)$
6:       Sample A uniformly at random from X, $|A| = \frac{n}{2}$
7:       Let $A' = $ RESET(A, k)
8:    Choose the $(k+1)$-th smallest element z from A' (by standard linear time selection)
9:    Let $\mathcal{C} = \{x \in X : x \leq z\}$
10:   **return** $\mathcal{C}$

Theorem 4. *Randomized selection is possible in expected fragile complexity $O(\log \log k)$ in the selected element.*

Proof. Let us show that the following procedure for selecting the k-th element in a set X with $|X| = n$, gives an expected fragile complexity $O(\log \log k)$ in the k-th element:

If $k > n^{\frac{1}{100}}$, then let $S' = X$. If $k \leq n^{\frac{1}{100}}$, then sample uniformly at random S from X, where $|S| = \frac{n}{k}$. Let $\mathcal{C} = $ RESET(S, k) and select the $k+1$-th smallest element z from $\mathcal{C}$ by standard linear time selection. Let $S' = \{x \in X : x \leq z\}$. Finally, apply to S' the randomized selection algorithm of [1].

Let x_k denote the k-th smallest element in X and let f_k denote the fragile complexity of x_k. Note that if $x_k \in S$, then, before constructing S', f_k is given by the fragile complexity of x_k in RESET(S, k) plus $O(|\mathcal{C}|)$ when finding the $(k+1)$-th smallest element in $\mathcal{C}$. Otherwise, x_k is not compared until S' is constructed. On the other hand, recall that the expected

f_k in the algorithm in [1] is $O(\log\log m)$ where m is the size of the input set. Hence, the expected f_k after selecting the $k+1$-th element in $\mathcal{C}$ is 1 when creating S' plus the expected f_k in the randomized selection algorithm in [1] that is $\sum_{|S'|} O(\log\log|S'|)\mathbb{P}[|S'|] = \mathbb{E}[O(\log\log|S'|)]$. Thus, $\mathbb{E}[f_k] = (\mathbb{E}[f_k \text{ in RESET}|x_k \in S] + \mathbb{E}[|\mathcal{C}|])\mathbb{P}[x_k \in S] + 0\mathbb{P}[x_k \notin S] + 1 + \mathbb{E}[O(\log\log|S'|)]$. Since the logarithm is a concave function, $\mathbb{E}[O(\log\log|S'|)] \le O(\log\log(\mathbb{E}[|S'|]))$. Therefore, if we prove that: (i) the expected fragile complexity of x_k before creating S' is $O(1)$ and (ii) $\mathbb{E}[|S'|] = c'k^c$ for some constants c and c'. Then, we obtain that $\mathbb{E}[f_k] \le O(1) + 1 + O(c\log\log k + \log c') = O(\log\log k)$, as desired. In order to prove (i) and (ii) we consider 2 cases:
(1) $k > n^{\frac{1}{100}}$, (2) $k \le n^{\frac{1}{100}}$.
Case 1) $S' = X$ and it makes no previous comparisons in any element, proving (i). In addition, S' has size less than k^{100}. Thus, (ii) holds.
Case 2) S is a sample of X with size $\frac{n}{k}$ and $S' = \text{RESET}(S, k)$.
First, let us show (i). If $x_k \notin S$, then there are no previous comparisons. Hence, the expected fragile complexity of x_k before constructing S' is given by $(\mathbb{E}[f_k \text{ in RESET}|x_k \in S'] + \mathbb{E}[|\mathcal{C}|])\mathbb{P}[x_k \in S] + 0$. Since S is an uniform random sample with size $\frac{n}{k}$, $\mathbb{P}[x_k \in S] = \frac{1}{k}$, it suffices to show that $\mathbb{E}[f_k \text{ in RESET}|x_k \in S'] + \mathbb{E}[|\mathcal{C}|] = O(k)$, which gives an expectation of $O(k)\frac{1}{k} = O(1)$, proving (i). So, let us show that $\mathbb{E}[f_k \text{ in RESET}|x_k \in S'] + \mathbb{E}[|\mathcal{C}|] = O(k)$. Let $A_0 = S$ and let A_1 be the sample of A_0 when passing through line 6 in RESET. Similarly, denote by A_i to the sample of A_{i-1} in the i-th recursive call of RESET and let $A'_i = \text{RESET}(A_i, k)$. Note that by definition $A'_0 = \mathcal{C}$. Let $\ell + 1$ be the number of recursive calls in $\text{RESET}(S, k)$.

Since A_i is a uniform random sample of size $\frac{|A_{i-1}|}{2}$ for all $i \ge 1$, $\mathbb{P}[x \in A_i | x \in A_{i-1}] = 2^{-1}$ and $\mathbb{P}[x \in A_i | x \notin A_{i-1}] = 0$. Hence, $\mathbb{P}[x_k \in A_i] = \mathbb{P}[x_k \in \cap_{i=0}^{i} A_i] = 2^{-i}$. Note that the number of comparisons of x_k in RESET is given by the number of times x_k is compared in lines 8 and 9. Thus, for each i-th recursive call: if $x_k \in A_i$, then x_k is compared once in line 9; and if $x_k \in A_i \cap A'_i$, then x_k is compared at most $|A'_i|$ times in line 8. Otherwise, x_k is not compared in that and the next iterations. Thus, $\mathbb{E}[f_k \text{ in RESET}|x_k \in S'] + \mathbb{E}[|\mathcal{C}|] \le \sum_{i=0}^{\ell}(1 + \mathbb{E}[|A'_i|])\mathbb{P}[x_k \in A_i] = \sum_{i=0}^{\ell} 2^{-i}(1 + \mathbb{E}[|A'_i|]) \le 2(1 + \mathbb{E}[|A'_i|])$. Let us compute $\mathbb{E}[|A'_i|]$. Since the $(\ell+1)$-th iteration $\text{RESET}(A_\ell, k)$ passes through the if in line 3, there is no new sample from A_ℓ. Thus, A'_ℓ is given by the $k+1$ smallest elements of A_ℓ. Therefore, $\mathbb{E}[|A'_\ell|] = k+1$ Denote by a'^i_j to the j-th smallest element of A'_i. For the case of $0 \le i < \ell$, we have $A'_i = \{x \in A_{i+1} : x \le a'^{i+1}_{k+1}\}$. Hence, $\mathbb{E}[|A'_i|] = \mathbb{E}[|\{x \in A_{i+1} : x \le a'^{i+1}_{k+1}\}|] = \mathbb{E}[|\{x \in A_{i+1} : x \le a'^{i+1}_1\}|] + \sum_{j=1}^{k} \mathbb{E}[|\{x \in A_{i+1} : a'^{i+1}_{j-1} < x \le a'^{i+1}_j\}|] \le \sum_{j=1}^{k+1} \sum_{t=1}^{\infty} t2^{-1}(2^{-(t-1)}) = 2(k+1)$. Therefore, $\mathbb{E}[f_k \text{ in RESET}|x_k \in S'] + \mathbb{E}[|\mathcal{C}|] = \sum_{i=0}^{\ell} 2^{-i}(1 + \mathbb{E}[|A'_i|]) \le 2 + 2\mathbb{E}[|A'_i|] = O(k)$ proving (i). Finally, let us show (ii): For simplicity, let c_j denote the j-th smallest element of $\mathcal{C}$. Then, $\mathbb{E}[|S'|] = \mathbb{E}[|\{x \in X : x \le c_1\}|] + \sum_{j=1}^{k} \mathbb{E}[|\{x \in X : c_j \le x \le c_{j+1}\}|] \le \sum_{j=1}^{k+1} \sum_{j=0}^{\infty} jk^{-1}(1 - k^{-1})^{j-1} = k(k+1) = O(k^2)$, proving (ii). $\square$

4   Sorting

When the input is known to have some amount of existing order, sorting can be done faster than $\Theta(n \log n)$. Quantifying the amount of existing order is traditionally done using measures of disorder [11], of which Inv and Runs are two classic examples.[2] A sorting algorithm is adaptive to a measure of disorder if it is faster for inputs with a smaller value of the measure. For the above measures, run times of $O(n \log(\text{Inv}/n))$ and $O(n \log(\text{Runs}))$ can be achieved. These results are best possible for comparison-based sorting, by standard information-theoretic arguments based on the number of different inputs having a given maximal value of the measure.

The fact [1,3] that we can sort all inputs in $\Theta(n \log n)$ time and $\Theta(\log n)$ fragile complexity can be viewed as being able to distribute the necessary comparisons evenly among the elements such that each element takes part in at most $\Theta(\log n)$ comparisons. Given the running times for adaptive sorting stated above, it is natural to ask if for an input with a given value of Inv or Runs we are able to sort in a way that distributes the necessary comparisons evenly among the elements, i.e., in a way such that each element takes part in at most $O(\log(\text{Inv}))$ or $O(\log(\text{Runs}))$ comparisons, respectively. In short, can we sort in fragile complexity $O(\log(\text{Inv}))$ and $O(\log(\text{Runs}))$? Or more generally, what problems can we solve with fragile complexity adaptive to Inv and Runs? In this section, we study the fragile complexity of deterministic algorithms for MINIMUM, MEDIAN, and SORTING and essentially resolve their adaptivity to Inv and Runs.

Due to space limitations, some proofs are deferred to the full paper [6].

Theorem 5. MINIMUM *has fragile complexity* $\Theta(\log(\text{Runs}))$.

Proof. For the upper bound: identify the runs in $O(1)$ fragile complexity by a scan of the input. Then, use a tournament on the heads of the runs since the minimum is the minimum of the heads of the runs. For the lower bound: apply the logarithmic lower bound for MINIMUM [1] on the heads of the runs.   □

Theorem 6. SORTING *has fragile complexity* $\Theta(\log n)$, *no matter what value of* Runs *is assumed for the input.*

Proof. The upper bound follows from general sorting. For the lower bound: the input consisting of a run R of length $n-1$ and one more element x has Runs $= 2$, but $\log n$ comparisons on x can be forced by an adversary before the position of x in R is determined.   □

Theorem 7. MEDIAN *has fragile complexity* $O(\log(\text{Runs}) + \log \log n)$.

---

[2] The measure Inv is defined as the total number of inversions in the input, where each of the $\binom{n}{2}$ pairs of elements constitute an inversion if the elements of the pair appear in the wrong order. The measure Runs is defined as the number of runs in the input, where a run is a maximal consecutive ascending subsequence.

Proof. Assume that $4 \cdot \text{Runs} \cdot \log n < n/2$, since otherwise the claimed fragile complexity is $O(\log n)$ for which we already have a median algorithm [1]. Consider the rank space $[1, n]$ (i.e., the indices of the input elements in the total sorted order) of the input elements and consider the rank interval $[a, b]$ around the median defined by $a = n/2 - 4 \cdot \text{Runs} \cdot \log n$ and $b = n/2 + 4 \cdot \text{Runs} \cdot \log n$. In each step of the algorithm, elements are removed in two ways: type A removals and type B removals. A removal of type A is a balanced removal, where a number of elements with ranks in $[1, a-1]$ are removed and the same number of elements with ranks in $[b+1, n]$ are removed. The key behind the type A removal is that the median element of the set prior to the removal is the same as the median of the set after the removal, if the median prior to the removal has a rank in $[a, b]$.

A removal of type B is a removal of elements with arbitrary rank. However, the total number of elements removed by type B removals is at most $7 \cdot \text{Runs} \cdot \log n$ during the entire run of the algorithm. Hence, repeated use of type A and type B removals will maintain the invariant that the median of the remaining elements has a rank in $[a, b]$.

We now outline the details of the algorithm. The first step is to identify all the runs in $O(1)$ fragile complexity by a scan. A run will be considered *short* if the run consists of fewer than $7 \cdot \log n$ elements and it will be considered *long* otherwise. A step of the algorithm proceeds by first performing a type B removal followed by a type A removal. A type B removal consists of removing all short runs that are present. The short runs that are removed will be reconsidered again at the end once the number of elements under consideration by the algorithm is less than $64 \cdot \text{Runs} \cdot \log n$.

Once a type B removal step is completed, only long runs remain under consideration. We now describe a type A removal step. Note that a long run may become short after a type A removal step, in which case it will be removed as part of the next type B removal step. Each run can become short (and be removed by a type B removal) only once, hence the total number of elements removed by type B removals will be at most $7 \cdot \text{Runs} \cdot \log n$, as claimed.

In the following, let $\mathcal{N}$ denote the elements under consideration just before a type A removal (i.e., the elements of the remaining long runs), and let $N = |\mathcal{N}|$. The algorithm stops when $N \leq 64 \cdot \text{Runs} \cdot \log n$.

To execute the type A removal step, the algorithm divides each long run R into blocks of length $\log n$. The blocks of a run are partitioned by a *partitioning block*. The partitioning block has the property that there are at least $|R|/7$ elements of R whose values are less than the values in the partitioning block and at least $5|R|/7$ elements of R whose value are greater than the elements in the partitioning block. One element x_R is selected from the partitioning block. We will refer to this element as a *partitioning* element. These partitioning elements are then sorted into increasing order, which incurs a cost of $O(\log(\text{Runs}))$ fragile complexity on each of the partitioning elements. The runs are then arranged in the same order as their partitioning elements. Label this sequence of runs as $R_1, R_2, \ldots, R_k$, and let t be the largest index such that $\sum_{i=1}^{t-1} |R_i| < N/8$.

Since the partitioning element x_{R_t} is smaller than all the elements in the blocks with values greater than their respective partitioning blocks in R_t, R_{t+1}, $\ldots, R_k$, we have that x_{R_t} is smaller than $(7/8)(5N/7) = 5N/8$ of the remaining elements. Hence in rank it is at least $N/8$ below the median of the remaining elements. By the invariant on the position in rank space of this median and the fact that $N > 64 \cdot \text{Runs} \cdot \log n$, we note that x_{R_t} has a rank below a. We also note that all the elements below the partitioning blocks in $R_1, R_2, \ldots, R_t$ have value less than x_{R_t}. This constitutes at least $(1/8)(N/7) = N/56$ elements in $\mathcal{N}$ with rank below a. Therefore, we can remove $N/56$ elements with rank below a. In a similar manner, we can find at least $N/56$ elements in $\mathcal{N}$ with rank above b. Removal of these $2N/56 = N/28$ elements in $\mathcal{N}$ constitutes a type A removal step.

Since the number of elements under consideration, i.e. N, decreases by a constant factor at each step, the algorithm performs $O(\log n)$ type A and type B removal steps before we have $N \leq 64 \cdot \text{Runs} \cdot \log n$. Since each block under consideration in a type A removal step has size $\log n$, we can guarantee that each element in a partitioning block only needs to be selected as a partitioning element $O(1)$ times. This implies that a total cost of $O(\log(\text{Runs}))$ fragile complexity is incurred on each element once we have that $N \leq 64 \cdot \text{Runs} \cdot \log n$.

We now describe the final step of the algorithm. At this point, the algorithm combines the last $\mathcal{N}$ elements with all the short runs removed during its execution up to this point, forming the set $\mathcal{S}$. This set is the original elements subjected to a series of type A removals, each of which are balanced and outside the rank interval $[a, b]$. Hence, the median of $\mathcal{S}$ is the global median. As $|\mathcal{S}| = O(\text{Runs} \cdot \log n)$, we can find this median in $O(\log(\text{Runs} \cdot \log n)) = O(\log(\text{Runs}) + \log \log n)$ fragile complexity [1], which dominates the total fragile complexity of the algorithm.

We note that for $\text{Runs} = 2$, we can improve the above result to $O(1)$ fragile complexity as follows. Let the two runs be R_1 and R_2, with $|R_1| \leq |R_2|$. Compare their middle elements x and y and assume $x \leq y$. Then the elements in the first half of R_1 are below $n/2$ other elements, and hence are below the median. Similarly, the elements in the last half of R_2 are above the median. Hence, we can remove $|R_1|/2$ elements on each side of the median by removing that many elements from one end of each run. The median of the remaining elements is equal to the global median. By recursion, we in $\log|R_1|$ steps end up with R_1 reduced to constant length. Then $O(1)$ comparisons with the center area of R_2 will find the median. Because both runs lose elements in each recursive step, both x and y will be new elements each time. The total fragile complexity of the algorithm is therefore $O(1)$.   □

Theorem 8. MINIMUM *has fragile complexity* $\Theta(\log(\text{Inv}))$.

Theorem 9. MEDIAN *has fragile complexity* $\Theta(\log(\text{Inv}))$.

Proof. As MEDIAN solves MINIMUM via padding with n elements of value $-\infty$, the lower bound follows from the lower bound on MINIMUM. For the upper bound, find R and I as in the upper bound for MINIMUM, sort I in fragile complexity $O(\log(\text{Inv}))$ and use the algorithm for MEDIAN for Runs = 2.   □

Theorem 10. SORTING *has fragile complexity* $\Theta(\log(\text{Inv}))$.

Acknowledgements. This material is based upon work performed while attending AlgoPARC Workshop on Parallel Algorithms and Data Structures at the University of Hawaii at Manoa, in part supported by the National Science Foundation under Grant No. CCF-1930579. We thank Timothy Chan and Qizheng He for their ideas improving the randomized selection algorithm.

P.B was partially supported by NSERC. P.C and J.I. were supported by F.R.S.-FNRS under Grant no MISU F 6001 1. R.F. was partially supported by the Independent Research Fund Denmark, Natural Sciences, grant DFF-7014-00041. J.I. was supported by NSF grant CCF-1533564. S.L. is Directeur de Recherches du F.R.S.-FNRS.

References

1. Afshani, P., et al.: Fragile complexity of comparison-based algorithms. In: Bender, M.A., Svensson, O., Herman, G. (eds.) 27th Annual European Symposium on Algorithms, ESA 2019, 9–11 September 2019, Munich/Garching, Germany. LIPIcs, vol. 144, pp. 2:1–2:19. Schloss Dagstuhl - Leibniz-Zentrum für Informatik (2019)
2. Ajtai, M., Komlós, J., Szemerédi, E.: An $O(n \log n)$ sorting network. In: Proceedings of the 15th Symposium on Theory of Computation, STOC 1983, pp. 1–9. ACM (1983)
3. Ajtai, M., Komlós, J., Szemerédi, E.: Sorting in $c \log n$ parallel steps. Combinatorica **3**(1), 1–19 (1983)
4. Alekseev, V.E.: Sorting algorithms with minimum memory. Kibernetika **5**(5), 99–103 (1969)
5. Batcher, K.E.: Sorting networks and their applications. In: Proceedings of AFIPS Spring Joint Computer Conference, pp. 307–314 (1968)
6. Bose, P., Cano, P., Fagerberg, R., Iacono, J., Jacob, R., Langerman, S.: Fragile complexity of adaptive algorithms (2021). To appear in arXiv
7. Brodal, G.S., Pinotti, M.C.: Comparator networks for binary heap construction. In: Arnborg, S., Ivansson, L. (eds.) SWAT 1998. LNCS, vol. 1432, pp. 158–168. Springer, Heidelberg (1998). https://doi.org/10.1007/BFb0054364
8. Bulánek, J., Koucký, M., Saks, M.E.: Tight lower bounds for the online labeling problem. SIAM J. Comput. **44**(6), 1765–1797 (2015)
9. Chvátal, V.: Lecture notes on the new AKS sorting network. Technical report DCS-TR-294, Department of Computer Science, Rutgers University, New Brunswick, NJ, October 1992
10. Dowd, M., Perl, Y., Rudolph, L., Saks, M.: The periodic balanced sorting network. J. ACM **36**(4), 738–757 (1989)
11. Estivill-Castro, V., Wood, D.: A survey of adaptive sorting algorithms. ACM Comput. Surv. **24**(4), 441–476 (1992)
12. Fredman, M.L.: Two applications of a probabilistic search technique: sorting x + y and building balanced search trees. In: Rounds, W.C., Martin, N., Carlyle, J.W., Harrison, M.A. (eds.) Proceedings of the 7th Annual ACM Symposium on Theory of Computing, Albuquerque, New Mexico, USA, 5–7 May 1975, pp. 240–244. ACM (1975)

13. Goodrich, M.T.: Zig-zag sort: a simple deterministic data-oblivious sorting algorithm running in $O(n \log n)$ time. In: Shmoys, D.B. (ed.) STOC 2014, pp. 684–693. ACM (2014)

14. Guibas, L.J., Sedgewick, R.: A dichromatic framework for balanced trees. In: 19th Annual Symposium on Foundations of Computer Science, Ann Arbor, Michigan, USA, 16–18 October 1978, pp. 8–21. IEEE Computer Society (1978)

15. Itai, A., Konheim, A.G., Rodeh, M.: A sparse table implementation of priority queues. In: Even, S., Kariv, O. (eds.) ICALP 1981. LNCS, vol. 115, pp. 417–431. Springer, Heidelberg (1981). https://doi.org/10.1007/3-540-10843-2_34

16. Jimbo, S., Maruoka, A.: A method of constructing selection networks with $O(\log n)$ depth. SIAM J. Comput. **25**(4), 709–739 (1996)

17. Parberry, I.: The pairwise sorting network. Parallel Process. Lett. **2**(2–3), 205–211 (1992)

18. Parker, B., Parberry, I.: Constructing sorting networks from k-sorters. Inf. Process. Lett. **33**(3), 157–162 (1989)

19. Paterson, M.S.: Improved sorting networks with $O(\log N)$ depth. Algorithmica **5**(1), 75–92 (1990)

20. Pippenger, N.: Selection networks. SIAM J. Comput. **20**(5), 878–887 (1991)

21. Pratt, V.R.: Shellsort and Sorting Networks. Outstanding Dissertations in the Computer Sciences, Garland Publishing, New York (1972)

22. Seiferas, J.I.: Sorting networks of logarithmic depth, further simplified. Algorithmica **53**(3), 374–384 (2009)

23. Hoory, S., Linial, N., Wigderson, A.: Expander graphs and their applications. BAMS Bull. Am. Math. Soc. **43**, 439–561 (2006)

24. Vadhan, S.P.: Pseudorandomness. Found. Trends Theoret. Comput. Sci. **7**(1–3), 1–336 (2012)

25. Willard, D.E.: Good worst-case algorithms for inserting and deleting records in dense sequential files. In: Zaniolo, C. (ed.) Proceedings of the 1986 ACM SIGMOD International Conference on Management of Data, Washington, DC, USA, 28–30 May 1986, pp. 251–260. ACM Press (1986)

26. Willard, D.E.: A density control algorithm for doing insertions and deletions in a sequentially ordered file in good worst-case time. Inf. Comput. **97**(2), 150–204 (1992)

27. Yao, A., Yao, F.F.: Lower bounds on merging networks. J. ACM **23**(3), 566–571 (1976)

FPT and Kernelization Algorithms for the Induced Tree Problem

Guilherme Castro Mendes Gomes[1]([✉]) [iD], Vinicius F. dos Santos[1] [iD],
Murilo V. G. da Silva[2] [iD], and Jayme L. Szwarcfiter[3,4] [iD]

[1] Departamento de Ciência da Computação, Universidade Federal de Minas Gerais,
Belo Horizonte, Brazil
{gcm.gomes,viniciussantos}@dcc.ufmg.br
[2] Departamento de Informática, Universidade Federal do Paraná, Curitiba, Brazil
murilo@inf.ufpr.br
[3] Universidade Federal do Rio de Janeiro, Rio de Janeiro, Brazil
[4] Universidade do Estado do Rio de Janeiro, Rio de Janeiro, Brazil
jayme@cos.ufrj.br

Abstract. The THREE-IN-A-TREE problem asks for an induced tree of
the input graph containing three mandatory vertices. In 2006, Chudnovsky and Seymour [Combinatorica, 2010] presented the first polynomial time algorithm for this problem, which has become a critical subroutine in many algorithms for detecting induced subgraphs, such as
beetles, pyramids, thetas, and even and odd-holes. In 2007, Derhy and
Picouleau [Discrete Applied Mathematics, 2009] considered the natural
generalization to k mandatory vertices with k being part of the input,
and showed that it is NP-complete; they named this problem INDUCED
TREE, and asked what is the complexity of FOUR-IN-A-TREE. Motivated
by this question and the relevance of the original problem, we study
the parameterized complexity of INDUCED TREE. We begin by showing
that the problem is W[1]-hard when jointly parameterized by the size
of the solution and minimum clique cover and, under the Exponential
Time Hypothesis, does not admit an $n^{o(k)}$ time algorithm. Afterwards,
we use Courcelle's Theorem to prove tractability under cliquewidth,
which prompts our investigation into which parameterizations admit single exponential algorithms; we show that such algorithms exist for the
unrelated parameterizations treewidth, distance to cluster, and vertex
deletion distance to co-cluster. In terms of kernelization, we present a linear kernel under feedback edge set, and show that no polynomial kernel
exists under vertex cover nor distance to clique unless NP ⊆ coNP/poly.
Along with other remarks and previous work, our tractability and kernelization results cover many of the most commonly employed parameters
in the graph parameter hierarchy.

Keywords: induced tree · parameterized complexity · FPT
algorithm · polynomial kernel · cross-composition

Work partially supported by CAPES, CNPq, and FAPEMIG. Full version permanently
available at https://arxiv.org/abs/2007.04468.

T. Calamoneri and F. Corò (Eds.): CIAC 2021, LNCS 12701, pp. 158–172, 2021.
https://doi.org/10.1007/978-3-030-75242-2_11

1   Introduction

Given a graph $G = (V, E)$ and a subset $K \subseteq V(G)$ of size three – here called the set of terminal vertices – the THREE-IN-A-TREE problem consists of finding an induced subtree of G that connects K. Despite the novelty of this problem, it has become an important tool in many detection algorithms, where it usually accounts for a significant part of the work performed during their executions. It was first studied by Chudnovsky and Seymour [13] in the context of theta and pyramid detection, the latter of which is a crucial part of perfect graph recognition algorithms [11] and the former was an open question of interest [10]. Across more than twenty pages, Chudnovsky and Seymour characterized all pairs (G, K) that do not admit a solution, which resulted in a $\mathcal{O}(mn^2)$ time algorithm for the problem on n-vertex, m-edge graphs. Since then, THREE-IN-A-TREE has shown itself as a powerful tool, becoming a crucial subroutine for the fastest known even-hole [9], beetle [9], and odd-hole [12] detection algorithms; to the best of our knowledge, these algorithms often rely on reductions to multiple instances of THREE-IN-A-TREE, e.g. the theta detection algorithm has to solve $\mathcal{O}(mn^2)$ THREE-IN-A-TREE instances to produce its output [34]. Despite its versatility, THREE-IN-A-TREE is not a silver bullet, and some authors discuss quite extensively why they think THREE-IN-A-TREE cannot be used in some cases [14,39]. Nevertheless, Lai et al. [34] very recently made a significant breakthrough and managed to reduce the complexity of Chudnovsky and Seymour's algorithm for THREE-IN-A-TREE to $\mathcal{O}(m \log^2 n)$, effectively speeding up many major detection algorithms, among other improvements to the number of THREE-IN-A-TREE instances required to solve some other detection problems.

As pondered by Lai et al. [34], the usage of THREE-IN-A-TREE as a go-to solution for detection problems may, at times, seem quite unnatural. In the aforementioned cases, one could try to tackle the problem by looking for constant sized minors or disjoint paths between terminal pairs and then resort to Kawarabayashi et al.'s [32] quadratic algorithm to finalize the detection procedure. The problem is that neither the minors nor the disjoint paths are guaranteed to be induced; to make the situation truly dire, this constraint makes even the most basic problems NP-hard. For instance, Bienstock [1,2] proved that TWO-IN-A-HOLE and THREE-IN-A-PATH are NP-complete. As such, it is quite surprising that THREE-IN-A-TREE can be solved in polynomial time and be of widespread importance. It is worth to note that the induced subgraph constraint is also troublesome from the parameterized point of view. MAXIMUM MATCHING, for instance, can be solved in polynomial time [23], but if we impose that the matching must be an induced subgraph, the problem becomes W[1]-hard when parameterized by the minimum number of edges in the matching [36].

Derhy and Picouleau [18] were the first to ponder how far we may push for polynomial time algorithms when considering larger numbers of terminal vertices, i.e. they were interested in the complexity of k-IN-A-TREE for $k \geq 4$. They were also the firsts to investigate the more general problem where k is not fixed, which they dubbed INDUCED TREE, for which we have the following formal definition:

INDUCED TREE
Instance: A graph G and a set $K \subseteq V(G)$.
Question: Is there an induced subtree of G that contains all vertices of K?

Derhy and Picouleau proved in [18] that INDUCED TREE is NP-complete even on planar bipartite cubic graphs of girth four, but solvable in polynomial time if the girth of the graph is larger than the number of terminals. A few years later, Derhy et al. [19] showed that FOUR-IN-A-TREE is solvable in polynomial time on triangle-free graphs, while Liu and Trotignon [35] proved that so is k-IN-A-TREE on graphs of girth at least k; their combined results imply that k-IN-A-TREE on graphs of girth at least k is solvable in polynomial time; it is important to remark that the running time of the algorithm of Liu and Trotignon [35] has no $n^{f(k)}$ term. In terms of the INDUCED PATH problem, Derhy and Picouleau [18] argued that their hardness reduction also applies to this problem and showed that THREE-IN-A-PATH is NP-complete even on graphs of maximum degree three. Fiala et al. [24] proved that k-IN-A-PATH, k-INDUCED DISJOINT PATHS, and k-IN-A-CYCLE can be solved in polynomial time on claw-free graphs for every fixed k, but that INDUCED PATH, INDUCED DISJOINT PATHS, and INDUCED CYCLE are NP-complete in fact, their positive results can be seen as XP algorithms for the latter three problems when parameterized by the number of terminals on claw-free graphs. Another related problem to INDUCED TREE is the well known STEINER TREE problem, where we want to find a subtree of the input with cost at most w connecting all terminals. Being one of Karp's 21 NP-hard problems [31], STEINER TREE has received a lot of attention over the decades. Relevant to our discussion, however, is its parameterized complexity. When parameterized by the number of terminals, it admits a single exponential time algorithm [22]; the same was proven to be true when treewidth [37] is the parameter [3]. On the other hand, when parameterized by cliquewidth [16], it is paraNP-hard since it is NP-hard even on cliques: we may reduce from STEINER TREE itself and by replacing each non-edge with an edge of cost $w + 1$. As we see below, the parameterized complexity of STEINER TREE and INDUCED TREE greatly differ.

Our Results. Given the hardness results for INDUCED TREE even on restricted graph classes, we focus our study on the parameterized complexity of the problem. We begin by presenting some algorithmic results for INDUCED TREE in Sect. 3, showing that the latter is W[1]-hard when simultaneously parameterized by the number of vertices in the solution ℓ and size of a minimum clique cover q and, moreover, does not admit an $n^{o(\ell+q)}$ time algorithm unless the Exponential Time Hypothesis [30] (ETH) fails. On the positive side, we prove tractability under cliquewidth using Courcelle's Theorem [15], which prompts us, in Sect. 4, to turn our attention to which parameters allow us to devise single exponential time algorithms for INDUCED TREE. Using Bodlaender et al.'s dynamic programming optimization machinery [3], we show that such algorithms exist under treewidth, distance to cluster, and distance to co-cluster, all of which are widely used in the literature [7,17,21,28,33,41] and were the smallest parameters for which we managed to obtain such algorithms. We conclude the study of vertex cover-related parameters in Sect. 5, where we prove that the problem does not admit a polynomial kernel when simultaneously parameterized by the size of the solution, diameter, and distance to any non-trivial graph class, including the class of independent sets; we also show no such kernel exists when parameterizing by bandwidth. All our negative kernelization results are obtained assuming NP $\not\subseteq$ coNP/poly. In the realm of structural parameters, a natural next step would be to consider the max leaf number parameter. In Sect. 6 we do so by: (i) showing that max leaf number and feedback edge set are equivalent parameterizations for INDUCED TREE, and (ii) presenting a kernel with $16q$ vertices and $17q$ edges when we parameterize by the size q of a minimum feedback edge set; aside from our contribution we only know of one other problem for which kernelization under feedback edge set was considered, namely the EDGE DISJOINT PATHS problem [27]. In terms of tractability and kernelization, our results encompass most of the commonly employed parameters of Sorge and Weller's graph parameter hierarchy [38]; we present a summary of our results in Fig. 1. To see why the distance to solution parameter sits between vertex cover and feedback vertex set, we refer to the end of Sect. 3. Missing proofs can be found in the full version of the paper.

2 Preliminaries

We refer the reader to [17,25] for basic background on parameterized complexity, and recall here only some basic definitions. A *parameterized problem* is a language $L \subseteq \Sigma^* \times \mathbb{N}$. For an instance $I = (x, q) \in \Sigma^* \times \mathbb{N}$, q is called the *parameter*. A parameterized problem is *fixed-parameter tractable* (FPT) if there exists an algorithm $\mathcal{A}$, a computable function f, and a constant c such that given an instance (x, q), $\mathcal{A}$ correctly decides whether $I \in L$ in time bounded by $f(q) \cdot |I|^c$; in this case, $\mathcal{A}$ is called an FPT *algorithm*. A kernelization algorithm, or just *kernel*, for a parameterized problem Π takes an instance (x, q) of the problem and, in time polynomial in $|x| + q$, outputs an instance (x', q') such that $|x'|, q' \leqslant g(q)$ for some function g, and $(x, q) \in \Pi$ if and only if $(x', q') \in \Pi$. Function g is called the *size* of the kernel and may be viewed as a measure of the "compressibility" of a problem using polynomial-time pre-processing rules. A kernel is called *polynomial* (resp. *quadratic, linear*) if $g(q)$ is a polynomial (resp. quadratic, linear) function in q. A breakthrough result of Bodlaender et al. [4] gave the first framework for proving that some parameterized problems do not admit polynomial kernels, by establishing so-called *composition algorithms*. Together with a result of Fortnow and Santhanam [26], this allows to exclude polynomial kernels under the assumption that NP $\not\subseteq$ coNP/poly, otherwise implying a collapse of the polynomial hierarchy to its third level [40].

All graphs in this work are finite and simple. We use standard graph theory notation and nomenclature for our parameters; for any undefined terminology in graph theory we refer to [6]. We denote the degree of vertex v on graph G by $\deg_G(v)$, and the set of natural numbers $\{1, 2, \ldots, t\}$ by $[t]$. A graph is a *cluster graph* if each of its connected components is a clique, while a *co-cluster graph* is

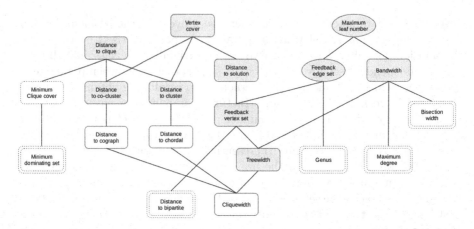

Fig. 1. Hasse diagram of graph parameters and associated results for INDUCED TREE. An edge from a lower parameter to a higher parameter indicates that the first is upper bounded by the latter; we remark that the edge between "Distance to solution" and "Vertex cover" exists when considering the reduction rule described in Sect. 3. Parameters surrounded by shaded ellipses have both single exponential time algorithms and polynomial kernels. Solid boxes represent parameters under which the problem is FPT but does not admit polynomial kernels; if the box is shaded, we have a single exponential time algorithm for that parameterization. A single dashed box corresponds to a W[1]-hard parameterization; double dashed boxes surround parameters under which the problem is paraNP-hard. Aside from the paraNP-hardness for genus, max degree, and distance to bipartite, all results are original contributions proposed in this work.

the complement of a cluster graph. The *vertex deletion distance to cluster* (*co-cluster*) of a graph G, is the size of the smallest set $U \subseteq V(G)$ such that $G \backslash U$ is a cluster (co-cluster) graph; when discussing these parameters, we refer to them simply as *distance to cluster* and *distance to co-cluster*. As defined in [8], a set $U \subseteq V(G)$ is an $\mathcal{F}$-*modulator* of G if $G \backslash U$ belongs to the graph class $\mathcal{F}$. When the context is clear, we omit the qualifier $\mathcal{F}$. For cluster and co-cluster graphs, one can decide if G admits a modulator of size q in time FPT on q [7].

3   Fixed-Parameter Tractability and Intractability

While it has been known for some time that INDUCED TREE is NP-complete even on planar bipartite cubic graphs, it is not known to be even in XP when parameterized by the natural parameter: the number of terminals. We take a first step with a negative result about this parameterization, ruling out the existence of an FPT algorithm unless FPT = W[1]; in fact, we show this for stronger parameterization: the maximum size of the induced tree that should contain the set of k terminal vertices K and the size of a minimum clique cover.

Theorem 1. INDUCED TREE *is* W[1]-*hard when simultaneously parameterized by the number of vertices of the induced tree and size of a minimum clique cover. Moreover, unless ETH fails, there is no $n^{o(k)}$ time algorithm for* INDUCED TREE.

Since the natural parameters offer little to no hope of fixed-parameter tractability, to obtain parameterized algorithms we turn our attention to the broad class of structural parameters. Our first positive result is a direct application of textbook MSO_1 formulae. We remark that the edge-weighted version of TWO-IN-A-TREE is strongly NP-complete [18].

Theorem 2. *When parameterized by cliquewidth*, INDUCED TREE *can be solved in* FPT *time. Moreover, the same holds true even for the edge-weighted version of* INDUCED TREE *where we want to minimize the weight of the tree.*

Towards showing that the minimum number of vertices we must delete to obtain a solution sits between feedback vertex set and vertex cover in Fig. 1, let $S \subset V(G)$ be such that $G \backslash S$ is a solution. First, note that S is a feedback vertex set of G; for the other inequality, take a vertex cover C of G and note that placing two vertices of $G \backslash C$ with the same neighborhood in C either generates a cycle in the solution or only one of them suffices – even if we have many terminals, we need to keep only two of them – so $|S| \leq |C| + 2^{|C|+1}$. In terms of paraNP-hardness results, we can easily show that INDUCED TREE is paraNP-hard when parameterized by bisection width[1]: to reduce from the problem to itself, we pick any terminal of the input and append to it a path with as many vertices as the original graph to obtain a graph with bisection width one. Similarly, when parameterizing by the size of a minimum dominating set and again reducing from INDUCED TREE to itself, we add a new terminal adjacent to any vertex of K and a universal vertex, which can never be part of the solution since it forms a triangle with the new terminal and its neighbor.

4   Single Exponential Time Algorithms

All results in this section rely on the *rank based approach* of Bodlaender et al. [3]. Even though our problem is unweighted, we found it convenient to solve the slightly more general weighted problem below when parameterizing INDUCED TREE by treewidth [37].

LIGHT CONNECTING INDUCED SUBGRAPH
Instance: A graph G, a set of k terminals $K \subseteq V(G)$, and two integers ℓ, f.
Question: Is there a connected induced subgraph of G on $\ell + k$ vertices and at most f edges that contains K?

Note that an instance (G, K) of INDUCED TREE is positive if and only if there is some integer ℓ where the LIGHT CONNECTING INDUCED SUBGRAPH instance $(G, K, \ell, \ell + k - 1)$ is positive. Our goal is to use the number of edges in the solution to LIGHT CONNECTING INDUCED SUBGRAPH as the cost of a

---

[1] The *width* of a bipartition (A, B) of $V(G)$ is the number of edges between the parts. The *bisection width* of G is equal to the minimum width among all bipartitions of $V(G)$ where $|A| \leq |B| \leq |A| + 1$.

partial solution in a dynamic programming algorithm. This shall be particularly useful for join nodes during our treewidth algorithm, as we may resort to the optimality of the solution to guarantee that the resulting induced subgraph of a join operation is acyclic. As usual, we assume that we are given a nice tree decomposition [17] in the input to our algorithm.

Theorem 3. *There is an algorithm for* LIGHT CONNECTING INDUCED SUBGRAPH *that, given a nice tree decomposition of width t of the n-vertex input graph G, runs in time $2^{\mathcal{O}(t)}n^{\mathcal{O}(1)}$.*

Corollary 1. *There is an algorithm for* INDUCED TREE *that, given a nice tree decomposition of width t of the n-vertex input graph G, runs in time $2^{\mathcal{O}(t)}n^{\mathcal{O}(1)}$.*

We also use the framework of Bodlaender et al. [3] for the next two results. For Theorem 4, we observe that a clique may have at most two vertices in any solution to INDUCED TREE. For Theorem 5, observe that at most two parts of a complete multipartite subgraph may intersect the solution and, if this is the case, one part has at most one vertex, otherwise we would have an induced C_4.

Theorem 4. *There is an algorithm for* INDUCED TREE *that runs in time $2^{\mathcal{O}(q)}n^{\mathcal{O}(1)}$ on graphs with distance to cluster at most q.*

Theorem 5. *There is an algorithm for* INDUCED TREE *that runs in time $2^{\mathcal{O}(q)}n^{\mathcal{O}(1)}$ where q is the distance to co-cluster.*

5   Kernelization Lower Bounds

In this section, we apply the cross-composition framework of Bodlaender et al. [5] to show that, unless $\mathsf{NP} \subseteq \mathsf{coNP/poly}$, INDUCED TREE does not admit a polynomial kernel under bandwidth[2], nor when parameterized by the distance to any graph class with at least one member with t vertices for each integer t, which we collectively call non-trivial classes. We say that an NP-hard problem R *OR-cross-composes* into a parameterized problem L if, given t instances $\{y_1, \ldots, y_t\}$ of R, we can construct, in time polynomial in $\sum_{i \in [t]} |y_i|$, an instance (x, k) of L that satisfies $k \leq p(\max_{i \in [t]} |y_i| + \log t)$ for some polynomial $p(\cdot)$ and admits a solution if and only if at least one instance y_i of R admits a solution; we say that R *AND-cross-composes* into L if the first two conditions hold but (x, k) has a solution if and only if all t instances of R admit a solution. We prove Theorem 6 using an AND-cross-composition from INDUCED TREE to itself. In our proof of Theorem 7 and Corollary 2, we exhibit an OR-cross-composition from HAMILTONIAN PATH to INDUCED TREE; as pointed by one of the reviewers, however, Corollary 2 is also a consequence of Theorem 1 and the fact that MULTICOLORED INDEPENDENT SET has no polynomial kernel when parameterized by the sum of the size of all but the largest color class [29].

---

[2] A *layout* of G is a bijection $f : V(G) \mapsto [n]$; the *width* of f is given by $\max_{uv \in E(G)} |f(u) - f(v)|$. The *bandwidth* of G is equal to the width of a minimum-width layout.

Theorem 6. *When parameterized by bandwidth,* INDUCED TREE *does not admit a polynomial kernel unless* NP $\subseteq$ coNP/poly.

Theorem 7. INDUCED TREE *does not admit a polynomial kernel when parameterized by the number of vertices of the induced tree, and size of a minimum vertex cover unless* NP $\subseteq$ coNP/poly.

Corollary 2. *For every non-trivial graph class* $\mathcal{G}$, INDUCED TREE *does not admit a polynomial kernel when parameterized by the number of vertices of the induced tree and size of a minimum* $\mathcal{G}$*-modulator unless* NP $\subseteq$ coNP/poly.

6   A Linear Kernel for Feedback Edge Set

In this section, we prove that INDUCED TREE admits a linear kernel when parameterized by the size q of a minimum feedback edge set. Throughout this section, we denote our input graph by G, the set of terminals by K, and the tree obtained by removing the edges of a minimum size feedback edge set F by $T(F)$. Note that, if G is connected and F is of minimum size, $G \backslash F$ is a tree; we may safely assume the first, otherwise we either have that (G, K) is a negative instance if K is spread across multiple connected components of G, or there must be some edge of F that merges two connected components of $G \backslash F$ and does not create a cycle, contradicting the minimality of F. The algorithm we describe takes as input a graph G and feedback edge set F and works in two steps: (i) it normalizes F into a feedback edge set that has a small number of edges incident to vertices of degree two in $T(F)$, then (ii) compresses long induced paths of G; we believe that (i) may be of independent interest to the community. We denote the set of leaves of a tree H by leaves (H).

Reduction Rule 1. *If G has a vertex v of degree one, remove v and, if $v \in K$, add the unique neighbor of v in G to K.*

Proof of Safeness of Rule 1. Safeness follows from the fact that a degree one vertex is in the solution if and only if its unique neighbor also is.   □

Observation 1. *After exhaustively applying Rule 1, for every minimum feedback edge set F of G, $T(F)$ has at most $2q$ leaves. Moreover, $T(F)$ has at most as many vertices of degree at least three as leaves.*

We begin with any minimum feedback edge set F of G. We partition $T(F) \backslash$ leaves $(T(F))$ into (D_2, D_*) according to the degree of the vertices of G in $T(F)$: $v \in D_2$ if and only if $\deg_{T(F)}(v) = 2$. For $u, v, f \in V(G)$, we say that u *F-links* v to f if $v = u$ or if $T(F) \backslash \{u\}$ has no $v - f$ path. We say that vertices u, f are an *F-pair* if the set of internal vertices of the unique $u - f$ path $P_F(u, f)$ of $T(F)$ is entirely contained in D_2; we denote the set of internal vertices by $P_F^*(u, f)$.

Normalization Rule 2. *Let $u, f, w_1, w_2 \in V(G)$ be such that u, f form an F-pair, $w_1 \neq w_2 \neq u \neq w_1$, w_2 is the unique neighbor of f that F-links it to u and w_1 F-links w_2 to u. If $fw_1 \in F$, remove edge fw_1 from F and add edge fw_2 to F.*

Proof of Safeness of Rule 2. Let $F' = F \backslash \{fw_1\}$ and note that w_2 F-links f to w_1; as such, edge fw_2 is in the unique cycle of $G \backslash F'$, so $F'' = F' \cup \{fw_2\}$ is a feedback edge set of G of size q. Furthermore, w_2 is the only vertex that has fewer neighbors in $T(F'')$ than in $T(F)$; since w_2 had two neighbors in $T(F)$, and F'' is a minimum feedback edge set of G, w_2 is a leaf of $T(F'')$, so it holds that $\mathsf{leaves}\,(T(F)) \subset \mathsf{leaves}\,(T(F''))$. □

Normalization Rule 2 guarantees that there are no edges in F between vertices of the paths between F-pairs, otherwise $|\mathsf{leaves}\,(T(F))|$ would not be maximal.

Normalization Rule 3. *Let $f, u, v \in V(G)$ be such that $v \notin \mathsf{leaves}\,(T(F)) \cup P_F(u, f)$, u, f form an F-pair, and $|P_F(u, f)| \geq 4$. If there are adjacent vertices $w_1, w_2 \in P_F^*(u, f)$ with $vw_1 \in F$ and w_2 F-linking v and w_1, remove edge vw_1 from F and add edge $w_1 w_2$ to F.*

Proof of Safeness of Rule 3. Let $F' = F \backslash \{w_1 w_2\}$. Since $G \backslash F'$ has one more edge than $T(F)$ and w_2 F-links v and w (see Figure 2), the unique cycle of $G \backslash F'$ contains edge $w_1 w_2$, so $F'' = F' \cup \{vw_1\}$ is a feedback edge set of G of size q. Since neither v nor w are leaves of $T(F)$ and $w_2 \in D_2$, $\deg_{T(F'')}(w_2) = 1$, so it holds that $|\mathsf{leaves}\,(T(F))| < |\mathsf{leaves}\,(T(F''))|$. □

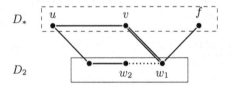

Fig. 2. Example for Normalization Rule 3, where the thick edge vw_1 is removed from F and the dotted edge $w_1 w_2$ added to F.

Note that, in Normalization Rule 3, the only properties that we exploit are that $v, w_1, w_2 \notin \mathsf{leaves}\,(T(F))$ and that there is at least one pair of vertices between v and f in D_2. So even if $v = u$, $u \in D_2$, or $f \in \mathsf{leaves}\,(T(F))$, we can apply Rule 3. Essentially, if Rule 3 is not applicable, for each edge $e \in F$ that contains a vertex w_1 of D_2 as an endpoint, either e has a leaf of $T(F)$ as its other endpoint, or w_1 is adjacent to two vertices not in D_2.

Normalization Rule 4. *Let $f, u, v, z, w_2 \in V(G)$ be such that $v \in \mathsf{leaves}\,((T(F)) \backslash \{f\}$, $w_2 \in D_2$ is the unique neighbor of v in $T(F)$, u, f and z, v are F-pairs, and u F-links f to z. If there is some $w_1 \in P_F^*(u, f)$ with $vw_1 \in F$, remove vw_2 from F and add vw_1.*

Proof of Safeness of Rule 4. Let $F' = F\backslash\{vw_1\}$. Since $T(F)$ is a tree and w_2 F-links w_1 and v, edge vw_2 is contained in the unique cycle of $G\backslash F'$; consequently, $F'' = F' \cup \{vw_2\}$ is a feedback edge set of G of size q, but it holds that the degrees of v and w_2 in $T(F'')$ are equal to one. Since $w_2 \notin$ leaves $(T(F))$, we have that leaves $(T(F)) \subset$ leaves $(T(F''))$. □

Our analysis for Rule 4 works even if $u = z$ or $z = w_2$: what is truly crucial is that $w_2 \in D_2$ and that $v \neq f$. We present an example of the general case in Fig. 3.

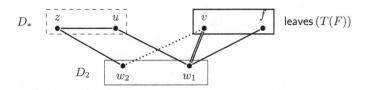

Fig. 3. Example for Normalization Rule 4, where the thick edge vw_1 is removed from F and the dotted edge vw_2 is added to F.

Normalization Rule 5. *Let $f, u, v, z, w_2, w_3 \in V(G)$ be such that $v \in$ leaves $((T(F))$, $w_3 \in N_{T(F)}(u) \cap P_F(u, f)$, $w_2w_3, vz \in E(T(F))$, u, f is an F-pair, $z \in D_*$, and u F-links f to v. If v is adjacent to some $w_1 \in P_F(u, f)\backslash\{w_2\}$ that F-links w_2 to f, remove vw_1 from F and add w_2w_3 to F.*

Proof of Safeness of Rule 5. Let $F' = F\backslash\{vw_3\}$. Since w_1 F-links w_2 to f, we have that w_2 F-links w_1 to v, so edge w_2w_3 belongs to the unique cycle of $G\backslash F'$ and, consequently, $F'' = F' \cup \{w_2w_3\}$ is a feedback edge set of G of size q. Since $\deg_{T(F)}(w_3) = \deg_{T(F)}(w_2) = 2$, $\deg_{T(F'')}(w_3) = \deg_{T(F'')}(w_2) = 1$, however, v is adjacent to both z and w_1 in $T(F'')$, so it holds that leaves $(T(F'')) =$ leaves $(T(F)) \cup \{w_2, w_3\}\backslash\{v\}$. □

We present an example of Normalization Rule 5 in Fig. 4. Our next lemma guarantees that the exhaustive application of rules 2 through 5 finds a set of paths in $T(F)$ that have many vertices of degree two in G; essentially, at this point, we are done minimizing the number of incident edges to vertices of D_2.

Lemma 1. *Let $a, b \in V(G)$ be an F-pair such that $a, b \notin D_2$, $|P_F^*(a, b)| \geq 5$, and let w be one of its inner vertices at distance at least three from both a and b. If none of the Rules 2 to 5 are applicable, then $\deg_G(w) = \deg_{T(F)}(w)$.*

At this point, paths between F-pairs are mostly the same as in G: only the two vertices closest to each endpoint may be adjacent to some leaves of $T(F)$, while all others have degree two in G. We say that u, f are a *strict F-pair* if for every $w \in P_F^*(u, f)$, $\deg_G(w) = 2$.

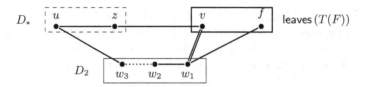

Fig. 4. Example for Normalization Rule 5, where the dotted edge w_2w_3 is added to F and the thick edge vw_1 is removed from F.

Reduction Rule 6. *Let $u, f \in V(G)$ be a strict F-pair. If there are adjacent vertices $w_1, w_2 \in P_F^*(u, f)$ such that either $w_1, w_2 \in K$ or $w_1, w_2 \notin K$, add a new vertex w^* to G that is adjacent to $N_G(w_1) \cup N_G(w_2) \backslash \{w_1, w_2\}$ and remove both w_1, w_2 from G. If $w_1, w_2 \in K$, set w^* as a terminal vertex.*

Proof of Safeness of Rule 6. Correctness follows directly from the hypotheses that $w_1 \in K$ if and only if $w_2 \in K$ and that both are degree two vertices. So, in a minimal solution H to (G, K), either both vertices are in H or neither is in H. For the converse, any minimal solution H' to the reduced instance (G', K') either has w^*, in which H is obtained by replacing w^* with both w_1 and w_2, or $w^* \notin V(H)$, in which case H' is a solution to (G, K).   □

Reduction Rule 7. *Let $u, f \in V(G)$ be a strict F-pair such that $|P_F^*(u, f)| \geq 4$. If Rule 6 is not applicable, replace $P_F^*(u, f)$ with three vertices a, t, b so that a is adjacent to u, b to f, t to both a and b, and t is a terminal of the new graph.*

Proof of Safeness of Rule 7. Let G' and K' be, respectively, the graph and set of terminals obtained after the application of the rule. Suppose H is a minimal solution to the INDUCED TREE instance (G, K), i.e. every vertex of H is contained in a path between two terminals. Note that, if $P_F^*(u, f) \cap V(H) = \emptyset$, H is also a solution to the instance (G', K'); as such, for the remainder of this paragraph, we may assume w.l.o.g. that $P_F^*(u, f) \cap V(H) \neq \emptyset$ and that $u \in V(H)$. If $P_F(u, f) \backslash \{u\} \nsubseteq V(H)$, $H' = H \cup \{a, t\} \backslash P_F^*(u, f)$ is a solution to (G', K'): since at least one vertex of $P_F(u, f)$ is not in $V(H)$ and every $w \in P_F^*(u, f)$ has degree two in G, the subpaths of $P_F(u, f)$ in H are used solely for the collection of terminal vertices of $P_F(u, f)$; consequently, H' is an induced tree of G' that contains all elements of K'. On the other hand, if $P_F(u, f) \subseteq V(H)$, $H' = H \cup \{a, t, b\} \backslash P_F^*(u, f)$ is a solution to (G', K'); to see that this is the case, note that $H \backslash P_F^*(u, f)$ is a forest with exactly two trees where u and f are in different connected components since $P_F(u, f)$ is the unique path between them in H, and $K' \subseteq V(H')$ since $P_F^* \subseteq V(H)$ and $K \backslash P_F^*(u, f) \subseteq V(H) \backslash P_F^*(u, f)$.

For the converse, let H' be a minimal solution to (G', K'). If $\{a, t, b\} \subseteq V(H')$, $H = H' \cup \{P_F^*(u, f)\} \backslash \{a, t, b\}$ is a solution of (G, K), as we are replacing one path consisting solely of degree two vertices with another that satisfies the same property. If $a \in V(H')$ but $b \notin V(H')$, then $t \in K'$ (recall that H' is minimal) and $u \in V(H)$, implying that there is at least one terminal vertex in $P_F^*(u, f)$. We branch our analysis in the following subcases, where $v \in P_F^*(u, f) \cap N_G(f)$:

- If $v \notin K$, then $H = H' \cup P_F^*(u,f)\backslash\{v\} \setminus \{a,t\}$ is a solution to (G,K): all terminals of $P_F^*(u,f)$ are contained in $P_F^*(u,v)$ and no cycle is generated since all vertices of the path $P_F^*(u,v)$ have degree two.
- If $v \in K$ but $f \notin V(H')$, $H = H' \cup P_F^*(u,f)\backslash\{a,t\}$ is a solution to (G,K): we cannot create any new cycle since $f \notin V(H)$ and u,f form a strict F-pair, moreover all terminals of $P_F^*(u,f)$ are contained in H.
- If $v \in K$ and $f \in V(H')$, there is at least one non-terminal vertex $w \in P_F^*(u,v)$ since Rule 6 is not applicable to $P_F(u,f)$. As such, we set $H = H' \cup P_F^*(u,w) \cup P_F^*(w,f)\backslash\{a,t\}$ and obtain a solution to (G,K).

Finally, if $\{a,t,b\} \cap V(H') = \emptyset$, it follows immediately from the assumption that H' is a solution to (G',K') that H' is also a solution to (G,K). Note that $P_F^*(u,f) \cap K \neq \emptyset$ since $|P_F^*(u,f)| \geq 4$ and Rule 6 is not applicable. $\qquad\square$

We are now ready to state our kernelization theorem.

Theorem 8. *When parameterized by the size q of a feedback edge set,* INDUCED TREE *admits a kernel with $16q$ vertices and $17q$ edges that can be computed in* $\mathcal{O}(q^2 + qn)$ *time.*

7 Concluding Remarks

In this work, we performed an extensive study of the parameterized complexity of INDUCED TREE and the existence of polynomial kernels for the problem, motivated by the relevance of THREE-IN-A-TREE in subgraph detection algorithms and related work on INDUCED TREE itself, specially the question on the complexity of FOUR-IN-A-TREE posed by Derhy and Picouleau [18]. We presented multiple positive and negative results for INDUCED TREE, of which we highlight its W[1]-hardness under the natural parameter, the linear kernel under feedback edge set, and the nonexistence of a polynomial kernel under vertex cover/distance to clique. The main question about the complexity of k-IN-A-TREE for *fixed k*, however, remains open; our hardness result showed that there is no $n^{o(k)}$ time algorithm under the ETH, but no XP algorithm is known to exist. There are also no known running time lower bounds for the parameters we study, and determining whether or not we can obtain $2^{o(q)}n^{\mathcal{O}(1)}$ time algorithms seems a feasible research direction; still in terms of algorithmic results, it would be quite interesting to see how we can avoid Courcelle's Theorem to get an algorithm when parameterizing by cliquewidth. The natural investigation of k-IN-A-TREE on different graph classes may provide some insights on how to tackle particular cases, such as FOUR-IN-A-TREE; this study has already been started in [20] and in others – such as cographs – may even be trivial, but many other cases may be quite challenging and much still remains to be done.

References

1. Bienstock, D.: On the complexity of testing for odd holes and induced odd paths. Discrete Math. **90**(1), 85–92 (1991). https://doi.org/10.1016/0012-365X(91)90098-M. http://www.sciencedirect.com/science/article/pii/0012365X9190098M

2. Bienstock, D.: On the complexity of testing for odd holes and induced odd paths. Discrete Math. **90**, 85–92 (1991). Discrete Math. **102**(1), 109 (1992). https://doi.org/10.1016/0012-365X(92)90357-L. http://www.sciencedirect.com/science/article/pii/0012365X9290357L

3. Bodlaender, H.L., Cygan, M., Kratsch, S., Nederlof, J.: Deterministic single exponential time algorithms for connectivity problems parameterized by treewidth. Inf. Comput. **243**, 86–111 (2015). 40th International Colloquium on Automata, Languages and Programming (ICALP 2013). https://doi.org/10.1016/j.ic.2014.12.008. http://www.sciencedirect.com/science/article/pii/S0890540114001606

4. Bodlaender, H.L., Downey, R.G., Fellows, M.R., Hermelin, D.: On problems without polynomial kernels. J. Comput. Syst. Sci. **75**(8), 423–434 (2009). https://doi.org/10.1016/j.jcss.2009.04.001. http://www.sciencedirect.com/science/article/pii/S0022000009000282

5. Bodlaender, H.L., Jansen, B.M.P., Kratsch, S.: Cross-composition: a new technique for kernelization lower bounds. In: Proceedings of the 28th International Symposium on Theoretical Aspects of Computer Science (STACS), Volume 9 of LIPIcs, pp. 165–176 (2011)

6. Bondy, J.A., Murty, U.S.R.: Graph Theory, 1st edn. Springer, Heidelberg (2008)

7. Boral, A., Cygan, M., Kociumaka, T., Pilipczuk, M.: A fast branching algorithm for cluster vertex deletion. Theory Comput. Syst. **58**(2), 357–376 (2015). https://doi.org/10.1007/s00224-015-9631-7

8. Cai, L.: Parameterized complexity of vertex colouring. Discrete Appl. Math. **127**(3), 415–429 (2003). https://doi.org/10.1016/S0166-218X(02)00242-1. http://www.sciencedirect.com/science/article/pii/S0166218X02002421

9. Chang, H.-C., Lu, H.-I.: A faster algorithm to recognize even-hole-free graphs. J. Comb. Theory Ser. B **113**, 141–161 (2015). https://doi.org/10.1016/j.jctb.2015.02.001. http://www.sciencedirect.com/science/article/pii/S0095895615000155

10. Chudnovsky, M., Kapadia, R.: Detecting a theta or a prism. SIAM J. Discrete Math. **22**(3), 1164–1186 (2008). arXiv:https://doi.org/10.1137/060672613. https://doi.org/10.1137/060672613

11. Chudnovsky, M., Robertson, N., Seymour, P., Thomas, R.: The strong perfect graph theorem. Ann. Math. **164**(1), 51–229 (2006). http://www.jstor.org/stable/20159988

12. Chudnovsky, M., Scott, A., Seymour, P., Spirkl, S.: Detecting an odd hole. J. ACM **67**(1) (2020). https://doi.org/10.1145/3375720

13. Chudnovsky, M., Seymour, P.: The three-in-a-tree problem. Combinatorica **30**(4), 387–417 (2010). https://doi.org/10.1007/s00493-010-2334-4

14. Chudnovsky, M., Seymour, P., Trotignon, N.: Detecting an induced net subdivision. J. Comb. Theory Ser. B **103**(5), 630–641 (2013). https://doi.org/10.1016/j.jctb.2013.07.005. http://www.sciencedirect.com/science/article/pii/S0095895613000531

15. Courcelle, B., Engelfriet, J.: Graph Structure and Monadic Second-Order Logic: A Language-Theoretic Approach. Encyclopedia of Mathematics and Its Applications. Cambridge University Press, Cambridge (2012). https://doi.org/10.1017/CBO9780511977619

16. Courcelle, B., Olariu, S.: Upper bounds to the clique width of graphs. Discrete Appl. Math. **101**(1), 77–114 (2000). https://doi.org/10.1016/S0166-218X(99)00184-5. http://www.sciencedirect.com/science/article/pii/S0166218X99001845

17. Cygan, M., et al.: Parameterized Algorithms, vol. 3. Springer, Cham (2015). https://doi.org/10.1007/978-3-319-21275-3

18. Derhy, N., Picouleau, C.: Finding induced trees. Discrete Appl. Math. **157**(17), 3552–3557 (2009). Sixth International Conference on Graphs and Optimization (2007). https://doi.org/10.1016/j.dam.2009.02.009. http://www.sciencedirect.com/science/article/pii/S0166218X09000663

19. Derhy, N., Picouleau, C., Trotignon, N.: The four-in-a-tree problem in triangle-free graphs. Graphs Comb. **25**(4), 489 (2009). https://doi.org/10.1007/s00373-009-0867-3

20. dos Santos, V.F., da Silva, M.V.G., Szwarcfiter, J.L.: The k-in-a-tree problem for chordal graphs. Matemática Contemporânea **44**, 1–10 (2015)

21. Doucha, M., Kratochvíl, J.: Cluster vertex deletion: a parameterization between vertex cover and clique-width. In: Rovan, B., Sassone, V., Widmayer, P. (eds.) MFCS 2012. LNCS, vol. 7464, pp. 348–359. Springer, Heidelberg (2012). https://doi.org/10.1007/978-3-642-32589-2_32

22. Dreyfus, S.E., Wagner, R.A.: The Steiner problem in graphs. Networks **1**(3), 195–207 (1971). http://dx.doi.org/10.1002/net.3230010302. https://onlinelibrary.wiley.com/doi/abs/10.1002/net.3230010302

23. Edmonds, J.: Paths, trees, and flowers. Can. J. Math. **17**, 449–467 (1965). https://doi.org/10.4153/CJM-1965-045-4

24. Fiala, J., Kamiński, M., Lidický, B., Paulusma, D.: The k-in-a-path problem for claw-free graphs. Algorithmica **62**(1), 499–519 (2012). https://doi.org/10.1007/s00453-010-9468-z

25. Fomin, F.V., Lokshtanov, D., Saurabh, S., Zehavi, M.: Kernelization: Theory of Parameterized Preprocessing. Cambridge University Press, Cambridge (2019). https://doi.org/10.1017/9781107415157

26. Fortnow, L., Santhanam, R.: Infeasibility of instance compression and succinct PCPs for NP. J. Comput. Syst. Sci. **77**(1), 91–106 (2011). Celebrating Karp's Kyoto Prize. https://doi.org/10.1016/j.jcss.2010.06.007. http://www.sciencedirect.com/science/article/pii/S0022000010000917

27. Ganian, R., Ordyniak, S.: The power of cut-based parameters for computing edge disjoint paths. In: Sau, I., Thilikos, D.M. (eds.) WG 2019. LNCS, vol. 11789, pp. 190–204. Springer, Cham (2019). https://doi.org/10.1007/978-3-030-30786-8_15

28. Gomes, G.C.M., Guedes, M.R., dos Santos, V.F.: Structural parameterizations for equitable coloring (2019). arXiv:1911.03297

29. Grüttemeier, N., Komusiewicz, C.: On the relation of strong triadic closure and cluster deletion. Algorithmica **82**(4), 853–880 (2019). https://doi.org/10.1007/s00453-019-00617-1

30. Impagliazzo, R., Paturi, R.: On the complexity of k-SAT. J. Comput. Syst. Sci. **62**(2), 367–375 (2001). https://doi.org/10.1006/jcss.2000.1727. http://www.sciencedirect.com/science/article/pii/S0022000000917276

31. Karp, R.M.: Reducibility among combinatorial problems. In: Miller, R.E., Thatcher, J.W., Bohlinger, J.D. (eds.) Complexity of Computer Computations. IRSS, pp. 85–103. Springer, Boston (1972). https://doi.org/10.1007/978-1-4684-2001-2_9

32. Kawarabayashi, K., Kobayashi, Y., Reed, B.: The disjoint paths problem in quadratic time. J. Comb. Theory Ser. B **102**(2), 424–435 (2012). https://doi.org/10.1016/j.jctb.2011.07.004. http://www.sciencedirect.com/science/article/pii/S0095895611000712

33. Komusiewicz, C., Kratsch, D., Le, V.B.: Matching cut: kernelization, single-exponential time FPT, and exact exponential algorithms. 283, 44–58 (2020). https://doi.org/10.1016/j.dam.2019.12.010. http://www.sciencedirect.com/science/article/pii/S0166218X19305530

34. Lai, K.-Y., Lu, H.-I., Thorup, M.: Three-in-a-tree in near linear time. In: Proceedings of the 52nd Annual ACM SIGACT Symposium on Theory of Computing, STOC 2020, pp. 1279–1292. Association for Computing Machinery, New York (2020). https://doi.org/10.1145/3357713.3384235

35. Liu, W., Trotignon, N.: The k-in-a-tree problem for graphs of girth at least k. Discrete Appl. Math. **158**(15), 1644–1649 (2010). https://doi.org/10.1016/j.dam.2010.06.005. http://www.sciencedirect.com/science/article/pii/S0166218X10002131

36. Moser, H., Sikdar, S.: The parameterized complexity of the induced matching problem. Discrete Appl. Math. **157**(4), 715–727 (2009). https://doi.org/10.1016/j.dam.2008.07.011. http://www.sciencedirect.com/science/article/pii/S0166218X08003211

37. Robertson, N., Seymour, P.D.: Graph minors. II. Algorithmic aspects of tree-width. J. Algorithms **7**(3), 309–322 (1986). https://doi.org/10.1016/0196-6774(86)90023-4. http://www.sciencedirect.com/science/article/pii/0196677486900234

38. Sorge, M., Weller, M.: The graph parameter hierarchy (2019, Unpublished manuscript)

39. Trotignon, N., Vušković, K.: A structure theorem for graphs with no cycle with a unique chord and its consequences. J. Graph Theory **63**(1), 31–67 (2010). https://doi.org/10.1002/jgt.20405. https://onlinelibrary.wiley.com/doi/abs/10.1002/jgt.20405. arXiv:https://onlinelibrary.wiley.com/doi/pdf/10.1002/jgt.20405

40. Yap, C.K.: Some consequences of non-uniform conditions on uniform classes. Theor. Comput. Sci. **26**(3), 287–300 (1983). https://doi.org/10.1016/0304-3975(83)90020-8. http://www.sciencedirect.com/science/article/pii/0304397583900208

41. Bonnet, É., Sikora, F.: The graph motif problem parameterized by the structure of the input graph. Discrete Appl. Math. **231**, 78–94 (2017). Algorithmic Graph Theory on the Adriatic Coast. https://doi.org/10.1016/j.dam.2016.11.016. http://www.sciencedirect.com/science/article/pii/S0166218X1630539X

The Multi-budget Maximum Weighted Coverage Problem

Francesco Cellinese[1]($\boxtimes$), Gianlorenzo D'Angelo[1], Gianpiero Monaco[2], and Yllka Velaj[3]

[1] Gran Sasso Science Institute, L'Aquila, Italy
{francesco.cellinese,gianlorenzo.dangelo}@gssi.it
[2] University of L'Aquila, L'Aquila, Italy
gianpiero.monaco@univaq.it
[3] University of Vienna, Vienna, Austria
yllka.velaj@univie.ac.at

Abstract. In this paper we consider the *multi-budget maximum weighted coverage problem*, a generalization of the classical maximum coverage problem, where we are given k budgets, a set X of elements, and a set $\mathcal{S}$ of bins where any $S \in \mathcal{S}$ is a subset of elements of X. Each bin S has its own cost, and each element its own weight. An outcome is a vector $O = (O_1, \ldots, O_k)$ where each budget b_i, for $i = 1, \ldots, k$, can be used to buy a subset of bins $O_i \subseteq \mathcal{S}$ of overall cost at most b_i. The objective is to maximize the total weight which is defined as the sum of the weights of the elements bought with the budgets.

We consider the classical combinatorial optimization problem of computing an outcome which maximizes the total weight and provide a $\left(1 - \frac{1}{\sqrt{e}}\right)$-approximation algorithm for the case when the maximum cost of a bin is upper-bounded by the minimum budget, i.e. the case in which each budget can be used to buy any bin. Moreover, we give a randomized Monte-Carlo algorithm for the general case that runs in polynomial time, satisfies the budget constraints in expectation, and guarantees an expected $1 - \frac{1}{e}$ approximation factor.

Keywords: Multi-Budget Coverage · Maximum Coverage · Approximation Algorithms

1 Introduction

Numerous problems with real-world applications in job scheduling, facility locations and resource allocations can be modeled through the classical maximum coverage problem [16, Ch. 3]. In this problem we are given a ground set X, a collection S of subsets of X with unit cost, and a budget k and the goal is selecting a sub-collection $S' \subseteq S$, such that $|S'| \leq k$, and the number of elements of X covered by S' is maximized. Motivated by the fact that applications in complex systems such as the Internet have spawned a recent interest in studying situations

© Springer Nature Switzerland AG 2021
T. Calamoneri and F. Corò (Eds.): CIAC 2021, LNCS 12701, pp. 173–186, 2021.
https://doi.org/10.1007/978-3-030-75242-2_12

involving multiple entities with different budgets, in this paper we consider the *Multi-budget Maximum Weighted Coverage problem* (MMWC), which is defined as follows. We are given a set X of n elements, and a set $\mathcal{S}$ of m bins, where any $S \in \mathcal{S}$ is a subset of elements of X (i.e. $S \subseteq X$). Each bin S has a cost that we denote as $c_S \in \mathbb{R}^+$, and each element $x \in X$ has a weight that we denote as $u_x \in \mathbb{R}^+$. We are also given a set of k budgets $B = \{b_1, \ldots, b_k\}$ such that $b_i > 0$ for $i = 1, \ldots, k$ (sometimes, when it is clear from the context that we refer to a budget, we denote it with i instead of b_i). An outcome of the problem is a vector $O = (O_1, \ldots, O_k)$ where, for any budget $i = 1, \ldots, k$, $O_i \subseteq \mathcal{S}$ and $\sum_{S \in O_i} c_S \leq b_i$. That is, each budget b_i can be used to buy a subset of elements of $\mathcal{S}$ which overall cost is at most b_i (sometimes, we say that a bin is *assigned to* or *covered by* a budget meaning that it is *bought* using that budget).

We denote with $C_i(O) = \bigcup_{S \in O_i} S$ the set of elements covered by budget b_i in the outcome O, and with $C(O) = \bigcup_{b_i \in B} C_i(O)$ the overall set of elements covered by the budgets in O. The *total weight* of an outcome O is defined as the sum of the weights of the covered elements, i.e., $R(O) = \sum_{x \in C(O)} u_x$. We consider the classical combinatorial optimization problem of computing an outcome which maximizes the total weight. Even if the model allows two budgets to be used for buying the same bin, we can assume, without loss of generality, that each bin is bought by at most one budget. Indeed, for any solution that does not satisfy this condition it is possible to define another solution that satisfies it and that does not decrease the value of the objective function. We notice that, when costs are constant and $k = 1$, the MMWC is exactly the maximum coverage problem.

The importance of our problem is underlined by the attention that variants of the classical coverage problem have attracted in the past years.

A greedy $1 - \frac{1}{e}$ approximation algorithm for the classical maximum coverage problem (MC) has been proposed in [23], and such result is tight [10]. A $\frac{5}{6}$ approximation algorithm has been showed in [3] for the case when each set in S has size at most three. The budgeted maximum coverage problem is a generalization of MC in which the subsets in S are associated with costs and the aim is to find a sub-collection $S' \subseteq S$ that maximizes the number of covered elements and whose overall cost does not exceed k. An algorithm that guarantees a $1 - \frac{1}{e}$ approximation factor for this case has been proposed in [19]. In [4] the authors consider a generalization of MC, where we are given a ground set of elements and a set of bins. Each bin has its own cost and the cost of each element depends on its associated bin. The goal is to find a subset of elements along with an associated set of bins, such that the overall cost is at most a given budget, and the profit, measured by a monotone submodular function over the elements, is maximized.

In the generalized maximum coverage problem we are given a budget L, a set of elements X, and a collections of bins B. Each element x has a positive utility $P(b, x)$ and a non negative weight $W(b, x)$ associated to each bin b. Also each bin has a weight. The goal is to find a subset of elements and bins where each element is associated to a bin such that the total weight is at most L and the overall utility is maximized. In [8] the authors present a $1 - \frac{1}{e} - \epsilon$ approximation algorithm for the problem.

Our Results. We first notice that, since our problem generalizes the classical maximum coverage problem, the hardness result of $1 - \frac{1}{e}$ for the maximum coverage problem proved in [10] also holds in our case. We provide a deterministic approximation algorithm for the case in which the maximum cost of a bin is upper-bounded by the minimum available budget, i.e. the case in which each budget can be used to buy any bin. The algorithm guarantees an approximation factor of $1 - \frac{1}{\sqrt{e}}$. Moreover, we provide a randomized Monte-Carlo algorithm for the general case that runs in polynomial time, satisfies the budget constraints in expectation, and guarantees an expected $1 - \frac{1}{e}$ approximation factor.

Further Related Work. The different budgets can represent different agents who are allowed to spend part of the whole budget. Our problem can capture this scenario, therefore in the following we review the literature related to coverage problems with multiple agents.

A relevant setting was first presented in [14]. The authors study a class of combinatorial problems with multi-agent submodular cost functions where we are given a set of elements X and a collection $C \subseteq 2^X$. We are also given k agents, where each agent i specifies a normalized monotone submodular cost function $f_i : 2^X \to \mathbb{R}^+$. The goal is to find a set $S \in C$ and a partition $S_1, \ldots, S_k$ of S such that $\sum_i f_i(S_i)$ is minimized. By fixing the collection C to any particular combinatorial structure, the authors define a subclass of fundamental optimization problems: vertex cover, shortest path, perfect matching and spanning tree. They give upper and lower bounds for all these problems. These results are then extended in [15] and in [24]. We note that the problem considered here is different from these previous works. In fact, in our setting, agents have budget constraints to satisfy. Moreover, we have bins and a single element can belong to more than one bin. Finally, we are interested in maximizing the sum of the weights of the covered elements.

Chekuri and Kumar in [6] introduce a variant of the maximum coverage problem called maximum coverage with group budget constraints (MCG). In this problem we are given a collection of sets $\mathcal{S} = \{S_1, \ldots, S_n\}$ where each set S_i is a subset of a given ground set X. The collection $\mathcal{S}$ is partitioned into groups $G_1, \ldots, G_l$. In MCG the goal is to pick k sets from $\mathcal{S}$ in order to maximize the cardinality of their union, with the additional constraint of picking at least one set from each group. For this setting the authors provide a $\frac{1}{2}$-approximation algorithm. Even if this is not a paper about multi-agent settings, MCG can model a variety of multi-agent problems related to coverage. The authors also present a cost version of the problem where each set S_i has an associated cost $c(S_i)$, there is a budget B_j for each group G_j, and an overall budget B. In the cost version the goal is to maximize the cardinality of the union of the selected sets, respecting also the budget constraints: the total cost of the sets selected in each group cannot exceed the group's budget and the overall cost of the selected sets cannot exceed B. The authors give a $\frac{1}{12}$-approximation algorithm for the cost version of the problem. This factor was improved to $\frac{1}{5}$ by Farbstein and Levin [9]. We notice that our problem can be reduced to the cost version of

MCG, however, the reduction would create instances where the groups are not disjoint. As shown in [9], if the groups are not disjoint, the MCG problem cannot be approximated within any constant factor.

Another interesting setting was presented in [5]. The authors consider the maximum budgeted allocation problem where we are given a set of indivisible items and agents. Each agent i is willing to pay b_{ij} on item j and has a budget b_i. The goal is to allocate items to agents to maximize the revenue, that is, the sum, over all the agents, of the price for the items she bought. The main results described in the paper are: a $\frac{3}{4}$-approximation algorithm that exploits a natural LP relaxation of the problem, and a $\frac{15}{16}$ hardness of approximation result. We notice that our problem differs from the one considered in this work. In fact, in our setting, we have bins and a single element can belong to more than one bin. Moreover, we have costs for the bins and weights for the elements.

A further relevant work with multiple agents is [27] that considers the submodular welfare problem: m items are to be distributed among n agents with utility functions $w_i : 2^{[m]} \to \mathbb{R}^+$. Assuming that agent i receives a set of items S_i, the goal is to maximize the total utility $\sum_{i=1}^{n} w_i(S_i)$. In this paper, the authors work in the value oracle model where the only access to the utility functions is through a black box returning $w_i(S)$ for a given set S. They develop a randomized continuous greedy algorithm which achieves a $\left(1 - \frac{1}{e}\right)$-approximation for their problem in the value oracle model. Our problem is different because we have bins with their own costs, a single element can belong to more than one bin, and agents have a budget constraint to satisfy.

The last relevant setting that we mention which is related to multiple agents maximum coverage problems is the general covering problem. We are given a set of elements X where each $x \in X$ is associated to a positive integer weight. Moreover, we are given n collections $S_1, \ldots, S_n$ of subsets of X where $S_i \subset 2^X$ is a subset of the power-set of the elements. The goal is to choose one subset s_i from each collection S_i such that their union $\cup_{i \in [n]} S_i$ has maximum total weight. In [13] the authors consider the general covering problem where the choice in each collection is made by an independent agent. For covering an element, the agents receive a revenue defined by a non-increasing revenue sharing function. This function defines the fraction that each covering agent receives from the elements. They study how to define a revenue sharing function such that every Nash equilibrium approximates the optimal solution by a factor of $1 - \frac{1}{e}$. They also show a centralized $1 - \frac{1}{e}$ approximation algorithm for the general covering problem. We notice that our MMWC can be modeled by the general covering problem. Indeed, for each agent i, S_i can be defined as the different subsets of elements that can be covered by using budgets b_i. However, notice that S_i can be exponential in $|X|$, which implies that the centralized algorithm proposed in [13] is not polynomial in our setting.

Another important research topic is the maximization of submodular set function: given a set of elements and a monotone submodular function, the goal is to find the subset of elements that gives the maximum value, subjected to some constraints. The case when the subset of elements must be an independent set of the matroid over the set of elements has been considered in [2], where the authors show an optimal randomized $\left(1 - \frac{1}{e}\right)$-approximation algorithm. A simpler algorithm has been proposed in [12]. The case of multiple k matroid constraints has been considered in [21], a $\frac{1}{k+\epsilon}$-approximation is given. An improved result appeared in [28]. Unconstrained non-monotone submodular maximization, has been considered in [1,11]. In the submodular set function subject to a knapsack constraint maximization problem we have a cost $c(x)$ for any element $x \in X$, and the goal is selecting a set $X' \subseteq X$ of elements that maximizes $f(X')$, where f is a monotone submodular function subject to the constraint that the sum of the costs of the selected elements is at most k. This problem admits a polynomial time algorithm that is $\left(1 - \frac{1}{e}\right)$-approximation [26]. In [20] the authors focus on multiple knapsack constraints: Given a d-dimensional budget vector $\mathbf{L}$, for some $d \geq 1$, and an oracle for a non-decreasing submodular set function f over a universe U, where each element $e \in U$ is associated with a d-dimensional cost vector, we seek a subset of elements $S \subseteq U$ whose total cost is at most $\mathbf{L}$, such that $f(S)$ is maximized.

In [7], the authors give a general framework that can be applied to the maximization of monotone and non-monotone submodular functions subject to the intersection of several independence constraints, including knapsack, matroid, and packing constraints. Their approach is based on approximately maximizing the multilinear extension of the submodular objective function and then round a fractional solution by means of suitable contention resolution schemes that depend on the type of constraints.

A different setting called submodular cost submodular knapsack was considered in [18]: given a set of elements $V = \{1, 2, \ldots, n\}$, two monotone non-decreasing submodular functions g and f $(f, g : 2^V \to \mathbb{R})$, and a budget b, the goal is to find a set of elements $X \subseteq V$ that maximizes the value $g(X)$ under the constraint that $f(X) \leq b$. The authors show that the problem cannot be approximated within any constant bound, give a $1/n$ approximation algorithm, and mainly focus on bi-criterion approximation.

In [25], the authors consider the setting of multivariate submodular optimization, where the argument of the function is a disjoint union of sets. They provide a $\alpha \left(1 - \frac{1}{e}\right)$ (resp. $\alpha \cdot 0.385$) for the maximization problem in the monotone (resp. non-monotone) case, where α is the approximation ratio for a multilinear formulation of the submodular maximization problem.

A last work that is worth mentioning is [22] where the authors consider a class of games, called distributed welfare games, that can be utilized to model the game theoretical version of the MMWC.

2   Single-Budget Case

The single-budget case corresponds to the budgeted maximum coverage problem for which an algorithm that guarantees a $1 - \frac{1}{e}$ approximation factor has been proposed in [19]. The problem generalizes the maximum coverage problem which is known to be NP-hard to approximate to within a factor greater than $1 - \frac{1}{e}$, therefore the algorithm in [19] is optimal from the approximation point of view.

Algorithm 1: Greedy algorithm for single-budget case.

   Input : $\mathcal{S}, X, b_1$

1 $O_1 := \emptyset$; $X' := X$; $\mathcal{S}' := \mathcal{S}$;

2 **repeat**

3   Select $S' \in \mathcal{S}$ that maximizes $\frac{\sum_{x \in S' \cap X'} u_x}{c_{S'}}$;

4   **if** $c_{S'} + \sum_{S \in O_1} c_S \leq b_1$ **then**

5         $O_1 := O_1 \cup \{S'\}$;

6         $X' := X' \setminus S'$;

7      $\mathcal{S}' := \mathcal{S}' \setminus \{S'\}$

8 **until** $\mathcal{S}' = \emptyset$;

9 **return** $O = (O_1)$;

In [19] the authors first notice that a natural greedy algorithm does not guarantee any bounded approximation factor; then they show that the algorithm which returns the maximum between the greedy solution and the single best bin achieves an approximation factor of $1 - \frac{1}{\sqrt{e}}$; moreover, they provide an algorithm which improves the approximation factor to $1 - \frac{1}{e}$ by first guessing three bins that are contained in an optimal solution, and then completing the solution with the greedy algorithm.

In the following we describe the greedy algorithm and show a property that will be used in the next section to prove an approximation bound on the multi-budget case. The pseudo-code is reported in Algorithm 1. The algorithm starts with an empty solution O_1 and, at each iteration selects the bin S' that maximizes the ratio between the weight of the elements in S' that are not yet covered and the cost of S' (line 3). If the sum of the cost of S' and that of the solution computed so far does not exceed the budget b_1, then S' is added to O_1 (lines 4–5).

Let O^* be an optimal solution. Let h be the number of iterations of Algorithm 1 until the first bin is not added to O_1 because it violates the budget constraint (i.e. either $h + 1$ is the first iteration in which the condition at line 4 is not satisfied, or all the bins in $\mathcal{S}$ are added to O_1). For each $j = 1, \ldots, h$, let S_j be the bin selected at iteration j and O_1^j be the set of bins O_1 computed at the end of iteration j. The next lemma is used in [19] to show the approximation ratio of Algorithm 1.

Lemma 1. *[19] For each $j = 1, \ldots, h$, the following holds:*

$$R(O_1^j) \geq \left[1 - \prod_{\ell=1}^{j} \left(1 - \frac{c_{S_\ell}}{b_1} \right) \right] R(O^*).$$

The following proposition will be used in the next section to prove an approximation guarantee in the multi-budget case.

Algorithm 2: Algorithm for the case $c_{\max} \leq b_{\min}$.

Input : $\mathcal{S}, X, B$

1 Run Algorithm 1 with input $(\mathcal{S}, X, \sum_{i=1}^{k} b_i)$;
2 Let $O_1' = \{S_1, \ldots, S_g\}$ be the output of Algorithm 1, where S_j is the bin chosen at iteration j, for $j = 1, \ldots, g$;
3 $O_i := \emptyset$, for each $i = 1, \ldots, k$;
4 **for** $j = 1, \ldots, g$ **do**
5   **if** There exists an index $i \leq k$ such that $c_{S_j} + \sum_{S \in O_i} c_S \leq b_i$ **then**
6   Let i be the smallest index such that $c_{S_j} + \sum_{S \in O_i} c_S \leq b_i$;
7        $O_i := O_i \cup \{S_j\}$;

8 **return** $O = (O_1, \ldots, O_k)$;

Proposition 1. *Let O^* be an optimal solution for the single-budget case and let $\alpha \in [0, 1]$. For each $j = 1, \ldots, h$, if the budget spent by iteration j is at least αb_1, then $R(O_1^j) \geq \left(1 - \frac{1}{e^\alpha}\right) R(O^*)$.*

Proof. By hypothesis, $\sum_{S \in O_1^j} c_S \geq \alpha b_1$. Then, by Lemma 1, we have:

$$R(O) \geq \left[1 - \prod_{\ell=1}^{j} \left(1 - \frac{c_{S_\ell}}{b_1}\right)\right] R(O^*) \geq \left[1 - \prod_{\ell=1}^{j} \left(1 - \frac{\alpha c_{S_\ell}}{\sum_{S \in O_1^j} c_S}\right)\right] R(O^*)$$

$$\geq \left[1 - \left(1 - \frac{\alpha}{j}\right)^j\right] R(O^*) \geq \left(1 - \frac{1}{e^\alpha}\right) R(O^*),$$

where the last two inequalities are due to the following observation (see [4]): For a sequence of numbers $a_1, \ldots, a_n$ such that $\sum_{\ell=1}^{n} a_\ell = A$, the function $\left[1 - \prod_{\ell=1}^{n} \left(1 - \frac{a_\ell \cdot \alpha}{A}\right)\right]$ achieves its minimum when $a_\ell = \frac{A}{n}$ and

$$\left[1 - \prod_{\ell=1}^{n} \left(1 - \frac{a_\ell \cdot \alpha}{A}\right)\right] \geq 1 - \left(1 - \frac{\alpha}{n}\right)^n \geq 1 - e^{-\alpha}.$$

$\square$

3 Smallest Budget Greater Than or Equal to the Highest Cost

In this section, we consider the multi-budget case (where we suppose that $k \geq 2$) in which all the budgets are large enough to be used to buy any bin in $\mathcal{S}$. Formally, if $c_{\max} = \max_{S \in \mathcal{S}} c_S$ and $b_{\min} = \min_{i \in B} b_i$, then $c_{\max} \leq b_{\min}$. Without loss of generality, we assume that the budgets are sorted in non-increasing order, that is $b_1 \geq b_2 \geq \cdots \geq b_k = b_{\min}$, ties are broken arbitrarily.

We give an algorithm that achieves a $1 - \frac{1}{\sqrt{e}}$ approximation ratio. The algorithm, whose pseudo-code is given in Algorithm 2, first defines an instance of the single-budget case which has the same elements X and bins $\mathcal{S}$, while the unique budget is equal to $\sum_{i=1}^{k} b_i$, which is the sum of all the budgets in the multi-budget instance. Algorithm 2 approximately solves this instance by means of Algorithm 1, and then assigns some of the bins returned to the k budgets in such a way that the budget constraints are not violated. In detail, let $\{S_1, \ldots, S_g\}$ be the bins returned by Algorithm 1, where S_j is the bin chosen at iteration j of Algorithm 1, for $j = 1, \ldots, g$. Iteratively, for each $j = 1, \ldots, g$, the algorithm assigns S_j to the budget b_i such that i is minimum (i.e. the budget is maximum) and the cost of S_j plus that of the partial solution O_i does not exceed budget b_i, if such a budget exists, otherwise it discards S_j (lines 4–7).

The next theorem shows a constant approximation bound for Algorithm 2. The assumption that $c_{\max} \leq b_k$ allows us to show that the cost of the bins bought using a budget is at least half of the entire budget $\sum_{i=1}^{k} b_i$. Moreover, these assigned bins satisfy the hypotheses of Proposition 1, therefore, we can exploit it with $\alpha = \frac{1}{2}$ to show the statement.

Theorem 1. *Let O^* be an optimal solution and O be the solution returned by Algorithm 2, then $R(O) \geq \left(1 - \frac{1}{\sqrt{e}}\right) R(O^*)$.*

Proof. Let us consider the last iteration h of Algorithm 1 for which the budget constraint is not violated (i.e. either $h + 1$ is the first iteration in which the condition at line 4 is not satisfied or Algorithm 1 includes all the bins in O_1').

We first show that the cost of bins $S_1, \ldots, S_h$ is at least a fraction $1 - \frac{1}{k}$ of the entire budget $\sum_{i=1}^{k} b_i$ or Algorithm 1 includes all the bins in $\mathcal{S}$ to O_1' by iteration h. Then, we show that there exists a $j \leq h$ such that Algorithm 2 is able to assign to the k budgets all the bins $S_1, \ldots, S_j$ and that the overall cost of these bins is at least half of the entire budget used by Algorithm 1. Finally, we exploit Proposition 1 to show the statement.

If Algorithm 1 does not include all the bins in $\mathcal{S}$ to O_1' by iteration h, then at iteration $h + 1$ the budget $\sum_{i=1}^{k} b_i$ is violated, that is $\sum_{j=1}^{h+1} c_{S_j} > \sum_{i=1}^{k} b_i$. Therefore, $\sum_{j=1}^{h} c_{S_j} > \sum_{i=1}^{k} b_i - c_{S_{h+1}} \geq \sum_{i=1}^{k} b_i - c_{\max} \geq \sum_{i=1}^{k} b_i - b_k \geq \left(1 - \frac{1}{k}\right) \sum_{i=1}^{k} b_i$, where the third inequality is due to the hypothesis that $c_{\max} \leq b_k$, and the last one is implied by the fact that b_k is the minimum budget, which implies that $b_k \leq \frac{1}{k} \sum_{i=1}^{k} b_i$.

Therefore, the cost of $S_1, \ldots, S_h$ is at least a fraction $1 - \frac{1}{k}$ of the entire budget $\sum_{i=1}^{k} b_i$ or Algorithm 1 includes all the bins in $\mathcal{S}$ to O_1' by iteration h.

We now show that there exists a $j \leq h$ such that Algorithm 2 is able to assign to the k budgets all the bins $S_1, \ldots, S_j$ and that the overall cost of these bins is at least half of the entire budget.

If Algorithm 2 is able to assign all the bins $S_1, \ldots, S_h$ to the budgets, then either the overall cost of these bins is at least $\left(1 - \frac{1}{k}\right) \sum_{i=1}^{k} b_i \geq \frac{1}{2} \sum_{i=1}^{k} b_i$, for $k \geq 2$, or we found an optimal solution which assigns all the bins in $\mathcal{S}$ to the budgets. In this case $j = h$.

Otherwise, let $j < h$ be an index such that bins $S_1, \ldots, S_j$ are all assigned to the budgets and $j + 1 \leq h$ is the first iteration of the cycle at lines 4–7 of

Algorithm 2 such that the condition at line 5 does not hold, that is S_{j+1} is the first bin in the greedy ordering that is not assigned to a budget because $\sum_{S \in O_i} c_S + c_{S_{j+1}} > b_i$, for each $i = 1, \ldots, k$, at iteration $j + 1$. Note that $j \geq 1$ because at least one bin is always assigned, assuming that $c_{\max} \leq b_k$.

We show that $\sum_{f=1}^{j} c_{S_f} \geq \frac{1}{2} \sum_{i=1}^{k} b_i$. Equivalently, we show that at the beginning of iteration $j + 1$ (at the end of iteration j), $\sum_{i=1}^{k} \sum_{S \in O_i} c_S \geq \frac{1}{2} \sum_{i=1}^{k} b_i$.

Since S_{j+1} is not assigned to any budget and $c_{S_{j+1}} \leq b_i$, for each budget b_i, then all the budgets are used to buy at least one bin before iteration $j + 1$, that is $O_i \neq \emptyset$, for each i. Formally, $\sum_{S \in O_i} c_S + c_{S_{j+1}} > b_i$ and $c_{S_{j+1}} \leq b_i$, imply $\sum_{S \in O_i} c_S > 0$, that is $O_i \neq \emptyset$, for each $i = 1, \ldots, k$.

If $\sum_{S \in O_i} c_S \geq \frac{1}{2} b_i$, for each budget b_i, then the statement holds. Otherwise, let ℓ be the smallest index such that $\sum_{S \in O_\ell} c_S < \frac{1}{2} b_\ell$. We analyze two cases:

o If $\ell = k$, then the bins assigned to O_k in previous iterations have not been assigned to any budget with $i < k$, which implies that $\sum_{S \in O_i} c_S + \sum_{S \in O_k} c_S > b_i \geq \frac{1}{2}(b_i + b_k)$. This holds in particular for budget $k - 1$. For any other budget $i < k - 1$, by hypothesis we have $\sum_{S \in O_i} c_S \geq \frac{1}{2} b_i$. Therefore, the overall cost of assigned bins is then $\sum_{i=1}^{k-2} \sum_{S \in O_i} c_S + \sum_{S \in O_{k-1}} c_S + \sum_{S \in O_k} c_S \geq \sum_{i=1}^{k-2} \frac{1}{2} b_i + \frac{1}{2}(b_{k-1} + b_k) = \frac{1}{2} \sum_{i=1}^{k} b_i$.

o If $\ell < k$, let us split the budgets different from ℓ into two groups according to their ordering. For each budget $i > \ell$, the bins assigned to O_i in previous iterations have not been assigned to budget ℓ, which implies that $\sum_{S \in O_\ell} c_S + \sum_{S \in O_i} c_S > b_\ell$. From $\sum_{S \in O_\ell} c_S < \frac{1}{2} b_\ell$, follows that $\sum_{S \in O_i} c_S > b_\ell - \sum_{S \in O_\ell} c_S > \frac{1}{2} b_\ell \geq \frac{1}{2} b_i$. In particular, let us consider budget $i = \ell + 1$ (note that this budget always exists, as $\ell < k$). From $\sum_{S \in O_\ell} c_S + \sum_{S \in O_i} c_S > b_\ell$ and $b_\ell \geq b_{\ell+1}$ we have that $\sum_{S \in O_\ell} c_S + \sum_{S \in O_{\ell+1}} c_S > \frac{1}{2}(b_{\ell+1} + b_\ell)$. Finally, for any budget $i < \ell$, by hypothesis we have $\sum_{S \in O_i} c_S \geq \frac{1}{2} b_i$. Therefore, the overall cost of the assigned bins is $\sum_{i=1}^{\ell-1} \sum_{S \in O_i} c_S + \sum_{S \in O_\ell} c_S + \sum_{S \in O_{\ell+1}} c_S + \sum_{i=\ell+2}^{k} \sum_{S \in O_i} c_S \geq \sum_{i=1}^{\ell-1} \frac{1}{2} b_i + \frac{1}{2}(b_\ell + b_{\ell+1}) + \sum_{i=\ell+2}^{k} \frac{1}{2} b_i = \frac{1}{2} \sum_{i=1}^{k} b_i$.

Let $O_1'' = \{S_1, \ldots, S_j\}$. We have just proved that the cost of O_1'' is a fraction $\frac{1}{2}$ of the budget given to Algorithm 1. Therefore, we can apply Proposition 1 with $\alpha = \frac{1}{2}$ to obtain $R(O_1'') \geq \left(1 - \frac{1}{\sqrt{e}}\right) R(O^{**})$. Where O^{**} is an optimal solution to the single-budget instance that has elements X, bins $\mathcal{S}$, and budget $\sum_{i=1}^{k} b_i$. Since this is a relaxation of the original multi-budget instance in which the assignment constraints are relaxed, then $R(O^{**}) \geq R(O^*)$. As Algorithm 2 assigns all the bins in O_1'' to the k budgets, the statement holds. □

We notice that Algorithm 2 fails to achieve a constant approximation if $c_{\max} > b_{\min}$. The problem is that, in this case, Algorithm 1 could select many sets that only few budgets can buy. Specifically, consider the following instance: given k budgets where $b_1 = 1 + \varepsilon$, for a small positive $\varepsilon > 0$, and $b_i = 1$ for any

$i = 2, \ldots, k$, the ground set $X = \{x_1, \ldots, x_{2k-2}\}$ contains $2k - 2$ elements where the first $k-1$ elements have weight $1+2\varepsilon$ and the last $k-1$ have weight 1, that is $u_{x_1} = u_{x_2} = \ldots = u_{x_{k-1}} = 1+2\varepsilon$ and $u_{x_k} = u_{x_{k+1}} = \ldots = u_{x_{2k-2}} = 1$. Moreover, $\mathcal{S} = \{S_1, \ldots, S_{2k-2}\}$ such that $S_j = \{x_j\}$ for any $j = 1, \ldots, 2k - 2$, where the cost of the first $k - 1$ sets is $1 + \varepsilon$ and the cost of the last $k - 1$ is 1, that is $c_{S_1} = c_{S_2} = \ldots = c_{S_{k-1}} = 1 + \varepsilon$ and $c_{S_k} = c_{S_{k+1}} = \ldots = c_{S_{2k-2}} = 1$. Notice that $c_{\max} > b_{\min}$. In such instance, for a sufficiently small ε, Algorithm 1 would select all the first $k - 1$ sets $S_1, \ldots, S_{k-1}$ and none of the sets $S_k, \ldots, S_{2k-2}$, and then Algorithm 2 would assign only S_1 to budget $_b 1$ for an overall weight of $1 + 2\varepsilon$. However, an optimal solution has total weight $k + 2\varepsilon$ where the set S_1 is assigned to budget b_1 and, for any $i = 2, \ldots, k$, the set S_{k-2+i} is assigned to budget b_i.

4   Randomized Algorithm for the General Case

In this section, we give a randomized algorithm for the general case that returns a solution that satisfies the budget constraints in expectation and achieves an expected approximation ratio of $1 - \frac{1}{e}$.

We start by defining an integer program for the general multi-budget case. Let D be all the pairs of a single bin and a single budged that satisfy the budget constraint, that is, $D := \{(S, i) \in \mathcal{S} \times B \mid c_S \leq b_i\}$. We define two types of binary variables: y_x indicates whether element x is covered by any budget, for each $x \in X$, and z_S^i, indicates whether bin S is assigned to budged i, for each $(S, i) \in D$.

The integer program is then as follows.

$$\max \sum_{x \in X} u_x y_x \qquad\qquad\qquad\qquad\qquad\qquad\qquad (\text{IP})$$

$$\sum_{(S,i) \in D} c_S z_S^i \leq b_i \qquad\qquad\qquad \text{for } i = 1, \ldots, k \qquad (1)$$

$$\sum_{(S,i) \in D} z_S^i \leq 1 \qquad\qquad\qquad\qquad \text{for all } S \in \mathcal{S} \qquad (2)$$

$$\sum_{S: x \in S} \sum_{(S,i) \in D} z_S^i \geq y_x \qquad\qquad\qquad \text{for all } x \in X \qquad (3)$$

$$y_x, z_S^i \in \{0, 1\} \qquad\qquad \text{for all } x \in X, (S, i) \in D \qquad (4)$$

Constraints (1) guarantee that the cost of the bins assigned to a budget does not exceed her budget. Constraints (2) guarantee that each bin is assigned to at most one budget. We recall that this is not required by the problem definition, however, in the centralized setting, we can assume that this condition holds for any solution and, in particular, for any optimal solution. Therefore we can add this constraint without loss of generality. Constraints (3), guarantee that, if an element x is covered (i.e. $y_x = 1$), then there exists at least a bin S that contains x and that is assigned to some budget i (i.e. $z_S^i = 1$, for some $(S, i) \in D$ such that $x \in S$).

The randomized algorithm is reported in Algorithm 3. It first solves the linear relaxation of (IP) where the integrality constraints are replaced with bounds

---

Algorithm 3: Randomized algorithm for the general case.

Input : $\mathcal{S}, X, B$

1 Solve the relaxation of (IP) where Constr. (4) are replaced with $y_x, z_S^i \in [0, 1]$, for $x \in X, (S, i) \in D$, and let (y^*, z^*) be an opt. fractional solution;

2 For each $S \in \mathcal{S}$, independently, select at most one variable $z_S^i, (S, i) \in D$, to be set to 1. Each variable is selected with probability z_S^{i*} and no variable is selected with probability $1 - \sum_{(S,i)\in D} z_S^{i*}$;

3 Let $O = (O_1, \ldots, O_k)$, where $O_i = \{S \mid z_S^i = 1\}$;

4 **return** $O = (O_1, \ldots, O_k)$;

---

between 0 and 1 on the variables. Let (y^*, z^*) be an optimal fractional solution of this relaxation. To round this fractional solution to binary values, variables z_S^i are interpreted as probabilities. In detail, let us consider each bin $S \in \mathcal{S}$, independently. Among all variables $z_S^i, (S, i) \in D$, we set at most a variable to 1 and all the other ones to 0. The probability that z_S^i is set to 1 is proportional to the value of the optimal fractional variable z_S^{i*}, while the probability that all variables $z_S^i, (S, i) \in D$, are set to 0 is $1 - \sum_{(S,i)\in D} z_S^{i*}$. Eventually, Algorithm 3 defines O_i as $O_i = \{S \mid z_S^i = 1\}$.

By the above process, we have that the probability that a bin S belongs to O_i, for each $(S, i) \in D$, is $\mathbf{P}[S \in O_i] = \mathbf{P}[z_S^i = 1] = z_S^{i*}$; the probability that a bin $S \in \mathcal{S}$ is assigned to some budget is equal to $\mathbf{P}[S \in \bigcup_{(S,i)\in D} O_i] = \sum_{(S,i)\in D} z_S^{i*}$; and the probability that S is not assigned to any budget is $\mathbf{P}[S \notin \bigcup_{(S,i)\in D} O_i] = 1 - \sum_{(S,i)\in D} z_S^{i*}$.

The solution $O = (O_1, \ldots, O_k)$ returned by Algorithm 3 satisfies the budget constraints in expectation. Indeed, $\mathbf{E}\left[\sum_{S\in O_i} c_S\right] = \sum_{(S,i)\in D} c_S \mathbf{P}[S \in O_i] = \sum_{(S,i)\in D} c_S z_S^{i*} \leq b_i$, where the last inequality is due to Constraint (1).

The following theorem shows the expected approximation ratio of Algorithm 3.

Theorem 2. *Let O^* be an optimal solution and O be the solution returned by Algorithm 3, then $\mathbf{E}[R(O)] \geq \left(1 - \frac{1}{e}\right) R(O^*)$.*

Proof. For each $x \in X$, let us denote by w_x the random binary variable that is equal to 1 if element x is covered. The probability that $w_x = 1$ is equal to the probability that x belongs to $\cup_{i\in B} C_i(O)$, that is,

$$\mathbf{P}[w_x = 1] = \mathbf{P}[x \in \cup_{i\in B} C_i(O)] = 1 - \mathbf{P}[x \notin \cup_{i\in B} C_i(O)].$$

The probability that x does not belong to $\cup_{i\in B} C_i(O)$ is equal to the probability that each bin S that contains x is not selected by Algorithm 3. Since bins are selected independently, this is equal to

$$\mathbf{P}[x \notin \cup_{i\in B} C_i(O)] = \prod_{S:x\in S} \mathbf{P}[S \notin \bigcup_{(S,i)\in D} O_i] = \prod_{S:x\in S} \left(1 - \sum_{(S,i)\in D} z_S^{i*}\right).$$

Since $1 - p \leq e^{-p}$, for any $p \in [0,1]$ and $\sum_{(S,i) \in D} z_S^{i*} \in [0,1]$ by Constraint (2), then this value is at most

$$\prod_{S:x \in S} \exp\left(-\sum_{(S,i) \in D} z_S^{i*}\right) = \exp\left(-\sum_{S:x \in S} \sum_{(S,i) \in D} z_S^{i*}\right).$$

By Constraint (3), $\sum_{S:x \in S} \sum_{(S,i) \in D} z_S^{i*} \geq y_x^*$ and then $\mathbf{P}[x \notin \cup_{i \in B} C_i(O)] \leq \exp(-y_x^*)$. Therefore,

$$\mathbf{P}[w_x = 1] \geq 1 - e^{-y_x^*} \geq \left(1 - e^{-1}\right) y_x^*,$$

where the last inequality is due to the fact that $1 - e^{-p} \geq (1 - e^{-1})p$, for any $p \in [0,1]$,[1] and $y_x^* \in [0,1]$. The expected value of $R(O)$, $\mathbf{E}[R(O)]$, is equal to

$$\mathbf{E}\left[\sum_{x \in X} u_x w_x\right] = \sum_{x \in X} u_x \mathbf{P}[w_x = 1] \geq \sum_{x \in X} u_x \left(1 - e^{-1}\right) y_x^* \geq \left(1 - e^{-1}\right) R(O^*),$$

since $\sum_{x \in X} u_x y_x^*$ is the optimum value for the linear relaxation of (IP). □

5   Future Work

The main open problems are that of finding a deterministic constant approximation algorithm for the general case and, in the case when $c_{\max} \leq b_{\min}$, closing the gap between our deterministic $\left(1 - \frac{1}{\sqrt{e}}\right)$-approximation algorithm and the $1 - \frac{1}{e}$ hardness of approximation for the maximum coverage problem [10]. One possible direction could be that of exploiting the framework of multilinear relaxation and contention resolution schemes given in [7]. We observe that MMWC cannot be directly reduced to any of the problems solved in [7], as the constraints of our problem cannot be directly modeled as knapsack, matroid, or sparse packing constraints (or their intersection). However, this does not exclude the existence of a suitable contention resolution scheme for our problem.

Moreover, extending the definition of multi-budget coverage to more general weight functions (e.g. general submodular functions) is worth further research.

References

1. Buchbinder, N., Feldman, M., Naor, J., Schwartz, R.: A tight linear time (1/2)-approximation for unconstrained submodular maximization. In: Proceedings of FOCS, pp. 649–658 (2012). https://doi.org/10.1109/FOCS.2012.73
2. Călinescu, G., Chekuri, C., Pál, M., Vondrák, J.: Maximizing a monotone submodular function subject to a matroid constraint. SIAM J. Comput. 40(6), 1740–1766 (2011). https://doi.org/10.1137/080733991

---

[1] To see this, observe that function $1 - e^{-p}$ is concave for $p > 0$ and it is equal to 0 when $p = 0$ and to $1 - e^{-1}$ when $p = 1$.

3. Caragiannis, I., Monaco, G.: A 6/5-approximation algorithm for the maximum 3-cover problem. J. Comb. Optim. **25**(1), 60–77 (2013). https://doi.org/10.1007/s10878-011-9417-z
4. Cellinese, F., D'Angelo, G., Monaco, G., Velaj, Y.: Generalized budgeted submodular set function maximization. In: Proceedings of MFCS. LIPIcs, vol. 117, pp. 31:1–31:14 (2018). https://doi.org/10.4230/LIPIcs.MFCS.2018.31
5. Chakrabarty, D., Goel, G.: On the approximability of budgeted allocations and improved lower bounds for submodular welfare maximization and gap. In: Proceedings of FOCS, pp. 687–696 (2008). https://doi.org/10.1137/080735503
6. Chekuri, C., Kumar, A.: Maximum coverage problem with group budget constraints and applications. In: Jansen, K., Khanna, S., Rolim, J.D.P., Ron, D. (eds.) APPROX/RANDOM-2004. LNCS, vol. 3122, pp. 72–83. Springer, Heidelberg (2004). https://doi.org/10.1007/978-3-540-27821-4_7
7. Chekuri, C., Vondrák, J., Zenklusen, R.: Submodular function maximization via the multilinear relaxation and contention resolution schemes. SIAM J. Comput. **43**(6), 1831–1879 (2014)
8. Cohen, R., Katzir, L.: The generalized maximum coverage problem. Inf. Process. Lett. **108**(1), 15–22 (2008). https://doi.org/10.1016/j.ipl.2008.03.017
9. Farbstein, B., Levin, A.: Maximum coverage problem with group budget constraints. J. Comb. Optim. **34**(3), 725–735 (2017). https://doi.org/10.1007/s10878-016-0102-0
10. Feige, U.: A threshold of ln n for approximating set cover. J. ACM **45**(4), 634–652 (1998). https://doi.org/10.1145/237814.237977
11. Feige, U., Mirrokni, V.S., Vondrák, J.: Maximizing non-monotone submodular functions. In: Proceedings of FOCS, pp. 461–471 (2007). https://doi.org/10.1137/090779346
12. Filmus, Y., Ward, J.: Monotone submodular maximization over a matroid via non-oblivious local search. SIAM J. Comput. **43**(2), 514–542 (2014). https://doi.org/10.1137/130920277
13. Gairing, M.: Covering games: approximation through non-cooperation. In: Proceedings of WINE, pp. 184–195 (2009). https://doi.org/10.1007/978-3-642-10841-9_18
14. Goel, G., Karande, C., Tripathi, P., Wang, L.: Approximability of combinatorial problems with multi-agent submodular cost functions. In: Proceedings of FOCS, pp. 755–764 (2009). https://doi.org/10.1109/FOCS.2009.81
15. Goel, G., Tripathi, P., Wang, L.: Combinatorial problems with discounted price functions in multi-agent systems. In: Proceedings of the FSTTCS, pp. 436–446 (2010). https://doi.org/10.4230/LIPIcs.FSTTCS.2010.436
16. Hochbaum, D.: Approximation Algorithms for NP-Hard Problems. PWS Publishing Company, Boston (1997). https://doi.org/10.1145/261342.571216
17. Iwata, S., Nagano, K.: Submodular function minimization under covering constraints. In: Proceedings of FOCS, pp. 671–680 (2009). https://doi.org/10.1109/FOCS.2009.31
18. Iyer, R.K., Bilmes, J.A.: Submodular optimization with submodular cover and submodular knapsack constraints. In: Proceedings of NIPS, pp. 2436–2444 (2013). https://doi.org/10.1145/1374376.1374389
19. Khuller, S., Moss, A., Naor, J.S.: The budgeted maximum coverage problem. Inf. Process. Lett. **70**(1), 39–45 (1999). https://doi.org/10.1016/S0020-0190(99)00031-9

20. Kulik, A., Shachnai, H., Tamir, T.: Approximations for monotone and nonmono-tone submodular maximization with knapsack constraints. Math. Oper. Res. **38**(4), 729–739 (2013)
21. Lee, J., Sviridenko, M., Vondrák, J.: Submodular maximization over multiple matroids via generalized exchange properties. Math. Oper. Res. **35**(4), 795–806 (2010). https://doi.org/10.1007/978-3-642-03685-9_19
22. Marden, J.R., Wierman, A.: Distributed welfare games. Oper. Res. **61**(1), 155–168 (2013). https://doi.org/10.1287/opre.1120.1137
23. Nemhauser, G.L., Wolsey, L.A., Fisher, M.L.: An analysis of approximations for maximizing submodular set functions-I. Math. Program. **14**(1), 265–294 (1978). https://doi.org/10.1007/BF01588971
24. Santiago, R., Shepherd, F.B.: Multi-agent submodular optimization. In: Proceed-ings of APPROX/RANDOM. LIPIcs, vol. 116, pp. 23:1–23:20 (2018). https://doi.org/10.4230/LIPIcs.APPROX-RANDOM.2018.23
25. Santiago, R., Shepherd, F.B.: Multivariate submodular optimization. In: Proceed-ings of ICML, vol. 97, pp. 5599–5609. PMLR (2019)
26. Sviridenko, M.: A note on maximizing a submodular set function subject to a knap-sack constraint. Oper. Res. Lett. **32**(1), 41–43 (2004). https://doi.org/10.1016/S0167-6377(03)00062-2
27. Vondrak, J.: Optimal approximation for the submodular welfare problem in the value oracle model. In: Proceedings of STOC, pp. 67–74. ACM (2008). https://doi.org/10.1145/1374376.1374389
28. Ward, J.: A (k+3)/2-approximation algorithm for monotone submodular k-set packing and general k-exchange systems. In: Proceedings of STACS, pp. 42–53 (2012). https://doi.org/10.4230/LIPIcs.STACS.2012.42

A Tight Lower Bound for Edge-Disjoint Paths on Planar DAGs

Rajesh Chitnis$^{(\boxtimes)}$

School of Computer Science, University of Birmingham, Birmingham, UK

Abstract. Given a graph G and a set $\mathcal{T} = \{(s_i, t_i) : 1 \leq i \leq k\}$ of k pairs, the VERTEX-DISJOINT PATHS (resp. EDGE-DISJOINT PATHS) problems asks to determine whether there exist pairwise vertex-disjoint (resp. edge-disjoint) paths $P_1, P_2, \ldots, P_k$ in G such that P_i connects s_i to t_i for each $1 \leq i \leq k$. Unlike their undirected counterparts which are FPT (parameterized by k) from Graph Minor theory, both the edge-disjoint and vertex-disjoint versions in directed graphs were shown by Fortune et al. (TCS '80) to be NP-hard for $k = 2$. This strong hardness for DISJOINT PATHS on general directed graphs led to the study of parameterized complexity on special graph classes, e.g., when the underlying undirected graph is planar. For VERTEX-DISJOINT PATHS on planar directed graphs, Schrijver (SICOMP '94) designed an $n^{O(k)}$ time algorithm which was later improved upon by Cygan et al. (FOCS '13) who designed an FPT algorithm running in $2^{2^{O(k^2)}} \cdot n^{O(1)}$ time. To the best of our knowledge, the parameterized complexity of EDGE-DISJOINT PATHS on planar (A directed graph is planar if its underlying undirected graph is planar) directed graphs is unknown.

We resolve this gap by showing that EDGE-DISJOINT PATHS is W[1]-hard parameterized by the number k of terminal pairs, even when the input graph is a planar directed acyclic graph (DAG). This answers a question of Slivkins (ESA '03, SIDMA '10). Moreover, under the Exponential Time Hypothesis (ETH), we show that there is no $f(k) \cdot n^{o(k)}$ algorithm for EDGE-DISJOINT PATHS on planar DAGs, where k is the number of terminal pairs, n is the number of vertices and f is any computable function. Our hardness holds even if both the maximum in-degree and maximum out-degree of the graph are at most 2. We now place our result in the context of previously known algorithms and hardness for EDGE-DISJOINT PATHS on special classes of directed graphs:

– **Implications for Edge-Disjoint Paths on DAGs**: Our result shows that the $n^{O(k)}$ algorithm of Fortune et al. (TCS '80) for EDGE-DISJOINT PATHS on DAGs is asymptotically tight, even if we add an extra restriction of planarity. The previous best lower bound (also under ETH) for EDGE-DISJOINT PATHS on DAGs was $f(k) \cdot n^{o(k/\log k)}$ by Amiri et al. (MFCS '16, IPL '19) which improved upon the $f(k) \cdot n^{o(\sqrt{k})}$ lower bound implicit in Slivkins (ESA '03, SIDMA '10).

Full version [3] is available at https://arxiv.org/abs/2101.10742.

© Springer Nature Switzerland AG 2021
T. Calamoneri and F. Corò (Eds.): CIAC 2021, LNCS 12701, pp. 187–201, 2021.
https://doi.org/10.1007/978-3-030-75242-2_13

- **Implications for Edge-Disjoint Paths on planar directed graphs**: As a special case of our result, we obtain that EDGE-DISJOINT PATHS on planar directed graphs is W[1]-hard parameterized by the number k of terminal pairs. This answers a question of Cygan et al. (FOCS '13) and Schrijver (pp. 417–444, Building Bridges II, '19), and completes the landscape (see Table 2) of the parameterized complexity status of edge and vertex versions of the DISJOINT PATHS problem on planar directed and planar undirected graphs.

1   Introduction

The DISJOINT PATHS problem is one of the most fundamental problems in graph theory: given a graph and a set of k terminal pairs, the question is to determine whether there exists a collection of k pairwise disjoint paths where each path connects one of the given terminal pairs? There are four natural variants of this problem depending on whether we consider undirected or directed graphs and the edge-disjoint or vertex-disjoint requirement. In undirected graphs, the edge-disjoint version is reducible to the vertex-disjoint version in polynomial time by considering the line graph. In directed graphs, the edge-disjoint version and vertex-disjoint version are known to be equivalent in terms of designing exact algorithms. Besides its theoretical importance, the DISJOINT PATHS problem has found applications in VLSI design, routing, etc. The interested reader is referred to the surveys [16] and [26, Chapter 9] for more details.

The case when the number of terminal pairs k is bounded is of special interest: given a graph with n vertices and k terminal pairs the goal is to try to design FPT algorithms, i.e., algorithms whose running time is $f(k) \cdot n^{O(1)}$ for some computable function f, or XP algorithms, i.e., algorithms whose running time is $n^{g(k)}$ for some computable function g. We now discuss some of the known results on exact[1] algorithms for different variants of the DISJOINT PATHS problem before stating our result.

Prior Work on Exact Algorithms for Disjoint Paths on Undirected Graphs: The NP-hardness for EDGE-DISJOINT PATHS and VERTEX-DISJOINT PATHS on undirected graphs was shown by Even et al. [14]. Solving the VERTEX-DISJOINT PATHS problem on undirected graphs is an important subroutine in checking whether a fixed graph H is a minor of a graph G. Hence, a core algorithmic result of the seminal work of Robertson and Seymour was their FPT algorithm [24] for VERTEX-DISJOINT PATHS (and hence also EDGE-DISJOINT PATHS) on general undirected graphs which runs in $O(g(k) \cdot n^3)$ time for some function g. The cubic dependence on the input size was improved to quadratic by Kawarabayashi et al. [19] who designed an algorithm running in $O(h(k) \cdot n^2)$ time for some function h. Both the functions g and h are quite large (at least quintuple exponential as per [1]). This naturally led to the search for faster FPT algorithms on planar graphs: Adler et al. [1] designed an algorithm for

---

[1] This paper focuses on exact algorithms for DISJOINT PATHS so we do not discuss here the many results regarding (in)approximability.

VERTEX-DISJOINT PATHS on planar graphs which runs in $2^{2^{O(k^2)}} \cdot n^{O(1)}$ time. Very recently, this was improved to an single-exponential time FPT algorithm which runs in $2^{O(k^2)} \cdot n^{O(1)}$ time by Lokshtanov et al. [20].

Prior Work on Exact Algorithms for Disjoint Paths on Directed Graphs: Unlike undirected graphs where both EDGE-DISJOINT PATHS and VERTEX-DISJOINT PATHS are FPT parameterized by k, the DISJOINT PATHS problem becomes significantly harder for directed graphs: Fortune et al. [15] showed that both EDGE-DISJOINT PATHS and VERTEX-DISJOINT PATHS on general directed graphs are NP-hard even for $k = 2$. The DISJOINT PATHS problem has also been extensively studied on special classes of digraphs:

- **Disjoint Paths on DAGs**: It is easy to show that VERTEX-DISJOINT PATHS and EDGE-DISJOINT PATHS are equivalent on the class of directed acyclic graphs (DAGs). Fortune et al. [15] designed an $n^{O(k)}$ algorithm for EDGE-DISJOINT PATHS on DAGs. Slivkins [28] showed W[1]-hardness for EDGE-DISJOINT PATHS on DAGs and a $f(k) \cdot n^{o(\sqrt{k})}$ lower bound (for any computable function f) under the Exponential Time Hypothesis [17,18] (ETH) follows from that reduction. Amiri et al. [2][2] improved the lower bound to $f(k) \cdot n^{o(k/\log k)}$ thus showing that the algorithm of Fortune et al. [15] is almost-tight.

- **Disjoint Paths on directed planar graphs**: Schrijver [25] designed an $n^{O(k)}$ algorithm for VERTEX-DISJOINT PATHS on directed planar graphs. This was improved upon by Cygan et al. [12] who designed an FPT algorithm running in $2^{2^{O(k^2)}} \cdot n^{O(1)}$ time. As pointed out by Cygan et al. [12], their FPT algorithm for VERTEX-DISJOINT PATHS on directed planar graphs does not work for the EDGE-DISJOINT PATHS problem. The status of parameterized complexity (parameterized by k) of EDGE-DISJOINT PATHS on directed planar graphs remained an open question. Table 1 gives a summary of known results for exact algorithms for DISJOINT PATHS on (subclasses of) directed graphs.

Our Result: We resolve this open question by showing a slightly stronger result: the EDGE-DISJOINT PATHS problem is W[1]-hard parameterized by k when the input graph is a planar DAG. First we define the EDGE-DISJOINT PATHS problem formally below, and then state our result:

EDGE-DISJOINT PATHS
Input: A directed graph $G = (V, E)$, and a set $\mathcal{T} \subseteq V \times V$ of k terminal pairs given by $\{(s_i, t_i) : 1 \le i \le k\}$.
Question: Do there exist k pairwise edge-disjoint paths $P_1, P_2, \ldots, P_k$ such that P_i is an $s_i \rightsquigarrow t_i$ path for each $1 \le i \le k$?
Parameter: k

Theorem 1.1. *The* EDGE-DISJOINT PATHS *problem on planar DAGs is W[1]-hard parameterized by the number k of terminal pairs. Moreover, under ETH, the*

---
[2] [2] considers a more general version than DISJOINT PATHS which allows congestion.

Table 1. The landscape of parameterized complexity results for DISJOINT PATHS on directed graphs. All lower bounds are under the Exponential Time Hypothesis (ETH). To the best of our knowledge, the entries marked with ???? have no known non-trivial results.

Graph class	Problem type	Algorithm	Lower Bound
General graphs	Vertex & edge-disjoint	????	NP-hard for $k = 2$
DAGs	Vertex & edge-disjoint	$n^{O(k)}$ [15]	$f(k) \cdot n^{o(\sqrt{k})}$ [28] $f(k) \cdot n^{o(k/\log k)}$ [2] $f(k) \cdot n^{o(k)}$ [this paper]
Planar graphs	Vertex-disjoint	$n^{O(k)}$ [25] $2^{2^{O(k^2)}} \cdot n^{O(1)}$ [12]	????
	Edge-disjoint	????	$f(k) \cdot n^{o(k)}$ [this paper]
Planar DAGs	Vertex-disjoint	$2^{2^{O(k^2)}} \cdot n^{O(1)}$ [12]	????
	Edge-disjoint	$n^{O(k)}$ [15]	$f(k) \cdot n^{o(k)}$ [this paper]

EDGE-DISJOINT PATHS *problem on planar DAGs cannot be solved* $f(k) \cdot n^{o(k)}$ *time where* f *is any computable function,* n *is the number of vertices and* k *is the number of terminal pairs. The hardness holds even if both the maximum in-degree and maximum out-degree of the graph are at most 2.*

Recall that the Exponential Time Hypothesis (ETH) states that n-variable m-clause 3-SAT cannot be solved in $2^{o(n)} \cdot (n + m)^{O(1)}$ time [17,18]. Prior to our result, only the NP-completeness of EDGE-DISJOINT PATHS on planar DAGs was known [29]. The reduction used in Theorem 1.1 is heavily inspired by some known reductions: in particular, the planar DAG structure (Fig. 1) is from [6] and the splitting operation (Fig. 2 and Definition 2.2) is from [4,5]. We view the simplicity of our reduction as evidence of success of the (now) established methodology of showing W[1]-hardness (and ETH-based hardness) for planar graph problems using GRID-TILING and its variants.

Placing Theorem 1.1 in the Context of Prior Work: Theorem 1.1 answers a question of Slivkins [28] regarding the parameterized complexity of EDGE-DISJOINT PATHS on planar DAGs. As a special case of Theorem 1.1, one obtains that EDGE-DISJOINT PATHS on planar directed graphs is W[1]-hard parameterized by the number k of terminal pairs: this answers a question of Cygan et al. [12] and Schrijver [27]. The W[1]-hardness result of Theorem 1.1 completes the landscape (see Table 2) of parameterized complexity of edge-disjoint and vertex-disjoint versions of the DISJOINT PATHS problem on planar directed and planar undirected graphs. Theorem 1.1 also shows that the $n^{O(k)}$ algorithm of Fortune et al. [15] for EDGE-DISJOINT PATHS on DAGS is asymptotically optimal, even if we add an extra restriction of planarity to the mix. Theorem 1.1 adds another problem (EDGE-DISJOINT PATHS on DAGs) to the relatively small[3] list of problems for which it is provably known that the planar version has the same asymptotic complexity as the problem on general graphs. This is in contrast to the fact that for several problems the planar version is easier by a square root factor in the exponent [21] as compared to general graphs.

Table 2. The landscape of parameterized complexity results for the four different versions (edge-disjoint vs vertex-disjoint & directed vs undirected) of DISJOINT PATHS on planar graphs.

Graph class	Problem type	Parameterized Complexity w.r.t by k
Planar undirected	Vertex-disjoint	FPT [1, 19, 20, 24]
	Edge-disjoint	
Planar directed	Vertex-disjoint	FPT [12]
	Edge-disjoint	W[1]-hard [**this paper**]

Organization of the Paper: In Sect. 2.1 we describe the construction of the instance $(G_2, \mathcal{T})$ of EDGE-DISJOINT PATHS. The two directions of the reduction are shown in Sect. 2.2 and Sect. 2.3 respectively. Finally, Sect. 2.4 contains the proof of Theorem 1.1. We conclude with some open questions in Sect. 3.

Notation: All graphs considered in this paper are directed and do not have self-loops or multiple edges. We use (mostly) standard graph theory notation [13]. The set $\{1, 2, 3, \ldots, M\}$ is denoted by $[M]$ for each $M \in \mathbb{N}$. A directed edge (resp. path) from s to t is denoted by $s \to t$ (resp. $s \rightsquigarrow t$). We use the **non-standard** notation: $s \rightsquigarrow s$ **does not** represent a self-loop but rather is to be viewed as *"just staying put"* at the vertex s. If $A, B \subseteq V(G)$ then we say that there is an $A \rightsquigarrow B$ path if and only if there exists two vertices $a \in A, b \in B$ such that there is an $a \rightsquigarrow b$ path. For $A \subseteq V(G)$ we define $N_G^+(A) = \{x \notin A : \exists\, y \in A \text{ such that } (y, x) \in E(G)\}$ and $N_G^-(A) = \{x \notin A : \exists\, y \in A \text{ such that } (x, y) \in E(G)\}$. For $A \subseteq V(G)$ we define $G[A]$ to be the graph induced on the vertex set A, i.e., $G[A] := (A, E_A)$ where $E_A := E(G) \cap (A \times A)$.

2   W[1]-Hardness of Edge-Disjoint Paths on Planar DAGs

To obtain W[1]-hardness for EDGE-DISJOINT PATHS, we reduce from the GRID-TILING-$\leq$ problem [23] which is defined below:

---

[3] The only such problems we are aware of are [5, 6, 22].

GRID-TILING-$\leq$

Input: Integers k, N, and a collection $\mathcal{S}$ of k^2 sets given by $\{S_{x,y} \subseteq [N] \times [N] : 1 \leq x, y \leq k\}$.

Question: For each $1 \leq x, y \leq k$ does there exist a pair $\gamma_{x,y} \in S_{x,y}$ such that

  - if $\gamma_{x,y} = (a, b)$ and $\gamma_{x+1,y} = (a', b')$ then $b \leq b'$, and
  - if $\gamma_{x,y} = (a, b)$ and $\gamma_{x,y+1} = (a', b')$ then $a \leq a'$

It is known [11, Theorem 14.30] that GRID-TILING-$\leq$ is W[1]-hard parameterized by k, and under the Exponential Time Hypothesis (ETH) has no $f(k) \cdot N^{o(k)}$ algorithm for any computable function f. We will exploit this result by reducing an instance $(k, N, \mathcal{S})$ of GRID-TILING-$\leq$ in poly(N, k) time to an instance $(G_2, \mathcal{T})$ of EDGE-DISJOINT PATHS such that G_2 is a planar DAG, number of vertices in G_2 is $|V(G_2)| = O(N^2 k^2)$ and number of terminal pairs is $|\mathcal{T}| = 2k$.

Remark 2.1. Our definition of GRID-TILING-$\leq$ above is slightly different than the one given in [11, Theorem 14.30]: there the constraints are first coordinate of $\gamma_{x,y}$ is $\leq$ first coordinate of $\gamma_{x+1,y}$ and second coordinate of $\gamma_{x,y}$ is $\leq$ second coordinate of $\gamma_{x,y+1}$. By rotating the axis by 90°, i.e., swapping the indices, our version of GRID-TILING-$\leq$ is equivalent to that from [11, Theorem 14.30].

2.1   Construction of the Instance $(G_2, \mathcal{T})$ of Edge-Disjoint Paths

Consider an instance $(N, k, \mathcal{S})$ of GRID-TILING-$\leq$. We now build an instance $(G_2, \mathcal{T})$ of EDGE-DISJOINT PATHS as follows: first in Sect. 2.1.1 we describe the construction of an intermediate graph G_1 (Fig. 1). The splitting operation is defined in Sect. 2.1.2, and the graph G_2 is obtained from G_1 by splitting each (black) grid vertex.

2.1.1   Construction of the Graph G_1

Given integers k and N, we build a directed graph G_1 as follows (refer to Fig. 1):

1. **Origin**: The origin is marked at the bottom left corner of Fig. 1. This is defined just so we can view the naming of the vertices as per the usual $X - Y$ coordinate system: increasing horizontally towards the right, and vertically towards the top.
2. **Grid (black) vertices and edges**: For each $1 \leq i, j \leq k$ we introduce a (directed) $N \times N$ grid $G_{i,j}$ where the column numbers increase from 1 to N as we go from left to right, and the row numbers increase from 1 to N as we go from bottom to top. For each $1 \leq q, \ell \leq N$ the unique vertex which is the intersection of the qth column and ℓ^{th} row of $G_{i,j}$ is denoted by $\mathbf{w}_{i,j}^{q,\ell}$. The vertex set and edge set of $G_{i,j}$ is defined formally as:
   - $V(G_{i,j}) = \{\mathbf{w}_{i,j}^{q,\ell} : 1 \leq q, \ell \leq N\}$

$$- \; E(G_{i,j}) = \left(\bigcup_{(q,\ell) \in [N] \times [N-1]} \mathbf{w}_{i,j}^{q,\ell} \to \mathbf{w}_{i,j}^{q,\ell+1} \right) \cup \left(\bigcup_{(q,\ell) \in [N-1] \times [N]} \mathbf{w}_{i,j}^{q,\ell} \to \mathbf{w}_{i,j}^{q+1,\ell} \right)$$

All vertices and edges of $G_{i,j}$ are shown in Fig. 1 using black color. Note that each horizontal edge of the grid $G_{i,j}$ is oriented to the right, and each vertical edge is oriented towards the top. We will later (Definition 2.2) modify the grid $G_{i,j}$ to *represent* the set $S_{i,j}$.

For each $1 \le i, j \le k$ we define the set of *boundary* vertices of the grid $G_{i,j}$ as follows:

$$\begin{aligned} \texttt{Left}(G_{i,j}) &:= \left\{ \mathbf{w}_{i,j}^{1,\ell} \; : \; \ell \in [N] \right\} ; \; \texttt{Right}(G_{i,j}) := \left\{ \mathbf{w}_{i,j}^{N,\ell} \; : \; \ell \in [N] \right\} \\ \texttt{Top}(G_{i,j}) &:= \left\{ \mathbf{w}_{i,j}^{\ell,N} \; : \; \ell \in [N] \right\} ; \; \texttt{Bottom}(G_{i,j}) := \left\{ \mathbf{w}_{i,j}^{\ell,1} \; : \; \ell \in [N] \right\} \end{aligned} \quad (1)$$

3. **Arranging the k^2 different $N \times N$ grids $\{G_{i,j}\}_{1 \le i,j \le k}$ into a large $k \times k$ grid:** We place the grids $G_{i,j}$ into a big $k \times k$ grid of grids left to right according to growing i and from bottom to top according to growing j.

4. **Blue vertices and red edges for horizontal connections:** For each $(i,j) \in [k-1] \times [k]$ we add a set of vertices $H_{i,j}^{i+1,j} := \left\{ \mathbf{h}_{i,j}^{i+1,j}(\ell) \; : \; \ell \in [N] \right\}$ shown in Fig. 1 using blue color. We also add the following three sets of edges (shown in Fig. 1 using red color):
   - a directed path of $N-1$ edges given by $\texttt{Path}(H_{i,j}^{i+1,j}) := \{ \mathbf{h}_{i,j}^{i+1,j}(\ell) \to \mathbf{h}_{i,j}^{i+1,j}(\ell+1) \; : \; \ell \in [N-1] \}$
   - a directed perfect matching from $\texttt{Right}(G_{i,j})$ to $H_{i,j}^{i+1,j}$ given by $\texttt{Matching}(G_{i,j}, H_{i,j}^{i+1,j}) := \left\{ \mathbf{w}_{i,j}^{N,\ell} \to \mathbf{h}_{i,j}^{i+1,j}(\ell) \; : \; \ell \in [N] \right\}$
   - a directed perfect matching from $H_{i,j}^{i+1,j}$ to $\texttt{Left}(G_{i+1,j})$ given by $\texttt{Matching}(H_{i,j}^{i+1,j}, G_{i+1,j}) := \{ \mathbf{h}_{i,j}^{i+1,j}(\ell) \to \mathbf{w}_{i+1,j}^{1,\ell} \; : \; \ell \in [N] \}$

5. **Blue vertices and red edges for vertical connections:** For each $(i,j) \in [k] \times [k-1]$ we add a set of vertices $V_{i,j}^{i,j+1} := \left\{ \mathbf{v}_{i,j}^{i,j+1}(\ell) \; : \; \ell \in [N] \right\}$ shown in Fig. 1 using blue color. We also add the following three sets of edges (shown in Fig. 1 using red color):
   - a directed path of $N-1$ edges given by $\texttt{Path}(V_{i,j}^{i,j+1}) := \{ \mathbf{v}_{i,j}^{i,j+1}(\ell) \to \mathbf{v}_{i,j}^{i,j+1}(\ell+1) \; : \; \ell \in [N-1] \}$
   - a directed perfect matching from $\texttt{Top}(G_{i,j})$ to $V_{i,j}^{i,j+1}$ given by $\texttt{Matching}(G_{i,j}, V_{i,j}^{i,j+1}) := \left\{ \mathbf{w}_{i,j}^{\ell,N} \to \mathbf{v}_{i,j}^{i,j+1}(\ell) \; : \; \ell \in [N] \right\}$
   - a directed perfect matching from $V_{i,j}^{i,j+1}$ to $\texttt{Bottom}(G_{i,j+1})$ given by $\texttt{Matching}(V_{i,j}^{i,j+1}, G_{i,j+1}) := \{ \mathbf{v}_{i,j}^{i,j+1}(\ell) \to \mathbf{w}_{i,j+1}^{\ell,1} \; : \; \ell \in [N] \}$

6. **Green (terminal) vertices and magenta edges:** For each $i \in [k]$ we add the following four sets of (terminal) vertices (shown in Fig. 1 using green color)

$$\begin{aligned} A &:= \{ a_i \; : \; i \in [k] \} \quad ; \quad B := \{ b_i \; : \; i \in [k] \} \\ C &:= \{ c_i \; : \; i \in [k] \} \quad ; \quad D := \{ d_i \; : \; i \in [k] \} \end{aligned} \quad (2)$$

For each $i \in [k]$ we add the edges (shown in Fig. 1 using magenta color)

$$\texttt{Source}(A) := \{ a_i \to \mathbf{w}_{i,1}^{\ell,1} \; : \; \ell \in [N] \} ; \; \texttt{Sink}(B) := \{ \mathbf{w}_{i,N}^{\ell,N} \to b_i \; : \; \ell \in [N] \} \quad (3)$$

For each $j \in [k]$ we add the edges (shown in Fig. 1 using magenta color)

$$\text{Source}(C) := \left\{ c_j \to \mathbf{w}_{1,j}^{1,\ell} \; : \; \ell \in [N] \right\} \; ; \; \text{Sink}(D) := \left\{ \mathbf{w}_{N,j}^{N,\ell} \to d_j \; : \; \ell \in [N] \right\} \quad (4)$$

This completes the construction of the graph G_1 (see Fig. 1).

Claim 2.1. G_1 *is a planar DAG*

Proof. Figure 1 gives a planar embedding of G_1. It is easy to verify from the construction of G_1 described at the start of Sect. 2.1.1 (see also Fig. 1) that G_1 is a DAG. $\square$

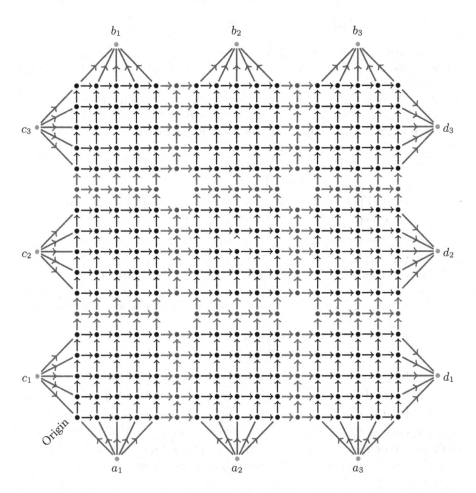

Fig. 1. The graph G_1 constructed for the input $k = 3$ and $N = 5$ via the construction described in Sect. 2.1.1. The final graph G_2 for the EDGE-DISJOINT PATHS instance is obtained from G_1 by the splitting operation (Definition 2.2) as described in Sect. 2.1.2. (Color figure online)

2.1.2 Obtaining the Graph G_2 from G_1 via the Splitting Operation

Observe (see Fig. 1) that every (black) grid vertex in G_1 has in-degree two and out-degree two. Moreover, the two in-neighbors and two out-neighbors do not appear alternately. For each (black) grid vertex $z \in G_1$ we set up the notation:

Definition 2.1. (four neighbors of each grid vertex in G_1) For each (black) grid vertex $\mathbf{z} \in G_1$ we define the following four vertices

- $\mathtt{west}(\mathbf{z})$ is the vertex to the left of $\mathbf{z}$ (as seen by the reader) which has an edge incoming into $\mathbf{z}$
- $\mathtt{south}(\mathbf{z})$ is the vertex below $\mathbf{z}$ (as seen by the reader) which has an edge incoming into $\mathbf{z}$
- $\mathtt{east}(\mathbf{z})$ is the vertex to the right of $\mathbf{z}$ (as seen by the reader) which has an edge outgoing from $\mathbf{z}$
- $\mathtt{north}(\mathbf{z})$ is the vertex above $\mathbf{z}$ (as seen by the reader) which has an edge outgoing from $\mathbf{z}$

We now define the splitting operation which allows us to obtain the graph G_2 from the graph G_1 constructed in Sect. 2.1.1.

Definition 2.2. (splitting operation) For each $i, j \in [k]$ and each $q, \ell \in [N]$

- If $(q, \ell) \notin S_{i,j}$, then we **split** the vertex $\mathbf{w}_{i,j}^{q,\ell}$ into two **distinct** vertices $\mathbf{w}_{i,j,\mathrm{LB}}^{q,\ell}$ and $\mathbf{w}_{i,j,\mathrm{TR}}^{q,\ell}$ and add the edge $\mathbf{w}_{i,j,\mathrm{LB}}^{q,\ell} \to \mathbf{w}_{i,j,\mathrm{TR}}^{q,\ell}$ (denoted by the dotted edge in Fig. 2). The 4 edges (see Definition 2.1) incident on $\mathbf{w}_{i,j}^{q,\ell}$ are now changed as follows (see Fig. 2):
  - Replace the edge $\mathtt{west}(\mathbf{w}_{i,j}^{q,\ell}) \to \mathbf{w}_{i,j}^{q,\ell}$ by the edge $\mathtt{west}(\mathbf{w}_{i,j}^{q,\ell}) \to \mathbf{w}_{i,j,\mathrm{LB}}^{q,\ell}$
  - Replace the edge $\mathtt{south}(\mathbf{w}_{i,j}^{q,\ell}) \to \mathbf{w}_{i,j}^{q,\ell}$ by the edge $\mathtt{south}(\mathbf{w}_{i,j}^{q,\ell}) \to \mathbf{w}_{i,j,\mathrm{LB}}^{q,\ell}$
  - Replace the edge $\mathbf{w}_{i,j}^{q,\ell} \to \mathtt{east}(\mathbf{w}_{i,j}^{q,\ell})$ by the edge $\mathbf{w}_{i,j,\mathrm{TR}}^{q,\ell} \to \mathtt{east}(\mathbf{w}_{i,j}^{q,\ell})$
  - Replace the edge $\mathbf{w}_{i,j}^{q,\ell} \to \mathtt{north}(\mathbf{w}_{i,j}^{q,\ell})$ by the edge $\mathbf{w}_{i,j,\mathrm{TR}}^{q,\ell} \to \mathtt{north}(\mathbf{w}_{i,j}^{q,\ell})$
- Otherwise, if $(q, \ell) \in S_{i,j}$ then the vertex $\mathbf{w}_{i,j}^{q,\ell}$ is **not split**, and we define $\mathbf{w}_{i,j,\mathrm{LB}}^{q,\ell} = \mathbf{w}_{i,j}^{q,\ell} = \mathbf{w}_{i,j,\mathrm{TR}}^{q,\ell}$. Note that the four edges (Definition 2.1) incident on $\mathbf{w}_{i,j}^{q,\ell}$ are unchanged.

Remark 2.2. To avoid case distinctions in the forthcoming proof of correctness of the reduction, we will use the following non-standard notation: the edge $s \rightsquigarrow s$ **does not** represent a self-loop but rather is to be viewed as *"just staying put"* at the vertex s. Note that this does not affect edge-disjointness.

We are now ready to define the graph G_2 and the set $\mathcal{T}$ of terminal pairs:

Definition 2.3. The graph G_2 is obtained by applying the splitting operation (Definition 2.2) to each (black) grid vertex of G_1, i.e., the set of vertices given by $\bigcup_{1 \leq i,j \leq k} V(G_{i,j})$. The set of terminal pairs is $\mathcal{T} := \big\{(a_i, b_i) : i \in [k]\big\} \cup \big\{(c_j, d_j) : j \in [k]\big\}$

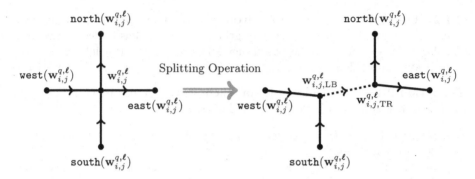

Fig. 2. The splitting operation for the vertex $\mathbf{w}_{i,j}^{q,\ell}$ when $(q,\ell) \notin S_{i,j}$. The idea behind this splitting is if we want edge-disjoint paths then we can go **either** left-to-right or bottom-to-top but not in **both** directions. On the other hand, if $(q,\ell) \in S_{i,j}$ then the picture on the right-hand side (after the splitting operation) would look exactly like that on the left-hand side.

The next claim shows that G_2 is also both planar and acyclic (like G_1).

Claim 2.2. $[\star]^4$ G_2 *is a planar DAG*

We now set up notation for the grids in G_2:

Definition 2.4. For each $i,j \in [k]$, we define $G_{i,j}^{\mathtt{split}}$ to be the graph obtained by applying the splitting operation (Definition 2.2) to each vertex of $G_{i,j}$. For each $i,j \in [k]$ and each $q,\ell \in [N]$ we define $\mathtt{split}(\mathbf{w}_{i,j}^{q,\ell}) := \left\{ \mathbf{w}_{i,j,\mathrm{LB}}^{q,\ell}, \mathbf{w}_{i,j,\mathrm{TR}}^{q,\ell} \right\}$.

2.2 Solution for Edge-Disjoint Paths $\Rightarrow$ Solution for Grid-Tiling-$\leq$

In this section, we show that if the instance $(G_2, \mathcal{T})$ of EDGE-DISJOINT PATHS has a solution then the instance $(k, N, \mathcal{S})$ of GRID-TILING-$\leq$ also has a solution.

Suppose that the instance $(G_2, \mathcal{T})$ of EDGE-DISJOINT PATHS has a solution, i.e., there is a collection of $2k$ pairwise edge-disjoint paths $\{P_1, P_2, \ldots, P_k, Q_1, Q_2, \ldots, Q_k\}$ in G_2 such that

$$P_i \text{ is an } a_i \rightsquigarrow b_i \text{ path } \forall i \in [k] \text{ and } Q_j \text{ is an } c_j \rightsquigarrow d_j \text{ path } \forall j \in [k] \quad (5)$$

To streamline the arguments of this section, we define the following subsets of vertices of G_2:

Definition 2.5. (horizontal & vertical levels) For each $j \in [k]$, we define

$$\text{HORIZONTAL}(j) = \{c_j, d_j\} \cup \left(\bigcup_{i=1}^{k} V(G_{i,j}^{\mathtt{split}}) \right) \cup \left(\bigcup_{i=1}^{k-1} H_{i,j}^{i+1,j} \right)$$

---

[4] Proofs of all results labeled with $[\star]$ are deferred to the full version [3].

For each $i \in [k]$, we define

$$\text{VERTICAL}(i) = \{a_i, b_i\} \cup \Big(\bigcup_{j=1}^{k} V(G_{i,j}^{\text{split}}) \Big) \cup \Big(\bigcup_{j=1}^{k-1} V_{i,j}^{i,j+1} \Big)$$

From Definition 2.5, it is easy to verify that $\text{VERTICAL}(i) \cap \text{VERTICAL}(i') = \emptyset = \text{HORIZONTAL}(i) \cap \text{HORIZONTAL}(i')$ for every $1 \le i \ne i' \le k$.

Definition 2.6. (boundary vertices in G_2) For each $1 \le i, j \le k$ we define the set of boundary vertices of the grid $G_{i,j}^{\text{split}}$ in the graph G_2 as follows:

$$\begin{aligned}
&\texttt{Left}(G_{i,j}^{\text{split}}) := \big\{ \mathbf{w}_{i,j,\text{LB}}^{1,\ell} \ : \ \ell \in [N] \big\} \ ; \ \texttt{Right}(G_{i,j}^{\text{split}}) := \big\{ \mathbf{w}_{i,j,\text{TR}}^{N,\ell} \ : \ \ell \in [N] \big\} \\
&\texttt{Top}(G_{i,j}^{\text{split}}) := \big\{ \mathbf{w}_{i,j,\text{TR}}^{\ell,N} \ : \ \ell \in [N] \big\} \ ; \ \texttt{Bottom}(G_{i,j}^{\text{split}}) := \big\{ \mathbf{w}_{i,j,\text{LB}}^{\ell,1} \ : \ \ell \in [N] \big\}
\end{aligned} \quad (6)$$

Lemma 2.1. [⋆] *For each $i \in [k]$ the path P_i satisfies the following two structural properties:*

- *every edge of the path P_i has both end-points in $\text{VERTICAL}(i)$*
- *P_i contains an $\texttt{Bottom}(G_{i,j}^{split}) \rightsquigarrow \texttt{Top}(G_{i,j}^{split})$ path for each $j \in [k]$.*

For each $j \in [k]$ the path Q_j satisfies the following two structural properties:

- *every edge of the path Q_j has both end-points in $\text{HORIZONTAL}(j)$*
- *Q_j contains an $\texttt{Left}(G_{i,j}^{split}) \rightsquigarrow \texttt{Right}(G_{i,j}^{split})$ path for each $i \in [k]$*

Lemma 2.2. [⋆] *For any $(i,j) \in [k] \times [k]$, let P', Q' be any $\texttt{Bottom}(G_{i,j}^{split}) \rightsquigarrow \texttt{Top}(G_{i,j}^{split})$, $\texttt{Left}(G_{i,j}^{split}) \rightsquigarrow \texttt{Right}(G_{i,j}^{split})$ paths in G_2 respectively. If P' and Q' are edge-disjoint then there exists $(\mu, \delta) \in S_{i,j}$ such that the vertex $\mathbf{w}_{i,j,LB}^{\mu,\delta} = \mathbf{w}_{i,j}^{\mu,\delta} = \mathbf{w}_{i,j,TR}^{\mu,\delta} = $ belongs to both P' and Q'*

Lemma 2.3. [⋆] *The instance $(k, N, \mathcal{S})$ of GRID-TILING-$\le$ has a solution.*

2.3 Solution for Grid-Tiling-$\le$ $\Rightarrow$ Solution for Edge-Disjoint Paths

In this section, we show that if the instance $(k, N, \mathcal{S})$ of GRID-TILING-$\le$ has a solution then the instance $(G_2, \mathcal{T})$ of EDGE-DISJOINT PATHS also has a solution.

Lemma 2.4. [⋆] *The instance $(G_2, \mathcal{T})$ of EDGE-DISJOINT PATHS has a solution.*

2.4 Proof of Theorem 1.1

Finally we are ready to prove our main theorem (Theorem 1.1).

Proof. Given an instance $(k, N, \mathcal{S})$ of GRID-TILING-$\leq$, we use the construction from Sect. 2.1 to build an instance $(G_2, \mathcal{T})$ of EDGE-DISJOINT PATHS such that G_2 is a planar DAG (Claim 2.2). It is easy to see that $n = |V(G_2)| = O(N^2 k^2)$ and G_2 can be constructed in poly(N, k) time.

It is known [11, Theorem 14.30] that GRID-TILING-$\leq$ is W[1]-hard parameterized by k, and under ETH cannot be solved in $f(k) \cdot N^{o(k)}$ time for any computable function f. Combining the two directions from Sect. 2.2 and Sect. 2.3, we get a parameterized reduction from GRID-TILING-$\leq$ to an instance of EDGE-DISJOINT PATHS which is a planar DAG and has $|\mathcal{T}| = 2k$ terminal pairs. Hence, it follows that EDGE-DISJOINT PATHS on planar DAGs is W[1]-hard parameterized by number k of terminal pairs, and under ETH cannot be solved in $f(k) \cdot n^{o(k)}$ time for any computable function f.

Finally we show how to edit G_2, without affecting the correctness of the reduction, so that both the out-degree and in-degree are at most 2. We present the argument for reducing out-degree: the argument for in-degree is analogous. Note that the only vertices in G_2 with out-degree > 2 are $A \cup C$. For each $c_j \in C$ we replace the directed star whose edges are from c_j to each vertex of Left$(G_{1,j})$ with a directed binary tree whose root is c_i, leaves are the set of vertices Left$(G_{1,j})$ and each edge is directed away from the root. It is easy to see that the edge-disjointness of paths from c_j to different vertices of Left$(G_{1,j})$ is preserved, and we have only increased the number of vertices by $O(k)$ while maintaining planarity and (directed) acyclicity. We do a similar transformation for each $a_i \in A$. It is easy to see that this editing adds $O(k^2)$ new vertices and takes poly(k) time, and therefore it is still true that $n = |V(G_2)| = O(N^2 k^2)$ and G_2 can be constructed in poly(N, k) time. □

3 Conclusion and Open Questions

In this paper we have shown that EDGE-DISJOINT PATHS on planar DAGs is W[1]-hard parameterized by k, and has no $f(k) \cdot n^{o(k)}$ algorithm under the Exponential Time Hypothesis (ETH) for any computable function f. The hardness holds even if both the maximum in-degree and maximum out-degree of the graph are at most 2. This answers a question of Slivkins [28] regarding the parameterized complexity of EDGE-DISJOINT PATHS on planar DAGS, and a question of Cygan et al. [12] and Schrijver [27] regarding the parameterized complexity of EDGE-DISJOINT PATHS on planar directed graphs.

We now propose some interesting open questions related to the complexity of the DISJOINT PATHS problem:

- What is the *correct* parameterized complexity of EDGE-DISJOINT PATHS on planar graphs parameterized by k? Can we design an $n^{O(k)}$ algorithm, or is it NP-hard even for $k = O(1)$ like the general version? Note that to prove the latter result, one would need to have directed cycles involved in the reduction since there is $n^{O(k)}$ algorithm of Fortune et al. [15] for EDGE-DISJOINT PATHS on DAGs.

- Is the half-integral version[5] of EDGE-DISJOINT PATHS FPT on directed planar graphs? It is easy to see that our W[1]-hardness reduction does not work for this problem.
- Given our W[1]-hardness result, can we obtain FPT (in)approximability results for the EDGE-DISJOINT PATHS problem on planar DAGs? To the best of our knowledge, there are no known (non-trivial) FPT (in)approximability results for any variants of the DISJOINT PATHS problem. This question might be worth considering even for those versions of the DISJOINT PATHS problem which are known to be FPT since the running times are astronomical (except maybe [20]). Some of the recent work [7–10] on polynomial time (in)approximability of the DISJOINT PATHS problem might be relevant.

Acknowledgements. We thank the anonymous reviewers of CIAC 2021 for their helpful comments. In particular, one of the reviewers suggested the strengthening of Theorem 1.1 for the case when the input graph has both in-degree and out-degree at most 2.

References

1. Adler, I., Kolliopoulos, S.G., Krause, P.K., Lokshtanov, D., Saurabh, S., Thilikos, D.M.: Irrelevant vertices for the planar disjoint paths problem. J. Comb. Theory Ser. B **122**, 815–843 (2017). https://doi.org/10.1016/j.jctb.2016.10.001
2. Amiri, S.A., Kreutzer, S., Marx, D., Rabinovich, R.: Routing with congestion in acyclic digraphs. Inf. Process. Lett. **151** (2019). https://doi.org/10.1016/j.ipl.2019.105836
3. Chitnis, R.: A tight lower bound for edge-disjoint paths on planar DAGs (2021). http://arxiv.org/abs/2101.10742
4. Chitnis, R., Feldmann, A.E.: A tight lower bound for Steiner orientation. In: Fomin, F.V., Podolskii, V.V. (eds.) CSR 2018. LNCS, vol. 10846, pp. 65–77. Springer, Cham (2018). https://doi.org/10.1007/978-3-319-90530-3_7
5. Chitnis, R., Feldmann, A.E., Suchý, O.: A tight lower bound for planar Steiner orientation. Algorithmica **81**(8), 3200–3216 (2019). https://doi.org/10.1007/s00453-019-00580-x
6. Chitnis, R.H., Feldmann, A.E., Hajiaghayi, M.T., Marx, D.: Tight bounds for planar strongly connected Steiner subgraph with fixed number of terminals (and extensions). SIAM J. Comput. **49**(2), 318–364 (2020). https://doi.org/10.1137/18M122371X
7. Chuzhoy, J., Kim, D.H.K., Li, S.: Improved approximation for node-disjoint paths in planar graphs. In: Proceedings of the 48th Annual ACM SIGACT Symposium on Theory of Computing, STOC 2016, Cambridge, MA, USA, 18–21 June 2016, pp. 556–569. ACM (2016). https://doi.org/10.1145/2897518.2897538

[5] Each edge can belong to at most two of the paths.

8. Chuzhoy, J., Kim, D.H.K., Nimavat, R.: New hardness results for routing on disjoint paths. In: Proceedings of the 49th Annual ACM SIGACT Symposium on Theory of Computing, STOC 2017, Montreal, QC, Canada, 19–23 June 2017, pp. 86–99. ACM (2017). https://doi.org/10.1145/3055399.3055411

9. Chuzhoy, J., Kim, D.H.K., Nimavat, R.: Almost polynomial hardness of node-disjoint paths in grids. In: Proceedings of the 50th Annual ACM SIGACT Symposium on Theory of Computing, STOC 2018, Los Angeles, CA, USA, 25–29 June 2018, pp. 1220–1233. ACM (2018). https://doi.org/10.1145/3188745.3188772

10. Chuzhoy, J., Kim, D.H.K., Nimavat, R.: Improved approximation for node-disjoint paths in grids with sources on the boundary. In: 45th International Colloquium on Automata, Languages, and Programming, ICALP 2018, 9–13 July 2018, Prague, Czech Republic. LIPIcs, vol. 107, pp. 38:1–38:14. Schloss Dagstuhl - Leibniz-Zentrum für Informatik (2018). https://doi.org/10.4230/LIPIcs.ICALP.2018.38

11. Cygan, M., et al.: Parameterized Algorithms. Springer, Cham (2015). https://doi.org/10.1007/978-3-319-21275-3

12. Cygan, M., Marx, D., Pilipczuk, M., Pilipczuk, M.: The planar directed k-vertex-disjoint paths problem is fixed-parameter tractable. In: 54th Annual IEEE Symposium on Foundations of Computer Science, FOCS 2013, 26–29 October 2013, Berkeley, CA, USA, pp. 197–206. IEEE Computer Society (2013). https://doi.org/10.1109/FOCS.2013.29

13. Diestel, R.: Graph theory. In: Graduate Texts in Mathematics, vol. 173, p. 7 (2012)

14. Even, S., Itai, A., Shamir, A.: On the complexity of timetable and multi-commodity flow problems. In: 16th Annual Symposium on Foundations of Computer Science, Berkeley, California, USA, 13–15 October 1975, pp. 184–193. IEEE Computer Society (1975). https://doi.org/10.1109/SFCS.1975.21

15. Fortune, S., Hopcroft, J.E., Wyllie, J.: The directed subgraph homeomorphism problem. Theor. Comput. Sci. **10**, 111–121 (1980). https://doi.org/10.1016/0304-3975(80)90009-2

16. Frank, A.: Packing paths, cuts, and circuits-a survey. In: Paths, Flows and VLSI-Layout, pp. 49–100 (1990)

17. Impagliazzo, R., Paturi, R.: On the complexity of k-SAT. J. Comput. Syst. Sci. **62**(2), 367–375 (2001). https://doi.org/10.1006/jcss.2000.1727

18. Impagliazzo, R., Paturi, R., Zane, F.: Which problems have strongly exponential complexity? J. Comput. Syst. Sci. **63**(4), 512–530 (2001). https://doi.org/10.1006/jcss.2001.1774

19. Kawarabayashi, K., Kobayashi, Y., Reed, B.A.: The disjoint paths problem in quadratic time. J. Comb. Theory Ser. B **102**(2), 424–435 (2012). https://doi.org/10.1016/j.jctb.2011.07.004

20. Lokshtanov, D., Misra, P., Pilipczuk, M., Saurabh, S., Zehavi, M.: An exponential time parameterized algorithm for planar disjoint paths. In: Proccedings of the 52nd Annual ACM SIGACT Symposium on Theory of Computing, STOC 2020, Chicago, IL, USA, 22–26 June 2020, pp. 1307–1316 (2020). https://doi.org/10.1145/3357713.3384250

21. Marx, D.: The square root phenomenon in planar graphs. In: Fomin, F.V., Freivalds, R., Kwiatkowska, M., Peleg, D. (eds.) ICALP 2013. LNCS, vol. 7966, p. 28. Springer, Heidelberg (2013). https://doi.org/10.1007/978-3-642-39212-2_4

22. Marx, D., Pilipczuk, M., Pilipczuk, M.: On subexponential parameterized algorithms for Steiner tree and directed subset TSP on planar graphs. In: 59th IEEE Annual Symposium on Foundations of Computer Science, FOCS 2018, Paris, France, 7–9 October 2018, pp. 474–484. IEEE Computer Society (2018). https://doi.org/10.1109/FOCS.2018.00052

23. Marx, D., Sidiropoulos, A.: The limited blessing of low dimensionality: when $1-1/d$ is the best possible exponent for d-dimensional geometric problems. In: 30th Annual Symposium on Computational Geometry, SOCG 2014, Kyoto, Japan, 08–11 June 2014, p. 67. ACM (2014). https://doi.org/10.1145/2582112.2582124
24. Robertson, N., Seymour, P.D.: Graph minors XIII. The disjoint paths problem. J. Comb. Theory Ser. B **63**(1), 65–110 (1995). https://doi.org/10.1006/jctb.1995.1006
25. Schrijver, A.: Finding k disjoint paths in a directed planar graph. SIAM J. Comput. **23**(4), 780–788 (1994). https://doi.org/10.1137/S0097539792224061
26. Schrijver, A.: Combinatorial Optimization: Polyhedra and Efficiency, vol. 24. Springer, Heidelberg (2003)
27. Schrijver, A.: Finding k partially disjoint paths in a directed planar graph. Building bridges II. Bolyai Soc. Math. Stud. **28**, 417–444 (2019). https://doi.org/10.1007/978-3-662-59204-5_13
28. Slivkins, A.: Parameterized tractability of edge-disjoint paths on directed acyclic graphs. SIAM J. Discret. Math. **24**(1), 146–157 (2010). https://doi.org/10.1137/070697781
29. Vygen, J.: NP-completeness of some edge-disjoint paths problems. Discret. Appl. Math. **61**(1), 83–90 (1995). https://doi.org/10.1016/0166-218X(93)E0177-Z

Upper Dominating Set: Tight Algorithms for Pathwidth and Sub-exponential Approximation

Louis Dublois[(✉)], Michael Lampis, and Vangelis Th. Paschos

Université Paris-Dauphine, PSL University, CNRS, LAMSADE, Paris, France
{louis.dublois,michail.lampis,paschos}@lamsade.dauphine.fr

Abstract. An upper dominating set is a minimal dominating set in a graph. In the UPPER DOMINATING SET problem, the goal is to find an upper dominating set of maximum size. We study the complexity of parameterized algorithms for UPPER DOMINATING SET, as well as its sub-exponential approximation. First, we prove that, under ETH, k-UPPER DOMINATING SET cannot be solved in time $O(n^{o(k)})$ (improving on $O(n^{o(\sqrt{k})})$), and in the same time we show under the same complexity assumption that for any constant ratio r and any $\varepsilon > 0$, there is no r-approximation algorithm running in time $O(n^{k^{1-\varepsilon}})$. Then, we settle the problem's complexity parameterized by pathwidth by giving an algorithm running in time $O^*(6^{pw})$ (improving the current best $O^*(7^{pw})$), and a lower bound showing that our algorithm is the best we can get under the SETH. Furthermore, we obtain a simple sub-exponential approximation algorithm for this problem: an algorithm that produces an r-approximation in time $n^{O(n/r)}$, for any desired approximation ratio $r < n$. We finally show that this time-approximation trade-off is tight, up to an arbitrarily small constant in the second exponent: under the randomized ETH, and for any ratio $r > 1$ and $\varepsilon > 0$, no algorithm can output an r-approximation in time $n^{(n/r)^{1-\varepsilon}}$. Hence, we completely characterize the approximability of the problem in sub-exponential time.

Keywords: FPT Algorithms · Sub-Exponential Approximation · Upper Domination

1 Introduction

In a graph $G = (V, E)$, a set $D \subseteq V$ is called a *dominating set* if all vertices of V are dominated by D, that is for every $u \in V$ either u belongs to D or u is a neighbor of some vertex in D. The well-known DOMINATING SET problem is studied with a minimization objective: given a graph, we are interested in finding the smallest dominating set. In this paper, we consider *upper dominating sets*, that is dominating sets that are minimal, where a dominating set D is minimal if no proper subset of it is a dominating set, that is if it does not contain any

© Springer Nature Switzerland AG 2021
T. Calamoneri and F. Corò (Eds.): CIAC 2021, LNCS 12701, pp. 202–215, 2021.
https://doi.org/10.1007/978-3-030-75242-2_14

redundant vertex. We study the problem of finding an upper dominating set of maximum size.

This problem is called UPPER DOMINATING SET, and is the Max-Min version of the DOMINATING SET problem. We call UPPER DOMINATING SET the considered optimization problem and k-UPPER DOMINATING SET the associated decision problem.

Studying Max-Min and Min-Max versions of some famous optimization problems is not a new idea, and it has recently attracted some interest in the literature: MINIMUM MAXIMAL INDEPENDENT SET [6,15,19] (also known as MINIMUM INDEPENDENT DOMINATING SET), MAXIMUM MINIMAL VERTEX COVER [5,25], MAXIMUM MINIMAL SEPARATOR [16], MAXIMUM MINIMAL CUT [12], MINIMUM MAXIMAL KNAPSACK [1,13,14] (also known as LAZY BUREAUCRAT PROBLEM), MAXIMUM MINIMAL FEEDBACK VERTEX SET [11]. In fact, the original motivation for studying these problems was to analyze the performance of naive heuristics compared to the natural Max and Min versions, but these Max-Min and Min-Max problems have gradually revealed some surprising combinatorial structures, which makes them as interesting as their natural Max and Min versions. The UPPER DOMINATING SET problem can be seen as a member of this framework, and studying it within this framework is one of our motivation.

This problem is also one of the six problems of the well-known *domination chain* (see [2,18]) and is somewhat one which has fewer results, compared to the famous DOMINATING SET and INDEPENDENT SET problems. Increasing our understanding of the UPPER DOMINATING SET problem compared to these two famous problems is another motivation.

UPPER DOMINATING SET was first considered in an algorithmic point of view by Cheston et al. [9], where they showed that the problem is NP-hard. In the more extensive paper considering this problem, Bazgan et al. [3] studied approximability, and classical and parameterized complexity of the UPPER DOMINATING SET problem. In the polynomial approximation paradigm, they proved that the problem does not admit an $n^{1-\varepsilon}$-approximation for any $\varepsilon > 0$, unless P=NP, making the problem as hard as INDEPENDENT SET, whereas there exists a greedy $\ln n$-approximation algorithm for the Min version DOMINATING SET.

Considering the parameterized complexity, they proved that the problem is as hard as the k-INDEPENDENT SET problem: k-UPPER DOMINATING SET is W[1]-hard parameterized by the standard parameter k. Nonetheless, in their reduction, there is an inherent quadratic blow-up in the size of the solution k, so they essentially proved that there is no algorithm solving k-UPPER DOMINATING SET in time $O(n^{o(\sqrt{k})})$. They also gave FPT algorithms parameterized by the pathwidth pw and the treewidth tw of the graph, in time $O^*((7^{pw})^1$ and $O^*(10^{tw})$, respectively.

Our Results: The state of the art summarized above motivates two basic questions: first, can we close the gap between the lower and upper bounds of

---
[1] O^* notation suppresses polynomial factors in the input size.

the complexity of the problem parameterized by pathwidth; second, since the polynomial approximation is essentially settled, can we design sub-exponential approximation algorithms which can reach any approximation ratio $r < n$? We answer these questions and along the way we give stronger FPT hardness results. In fact, we prove the following:

(i) In Sect. 3, we show the following: under ETH, there is no algorithm solving k-UPPER DOMINATING SET in time $O(n^{o(k)})$; and under the same complexity assumption, for any ratio r and any $\varepsilon > 0$, there is no algorithm for this problem that outputs an r-approximation in time $O(n^{k^{1-\varepsilon}})$.

(ii) In Sect. 4, we give a dynamic programming algorithm parameterized by pathwidth that solves UPPER DOMINATING SET in time $O^*(6^{pw})$. Surprisingly, this result is obtained by slightly modifying the algorithm of Bazgan et al. [3]. We then prove the following: under SETH, and for any $\varepsilon > 0$, UPPER DOMINATING SET cannot be solved in time $O^*((6-\varepsilon)^{pw})$. This is our main result, and it shows that our algorithm for pathwidth is optimal.

(iii) In Sect. 5, we give a simple time-approximation trade-off: for any ratio $r < n$, there exists an algorithm for UPPER DOMINATING SET that ouputs an r-approximation in time $n^{O(n/r)}$. We also give a matching lower bound: under the randomized ETH, for any ratio $r > 1$ and any $\varepsilon > 0$, there is no algorithm that outputs an r-approximation running in time $n^{(n/r)^{1-\varepsilon}}$.

2   Preliminaries

We use standard graph-theoretic notation and we assume familiarity with the basics of parameterized complexity (e.g. pathwidth, the SETH and FPT algorithms), as given in [10]. Let $G = (V, E)$ be a graph with $|V| = n$ vertices and $|E| = m$ edges. For a vertex $u \in V$, the set $N(u)$ denotes the set of neighbors of u, $d(u) = |N(u)|$, and $N[u]$ the closed neighborhood of u, i.e. $N[u] = N(u) \cup \{u\}$. For a subset $U \subseteq V$ and a vertex $u \in V$, we note $N_U(u) = N(u) \cap U$. Furthermore, for $U \subseteq V$, we note $N(U) = \bigcup_{u \in U} N(u)$. For an edge set $E' \subseteq E$, we use $V(E')$ to denote the set of its endpoints. For $V' \subseteq V$, we note $G[V']$ the subgraph of G induced by V'.

An *upper dominating set* $D \subseteq V$ of a graph $G(V, E)$ is a set of vertices that dominates all vertices of G, and which is minimal. Note that D is minimal if we have the following: for every vertex $u \in D$, either u has a private neighbor, that is a neighbor that is dominated only by u, or u is its own private vertex, that is u is only dominated by itself. We note an upper dominating set $D = S \cup I$, where S is the set of vertices of D which have at least one private neighbor, and I is the set of vertices of D which forms an independent set, that is the set of vertices which are their own private vertices.

Note that a maximal independent set I (also known as an independent dominating set) is an upper dominating set since it is a set of vertices which dominates the whole graph and such that every vertex $u \in I$ is its own private vertex.

3   FPT and FPT-Approximation Hardness

In this section, we present two hardness results for the k-UPPER DOMINATING SET problem in the parameterized paradigm: we prove first that the considered problem cannot be solved in time $O(n^{o(k)})$ under the ETH ; and we prove then under the same complexity assumption that for any constant approximation ratio $0 < r < 1$ and any $\varepsilon > 0$, there is no FPT algorithm giving an r-approximation for the k-UPPER DOMINATING SET problem running in time $O(n^{k^{1-\varepsilon}})$.

Note that k-UPPER DOMINATING SET being W[1]-hard was already proved by Bazgan et al. [3]. To get this result, they made a reduction from the k-MULTICOLORED CLIQUE problem. Nonetheless, in this reduction, the size of the solution of the k-UPPER DOMINATING SET problem was quadratic compared to the size of the solution of the k-MULTICOLORED CLIQUE problem. Thus, they proved essentially the next result: k-UPPER DOMINATING SET problem cannot be solved in time $O(n^{o(\sqrt{k})})$.

To obtain our desired negative results, we will make a reduction from the k-INDEPENDENT SET problem to our problem. So recall that we have the following hardness results for the k-INDEPENDENT SET problem:

Lemma 1 (Theorem 5.5 from [8]). *Under ETH, k-INDEPENDENT SET cannot be solved in time $O(n^{o(k)})$.*

Lemma 2 (Corollary 2 from [4]). *Under ETH, for any constant $r > 0$ and any $\varepsilon > 0$, there is no r-approximation algorithm for k-INDEPENDENT SET running in time $O(n^{k^{1-\varepsilon}})$.*

We will obtain similar results for the k-UPPER DOMINATING SET by doing a reduction from k-INDEPENDENT SET. This reduction will linearly increase the size of the solutions between the two problems, so these two hardness results for the latter problem will hold for the former problem.

Before we proceed further in the description of our reduction, note that we will use a variant of the k-INDEPENDENT SET problem. In this variant, the graph G contains k cliques which are connected to each other, and if a solution of size k exists, then this solution takes exactly one vertex per clique. Note that the Lemmas 1 and 2 hold on this particular instance, since this is a case where the problem remains hard to solve in FPT time and to approximate in FPT time. So we will use this variant.

Let us now present our reduction. We are given a k-INDEPENDENT SET instance G with n vertices and m edges, where the n vertices are partitioned in k distinct cliques $V_1, \ldots, V_k$ connected to each other. We define the following number: $A = 5$. We set our budget to be $k' = Ak$.

We construct our instance G' of k'-UPPER DOMINATING SET as follows:

1. For any vertex $u \in V(G)$, create an independent set Z_u of size A.
2. For any edge $(u, v) \in E(G)$, add all edges between the vertices of Z_u and the vertices of Z_v.

3. For any $i \in \{1, \ldots, n\}$, let W_i be the *group* associated to the clique V_i, which contains all vertices of all independent sets Z_u such that the vertex u belongs to the clique V_i. For any $i \in \{1, \ldots, k\}$, create a vertex z_i connected to all vertices of the group W_i.

Now that we have presented our reduction, we argue that it is correct. Recall that the target size of an optimal solution in G' is k' as defined above. We can prove that, given an independent set I of size at least k in G, we can construct an upper dominating set of size at least Ak in G' by taking the A vertices of the independent set Z_u for any vertex $u \in I$.

Lemma 3. *If G has an independent set of size at least k, then G' has an upper dominating set of size at least k'.*

The idea of the following proof is the following: if an upper dominating set in G' of size at least k' has not the form described in Lemma 3, then it cannot have size at least k', enabling us to construct an independent set of size at least k in G from an upper dominating set which has the desired form.

Lemma 4. *If G' has an upper dominating set of size at least k', then G has an independent set of size at least k.*

Now that we have proved the correctness of our reduction and since the blow-up of the reduction is linear in both the size of the instance and the size of the solution, we can now present one of the main results of this section:

Theorem 1. *Under ETH, k-UPPER DOMINATING SET cannot be solved in time $O(n^{o(k)})$.*

From now one and to obtain the FPT-approximation hardness result, we now consider our reduction above with A being sufficiently large. Note that all the properties we have found before still hold since A remains a constant.

Let $0 < r < 1$. To obtain the FPT-approximation hardness result for the k-INDEPENDENT SET problem (see Lemma 2), Bonnet et al. [4] made a gap-amplification reduction from an instance ϕ of 3-SAT to an instance (G, k) of k-INDEPENDENT SET problem. Essentially, this reduction gives the following gap:

- YES-instance: If ϕ is satisfiable, then $\alpha(G) = k$.
- NO-instance: If ϕ is not satisfiable, then $\alpha(G) \leq rk$.

In this gap, $\alpha(G)$ is the size of a maximum independent set in G, and k corresponds in fact to a value which depends on the reduction, but designating it by k ease our purpose.

To obtain a similar result for the k-UPPER DOMINATING SET problem, and by using our reduction above, we have to prove that our reduction keep a gap of value r. Thus, we need to prove the following:

- YES-instance: If ϕ is satisfiable, then $\alpha(G) = k$ and $\Gamma(G') = Ak$.

– NO-instance: If ϕ is satisfiable, then $\alpha(G) \leq rk$ and $\Gamma(G') \leq rAk$.

where $\Gamma(G')$ is the size of a maximum upper dominating set in G'.

Note that we have proved the first condition in Lemma 3, since an independent set of size at least k in G necessarily has size exactly k.

Thus, we just need to prove the second condition. To prove it, we will in fact prove the contraposition, to ease our proof. This is given in the following Lemma. The proof of this Lemma uses some arguments made in the proof of Lemma 4, and by choosing carefully which vertices we can put in the independent set we want to construct.

Lemma 5. *If there exists an upper dominating set in G' of size $> rAk$, then there exists an independent set in G of size $> rk$.*

Now that we have proved the correctness of the gap-amplification of our reduction, we can present the second main result of this section:

Theorem 2. *Under ETH, for any constant $r > 0$ and any $\varepsilon > 0$, there is no r-approximation algorithm for k-UPPER DOMINATING SET running in time $O(n^{k^{1-\varepsilon}})$.*

4   Pathwidth

4.1   FPT Algorithm Parameterized by Pathwidth

In this section, we present an algorithm for the UPPER DOMINATING SET problem parameterized by the pathwidth pw of the given graph. We prove that, given a graph $G = (V, E)$ and a path decomposition $(T, \{X_t\}_{t \in V(T)})$ of width pw, there exists a dynamic programming algorithm that solves UPPER DOMINATING SET in time $O(6^{pw} \cdot pw)$. Due to space constraints, we only sketch the basic ideas and explain, on a high level, how we manage to get the desired complexity.

Note that Bazgan et al. have designed an FPT algorithm for UPPER DOMINATING SET running in time $O^*(7^{pw})$ [3]. Our algorithm essentially works as their algorithm: we have the same set of colors to give to the vertices; and our Initialization and Forget nodes are similar to theirs.

Nonetheless, we have modified the Introduce nodes in order to lower the complexity to $O(6^{pw} \cdot pw)$. For an Introduce node $X_t = X_{t'} \cup \{v\}$ (and a vertex $v \notin X_{t'}$), Bazgan et al. did the following: they go through all possible colorings of the bag X_t and consider every subset of the neighborhood of v to give the right color to the vertices of this subset. Thus, since they consider every subset of the neighborhood of v, they get an algorithm running in time $O^*(7^{pw})$.

In our algorithm, we do the following: for an Introduce node $X_t = X_{t'} \cup \{v\}$, we go through all possible colorings of the bag $X_{t'}$ and through all colorings of the vertex v, and we update the value in the table depending on the corresponding colorings of $X_{t'}$ and v. Doing so, and by being careful on the color given to v, it enables us to get an algorithm running in time $O(6^{pw} \cdot pw)$. We obtain the following Theorem:

Theorem 3. *The* UPPER DOMINATING SET *problem can be solved in time* $O(6^{pw} \cdot pw)$, *where* pw *is the input graph's pathwidth.*

4.2   Lower Bound

In this section, we present a lower bound on the complexity of any FPT algorithm for the UPPER DOMINATING SET problem parameterized by the pathwidth of the graph matching our previous algorithm. More precisely, we prove that, under SETH, for any $\varepsilon > 0$, there is no algorithm for UPPER DOMINATING SET running in time $O^*((6 - \varepsilon)^{pw})$, where pw is the pathwidth of the input graph.

To get this result, we will do a reduction from the q-CSP-6 problem (see [23]) to the UPPER DOMINATING SET problem. In the former problem, we are given a CONSTRAINT SATISFACTION (CSP) instance with n variables and m constraints. The variables take values over a set of size 6. Without loss of generality, let $\{0, 1, 2, 3, 4, 5\}$ be this set. Each constraint involves at most q variables, and is given as a list of acceptable assignments for these variables. Without loss of generality, we force the following condition: each constraint involves exactly q variables, because if it has fewer, we can add to it new variables and augment the list of satisfying assignments so that the value of the new variables is irrelevant.

The following result, shown in [23], is a natural consequence of the SETH, and will be the starting point to obtain the desired lower bound:

Lemma 6 (Lemma 2 from [23]). *If the SETH is true, then, for all $\varepsilon > 0$, there exists a q such that n-variables q-CSP-6 cannot be solved in time $O^*((6 - \varepsilon)^n)$.*

We note that in [23], it was shown that for any constant B, q-CSP-B cannot be solved in time $O^*((B - \varepsilon)^n)$ under the SETH. For our purpose, only the case where $B = 6$ is relevant because this corresponds to the base of our target lower bound.

We will produce a polynomial time reduction from an instance of q-CSP-6 with n variables to an equivalent instance of UPPER DOMINATING SET whose pathwidth is bounded by $n + O(1)$. Thus, any algorithm for the latter problem running faster than $O^*((6 - \varepsilon)^{pw})$ would give a $O^*((6 - \varepsilon)^n)$ algorithm for the former problem, contradicting SETH.

Before we proceed further in the description of our reduction, let us give the basic ideas, which look like other SETH-based lower bounds from the literature [17, 20–22, 24]. The constructed graph consists of a main part of n paths of length $4\,m$, each divided into m sections. The idea is that an optimal solution will verify, for each path, a specific pattern in the whole graph. For four consecutive vertices, there are six ways for taking exactly two vertices among the four and dominating the two others. These six ways for each path will represent all possible assignments for all variables. Then, we will add some *verification* gadgets for each constraint and attach it to the corresponding section, in order to check that the selected assignment satisfies the constraint or not.

A first difficulty of this reduction is to prove that an optimal solution of the UPPER DOMINATING SET instance has the desired form, and more precisely that the pattern selected for a variable is constant throughout the graph. To answer this difficulty, and by using a technique introduced in [24], we make a polynomial number of copies of this construction and we connect them together, enabling us to have a sufficiently large copy where the patterns are kept constant in this copy.

Moreover, we need to be careful in our verification gadgets in order to have the following conditions: the vertices of the paths taken in the solution must not have any private neighbor in the corresponding verification gadget, because otherwise it would be impossible to keep the patterns constant in a sufficiently large copy of the graph; and the vertices of the paths not taken in the solution must not be dominated by the corresponding verification gadget, because otherwise there can be some vertices of the paths taken in the solution that have no private neighbor.

Construction. Let us now present our reduction. We are given a q-CSP-6 instance φ with n variables $x_1, \ldots, x_n$ taking values over the set $\{0, 1, 2, 3, 4, 5\}$, and m constraints $c_0, \ldots, c_{m-1}$, each containing exactly q variables and C_j possible assignments over these q variables, for each $j \in \{0, \ldots, m-1\}$. We define the following numbers: $A = 4q + 2$ and $F = (2n+1)(4n+1)$. We set our budget to be $k = Fm(2n + A) + 2n$.

We construct our instance of UPPER DOMINATING SET as follows:

1. For $i \in \{1, \ldots, n\}$, we construct a path P_i of $4Fm + 6$ vertices: the vertices are labeled $u_{i,j}$ for $j \in \{-3, \ldots, 4Fm+2\}$; and for each i, j the vertex $u_{i,j}$ is connected to $u_{i,j+1}$. We call these paths the *main* part of our graph.
2. For each section $j \in \{0, \ldots, Fm-1\}$, let $j' = j \mod m$. We construct a verification gadget H_j as follows:
   (a) A clique K_j of size $AC_{j'}$ such that the $AC_{j'}$ vertices are partitioned into $C_{j'}$ cliques $K_j^1, \ldots, K_j^{C_{j'}}$, each corresponding to a satisfying assignment σ_l in the list of $c_{j'}$, for $l \in \{1, \ldots, C_{j'}\}$, and each containing exactly A vertices.
   (b) A clique L_j of size $AC_{j'}$ such that the $AC_{j'}$ vertices are partitioned in $C_{j'}$ cliques $L_j^1, \ldots, L_j^{C_{j'}}$, each containing exactly A vertices.
   (c) For each $i \in \{1, \ldots, n\}$ such that x_i is involved in $c_{j'}$, and for each satisfying assignment σ_l in the list of $c_{j'}$: if σ_l sets x_i value 0, connect the two vertices $u_{i,4j+2}$ and $u_{i,4j+3}$ to the A vertices of the clique K_j^l; if σ_l sets x_i value 1, connect the two vertices $u_{i,4j+3}$ and $u_{i,4j}$ to the A vertices of the clique K_j^l; if σ_l sets x_i value 2, connect the two vertices $u_{i,4j}$ and $u_{i,4j+1}$ to the A vertices of the clique K_j^l; if σ_l sets x_i value 3, connect the two vertices $u_{i,4j+1}$ and $u_{i,4j+2}$ to the A vertices of the clique K_j^l; if σ_l sets x_i value 4, connect the two vertices $u_{i,4j+1}$ and $u_{i,4j+3}$ to the A vertices of the clique K_j^l; if σ_l sets x_i value 5, connect the two vertices $u_{i,4j}$ and $u_{i,4j+2}$ to the A vertices of the clique K_j^l.
   (d) For each satisfying assignment σ_l in the list of $c_{j'}$, do the following: add a matching between the vertices of K_j^l and the vertices of L_j^l; for any

$l' \in \{1, \ldots, C_{j'}\}$ with $l' \neq l$, add all the edges between the vertices of K_j^l and the vertices of $L_j^{l'}$.

(e) Add a vertex w connected to all the vertices of the clique L_j.

Now that we have presented our reduction, we argue that it is correct and that the obtained graph G has the desired pathwidth. Recall that the target size of an optimal solution in G is k as defined above.

Lemma 7. *If φ is satisfiable, then there exists an upper dominating set in G of size at least k.*

Proof. Assume φ admits some satisfying assignment $\rho : \{x_1, \ldots, x_n\} \to \{0, 1, 2, 3, 4, 5\}$. We construct a solution S of the instance G of UPPER DOMINATING SET as follows:

1. For each $i \in \{1, \ldots, n\}$, let α and β be the following numbers: if $\rho(x_i) = 0$, let $\alpha = 2$ and $\beta = 3$; if $\rho(x_i) = 1$, let $\alpha = 3$ and $\beta = 0$; if $\rho(x_i) = 2$, let $\alpha = 0$ and $\beta = 1$; if $\rho(x_i) = 3$, let $\alpha = 1$ and $\beta = 2$; if $\rho(x_i) = 4$, let $\alpha = 1$ and $\beta = 3$; if $\rho(x_i) = 5$, let $\alpha = 0$ and $\beta = 2$. Let $U = \bigcup_{j=0}^{Fm-1} \{u_{i,4j+\alpha}, u_{i,4j+\beta}\}$. We add to the solution all vertices of $(V(P_i) \setminus \{u_{i,-3}, u_{i,-2}, u_{i,-1}, u_{i,4Fm}, u_{i,4Fm+1}, u_{i,4Fm+2}\}) \setminus U$.
2. For each $j \in \{0, \ldots, Fm - 1\}$, let $j' = j \mod m$. Consider the unique possible assignment σ_{l^*} in the list of $c_{j'}$ satisfied by ρ (such a unique possible assignment must exist since ρ satisfies φ), and take the A vertices of the clique $L_j^{l^*}$.
3. For each $i \in \{1, \ldots, n\}$, do the following: if $\rho(x_i) = 0$, then add $u_{i,-3}, u_{i,4Fm}$ and $u_{i,4Fm+1}$ to S; if $\rho(x_i) = 1$, then add $u_{i,-2}$ and $u_{i,4Fm+1}$ to S; if $\rho(x_i) = 2$, then add $u_{i,-2}, u_{i,-1}$ and $u_{i,4Fm+2}$ to S; if $\rho(x_i) = 3$, then add $u_{i,-3}$ and $u_{i,4Fm+2}$ to S; if $\rho(x_i) = 4$, then add $u_{i,-3}$ and $u_{i,4Fm+1}$ to S; if $\rho(x_i) = 5$, then add $u_{i,-2}$ and $u_{i,4Fm+2}$ to S.

Let us now argue why this solution has size at least k. In the first step, we have selected $2Fmn$ vertices. To see this, let $Q_{i,j}$ be the sub-path of P_i corresponding to the section j ($j \in \{0, \ldots, Fm-1\}$), i.e. $Q_{i,j} = \{u_{i,4j}, u_{i,4j+1}, u_{i,4j+2}, u_{i,4j+3}\}$. Observe that we have put exactly two vertices of $Q_{i,j}$ in U, which leaves two vertices in the solution, for all i and all j. Consider now any $j \in \{0, \ldots, Fm-1\}$ and the corresponding verification gadget H_j. In this gadget, we have selected all the vertices of the clique $L_j^{l^*}$, corresponding to the satisfied assignment σ_{l^*}. So we have selected AFm vertices for all the verification gadgets. Finally, at least $2n$ vertices have been added to the solution at step 3. So the total size is at least $2Fmn + AFm + 2n = k$.

Let us now argue why the solution is a valid upper dominating set.

Consider any $j \in \{0, \ldots, Fm - 1\}$ and let $j' = j \mod m$. We have selected the A vertices of the clique $L_j^{l^*}$ corresponding to the unique possible assignment σ_{l^*} in the list of $c_{j'}$ satisfied by ρ (such a unique possible assignment must exist since ρ satisfies φ). Since L_j is a clique, since the vertices of $L_j^{l^*}$ are connected to all vertices of $K_j^{l'}$, for any $l' \in \{1, \ldots, C_{j'}\}$ with $l' \neq l^*$, since there is a matching between the vertices of $L_j^{l^*}$ and the vertices of $K_j^{l^*}$, and since the vertex w is connected to all vertices of L_j, we have that all the vertices of H_j are dominated by S.

Now, observe that, since σ_{l^*} is satisfied by ρ, it means that the values given by ρ to the variables appearing in the constraint $c_{j'}$ satisfy σ_{l^*}, so by the construction it follows that the neighbors of the vertices of $K_j^{l^*}$ in the paths all belongs to U. Indeed, consider any variable x_i appearing in $c_{j'}$: if σ_{l^*} sets value 0 to x_i, then $\rho(x_i) = 0$, and then, for $\alpha = 2$ and $\beta = 3$, we have that $u_{i,4j+\alpha}$ and $u_{i,4j+\beta}$ are in U and are the only vertices of $Q_{i,j}$ neighbors of the vertices of $K_j^{l^*}$; it remains true whether σ_{l^*} sets value $1, 2, 3, 4$ or 5 to x_i with the convenient α and β. So all the neighbors of $K_j^{l^*}$ in the main part of the graph are not in S. Moreover, no vertex of K_j is taken in the solution, and no vertex of $L_j \setminus L_j^{l^*}$ is taken in the solution. By these facts, and since the only edges between $L_j^{l^*}$ and $K_j^{l^*}$ is a perfect matching between the vertices of these two sets, it follows that each vertex of $L_j^{l^*}$ has a private neighbor, namely its unique neighbor in $K_j^{l^*}$.

Consider now any $i \in \{1, \ldots, n\}$. The set U never takes three consecutive vertices in the path P_i, so $(V(P_i) \setminus \{u_{i,-3}, u_{i,-2}, u_{i,-1}, u_{i,4Fm}, u_{i,4Fm+1}, u_{i,4Fm+2}\}) \setminus U$ is a dominating set in the path $(V(P_i) \setminus \{u_{i,-3}, u_{i,-2}, u_{i,-1}, u_{i,4Fm}, u_{i,4Fm+1}, u_{i,4Fm+2}\})$. Observe now that, for any $j \in \{0, \ldots, Fm - 1\}$, the vertices of the clique K_j in the gadget H_j are never taken by the solution, so the vertices of the path P_i are only dominated by the vertices of P_i, whether the variable x_i appears in $c_{j'}$ or not (for $j' = j \mod m$). Moreover, by the same argument, the neighbors in the verification gadgets of the vertices of the path P_i taken in the solution are never taken in the solution.

If $\rho(x_i) \in \{0, 1, 2, 3\}$, then U takes two consecutive vertices, leaves two consecutive vertices in S, takes again two consecutive vertices, and so on. In these cases, the two vertices of S each have a private neighbor, namely their other neighbor in the path. If $\rho(x_i) \in \{4, 5\}$, then U takes a vertex, leaves a vertex in S, takes a vertex, and so on. In these cases, the vertices of S are their own private vertex. So all the vertices of the path either have a private neighbor, or are their own private vertices.

Nonetheless, we have to be more careful for the first and last sections (for $j = 0$ and $j = Fm - 1$). By the step 3 of our construction of the solution S, and by some simple observations, we have that all vertices of the main part are dominated, and that the vertices of the main part which belong to the solution either have a private neighbor in the corresponding path, or are their own private vertices. □

Let us now prove the other direction of our reduction. The idea of this proof is the following: by partitioning the graph into different parts and upper bound the cost of these parts, we prove that if an upper dominating set in G has not the same form as in Lemma 7 in a sufficiently large copy, then it has size strictly less than k, enabling us to produce a satisfiable assignment for φ using the copy where the upper dominating set has the desired form.

Lemma 8. *If there exists an upper dominating set of size at least k in G, then φ is satisfiable.*

We can now show that the pathwidth of G is bounded by $n + O(1)$.

Lemma 9. *The pathwidth of G is at most $n + O(1)$.*

We are now ready to present the main result of this section:

Theorem 4. *Under SETH, for all $\varepsilon > 0$, no algorithm solves UPPER DOMINATING SET in time $O^*((6 - \varepsilon)^{pw})$, where pw is the input graph's pathwidth.*

5   Sub-exponential Approximation

5.1   Sub-exponential Approximation Algorithm

In this section, we present a sub-exponential approximation algorithm for the UPPER DOMINATING SET problem. We prove the following: for any $r < n$, there exists an r-approximation algorithm for the UPPER DOMINATING SET problem running in time $n^{O(n/r)}$.

To show this result, we use a common tool to design sub-exponential algorithms: partitioning the set of vertices $V(G)$ of the input graph into a convenient number of subsets of the same size. On each subset, we create a number of solutions: all maximal independent sets I in the subgraph induced by the considered set of vertices; and all subsets S of the considered subset. For each maximal independent set I, we extend it to the whole graph. For each subset S, we first go through all subsets of neighbors of vertices of S in order to find the correct set of private neighbors, and then we extend the solution to the whole graph. At the end, we output the best solution encountered. By computing all maximal independent sets I and by going through all subsets S, we prove that there exists at least one valid upper dominating set which has the desired size. Note that, given a subset of an upper dominating set whose vertices have private neighbors, it may be impossible to extend the partial solution if we do not know their private vertices. This is why we need to find the private vertices of the subset S we consider, since in our proof the solution which has the desired size may come from such a subset S. We prove the following:

Theorem 5. *For any $r < n$, UPPER DOMINATING SET is r-approximable in time $n^{O(n/r)}$.*

5.2   Sub-exponential Inapproximability

In this section, we give a lower bound on the complexity of any r-approximation algorithm, matching our algorithm of the previous section. We get the following result: for any $r < n$ and any $\varepsilon > 0$, there is no algorithm that outputs an r-approximation for the UPPER DOMINATING SET problem running in time $n^{(n/r)^{1-\varepsilon}}$.

To obtain this result, we will first prove the desired lower bound for the MAXIMUM MINIMAL HITTING SET problem. In this problem, we are given an hypergraph and we want to find a set of vertices which cover all hyper-edges. Moreover, we need that this set is minimal, i.e. every vertex in the solution covers a private hyper-edge, and we want the solution to be of maximum size.

To obtain this lower bound for the MAXIMUM MINIMAL HITTING SET problem, we will do a reduction from the MAXIMUM INDEPENDENT SET problem. Then, we will make a reduction from the MAXIMUM MINIMAL HITTING SET problem to the UPPER DOMINATING SET problem to transfer this lower bound to our problem.

Recall that we have the following lower bound by Chalermsook et al. [7] for the MAXIMUM INDEPENDENT SET problem:

Theorem 6 (Theorem 1.2 from [7]). *For any $\varepsilon > 0$ and any sufficiently large $r > 1$, if there exists an r-approximation algorithm for* MAXIMUM INDEPENDENT SET *running in time $2^{(n/r)^{1-\varepsilon}}$, then the randomized ETH is false.*

We note that making a reduction from the MAXIMUM MINIMAL HITTING SET problem to derive hardness result for the UPPER DOMINATING SET problem has already be done by Bazgan et al. [3]. Indeed, to get the $n^{1-\varepsilon}$-inapproximability result for the UPPER DOMINATING SET problem, they first derive this bound of the MAXIMUM MINIMAL HITTING SET problem and then they designed an approximation-preserving reduction between these two problems, enabling them to transfer this hardness result to the UPPER DOMINATING SET problem.

In fact, to obtain the hardness result for the MAXIMUM MINIMAL HITTING SET problem, they made a reduction from the MAXIMUM INDEPENDENT SET problem. Our first reduction is similar to this reduction and will allows us to get the desired hardness result for the MAXIMUM MINIMAL HITTING SET problem. Our second reduction, from MAXIMUM MINIMAL HITTING SET to UPPER DOMINATING SET is the approximation-preserving reduction designed by Bazgan et al. [3].

Note that our reduction from MAXIMUM INDEPENDENT SET to MAXIMUM MINIMAL HITTING SET create a quadratic (in n) blow-up of the size of the instance of the latter problem. Such a blow-up does not allow us to derive the desired running-time. To answer this difficulty, we make another step in the reduction where we "sparsify" the instance of MAXIMUM MINIMAL HITTING SET in order to keep the blow-up under control. To prove that the inapproximability gap stays the same, we use a probabilistic analysis with Chernoff bounds.

We will first prove the following hardness result:

Theorem 7. *For any $\varepsilon > 0$ and any sufficiently large $r > 1$, if there exists an r-approximation algorithm for* MAXIMUM MINIMAL HITTING SET *running in time $n^{(n/r)^{1-\varepsilon}}$, then the randomized ETH is false.*

With this hardness result for MAXIMUM MINIMAL HITTING SET, and by using the reduction of Bazgan et al. [3], we get the following hardness result for UPPER DOMINATING SET:

Theorem 8. *For any $\varepsilon > 0$ and any sufficiently large $r > 1$, if there exists an r-approximation for* UPPER DOMINATING SET *running in time $n^{(n/r)^{1-\varepsilon}}$, then the randomized ETH is false.*

References

1. Arkin, E.M., Bender, M.A., Mitchell, J.S.B., Skiena, S.: The lazy bureaucrat scheduling problem. Inf. Comput. **184**(1), 129–146 (2003). https://doi.org/10.1016/S0890-5401(03)00060-9
2. Bazgan, C., Brankovic, L., Casel, K., Fernau, H.: Domination chain: characterisation, classical complexity, parameterised complexity and approximability. Discrete Appl. Math. **280**, 23–42 (2019)
3. Bazgan, C., et al.: The many facets of upper domination. Theor. Comput. Sci. **717**, 2–25 (2018). https://doi.org/10.1016/j.tcs.2017.05.042
4. Bonnet, E., Escoffier, B., Kim, E.J., Paschos, V.T.: On subexponential and FPT-time inapproximability. In: Parameterized and Exact Computation - 8th International Symposium, IPEC 2013, Sophia Antipolis, France, 4–6 September 2013, Revised Selected Papers, pp. 54–65 (2013). https://doi.org/10.1007/978-3-319-03898-8_6
5. Boria, N., Croce, F.D., Paschos, V.T.: On the max min vertex cover problem. In: Approximation and Online Algorithms - 11th International Workshop, WAOA 2013, Sophia Antipolis, France, 5–6 September 2013, Revised Selected Papers, pp. 37–48 (2013). https://doi.org/10.1007/978-3-319-08001-7_4
6. Bourgeois, N., Croce, F.D., Escoffier, B., Paschos, V.T.: Fast algorithms for min independent dominating set. Discret. Appl. Math. **161**(4–5), 558–572 (2013). https://doi.org/10.1016/j.dam.2012.01.003
7. Chalermsook, P., Laekhanukit, B., Nanongkai, D.: Independent set, induced matching, and pricing: Connections and tight (subexponential time) approximation hardnesses. In: 54th Annual IEEE Symposium on Foundations of Computer Science, FOCS 2013, 26–29 October 2013, Berkeley, CA, USA, pp. 370–379 (2013). https://doi.org/10.1109/FOCS.2013.47
8. Chen, J., Huang, X., Kanj, I.A., Xia, G.: Strong computational lower bounds via parameterized complexity. J. Comput. Syst. Sci. **72**(8), 1346–1367 (2006). https://doi.org/10.1016/j.jcss.2006.04.007
9. Cheston, G.A., Fricke, G., Hedetniemi, S.T., Jacobs, D.P.: On the computational complexity of upper fractional domination. Discret. Appl. Math. **27**(3), 195–207 (1990). https://doi.org/10.1016/0166-218X(90)90065-K
10. Cygan, M., et al.: Parameterized Algorithms. Springer, Cham (2015). https://doi.org/10.1007/978-3-319-21275-3
11. Dublois, L., Hanaka, T., Ghadikolaei, M.K., Lampis, M., Melissinos, N.: (In)approximability of maximum minimal FVS. CoRR abs/2009.09971 (2020). https://arxiv.org/abs/2009.09971
12. Eto, H., Hanaka, T., Kobayashi, Y., Kobayashi, Y.: Parameterized algorithms for maximum cut with connectivity constraints. In: 14th International Symposium on Parameterized and Exact Computation, IPEC 2019, 11–13 September 2019, Munich, Germany, pp. 13:1–13:15 (2019). https://doi.org/10.4230/LIPIcs.IPEC.2019.13
13. Furini, F., Ljubic, I., Sinnl, M.: An effective dynamic programming algorithm for the minimum-cost maximal knapsack packing problem. Eur. J. Oper. Res. **262**(2), 438–448 (2017). https://doi.org/10.1016/j.ejor.2017.03.061

14. Gourvès, L., Monnot, J., Pagourtzis, A.: The lazy bureaucrat problem with common arrivals and deadlines: approximation and mechanism design. In: Fundamentals of Computation Theory - 19th International Symposium, FCT 2013, Liverpool, UK, 19–21 August 2013. Proceedings, pp. 171–182 (2013). https://doi.org/10.1007/978-3-642-40164-0_18
15. Halldórsson, M.M.: Approximating the minimum maximal independence number. Inf. Process. Lett. **46**(4), 169–172 (1993). https://doi.org/10.1016/0020-0190(93)90022-2
16. Hanaka, T., Bodlaender, H.L., van der Zanden, T.C., Ono, H.: On the maximum weight minimal separator. Theor. Comput. Sci. **796**, 294–308 (2019). https://doi.org/10.1016/j.tcs.2019.09.025
17. Hanaka, T., Katsikarelis, I., Lampis, M., Otachi, Y., Sikora, F.: Parameterized orientable deletion. In: Eppstein, D. (ed.) 16th Scandinavian Symposium and Workshops on Algorithm Theory, SWAT 2018, 18–20 June 2018, Malmö, Sweden. LIPIcs, vol. 101, pp. 24:1–24:13. Schloss Dagstuhl - Leibniz-Zentrum für Informatik (2018). https://doi.org/10.4230/LIPIcs.SWAT.2018.24
18. Haynes, T.W., Hedetniemi, S.T., Slater, P.J.: Fundamentals of Domination in Graphs, Pure and Applied Mathematics, vol. 208. Dekker, New York (1998)
19. Hurink, J.L., Nieberg, T.: Approximating minimum independent dominating sets in wireless networks. Inf. Process. Lett. **109**(2), 155–160 (2008). https://doi.org/10.1016/j.ipl.2008.09.021
20. Jaffke, L., Jansen, B.M.P.: Fine-grained parameterized complexity analysis of graph coloring problems. In: Fotakis, D., Pagourtzis, A., Paschos, V.T. (eds.) CIAC 2017. LNCS, vol. 10236, pp. 345–356. Springer, Cham (2017). https://doi.org/10.1007/978-3-319-57586-5_29
21. Katsikarelis, I., Lampis, M., Paschos, V.T.: Structural parameters, tight bounds, and approximation for (k, r)-center. In: Okamoto, Y., Tokuyama, T. (eds.) 28th International Symposium on Algorithms and Computation, ISAAC 2017, 9–12 December 2017, Phuket, Thailand. LIPIcs, vol. 92, pp. 50:1–50:13. Schloss Dagstuhl - Leibniz-Zentrum für Informatik (2017). https://doi.org/10.4230/LIPIcs.ISAAC.2017.50
22. Katsikarelis, I., Lampis, M., Paschos, V.T.: Structurally parameterized d-scattered set. In: Brandstädt, A., Köhler, E., Meer, K. (eds.) Graph-Theoretic Concepts in Computer Science - 44th International Workshop, WG 2018, Cottbus, Germany, 27–29 June 2018, Proceedings. Lecture Notes in Computer Science, vol. 11159, pp. 292–305. Springer, Heidelberg (2018). https://doi.org/10.1007/978-3-030-00256-5_24
23. Lampis, M.: Finer tight bounds for coloring on clique-width. In: Chatzigiannakis, I., Kaklamanis, C., Marx, D., Sannella, D. (eds.) 45th International Colloquium on Automata, Languages, and Programming, ICALP 2018, 9–13 July 2018, Prague, Czech Republic. LIPIcs, vol. 107, pp. 86:1–86:14. Schloss Dagstuhl - Leibniz-Zentrum für Informatik (2018). https://doi.org/10.4230/LIPIcs.ICALP.2018.86
24. Lokshtanov, D., Marx, D., Saurabh, S.: Known algorithms on graphs of bounded treewidth are probably optimal. ACM Trans. Algorithms **14**(2), 13:1–13:30 (2018). https://doi.org/10.1145/3170442
25. Zehavi, M.: Maximum minimal vertex cover parameterized by vertex cover. In: Mathematical Foundations of Computer Science 2015–40th International Symposium, MFCS 2015, Milan, Italy, 24–28 August 2015, Proceedings, Part II. pp. 589–600 (2015). https://doi.org/10.1007/978-3-662-48054-0_49

On 2-Clubs in Graph-Based Data Clustering: Theory and Algorithm Engineering

Aleksander Figiel, Anne-Sophie Himmel, André Nichterlein$^{(\boxtimes)}$,
and Rolf Niedermeier

TU Berlin, Faculty IV, Algorithmics and Computational Complexity,
Berlin, Germany
{a.figiel,anne-sophie.himmel,andre.nichterlein,
rolf.niedermeier}@tu-berlin.de

Abstract. Editing a graph into a disjoint union of clusters is a standard optimization task in graph-based data clustering. Here, complementing classic work where the clusters shall be cliques, we focus on clusters that shall be 2-clubs, that is, subgraphs of diameter at most two. This naturally leads to the two NP-hard problems 2-CLUB CLUSTER EDITING (the editing operations are edge insertion and edge deletion) and 2-CLUB CLUSTER VERTEX DELETION (the editing operations are vertex deletions). Answering an open question, we show that 2-CLUB CLUSTER EDITING is W[2]-hard with respect to the number of edge modifications, thus contrasting the fixed-parameter tractability result for the classic CLUSTER EDITING problem (considering cliques instead of 2-clubs). Then, focusing on 2-CLUB CLUSTER VERTEX DELETION, which is easily seen to be fixed-parameter tractable, we show that under standard complexity-theoretic assumptions it does not have a polynomial-size problem kernel when parameterized by the number of vertex deletions. Nevertheless, we develop several effective data reduction and pruning rules, resulting in a competitive solver, outperforming a standard CPLEX solver in most instances of an established biological test data set.

1 Introduction

Graph-based data clustering is one of the most important application domains for graph modification problems [31]. Roughly speaking, the goal herein is to transform a given graph into (usually disjoint) clusters, thereby performing as few modification operations (edge deletions, edge insertions, vertex deletions) as possible. This type of problems typically is NP-hard. The perhaps most prominent problem herein is CLUSTER EDITING (also known as CORRELATION CLUSTERING), where the clusters are requested to be cliques and one is allowed to perform both edge insertions and edge deletions. There has been a lot of work

A. Figiel—Partially supported by DFG project NI 369/18.
A.-S. Himmel—Supported by DFG project NI 369/16.

T. Calamoneri and F. Corò (Eds.): CIAC 2021, LNCS 12701, pp. 216–230, 2021.
https://doi.org/10.1007/978-3-030-75242-2_15

on CLUSTER EDITING, e.g., see the surveys by Böcker and Baumbach [2] and by Crespelle et al. [7]. However, also the variant where one modifies the input graph by vertex deletions received significant interest [6,12,20,32].

Arguably, for many data science applications the request that the clusters have to be cliques is too rigid. Hence, the consideration of clique relaxations for defining clusters gained attention in graph-based data clustering [1,17,25,27]. In this work, we focus on so-called *2-clubs* as clusters [25,27]: these are diameter-at-most-two graphs (hence, cliques are 1-clubs). Other than finding cliques, finding 2-clubs of size at least k is fixed-parameter tractable with respect to k [19,30]. Note that 2-clubs have been used in the context of biological data analysis [21, 28]. Moreover, 2-clubs have been studied in the context of covering vertices in a graph [9–11].

Now, continuing and complementing previous work of Liu et al. [25], we study both the edge editing variant (referred to as 2-CLUB CLUSTER EDITING) and the vertex deletion variant (referred to as 2-CLUB CLUSTER VERTEX DELETION). We contribute the following three main results:

1. Answering an open question of Liu et al. [25], in Sect. 2 we show that 2-CLUB CLUSTER EDITING is W[2]-hard with respect to the number of modified edges (deletions and insertions), hence most likely not fixed-parameter tractable. This stands in sharp contrast to the problems CLUSTER EDITING [16] and the more general s-PLEX CLUSTER EDITING [17][1], both known to be fixed-parameter tractable for the parameter number of edge modifications. The W[2]-hardness seems surprising considering the fact that while CLUSTER EDITING is fixed-parameter tractable [2] and 2-CLUB CLUSTER EDITING is presumably not, by way of contrast finding cliques is presumably not fixed-parameter tractable while finding 2-clubs is.

2. Complementing fixed-parameter tractability and kernelization results for CLUSTER VERTEX DELETION [6,20,32] and s-PLEX CLUSTER VERTEX DELETION [1], in Sect. 2 we show that, other than these related problems and despite being easily seen to be fixed-parameter tractable for the parameter solution size, 2-CLUB CLUSTER VERTEX DELETION is unlikely to have a polynomial-size problem kernel.[2]

3. In Sects. 3 and 4, we explore the fixed-parameter tractability of 2-CLUB VERTEX DELETION from a more practical angle and develop several efficient data reduction rules together with effective search-tree pruning rules. Performing an empirical evaluation with standard biological data, we show that our tuned algorithmic approach (based on branching and data reduction) in most relevant cases clearly outperforms a standard ILP formulation solved CPLEX,

---

[1] This is the generalization of CLUSTER EDITING where clusters are requested to be s-plexes (and not cliques); an s-plex is a subgraph where each vertex is connected to all other vertices of the s-plex except for at most $s - 1$ vertices. Notably, a clique is a 1-plex.

[2] It has been featured as an open problem whether the edge deletion variant s-CLUB CLUSTER EDGE DELETION has a polynomial-size problem kernel [7,25].

thus providing a state-of the art software tool for the vertex deletion variant
of graph-based data clustering with 2-clubs.

Due to the lack of space, several details had to be deferred to the full version [14].

Preliminaries. All graphs considered in our work are undirected and simple. For
a graph $G = (V, E)$ we set $n := |V|$ and $m := |E|$. We denote with $\binom{V}{2}$ the set of
all two-element subsets of V. For a vertex $v \in V$, we denote by $N_G(v) := \{w \in V \mid \{v, w\} \in E\}$ the *open neighborhood* of v and by $N_G[v] := N_G(v) \cup \{v\}$ the
closed neighborhood of v. The *degree* of v is $\deg_G(v) := |N_G(v)|$. For a vertex
subset $V' \subseteq V$, let $N_G[V'] := \bigcup_{v \in V'} N_G[v]$. If it is clear from the context, then
we omit G from the subscripts. We denote by $G[V']$ the subgraph of G induced
by the vertex set $V' \subseteq V$ and by $G[E']$ the subgraph of G with edge set $E' \subseteq E$,
that is, $G[E'] := (V, E')$. The graph $G - v$ is obtained by deleting $v \in V$ from G,
that is $G - v := G[V \setminus \{v\}]$.

    A path P in G is an ordered sequence of pairwise distinct vertices v_1,
$v_2, \ldots, v_{k+1} \in V$ such that $\{v_i, v_{i+1}\} \in E$ for all $i \in \{1, \ldots, k\}$. It is also an
induced path if these are the only edges between its vertices. The length of P
is k. We denote by P_n a path on n vertices. The *distance* of two vertices $s, t \in V$,
denoted by $\text{dist}_G(s, t)$, is the length of a shortest path connecting s and t if one
exists, and ∞ otherwise. The *diameter* of a graph is the maximum distance of
any two vertices, formally $\max_{s,t \in V} \text{dist}_G(s, t)$. A graph is said to be *connected*
if there exists a path between all pairs of its vertices. A (connected) *component*
of a graph G is a maximal vertex set $S \subseteq V$ such that $G[S]$ is connected.

s-Club. An *s-club* is a graph of diameter at most s. A *clique* is a 1-club.
Furthermore, an *s-club cluster graph* is a graph in which each component
is an s-club. In this paper, we consider the following two problems, where
$E \triangle F := (E \setminus F) \cup (F \setminus E)$ denotes the *symmetric difference* of two sets E
and F.

s-CLUB CLUSTER EDITING
Input: An undirected graph $G = (V, E)$ and an integer $k \in \mathbb{N}$.
Question: Is there an edge set $F \subseteq \binom{V}{2}$ with $|F| \leq k$ such that $G[E \triangle F]$
                is an s-club cluster graph?

s-CLUB CLUSTER VERTEX DELETION
Input: An undirected graph $G = (V, E)$ and an integer $k \in \mathbb{N}$.
Question: Is there a vertex subset $S \subseteq V$ with $|S| \leq k$ such that $G[V \setminus S]$
                is an s-club cluster graph?

An edge set $F \subseteq \binom{V}{2}$ such that $G[E \triangle F]$ is an s-club cluster graph is called an
s-club editing set and a vertex set $S \subseteq V$ such that $G[V \setminus S]$ is an s-club cluster
graph is called an *s-club vertex deletion set*.

2-Club. A 2-club is a graph with diameter at most two. This means that for all pairs of vertices $u, v \in V$ it holds that u and v are adjacent or have at least one common neighbor. Note that 2-clubs are *non-hereditary*, that is, if G is a 2-club, then deleting vertices from G may destroy this property. This is a significant difference in comparison with cliques.

Using terminology of Liu et al. [25], we call a path $stuv$ in G a *restricted P_4* if $\text{dist}_G(s, v) = 3$. That is, a restricted P_4 is a shortest path connecting s and v and is thus also an induced P_4. The following characterization is easy to verify:

Observation 1 ([25, Lemma 3]). *A graph G is a 2-club cluster graph if and only if it contains no restricted P_4.*

Parameterized Algorithmics [8]. A parameterized problem $\Pi \subseteq \Sigma^* \times \mathbb{N}$ is a set of pairs (I, k), where I denotes the problem instance and k is the parameter. Problem Π is *fixed-parameter tractable* (FPT) if there exists an algorithm solving any instance of Π in $f(k) \cdot |I|^c$ time, where f is some computable function and c is some constant. A *parameterized reduction* from a parameterized problem $\Pi \subseteq \Sigma^* \times \mathbb{N}$ to a parameterized problem $\Pi' \subseteq \Sigma^* \times \mathbb{N}$ is a function which maps any instance $(I, k) \in \Sigma^* \times \mathbb{N}$ to another instance $(I', k') \in \Sigma^* \times \mathbb{N}$ such that (1) (I', k') can be computed from (I, k) in FPT time, (2) $k' \leq g(k)$ for some computable function g, and (3) $(I, k) \in \Pi \iff (I', k') \in \Pi'$. If Π is W[i]-hard, $i \geq 1$, then such a parameterized reduction shows that also Π' is W[i]-hard, that is, presumably not fixed-parameter tractable. A *reduction to a problem kernel* is a parameterized self-reduction (from Π to Π) such that (I', k') can be computed in polynomial time and $|I'| \leq g(k)$. If g is a polynomial, then (I', k') is called a *polynomial kernel*. Problem kernels are usually achieved by applying *data reduction rules*. Given an instance (I, k), a data reduction rule computes in polynomial time a new instance (I', k'). We call a data reduction rule *safe* if $(I, k) \in \Pi \iff (I', k') \in \Pi$.

2   Hardness Results

It is easy to see that 2-CLUB CLUSTER VERTEX DELETION is fixed-parameter tractable with respect to solution size k [25]: By Observation 1, it is enough to recursively search for a restricted P_4 $stuv$ and delete a vertex to separate s and v. In contrast, we subsequently show that 2-CLUB CLUSTER EDITING is W[2]-hard with respect to solution size k answering an open question of Liu et al. [25]. Intuitively, the hardness is due to the fact that there is a "non-local" way of destroying a restricted P_4 with edge insertions, see Fig. 1 for an illustration.

The basic idea of our parameterized reduction from DOMINATING SET[3] is inspired by a parameterized reduction by Gao et al. [15, Theorem 1] who showed hardness for the problem of reducing the diameter of a given graph to two by inserting at most k edges. In our reduction we need to take care of the possibility

---

[3] Given an undirected graph $G = (V, E)$ and an integer k, the question is whether there is a dominating set $V' \subseteq V$ (that is, $N[V'] = V$) of size at most k.

six local
modifications:

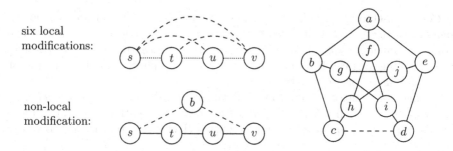

non-local
modification:

Fig. 1. *Left (top and bottom):* All possible modifications to destroy a restricted P_4. Top: The six "local" modifications; that is, any edge which is inserted (dashed edges) or deleted (dotted edges) has both its ends in the P_4. Bottom: A "non-local" modification (the two inserted edges are dashed), where b can be any vertex other than s, t, u and v. *Right side:* The dashed edge indicates the single optimal solution (inserting the edge, the resulting Petersen graph has diameter two) which is a non-local modification. Since the distance between c and d was four, the insertion of edge $\{c, d\}$ is not part of any local modification.

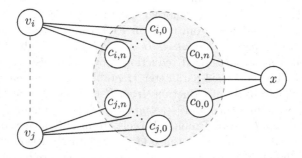

Fig. 2. A schematic picture of the construction of G' in the proof of Theorem 1. The vertices in the gray circle form a clique, but only the vertices connected to v_i, v_j, or x are shown. The dashed gray edge between v_i and v_j exists if $\{v_i, v_j\} \in E(G)$.

to delete edges, which changes many details of the construction. DOMINATING SET remains W[2]-hard with respect to k for graphs with diameter two [26]. Thus we can assume that the DOMINATING SET instance has diameter two.

Theorem 1. 2-CLUB CLUSTER EDITING *is W[2]-hard with respect to k.*

Proof. Let $G = (V, E)$ be a graph with diameter two. We construct a graph G' in such a way that G has a dominating set of size at most k if and only if G' has a 2-club editing set of size at most k. The graph $G' = (V', E')$ can be broken down into the following parts: the original graph G, a clique $C \subseteq V'$ of cardinality $(n + 1)^2$, and a single vertex x. We assign two indices for the vertices $c_{i,j} \in C$ such that $i, j \in \{0, \dots, n\}$. The vertices in V only have one index: $v_i \in V$, $i \in \{1, \dots, n\}$. In addition to the existing edges in G and C, add the following edges: for each $j \in \{0, \dots, n\}$ add $\{x, c_{0,j}\}$ and for each $c_{i,j} \in$

$C, i \neq 0$, add $\{v_i, c_{i,j}\}$. The graph G' has $\mathcal{O}(n^4)$ edges and $\mathcal{O}(n^2)$ vertices and can be constructed in polynomial time. For a schematic picture of G' see Fig. 2. Note that the only pairs of vertices with distance three are x and $v_i \in V$, all others have distance at most two.

We claim that there exists a dominating set of size at most k for G if and only if there exists a 2-club editing set of size at most k for G' (which only inserts edges).

"$\Rightarrow$": Let D be a dominating set for G with $|D| \leq k$, and $F := \{\{x, v\} \mid v \in D\}$. Let $H := G'[E' \triangle F]$. For every $v_i \in V$, either $v_i \in D$ and then $\text{dist}_H(x, v_i) = 1$, or $v_i \notin D$ and then v_i has a neighbor in D and thus $\text{dist}_H(x, v_i) = 2$. This means that H is a 2-club cluster graph and F is a 2-club editing set for G' with $|F| \leq k$.

"$\Leftarrow$": Let F be a 2-club editing set for G' with $|F| \leq k$ and $H = G'[E' \triangle F]$ be the resulting 2-club cluster graph. Assume without loss of generality that F is minimal. Removing any edge would only be optimal if H contained more than one 2-club cluster. Note that the size of a minimum cut of G' is $n + 1$ and that $k < n$. Hence, there is only one 2-club cluster in a solution and no edge is removed.

For any inserted edge $\{a, b\} \in F$ exactly one of the following cases applies, since the distance between x and some $v_i \in V$ has to be reduced by means of inserting $\{a, b\}$.

- $\{a, b\} = \{v_i, x\}$: Then $\text{dist}_H(x, v_i) = 1$ and for $a \in N_G(v_i)$ $\text{dist}_H(x, a) \leq 2$. We interpret this as v_i being a dominating vertex in G.
- $\{a, b\} = \{v_i, c_{0,j}\}$: This edge enables a path of length two from v_i to x via $c_{0,j}$. This means that this edge is only of benefit to v_i. Then $F' = (F \setminus \{v_i, c_{0,j}\}) \cup \{x, v_i\}$ is also a 2-club editing set with $|F| = |F'|$.
- $\{a, b\} = \{v_i, v_j\}$: This means that one of the vertices has an edge to x. Without loss of generality assume that $\{x, v_i\} \in F$. Note that F is only minimal if $\{x, v_j\} \notin F$, as the edge $\{v_i, v_j\}$ is only of benefit to v_j and no other vertices since it enables a path of length two from v_j to x via v_i. Then $F' = (F \setminus \{v_i, v_j\}) \cup \{x, v_j\}$ is also a 2-club editing set with $|F| = |F'|$.
- $\{a, b\} = \{v_i, c_{j,k}\}, j \neq i, j \neq 0$: This means that there is an edge $\{x, c_{j,k}\} \in F$, otherwise F would not be minimal. The edge $\{v_i, c_{j,k}\}$ enables a path of length two from v_i to x via $c_{j,k}$. This means that the edge is of no benefit to any other vertices. Then $F' = (F \setminus \{v_i, c_{j,k}\}) \cup \{x, v_i\}$ is also a 2-club editing set with $|F| = |F'|$.
- $\{a, b\} = \{x, c_{i,j}\}, i \neq 0$: This edge enables a path of length two from v_i to x via $c_{i,j}$. In the previous case, we have seen that there exists an F' with $\{x, c_{i,j}\} \in F'$ such that there exists no edge $\{v_\ell, c_{i,j}\} \in F'$ with $\ell \neq i$. This means that the edge $\{x, c_{i,j}\}$ is of no benefit to any other vertices. Then $F'' = (F' \setminus \{x, c_{i,j}\}) \cup \{x, v_i\}$ is also a 2-club editing set with $|F| = |F''|$.

Altogether, we know that there exists an F' with $|F'| = |F|$ such that F' is a 2-club editing set of the form $\{\{x, v\} \mid v \in D\}$ for some $D \subseteq V$. This means that D is a dominating set for G with $|D| \leq k$.

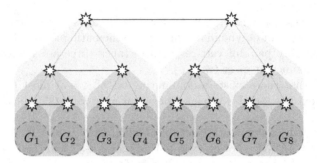

Fig. 3. Illustration of the construction for Theorem 2 exemplified for $\ell = 8$. Star-shaped vertices have $k' + 1$ additional leaves and are connected to all vertices in the gray-shaded area below them.

Summarizing, the reduction from (G, k) to (G', k) is a valid parameterized reduction from DOMINATING SET for graphs with diameter two to 2-CLUB CLUSTER EDITING. Since DOMINATING SET is W[2]-hard for graphs of diameter two [26], this yields that 2-CLUB CLUSTER EDITING is also W[2]-hard.   □

Next, we use the OR-cross-composition framework of Bodlaender et al. [5] to show that 2-CLUB CLUSTER VERTEX DELETION does not admit a polynomial kernel with respect to the solution size k.

Theorem 2. 2-CLUB CLUSTER VERTEX DELETION *does not admit a polynomial kernel with respect to k unless* NP $\subseteq$ coNP / poly.

Proof (Sketch). Given ℓ instances $(G_1 = (V_1, E_1), k), \ldots, (G_\ell = (V_\ell, E_\ell), k)$, we subsequently construct in polynomial time a new instance $(G' = (V', E'), k')$ that is a yes-instance if and only if at least one of the ℓ instances $(G_i = (V_i, E_i), k)$ is a yes-instance. The theorem then follows from applying the OR-cross-composition framework of Bodlaender et al. [5].

Without loss of generality, assume that ℓ is a power of two (otherwise copy instances until ℓ is a power of two). We set $k' := k + \log \ell$. To describe G', we need a simple *selection-gadget* consisting of two stars with $k' + 1$ leaves each where the two center vertices are adjacent. Observe that in the selection-gadget the leaves of one star are at distance three to the leaves of the other star. Moreover, since each star has more than k' leaves, the only possibility to transform a selection-gadget into a 2-club cluster graph is to delete one of the two center vertices.

We can now define G': To this end, we recursively create an "instance-selector" that forces the selection of exactly one instance G_i as shown in Fig. 3. First, add a selection-gadget with the two center vertices c_L and c_R (left and right). Second, recursively build the two graphs G_L, G_R composing $G_1, \ldots, G_{\ell/2}$ and $G_{\ell/2+1}, \ldots, G_\ell$ respectively until G_L, G_R consist of only one input instance. Make every vertex in G_L (in G_R) adjacent to c_L (to c_R). Note that this recursive procedure has recursion depth $\log \ell$. The construction of (G', k') can clearly be done in polynomial time.   □

3   Algorithms for 2-Club Cluster Vertex Deletion

In this section, we first formulate 2-CLUB CLUSTER VERTEX DELETION as an Integer Linear Program (ILP) and then introduce a branch&bound-algorithm solving a generalization of 2-CLUB CLUSTER VERTEX DELETION. We use the ILP-formulation in our experiments to evaluate our branch&bound algorithm.

ILP Formulation. By Observation 1, a graph is a 2-club cluster graph if and only if it contains no restricted P_4. Recall that a restricted P_4 is an induced P_4 *stuv* that is also a shortest path between s and t. Thus, there exists no vertex $w \in N(s) \cap N(v)$ in the common neighborhood of s and v. The deletion of a vertex cannot create any new induced path but it might "promote" an induced P_4 to a restricted P_4. Hence, if $N(s) \cap N(v) = \emptyset$ for any induced P_4 *stuv* in G, then at least one vertex from *stuv* must be deleted.

We introduce a variable x_v for each vertex $v \in V$. This variable has a value of 1 if and only if v is in the 2-club vertex deletion set. This leads to the following ILP formulation:

$$\text{min:} \quad \sum_{v \in V} x_v$$

$$\text{s.t.} \quad x_s + x_t + x_u + x_v + \sum_{b \in N(s) \cap N(v)} (1 - x_b) \geq 1 \quad \text{for all induced } P_4\text{'s } stuv \text{ in } G$$

$$x_v \in \{0, 1\} \quad \text{for all } v \in V.$$

Branch&Bound Algorithm. We describe an algorithm for the following generalization of 2-CLUB CLUSTER VERTEX DELETION as this variant allows more flexibility in the design of data reduction rules and for deriving lower bounds.

GENERALIZED 2-CLUB CLUSTER VERTEX DELETION (GEN2CVD)

Input: An undirected graph $G = (V, E)$, an integer $k \in \mathbb{N}$, a set $F \subseteq V$ of permanent vertices, and a weight function $w : V \to \mathbb{N}^+$.

Question: Is there an $S \subseteq V$ with $w(S) \leq k$ and $S \cap F = \emptyset$ such that $G[V \setminus S]$ is a 2-club cluster graph?

Note that an instance (G, k) of 2-CLUB CLUSTER VERTEX DELETION is clearly equivalent to the instance $(G, k, \emptyset, w \equiv 1)$ of GEN2CVD.

Our algorithm uses a simple branching rule that takes a restricted P_4 and branches into all four cases of deleting one vertex which implies updates of the set F of permanent vertices in each branch. If some vertex of the restricted P_4 *stuv* is already in F, then we skip the corresponding case in the branching. Thus, the branching itself "grows" the set F of permanent vertices that will reduce the cases to be considered later in the branching. Moreover, if more than one restricted P_4 exists, then the algorithm chooses one with most vertices in F and uses the weights of the vertices as tiebreaker.

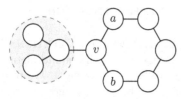 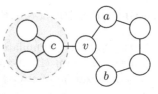

(a) Reduction Rule 1 can be applied on vertex v. The gray area contains vertices that can be marked as permanent.

(b) Reduction Rule 1 cannot be applied on vertex v because v is a bridge for vertices a and b (in fact deleting c is the unique optimal solution).

Fig. 4. Examples of graphs with a cut vertex v (all vertex weights are one). The gray area is a 2-club that is isolated from the rest of the graph after deleting v. Note that in the first graph the removal of v increases the distance of a and b to four. Thus no induced P_4 exists between a and b.

We enrich the basic branching algorithm with lower bounds and data reduction rules. A simple lower bound is obtained by (heuristically) computing a set $\mathcal{P}$ of vertex disjoint restricted P_4's. Then it is easy to see that any solution has to delete at least $|\mathcal{P}|$ many vertices. Moreover, for each restricted P_4 one has to pay at least the weight of the lightest vertex. Thus, $\sum_{stuv \in \mathcal{P}} \min_{v \in \{s,t,u,v\}} \{w(v)\}$ is a (better) lower bound.

We implemented several data reductions rules. We exemplify one rule below, illustrating the issues arising from the non-hereditary nature of the problem. A vertex $b \in V$ is a cut vertex if the deletion of b increases the number of connected components. A vertex $b \in V$ is a bridge vertex if for some $s, v \in N(b)$ there exists an induced P_4 $stuv$. (Deleting b might turn $stuv$ into a restricted P_4.) For ease of notation, assume that all vertex weights are one. (The weighted case requires some more case distinctions.) The rule is exemplified in Fig. 4, where the right side (b) demonstrates the necessity of the requirement that v is not a bridge vertex.

Reduction Rule 1. *Let $v \in V$ with $w(v) = 1$ be a non-permanent cut vertex that is not a bridge. For each component C in $G - v$ that is a 2-club mark the vertices in C as permanent.*

While these (and more) lower bounds and data reduction rules give a significant speedup in practice (also cf. Komusiewicz et al. [23]), we could not show an improved theoretical worst-case bound: The branching into four cases leads to the following result (note that a restricted P_4 can be found in $O(nm)$ time).

Proposition 3. GENERALIZED 2-CLUB CLUSTER VERTEX DELETION *can be solved in $O(4^k \cdot nm)$ time.*

4   Experimental Evaluation

4.1   Setup

We implemented our branch&bound algorithm (see Sect. 3) for GENERALIZED 2-CLUB CLUSTER VERTEX DELETION in C++ (we use the algorithm to solve 2-CLUB CLUSTER VERTEX DELETION).[4] This solver (called solverALL in the following) computes a 2-club vertex deletion set size of minimum cost and outputs the solution set. It uses all data reduction rules described in the full version [14]. Note that for the implementation of some data reduction rules and lower bounds we use heuristics.

We will compare the performance of our solver against the ILP formulation from Sect. 3 solved using CPLEX (we will refer to this solver as CPLEX). All experiments were run on a machine with an Intel Xeon W-2125 8-core, 4.0 GHz CPU and 256 GB of RAM running Ubuntu 18.04. We used a recent version of CPLEX, 12.8, for our experiments. Analogously to previous work for the related CLUSTER EDITING [4], we focus on implementations computing optimal solutions. Thus, we also require CPLEX to compute optimal solutions.

For our analysis we used a real-world biological dataset[5] that has been used for the evaluation of WEIGHTED CLUSTER EDITING solvers [3,29]. We created two sets of unweighted graphs (called Bio33 and Bio50; each with $\approx$ 430 graphs) as described by Hartung and Hoos [18]. These instances had up to 250 vertices and 15,000 edges.

4.2   Results

We next analyze the performance of our solver in detail. To this end, we start with comparing the theoretical bounds with the results of our experiments. The number of branches in our search tree is far below the theoretical worst-case bound of 4^k (even far below the 3.31^k bound of the search tree of Liu et al. [25]) given in Proposition 3. This is a clear indication that the data reduction rules and lower bounds have a strong impact on our solver. Another observation is that the impact of the number of input graph vertices on the running time is quite significant. The reason for this is the high polynomial running time for executing the data reduction rules and computing lower bounds: Our best upper bound on the running time (in terms of n) of one recursive step (including data reduction and lower bounds) is $O(n^4)$. We show subsequently that the high running-time cost for the data reduction rules is justified.

The Bio33 instances are in general harder for our solver than the Bio50 instances. The reason for this is that the Bio50 instances are more dense and allow to cluster in less 2-clubs of larger size with fewer vertex removals.

---

[4] The source code is available at https://fpt.akt.tu-berlin.de/software/two-club-editing/two-club-vertex-deletion.zip and includes the source code for the ILP formulation using CPLEX.

[5] The dataset is available at https://bio.informatik.uni-jena.de/data/#cluster_editing_data.

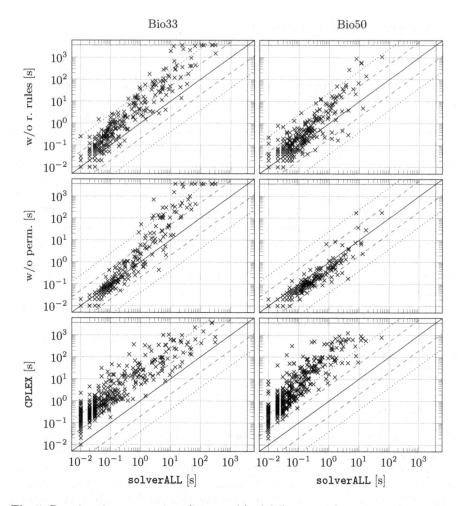

Fig. 5. Running time comparison (in seconds) of different configurations of our solver and CPLEX on two datasets (left: Bio33, right: Bio50). Each dot represents one instance with the x and y coordinates indicating the running time of the respective solver (in seconds). Hence, a dot above (below) the solid diagonal indicates the solver on the x-axis (y-axis) is faster on the corresponding instance. The diagonal lines mark running time factors of 1 (solid), 5 (dashed) and 25 (dotted). Dots on the solid horizontal and vertical red lines (at 3600 s) indicate timeouts. In each plot the running time of our `solverALL` is displayed at the x-axis. The y-axis shows in each row of the plots a different solver; these are from top to bottom: Three configurations of our solver where certain features are disabled (first all data reduction rules, then permanent vertices with the corresponding data reduction rules that require permanent vertices). The last row shows the comparison against CPLEX.

Comparisons. We next compare our solver `solverALL` to several variants of it where we deactivate key features and to CPLEX. The comparisons are illustrated in Fig. 5.

The first row of plots in Fig. 5 shows that if we deactivate the data reduction rules, then the performance becomes much worse, especially on the harder instances that require more than 10 s to solve. On average, `solverALL` (with all data reduction rules) is 6.7 times faster on the Bio33 instances and 3 times faster on the Bio50 instances. This is in stark contrast to the kernelization lower bound given in Theorem 2 and gives hope to find small parameters based on which one may perform a mathematical analysis yielding polynomial kernel sizes.

The plots in the second row of Fig. 5 show the effect of turning off permanent vertices and the corresponding data reduction rules. Note that in the Bio50 dataset the variant without permanent vertices is faster on most instances, very likely due to respective data reduction rule being expensive. However, the results for the Bio33 dataset show a different picture. In fact, one can see in both data sets that turning off the feature of permanent vertices solves the easier instances even faster and slows down the solver on the harder instances. The lack of "hard" instances in the Bio50 dataset is the reason for the variant without permanent vertices being faster there. On average, `solverALL` (with permanent vertices) is 5 times faster on the Bio33 instances but 1.6 times slower on the Bio50 instances.

The plots in the last row of Fig. 5 show that our solver is almost always faster than CPLEX by a factor of 5–25 for Bio33 and a factor of 25–100 for Bio50. On average, `solverALL` is 29.3 times faster on the Bio33 instances and 103.6 times faster on the Bio50 instances. For Bio33 it appears that for harder instances CPLEX is not much slower than our solver. On Bio50, CPLEX does very poorly compared to our solver. This is likely due to the minimum 2-club vertex deletion set size (the parameter in our FPT-algorithm) on these graphs being smaller than for Bio33. Moreover, the process for building the ILP model for CPLEX is usually fairly fast, but for larger instances it can take up to 60 s. For example, in Bio50 there is a graph with 205 vertices and 10455 edges which is already a 2-club cluster graph. It takes about 50 s to create the ILP model, and when exported to a file it takes up to 1.6 GB (uncompressed) and includes 5.8 million constraints, whereas the original graph only takes up 72 kB stored in an edge list format.

Summarizing, our solver outperforms a standard ILP-formulation solved with CPLEX. To this end, good data reduction rules are crucial to the practical performance of our solver.

5   Conclusion

We investigated the problem of modifying graphs into 2-club cluster graphs. We have shown that 2-CLUB CLUSTER EDITING is W[2]-hard for the parameter solution size k. Furthermore, we developed and engineered a competitive branch&bound algorithm for the fixed-parameter tractable 2-CLUB CLUSTER VERTEX DELETION problem.

On the theoretical side, we left open whether our "no-poly-kernel" result for 2-CLUB CLUSTER VERTEX DELETION parameterized by solution size transfers to the closely related 2-CLUB CLUSTER EDGE DELETION problem, a further open problem from the literature [7,25]. Moreover, it would be interesting to see whether our results also generalize to using s-clubs with $s \geq 3$. For other 2-club related graph modification problems to be studied one could consider overlapping clusters [13] or use stricter 2-club models such as well-connected 2-clubs [24]. Limiting the number of local manipulations [22] is another restriction worthwhile investigations. On the empirical and algorithm engineering side, note that while our solver showed strong performance when working with biological data, preliminary experiments showed that this is less so when attacking social network data. The reasons for this call for an explanation.

Acknowledgment. We thank anonymous reviewers for their valuable feedback.

References

1. van Bevern, R., Moser, H., Niedermeier, R.: Approximation and tidying - a problem kernel for s-plex cluster vertex deletion. Algorithmica **62**(3–4), 930–950 (2012). https://doi.org/10.1007/s00453-011-9492-7
2. Böcker, S., Baumbach, J.: Cluster editing. In: Bonizzoni, P., Brattka, V., Löwe, B. (eds.) CiE 2013. LNCS, vol. 7921, pp. 33–44. Springer, Heidelberg (2013). https://doi.org/10.1007/978-3-642-39053-1_5
3. Böcker, S., Briesemeister, S., Bui, Q.B.A., Truß, A.: Going weighted: parameterized algorithms for cluster editing. Theoret. Comput. Sci. **410**(52), 5467–5480 (2009). https://doi.org/10.1016/j.tcs.2009.05.006
4. Böcker, S., Briesemeister, S., Klau, G.W.: Exact algorithms for cluster editing: evaluation and experiments. Algorithmica **60**(2), 316–334 (2011). https://doi.org/10.1007/s00453-009-9339-7
5. Bodlaender, H.L., Jansen, B.M.P., Kratsch, S.: Kernelization lower bounds by cross-composition. SIAM J. Discrete Math. **28**(1), 277–305 (2014). https://doi.org/10.1137/120880240
6. Boral, A., Cygan, M., Kociumaka, T., Pilipczuk, M.: A fast branching algorithm for cluster vertex deletion. Theory Comput. Syst. **58**(2), 357–376 (2016). https://doi.org/10.1007/s00224-015-9631-7
7. Crespelle, C., Drange, P.G., Fomin, F.V., Golovach, P.A.: A survey of parameterized algorithms and the complexity of edge modification. CoRR, abs/2001.06867 (2020). https://arxiv.org/abs/2001.06867
8. Cygan, M., et al.: Parameterized Algorithms. Springer, Heidelberg (2015). https://doi.org/10.1007/978-3-319-21275-3
9. Dondi, R., Lafond, M.: On the tractability of covering a graph with 2-clubs. In: Gasieniec, L.A., Jansson, J., Levcopoulos, C. (eds.) FCT 2019. LNCS, vol. 11651, pp. 243–257. Springer, Cham (2019). https://doi.org/10.1007/978-3-030-25027-0_17
10. Dondi, R., Mauri, G., Sikora, F., Zoppis, I.: Covering a graph with clubs. J. Graph Algorithms Appl. **23**(2), 271–292 (2019). https://doi.org/10.7155/jgaa.00491
11. Dondi, R., Mauri, G., Zoppis, I.: On the tractability of finding disjoint clubs in a network. Theoret. Comput. Sci. **777**, 243–251 (2019). https://doi.org/10.1016/j.tcs.2019.03.045

12. Doucha, M., Kratochvíl, J.: Cluster vertex deletion: a parameterization between vertex cover and clique-width. In: Proceedings of the 37th International Symposium on Mathematical Foundations of Computer Science (MFCS 2012). LNCS, vol. 7464, pp. 348–359. Springer, Heidelberg (2012). https://doi.org/10.1007/s00453-011-9492-7

13. Fellows, M.R., Guo, J., Komusiewicz, C., Niedermeier, R., Uhlmann, J.: Graph-based data clustering with overlaps. Discrete Optim. **8**(1), 2–17 (2011)

14. Figiel, A., Himmel, A., Nichterlein, A., Niedermeier, R.: On 2-clubs in graph-based data clustering: theory and algorithm engineering. CoRR, abs/2006.14972 (2020). https://arxiv.org/abs/2006.14972

15. Gao, Y., Hare, D.R., Nastos, J.: The parametric complexity of graph diameter augmentation. Discrete Appl. Math. **161**(10–11), 1626–1631 (2013). https://doi.org/10.1016/j.dam.2013.01.016

16. Gramm, J., Guo, J., Hüffner, F., Niedermeier, R.: Graph-modeled data clustering: exact algorithms for clique generation. Theory Comput. Syst. **38**(4), 373–392 (2005). https://doi.org/10.1007/s00224-004-1178-y

17. Guo, J., Komusiewicz, C., Niedermeier, R., Uhlmann, J.: A more relaxed model for graph-based data clustering: s-plex cluster editing. SIAM J. Discrete Math. **24**(4), 1662–1683 (2010). https://doi.org/10.1137/090767285

18. Hartung, S., Hoos, H.H.: Programming by optimisation meets parameterised algorithmics: a case study for cluster editing. In: Dhaenens, C., Jourdan, L., Marmion, M.-E. (eds.) LION 2015. LNCS, vol. 8994, pp. 43–58. Springer, Cham (2015). https://doi.org/10.1007/978-3-319-19084-6_5

19. Hartung, S., Komusiewicz, C., Nichterlein, A.: Parameterized algorithmics and computational experiments for finding 2-clubs. J. Graph Algorithms Appl. **19**(1), 155–190 (2015). https://doi.org/10.1007/978-3-642-33293-7_22

20. Hüffner, F., Komusiewicz, C., Moser, H., Niedermeier, R.: Fixed-parameter algorithms for cluster vertex deletion. Theory Comput. Syst. **47**(1), 196–217 (2010). https://doi.org/10.1007/s00224-008-9150-x

21. Jia, S., et al.: Viewing the meso-scale structures in protein-protein interaction networks using 2-clubs. IEEE Access **6**, 36780–36797 (2018). https://doi.org/10.1109/ACCESS.2018.2852275

22. Komusiewicz, C., Uhlmann, J.: Cluster editing with locally bounded modifications. Discrete Appl. Math. **160**(15), 2259–2270 (2012). https://doi.org/10.1016/j.dam.2012.05.019

23. Komusiewicz, C., Nichterlein, A., Niedermeier, R.: Parameterized algorithmics for graph modification problems: on interactions with heuristics. In: Mayr, E.W. (ed.) WG 2015. LNCS, vol. 9224, pp. 3–15. Springer, Heidelberg (2016). https://doi.org/10.1007/978-3-662-53174-7_1

24. Komusiewicz, C., Nichterlein, A., Niedermeier, R., Picker, M.: Exact algorithms for finding well-connected 2-clubs in sparse real-world graphs: theory and experiments. Eur. J. Oper. Res. **275**(3), 846–864 (2019). https://doi.org/10.1016/j.ejor.2018.12.006

25. Liu, H., Zhang, P., Zhu, D.: On editing graphs into 2-club clusters. In: Snoeyink, J., Lu, P., Su, K., Wang, L. (eds.) AAIM/FAW 2012. LNCS, vol. 7285, pp. 235–246. Springer, Heidelberg (2012). https://doi.org/10.1007/978-3-642-29700-7_22

26. Lokshtanov, D., Misra, N., Philip, G., Ramanujan, M.S., Saurabh, S.: Hardness of r-dominating set on graphs of diameter $(r + 1)$. In: Gutin, G., Szeider, S. (eds.) IPEC 2013. LNCS, vol. 8246, pp. 255–267. Springer, Cham (2013). https://doi.org/10.1007/978-3-319-03898-8_22

27. Misra, N., Panolan, F., Saurabh, S.: Subexponential algorithm for d-cluster edge deletion: exception or rule? J. Comput. Syst. Sci. **113**, 150–162 (2020). https://doi.org/10.1016/j.jcss.2020.05.008

28. Pasupuleti, S.: Detection of protein complexes in protein interaction networks using n-clubs. In: Marchiori, E., Moore, J.H. (eds.) EvoBIO 2008. LNCS, vol. 4973, pp. 153–164. Springer, Heidelberg (2008). https://doi.org/10.1007/978-3-540-78757-0_14

29. Rahmann, S., Wittkop, T., Baumbach, J., Martin, M., Truss, A., Böcker, S.: Exact and heuristic algorithms for weighted cluster editing. In: Proceedings of the 6th Computational Systems Bioinformatics Conference (CSB 2007), pp. 391–401. World Scientific (2007). https://doi.org/10.1142/9781860948732_0040

30. Schäfer, A., Komusiewicz, C., Moser, H., Niedermeier, R.: Parameterized computational complexity of finding small-diameter subgraphs. Optim. Lett. **6**(5), 883–891 (2012). https://doi.org/10.1007/s11590-011-0311-5

31. Shamir, R., Sharan, R., Tsur, D.: Cluster graph modification problems. Discrete Appl. Math. **144**(1–2), 173–182 (2004)

32. Tsur, D.: Faster parameterized algorithm for cluster vertex deletion. CoRR, abs/1901.07609 (2019)

A Multistage View on 2-Satisfiability

Till Fluschnik[✉][iD]

Technische Universität Berlin, Faculty IV, Algorithmics and Computational
Complexity, Berlin, Germany
till.fluschnik@tu-berlin.de

Abstract. We study q-SAT in the multistage model, focusing on the
linear-time solvable 2-SAT. Herein, given a sequence of q-CNF formulas
and a non-negative integer d, the question is whether there is a sequence
of satisfying truth assignments such that for every two consecutive truth
assignments, the number of variables whose values changed is at most d.
We prove that MULTISTAGE 2-SAT is NP-hard even in quite restricted
cases. Moreover, we present parameterized algorithms (including kernelization) for MULTISTAGE 2-SAT and prove them to be asymptotically
optimal.

Keywords: temporal problems · symmetric difference · parameterized
complexity · problem kernelization

1 Introduction

q-SATISFIABILITY (q-SAT) is one of the most basic and best studied decision
problems in computer science: It asks whether a given boolean formula in conjunctive normal form, where each clause consists of at most q literals, is satisfiable. q-SAT is NP-complete for $q \geq 3$, while 2-SATISFIABILITY (2-SAT) is
linear-time solvable [1]. The recently introduced multistage model [15,22] takes
a sequence of instances of some decision problem (e.g., modeling one instance
that evolved over time), and asks whether there is a sequence of solutions to
them such that, roughly speaking, any two consecutive solutions do not differ
too much. We introduce q-SAT in the multistage model, defined as follows.[1]

MULTISTAGE q-SAT (MqSAT)
Input: A set X of variables, a sequence $\Phi = (\phi_1, \ldots, \phi_\tau)$, $\tau \in \mathbb{N}$, of q-CNF
  formulas over literals over X, and an integer $d \in \mathbb{N}_0$.
Question: Are there τ truth assignments $f_1, \ldots, f_\tau \colon X \to \{\bot, \top\}$ such that
  (i) for each $i \in \{1, \ldots, \tau\}$, f_i is a satisfying truth assignment for ϕ_i, and
  (ii) for each $i \in \{1, \ldots, \tau - 1\}$, it holds that $|\{x \in X \mid f_i(x) \neq f_{i+1}(x)\}| \leq d$?

---

[1] We identify `false` and `true` with $\bot$ and $\top$, respectively.

T. Fluschnik—Supported by DFG, project TORE (NI/369-18).

© Springer Nature Switzerland AG 2021
T. Calamoneri and F. Corò (Eds.): CIAC 2021, LNCS 12701, pp. 231–244, 2021.
https://doi.org/10.1007/978-3-030-75242-2_16

Constraint (ii) of MqSAT can also be understood as that the Hamming distance of two consecutive truth assignments interpreted as n-dimensional vectors over $\{\bot, \top\}$ is at most d, or when considering the sets of variables set true, then the symmetric difference of two consecutive sets is at most d.

In this work, we focus on M2SAT yet relate most of our results to MqSAT. We study M2SAT in terms of classic computational complexity and parameterized algorithmics [12].

Motivation. In theory as well as in practice, it is common to model problems as q-SAT- or even 2-SAT-instances. Once being modeled, established solvers specialized on q-SAT are employed. In some cases, a sequence of problem instances (e.g., modeling a problem instance that changes over time) is to solve such that any two consecutive solutions are similar in some way (e.g., when costs are inferred for setup changes). Hence, when following the previously described approach, each problem instance is first modeled as a q-SAT instance such that a sequence of q-SAT-instances remains to be solved. Comparably to the single-stage setting, understanding the multistage setting could give raise to a general approach for solving different (multistage) problems. With MqSAT we introduce the first problem that models the described setup. Note that, though a lot of variants of q-SAT exist, MqSAT is one of the very few variants that deal with a sequence of q-SAT-instances [31].

Our Contributions. Our results for MULTISTAGE 2-SAT are summarized in Fig. 1. We prove MULTISTAGE 2-SAT to be NP-hard, even in fairly restricted cases: (i) if $d = 1$ and the maximum number m of clauses in any stage is six, or (ii) if there are only two stages. These results are tight in the sense that M2SAT is linear-time solvable when $d = 0$ or $\tau = 1$. While NP-hardness for $d = 1$ implies that there is no $(n + m + \tau)^{f(d)}$-time algorithm for any function f unless P = NP, where n denotes the number of variables, we prove that when parameterized by the dual parameter $n-d$ (the minimum number of variables not changing between any two consecutive layers), M2SAT is W[1]-hard and solvable in $\mathcal{O}^*(n^{\mathcal{O}(n-d)})$ time.[2] We prove this algorithm to be tight in the sense that, unless the Exponential Time Hypothesis (ETH) breaks, there is no $\mathcal{O}^*(n^{o(n-d)})$-time algorithm. Further, we prove that M2SAT is solvable in $\mathcal{O}^*(2^{\mathcal{O}(n)})$ time but not in $\mathcal{O}^*(2^{o(n)})$ time unless the ETH breaks. Likewise, we prove that M2SAT is solvable in $\mathcal{O}^*(n^{\mathcal{O}(\tau \cdot d)})$ time but not in $\mathcal{O}^*(n^{o(d) \cdot f(\tau)})$ time for any function f unless the ETH breaks. As to efficient and effective data reduction, we prove M2SAT to admit problem kernelizations of size $\mathcal{O}(m \cdot \tau)$ and $\mathcal{O}(n^2 \tau)$, but none of size $(n+m)^{\mathcal{O}(1)}$, $\mathcal{O}((n+m+\tau)^{2-\varepsilon})$, or $\mathcal{O}(n^{2-\varepsilon}\tau)$, $\varepsilon > 0$, unless NP $\subseteq$ coNP / poly.

Related Work. q-SAT is one of the most famous decision problems with a central role in NP-completeness theory [11,27], for the (Strong) Exponential Time Hypothesis [25,26], and in the early theory on kernelization lower bounds [6,21], for instance. In contrast to q-SAT with $q \geq 3$, 2-SAT is proven to be polynomial-[28], even linear-time [1] solvable. Several applications of 2-SAT are known (see,

---

[2] The $\mathcal{O}^*$-notation suppresses factors polynomial in the input size.

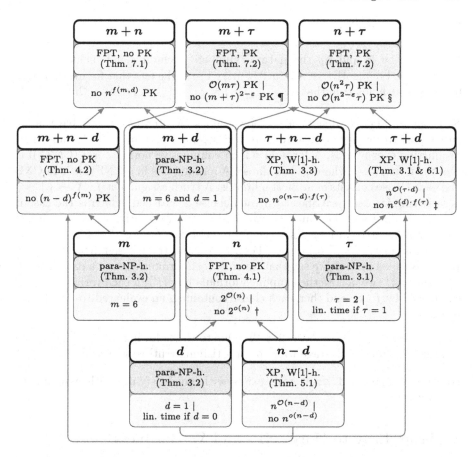

Fig. 1. Our results for MULTISTAGE 2-SAT. Each box gives, regarding to a parameterization (top layer), our parameterized classification (middle layer) with additional details on the corresponding result (bottom layer). Arrows indicate the parameter hierarchy: An arrow from parameter p_1 to p_2 indicates that $p_1 \leq p_2$. "PK" and "no PK" stand for "polynomial problem kernel" and "no polynomial problem kernel unless NP $\subseteq$ coNP / poly", respectively. †: unless the ETH breaks (Theorem 3.2). ‡: unless the ETH breaks (Theorem 3.1). ¶: unless NP $\subseteq$ coNP / poly (Theorem 7.3) §: unless NP $\subseteq$ coNP / poly (Theorem 3.1).

e.g., [10,16,23,30]). In the multistage model, various problems from different fields were studied, e.g. graph theory [2,3,9,19,20,22], facility location [15], knapsack [5], or committee elections [7]. Also variations to the multistage model were studied, e.g. with a global budget [24], an online-version [4], or using different distance measures for consecutive stages [7,20].

2   Preliminaries

We denote by $\mathbb{N}$ and $\mathbb{N}_0$ the natural numbers excluding and including zero, respectively. Frequently, we will tacitly make use of the fact that for every $n \in \mathbb{N}$, $0 \le k \le n$, it holds true that $1 + \sum_{i=1}^{k} \binom{n}{i} = \sum_{i=0}^{k} \binom{n}{i} \le 1 + n^k \le 2n^k$.

Satisfiability. Let X denote a set of variables. A literal is a variable that is either positive or negated (we denote the negation of x by $\neg x$). A clause is a disjunction over literals. A formula ϕ is in conjunctive normal form (CNF) if it is of the form $\bigwedge_i C_i$, where C_i is a clause. A formula ϕ is in q-CNF if it is in CNF and each clause consists of at most q literals. A truth assignment $f \colon X \to \{\bot, \top\}$ is satisfying for ϕ (or satisfies ϕ) if each clause is satisfied, which is the case if at least one literal in the clause is evaluated to true (a positive variable assigned true, or a negated variable assigned false). For $a, b \in \{\bot, \top\}$, let $a \oplus b := \bot$ if $a = b$, and $a \oplus b := \top$ otherwise. For $X' \subset X$, an truth assignment $f' \colon X' \to \{\bot, \top\}$ is called *partial*. We say that we *simplify* a formula ϕ given a partial truth assignment f' (we denote the simplified formula by $\phi[f']$) if each variable $x \in X'$ is replaced by $f'(x)$, and then each clause containing an evaluated-to-true literal is deleted.

Preprocessing on MULTISTAGE 2-SAT. Due to the following data reduction, we can safely assume each stage to admit a satisfying truth assignment.

Reduction Rule 2.1. *If a stage exists with no satisfying truth assignment, then return* no.

3   From Easy to Hard: NP- and W-Hardness

MULTISTAGE 2-SAT is linear-time solvable if the input consists of only one stage, or if all or none variables are allowed to change its truth assignment between two consecutive stages.

Observation 3.1 ($\bigstar^3$). MULTISTAGE 2-SAT *is linear-time solvable if (i) $\tau = 1$, (ii) $d = 0$, or (iii) $d = n$.*

We will prove that the cases (i) and (ii) in Observation 3.1 are tight: MULTISTAGE 2-SAT becomes NP-hard if $\tau \ge 2$ (Sects. 3.1 & 3.3) or $d = 1$ (Sect. 3.2). For the case (iii) in Observation 3.1 the picture looks different: we prove MULTISTAGE 2-SAT to be polynomial-time solvable if $n - d \in \mathcal{O}(1)$ (Sect. 5).

3.1   From One to Two Stages

In this section, we prove that MULTISTAGE 2-SAT becomes NP-hard if $\tau \ge 2$. In fact, we prove the following.

---

[3] Details and proofs (marked with $\bigstar$) are deferred to the appendix.

Theorem 3.1 (★). MULTISTAGE 2-SAT *is NP-hard, even for two stages, where the variables appear all negated in one and all positive in the other stage. Moreover,* MULTISTAGE 2-SAT

(i) *is* W[1]*-hard when parameterized by d even if* $\tau = 2$,
(ii) *admits no* $n^{o(d) \cdot f(\tau)}$*-time algorithm for any function f unless the ETH breaks, and*
(iii) *admits no problem kernelization of size* $\mathcal{O}(n^{2-\varepsilon} \cdot f(\tau))$ *for any* $\varepsilon > 0$ *and function f, unless* NP $\subseteq$ coNP / poly.

We will reduce from the following NP-hard problem:

WEIGHTED 2-SAT
Input: A set of variables X, a 2-CNF ϕ over X, and an integer k.
Question: Is there a satisfying truth assignment for ϕ with at most k variables set true?

When parameterized by the number k of set-to-true variables, WEIGHTED 2-SAT is W[1]-complete [14,18]. Moreover, WEIGHTED 2-SAT admits no $n^{o(k)}$-time algorithm unless the ETH breaks [8] and no problem bikernelization of size $\mathcal{O}(n^{2-\varepsilon})$, $\varepsilon > 0$, unless NP $\subseteq$ coNP / poly [13].

Construction 3.1. Let (X, ϕ, k) be an instance of WEIGHTED 2-SAT, where $\phi = \bigwedge_{i=1}^{m} C_i$. Construct $\Phi = (\phi_1, \phi_2)$, where $\phi_2 := \phi$ and $\phi_1 := \bigwedge_{x \in X}(\neg x)$ consists of n size-one clauses, where each variable appears negated in one clause. Finally, set $d := k$.   ◇

Remark 3.1. Theorem 3.1(iii) can be generalized to MULTISTAGE q-SAT: Instead from WEIGHTED 2-SAT, we reduce (in an analogous way) from WEIGHTED q-SAT which admits no problem bikernelization of size $\mathcal{O}(n^{q-\varepsilon})$, $\varepsilon > 0$, unless NP $\subseteq$ coNP / poly [13]. Thus, unless NP $\subseteq$ coNP / poly, MULTISTAGE q-SAT admits no problem kernel of size $\mathcal{O}(n^{q-\varepsilon} \cdot f(\tau))$ for any $\varepsilon > 0$ and function f.

3.2   From Zero to One Allowed Change

In this section, we prove that MULTISTAGE 2-SAT becomes NP-hard if $d = 1$ and the maximum number m of clauses in any stage is six. In fact, we prove the following.

Theorem 3.2. MULTISTAGE 2-SAT *is NP-hard, even if the number of clauses in each stage is at most six and* $d = 1$. *Moreover, unless the ETH breaks,* MULTISTAGE 2-SAT *admits no* $\mathcal{O}^*(2^{o(n)})$*-time algorithm.*

Construction 3.2. Let (X, ϕ) be an instance of 3-SAT, where $\phi = \bigwedge_{i=1}^{m} C_i$ and each clause consists of exactly three literals. Let ℓ_j^i, $j \in \{1, 2, 3\}$, denote the literals in C_i for each $i \in \{1, \ldots, m\}$. Construct instance (X', Φ, d) of M2SAT as follows. First, construct $X' := X \cup B$, where $B := \{b_1, b_2, b_3\}$. Let

$$\phi_B := (b_1 \vee b_2) \wedge (b_1 \vee b_3) \wedge (b_2 \vee b_3), \text{ and}$$
$$\phi_{\neg B} := (\neg b_1 \vee \neg b_2) \wedge (\neg b_1 \vee \neg b_3) \wedge (\neg b_2 \vee \neg b_3).$$

Next, construct $\Phi := (\phi_i, \ldots, \phi_{2m})$ as follows. For each $i \in \{1, \ldots, m\}$, construct

$$\phi_{2i-1} := \phi_{\neg B}, \quad \text{and} \quad \phi_{2i} := (\ell_1^i \vee b_1) \wedge (\ell_2^i \vee b_2) \wedge (\ell_3^i \vee b_3) \wedge \phi_B$$

Finally, set $d := 1$. $\diamond$

Observation 3.2 ($\bigstar$). *In every solution to an instance obtained from Construction 3.2, in each odd stage exactly two b_j are set to false and in each even stage exactly two b_j are set to true.*

Lemma 3.1 ($\bigstar$). *Let $\mathcal{I} = (X, \phi)$ be an instance of 3-SAT, and let $\mathcal{I}' = (X', \Phi, d)$ be an instance of MULTISTAGE 2-SAT obtained from $\mathcal{I}$ using Construction 3.2. Then, $\mathcal{I}$ is a yes-instance if and only if $\mathcal{I}'$ is a yes-instance.*

Proof (Proof of Theorem 3.2). Construction 3.2 forms a polynomial-time many-one reduction to an instance with $d = 1$, $m = 6$, and $n = |X| + 3$. Hence, M2SAT is NP-hard, even if $d = 1$ and $m = 6$, and, unless the ETH breaks, admits no $\mathcal{O}^*(2^{o(n)})$-time algorithm since no $\mathcal{O}^*(2^{o(|X|)})$-time algorithm exists for 3-SAT [8]. $\square$

3.3   From All to All But k Allowed Changes

In this section, we prove that MULTISTAGE 2-SAT is W[1]-hard when parameterized by the lower bound $n - d$ on the number of unchanged variables between any two consecutive stages.

Theorem 3.3. MULTISTAGE 2-SAT *is W[1]-hard when parameterized by $n - d$ even if $\tau = 2$, and, unless the ETH breaks, admits no $\mathcal{O}^*(n^{o(n-d) \cdot f(\tau)})$-time algorithm for any function f.*

We reduce from the following NP-hard problem:

MULTICOLORED INDEPENDENT SET (MIS)
Input: An undirected, k-partite graph $G = (V^1, \ldots, V^k, E)$.
Question: Is there an independent set S such that $|S \cap V^i| = 1$ for all $i \in \{1, \ldots, k\}$?

MIS is W[1]-hard with respect to k [17] and unless the ETH breaks, there is no $f(k) \cdot n^{o(k)}$-time algorithm [29].

Construction 3.3. Let $\mathcal{I} = (G = (V^1, \ldots, V^k, E))$ be an instance of MIS and let $V := V^1 \uplus \cdots \uplus V^k$, $n := |V|$, and $V^i = \{v_1^i, \ldots, v_{|V_i|}^i\}$ for all $i \in \{1, \ldots, k\}$. We construct an instance $\mathcal{I}' = (X, (\phi_1, \phi_2), d)$ with $d := n - k$ as follows. Let $X := X^1 \cup \cdots \cup X^k$ with $X^i = \{x_j^i \mid v_j^i \in V_i\}$ for all $i \in \{1, \ldots, k\}$. Let for all $i \in \{1, \ldots, k\}$

$$\phi_i^* := \bigwedge_{j,j' \in \{1, \ldots, |V^i|\}, j \neq j'} (\neg x_j^i \vee \neg x_{j'}^i), \quad \text{and let} \quad \phi_E := \bigwedge_{\{v_j^i, v_{j'}^{i'}\} \in E} (\neg x_j^i \vee \neg x_{j'}^{i'}).$$

Let

$$\phi_1 := \bigwedge_{x \in X} (x) \qquad \text{and} \qquad \phi_2 := \phi_E \wedge \bigwedge_{i \in \{1,\ldots,k\}} \phi_i^*.$$

This finishes the construction. ◇

Lemma 3.2 (★). *Let* $\mathcal{I} = (G = (V^1, \ldots, V^k, E))$ *be an instance of* MIS, *and let* $\mathcal{I}' = (X, (\phi_1, \phi_2), d)$ *be an instance of* MULTISTAGE 2-SAT *obtained from* $\mathcal{I}$ *using Construction 3.3. Then,* $\mathcal{I}$ *is a* **yes**-*instance if and only if* $\mathcal{I}'$ *is a* **yes**-*instance.*

Proof (Proof of Theorem 3.3). Construction 3.3 runs in polynomial time and outputs an equivalent instance (Lemma 3.2) with two stages and $d = n - k$. As Construction 3.1 also forms a parametric transformation, M2SAT is W[1]-hard when parameterized by $n - d$ even if $\tau = 2$. Moreover, unless the ETH breaks, M2SAT admits no $n^{o(n-d) \cdot f(\tau)}$-time algorithm for any function f since no $n^{o(k)}$-time algorithm exists for MIS. □

4 Fixed-Parameter Tractability Regarding the Number of Variables and $m + n - d$

In this section, we prove that MULTISTAGE 2-SAT is fixed-parameter tractable regarding the number of variables (Sect. 4.1) and regarding the parameter $m + n - d$, the maximum number of clauses over all input formulas and the minimum number of variables not changing between any two consecutive stages (Sect. 4.2).

4.1 Fixed-Parameter Tractability Regarding the Number of Variables

We prove that MULTISTAGE 2-SAT is fixed-parameter tractable regarding the number of variables.

Theorem 4.1 (★). MULTISTAGE 2-SAT *is solvable in* $\mathcal{O}(\min\{2^n n^d, 4^n\} \cdot \tau \cdot (n + m))$ *time.*

Remark 4.1. Theorem 4.1 is asymptotically optimal regarding n unless the ETH breaks (Theorem 3.2). Moreover, Theorem 4.1 is easily adaptable to MULTI-STAGE q-SAT with $q \geq 3$ as, for every $q \geq 3$, the number of truth assignments is 2^n and each is verifiable in linear time.

4.2 Fixed-Parameter Tractability Regarding $m + n - d$

We prove that MULTISTAGE 2-SAT is fixed-parameter tractable regarding the parameter $m + n - d$.

Theorem 4.2. MULTISTAGE 2-SAT *is solvable in* $\mathcal{O}(4^{2(m+n-d)} \tau(n+m))$ *time.*

To prove Theorem 4.2, we will show that either Theorem 4.1 applies with $n \leq 2(m + n - d)$ or the following.

Lemma 4.1. MULTISTAGE 2-SAT *solvable in* $\mathcal{O}(\tau(n + m))$ *time if* $2m < d$.

Proof. Let $\mathcal{I} = (X, \Phi, d)$ be an instance of M2SAT with $\Phi = (\phi_1, \ldots, \phi_\tau)$ on n variables and each formula contains at most m clauses. Due to Reduction Rule 2.1, we can safely assume that each formula of Φ admits a satisfying truth assignment. Let $X_i \subseteq X$ be the set of variables appearing as literals in ϕ_i for each $i \in \{1, \ldots, \tau\}$. Note that $|X_i| \leq 2m$ for each $i \in \{1, \ldots, \tau\}$. Compute in linear time a satisfying truth assignment $f_1 \colon X \to \{\bot, \top\}$ for ϕ_1. Compute for each $i \in \{2, \ldots, \tau\}$ in linear time a satisfying truth assignment $f_i' \colon X_i \to \{\bot, \top\}$ for ϕ_i. Next, iteratively for $i = 2, \ldots, \tau$, set for all $x \in X$

$$f_i(x) = \begin{cases} f_i'(x), & \text{if } x \in X_i, \\ f_{i-1}(x), & \text{if } x \in X \setminus X_i. \end{cases}$$

Clearly, truth assignment f_i satisfies ϕ_i. Moreover, for all $i \in \{2, \ldots, \tau\}$ it holds that $|\{x \in X \mid f_{i-1}(x) \neq f_i(x)\}| \leq |X_i| \leq 2m < d$, and hence $(f_1, \ldots, f_\tau)$ is a solution to $\mathcal{I}$. □

Proof (Proof of Theorem 4.2). Let $\mathcal{I} = (X, \Phi, d)$ be an instance of M2SAT with $\Phi = (\phi_1, \ldots, \phi_\tau)$ on n variables and each formula contains at most m clauses. We distinguish how $2(m + n - d)$ relates to $2n - d$.

Case 1: $2(m + n - d) \geq 2n - d$. Since $d \leq n$, it follows that $2(m + n - d) \geq n$. Due to Theorem 4.1, we can solve $\mathcal{I}$ in $\mathcal{O}(\min\{2^n n^d, 4^n\} \tau(n + m)) \subseteq \mathcal{O}(4^{2(m+n-d)} \tau(n + m))$ time.

Case 2: $2(m+n-d) < 2n-d$. We have that $2(m+n-d) < 2n-d \iff 2m < d$. Due to Lemma 4.1, we can solve $\mathcal{I}$ in $\mathcal{O}(\tau(n + m))$ time. □

Remark 4.2. Theorem 4.2 can be adapted for MULTISTAGE q-SAT for every $q \geq 3$, where Lemma 4.1 is restated for $qm < d$ and we check for a satisfying truth assignment for each stage in $\mathcal{O}^*(2^{qm})$ time. To adapt the proof of Theorem 4.2, we then relate $q(m + n - d)$ with $qn - (q - 1)d$ and either employ the adapted Theorem 4.1 (see Remark 4.1), or the adapted Lemma 4.1.

5   XP Regarding the Number of Consecutive Non-changes

We prove that MULTISTAGE 2-SAT is in XP when parameterized by the lower bound $n - d$ on non-changes between consecutive stages, the parameter "dual" to d.

Theorem 5.1. MULTISTAGE 2-SAT *is solvable in* $\mathcal{O}(n^{4(n-d)+1} \cdot 2^{4(n-d)} \tau(n + m))$ *time.*

Let $\mathcal{I} = (X, \Phi = (\phi_1, \dots, \phi_\tau), d)$ be a fixed yet arbitrary instance with n variables. Two partial truth assignments $f_Y \colon Y \to \{\bot, \top\}$ and $f_Z \colon Z \to \{\bot, \top\}$ with $Y, Z \subseteq X$ are called *compatible* if for all $x \in Y \cap Z$ it holds that $f_Y(x) = f_Z(x)$. For two compatible assignments f_Y, f_Z, let

$$f_Y \cup f_Z \colon Y \cup Z \to \{\bot, \top\}, \qquad f_Y \cup f_Z(x) := \begin{cases} f_Y(x), & \text{if } x \in Y, \\ f_Z(x), & \text{if } x \in Z \setminus Y. \end{cases}$$

With a similar idea as in the proof of Theorem 4.1, we will construct a directed graph with terminals s and t such that there is an s-t path in G if and only if $\mathcal{I}$ is a yes-instance.

Construction 5.1. Given $\mathcal{I}$, we construct a graph $G = (V, E)$ with vertex set $V := V^{1 \to 3} \cup V^{2 \to 4} \cup \dots \cup V^{\tau-2 \to \tau} \cup \{s, t\}$, where for each $Y, Z \in \binom{X}{n-d}$, we have that $(f_Y, f_Z) \in V^{i \to i+2}$ if and only if f_Y, f_Z are compatible and each of $\phi_i[f_Y]$, $\phi_{i+1}[f_Y \cup f_Z]$, and $\phi_{i+2}[f_Z]$ is satisfiable, and the following arcs: (i) (s, v) for all $v \in V^{1 \to 3}$, (ii) (v, t) for all $v \in V^{\tau-2 \to \tau}$, and (iii) $((f_Y, f_Z), (f_{Y'}, f_{Z'})) \in V^{i \to i+2} \times V^{j \to j+2}$ if $j = i+1$ and $f_Z = f_{Y'}$ (implying that $Z = Y'$).   ◇

Lemma 5.1 (★). *Construction 5.1 computes a graph of size $\mathcal{O}(n^{4(n-d)+1} \cdot 2^{4(n-d)}\tau)$ and can be done in $\mathcal{O}(n^{4(n-d)+1} \cdot 2^{4(n-d)}\tau(n+m))$ time.*

Lemma 5.2 (★). *Let $\mathcal{I}$ be an instance of* MULTISTAGE 2-SAT *and let G be the graph obtained from applying Construction 5.1 to $\mathcal{I}$. Then, $\mathcal{I}$ is a yes-instance if and only if G admits an s-t paths.*

Proof (Proof of Theorem 5.1). Given an instance $\mathcal{I} = (X, \Phi = (\phi_1, \dots, \phi_\tau), d)$ of M2SAT, apply Construction 5.1 in $\mathcal{O}(n^{4(n-d)+1} \cdot 2^{4(n-d)}\tau(n+m))$ time to obtain graph G with terminals s and t of size $\mathcal{O}(n^{4(n-d)+1} \cdot 2^{4(n-d)}\tau)$ (Lemma 5.1). Return, in time linear in the size of G, yes if G admits an s-t path, and no otherwise (Lemma 5.2).   □

Remark 5.1. Theorem 5.1 is asymptotically optimal regarding $n-d$ unless the ETH breaks (Theorem 3.3). Moreover, Theorem 5.1 does not generalize to MULTISTAGE q-SAT for $q \geq 3$, as MqSAT is already NP-hard for one stage and hence for any number $n-d$.

6 XP Regarding Number of Stages and Consecutive Changes

In this section, we prove that MULTISTAGE 2-SAT is in XP when parameterized by $\tau + d$.

Theorem 6.1. MULTISTAGE 2-SAT *is solvable in $\mathcal{O}(n^{2\tau \cdot d} \cdot 2^{\tau \cdot d+1} \cdot \tau \cdot (n+m))$ time.*

Algorithm 1: XP-algorithm on input instance (X, ϕ, d).

---

1 **foreach** $X' \subseteq X : |X'| \leq \tau \cdot d$ **do**   // $1 + n^{\tau \cdot d}$ **many**
2   **foreach** $f_1 \colon X' \to \{\bot, \top\}$ **do**   // $2^{|X'|}$ **many**
3         $\phi_1^* \leftarrow$ **simplify**(ϕ_1, f_1);
4   **foreach** $g_2, g_3, \ldots, g_\tau : g_i \in \mathcal{F}(X') \; \forall i \in \{2, \ldots, \tau\}$ **do** // $2^\tau |X'|^{\tau \cdot d}$ **many**
5   **foreach** $i \in \{2, \ldots, \tau\}$ **do**
6                 $f_i(x) \leftarrow f_{i-1}(x) \oplus g_i(x) \; \forall x \in X'$; $\phi_i^* \leftarrow$ **simplify**(ϕ_i, f_i);
7   **if** $(X \setminus X', (\phi_1^*, \ldots, \phi_\tau^*), 0)$ *is a yes-instance of* M2SAT **then**
8   **return** *yes*   // **decidable in linear time**
                 (Observation 3.1)

9 **return** *no*

Let $\mathcal{I} = (X, \Phi = (\phi_1, \ldots, \phi_\tau), d)$ be a fixed yet arbitrary instance with $\tau \cdot d < n$, as otherwise Theorem 4.1 applies. On a high level, our Algorithm 1 works as follows:

(1) Guess $q \leq \tau \cdot d$ variables $X' \subseteq X$ that will change over time.
(2) Guess an initial truth assignment of the variables in X'.
(3) For each but the first stage, guess the at most $\min\{q, d\}$ possible variables to change.
(4) Set the variables to the guessed true or false values, delete clauses which are set to true.
(5) Return **yes** if the resulting instance with $d = 0$ is a **yes**-instance (linear-time checkable).
(6) If the algorithm never (for all possible guesses) returned **yes**, then return **no**.

For any $X' \subseteq X$, define the set of all truth assignments to variables of X' with at most $\min\{|X'|, d\}$ true values by $\mathcal{F}(X') := \{f \colon X' \to \{\bot, \top\} \mid |\{x \in X' \mid f(x) = \top\}| \leq \min\{|X'|, d\}\}$. With the next two lemmas, we prove that Algorithm 1 is correct and runs in XP-time regarding $\tau + d$.

Lemma 6.1 (★). *Algorithm 1 returns* **yes** *if and only if the input instance is a* **yes**-*instance.*

Lemma 6.2 (★). *Algorithm 1 runs in* $\mathcal{O}(n^{2\tau \cdot d} \cdot 2^{\tau \cdot d + 1} \tau(n + m))$ *time.*

We are set to prove the main result from this section.

Proof (Proof of Theorem 6.1). Let $\mathcal{I} = (X, \Phi = (\phi_1, \ldots, \phi_\tau), d)$ be an instance of M2SAT with n variables and at most m clauses in each stage's formula. If $\tau \cdot d \geq n$, then, by Theorem 4.1, we know that M2SAT is solvable in $\mathcal{O}(2^{2\tau \cdot d} \cdot \tau(n + m))$ time. Otherwise, if $\tau \cdot d < n$, then Algorithm 1 runs in $\mathcal{O}(n^{2\tau \cdot d} \cdot 2^{\tau \cdot d + 1} \tau(n + m))$ time (Lemma 6.2) and correctly decides $\mathcal{I}$ (Lemma 6.1). $\square$

Remark 6.1. Theorem 6.1 is asymptotically optimal regarding d unless the ETH breaks (Theorem 3.1). Moreover, Theorem 6.1 is not adaptable to MULTISTAGE q-SAT with $q \geq 3$ unless P = NP since MULTISTAGE q-SAT with $q \geq 3$ is NP-hard even with $\tau + d \in \mathcal{O}(1)$.

7   Efficient and Effective Data Reduction

In this section, we study efficient and provably effective data reduction for MUL-TISTAGE 2-SAT in terms of problem kernelization. We focus on the parameter combinations $n+m$, $n+\tau$, and $m+\tau$. We prove that no problem kernelization of size polynomial in $n + m$ exists unless NP $\subseteq$ coNP / poly (Sect. 7.1), and that a problem kernelization of size quadratic in $m + \tau$ and of size cubic in $n + \tau$ exists (Sect. 7.2). Finally, we prove that no problem kernel of size truly subquadratic in $m + \tau$ exists unless NP $\subseteq$ coNP / poly.

7.1   No Time-Independent Polynomial Problem Kernelization

When parameterized by $n + m$, efficient and effective data reduction appears unlikely.

Theorem 7.1 (★). *Unless* NP $\subseteq$ coNP / poly, MULTISTAGE 2-SAT *admits no problem kernel of size polynomial in* $n^{f(m,d)}$, *for any function f only depending on m and d.*

Remark 7.1. Due to Theorem 4.1, MULTISTAGE 2-SAT yet admits a problem kernel of size $2^{\mathcal{O}(n)}$.

7.2   Polynomial Problem Kernelizations

We prove problem kernelizations of size polynomial in $n + \tau$ and $m + \tau$.

Theorem 7.2. MULTISTAGE 2-SAT *admits a linear-time computable problem kernelization of size $\mathcal{O}(n^2\tau)$ and of size $\mathcal{O}(m \cdot \tau)$.*

We employ the following two immediate reduction rules (each is clearly correct and applicable in linear time):

Reduction Rule 7.1. In each stage, delete all but one appearances of a clause in the formula.

Reduction Rule 7.2. Delete a variable that appears in no stage's formula as a literal.

Proof (Proof of Theorem 7.2). Observe that there are at most $N := 2n + \binom{2n}{2} \in \mathcal{O}(n^2)$ many pairwise different clauses. After exhaustively applying Reduction Rule 7.1, we have $m \leq N \in \mathcal{O}(n^2)$. After exhaustively applying Reduction Rule 7.2, it follows that for each variable, there is at least one clause, and hence, $n \leq 2 \cdot m \cdot \tau$. ☐

Remark 7.2. Theorem 7.2 adapts easily to MULTISTAGE q-SAT. Herein, the problem kernel sizes are $\mathcal{O}(n^q \cdot \tau)$ and $\mathcal{O}(q \cdot m \cdot \tau)$.

Subsequently, we prove that a linear kernel appears unlikely.

Theorem 7.3. *Unless* $\mathrm{NP} \subseteq \mathrm{coNP} / \mathrm{poly}$, MULTISTAGE 2-SAT *admits no problem kernel of size* $\mathcal{O}((m + n + \tau)^{2-\varepsilon})$ *for any* $\varepsilon > 0$.

To prove Theorem 7.3, we show that there is a linear parametric transformation from VERTEX COVER parameterized by $|V|$ to MULTISTAGE 2-SAT parameterized by $n + m + \tau$.

Construction 7.1. Let $\mathcal{I} = (G, k)$ with $G = (V, E)$ be an instance of VERTEX COVER. Denote the vertices $V = \{v_1, \dots, v_n\}$. We construct the instance $\mathcal{I}' = (X, \Phi, d)$ of M2SAT with $d = k$ and $\Phi = (\phi_0, \phi_1, \dots, \phi_n)$ as follows. Let $X = X_V \cup B$ with $X_V = \{x_i \mid v_i \in V\}$ and $B = \{b_1, \dots, b_k\}$. Let

$$\phi_0 := \bigwedge_{i=1}^{n} (\neg x_i) \wedge \bigwedge_{j=1}^{k} (\neg b_j) \quad \text{and}$$

$$\phi_i := \bigwedge_{\{v_i, v_j\} \in E} (x_i \vee x_j) \wedge \begin{cases} \bigwedge_{j=1}^{k} (b_j) & \text{if } i \bmod 2 = 0, \\ \bigwedge_{j=1}^{k} (\neg b_j) & \text{if } i \bmod 2 = 1, \end{cases} \quad \forall i \in \{1, \dots, n\}.$$

Note that $\tau + m + |X| \in \mathcal{O}(n)$, since each vertex degree is at most $n - 1$.   ◇

Lemma 7.1 (★). *Let* $\mathcal{I} = (G, k)$ *be an instance of* VERTEX COVER, *and let* $\mathcal{I}' = (X', \Phi', d)$ *be the instance of* MULTISTAGE 2-SAT *obtained from* $\mathcal{I}$ *using Construction 7.1. Then,* $\mathcal{I}$ *is a* **yes**-*instance if and only if* $\mathcal{I}'$ *is a* **yes**-*instance.*

Proof (Proof of Theorem 7.3). Construction 7.1 is a linear parametric transformation (Lemma 7.1) such that $\tau + m + |X| \in \mathcal{O}(|V|)$. Since VERTEX COVER admits no problem bikernelization of size $\mathcal{O}(|V|^{2-\varepsilon})$, $\varepsilon > 0$ [13], the statement follows.   □

Remark 7.3. Theorem 7.3 can be easily adapted to MULTISTAGE q-SAT when taking q-HITTING SET as source problem [13], ruling out problem kernelizations of size $\mathcal{O}((n + m + \tau)^{q-\varepsilon})$, $\varepsilon > 0$ (unless $\mathrm{NP} \subseteq \mathrm{coNP} / \mathrm{poly}$).

8   Conclusion

While 2-SAT is linear-time solvable, its multistage model MULTISTAGE 2-SAT is intractable in even surprisingly restricted cases. This is also reflected by the fact that several of our direct upper bounds are already asymptotically optimal. By our results, the most interesting difference between MULTISTAGE 2-SAT and MULTISTAGE q-SAT, with $q \geq 3$, is that the former is efficiently solvable if the numbers of stages and allowed consecutive changes are constant, which is not the case for the latter (unless $\mathrm{P} = \mathrm{NP}$). Finally, our results show that exact solutions are far from practical, waving the path for randomized or heuristic approaches.

Acknowledgements. I thank Hendrik Molter and Rolf Niedermeier for their constructive feedbacks.

References

1. Aspvall, B., Plass, M.F., Tarjan, R.E.: A linear-time algorithm for testing the truth of certain quantified boolean formulas. Inf. Process. Lett. **8**(3), 121–123 (1979). https://doi.org/10.1016/0020-0190(79)90002-4
2. Bampis, E., Escoffier, B., Kononov, A.V.: LP-based algorithms for multistage minimization problems. CoRR abs/1909.10354 (2019). http://arxiv.org/abs/1909.10354
3. Bampis, E., Escoffier, B., Lampis, M., Paschos, V.T.: Multistage matchings. In: Proceedings of 16th SWAT. LIPIcs, vol. 101, pp. 7:1–7:13. Schloss Dagstuhl–Leibniz-Zentrum für Informatik (2018). https://doi.org/10.4230/LIPIcs.SWAT.2018.7
4. Bampis, E., Escoffier, B., Schewior, K., Teiller, A.: Online multistage subset maximization problems. In: Proceedings of 27th ESA. LIPIcs, vol. 144, pp. 11:1–11:14. Schloss Dagstuhl–Leibniz-Zentrum für Informatik (2019). https://doi.org/10.4230/LIPIcs.ESA.2019.11
5. Bampis, E., Escoffier, B., Teiller, A.: Multistage knapsack. In: Proceedings of 44th MFCS. LIPIcs, vol. 138, pp. 22:1–22:14. Schloss Dagstuhl–Leibniz-Zentrum für Informatik (2019). https://doi.org/10.4230/LIPIcs.MFCS.2019.22
6. Bodlaender, H.L., Downey, R.G., Fellows, M.R., Hermelin, D.: On problems without polynomial kernels. J. Comput. Syst. Sci. **75**(8), 423–434 (2009). https://doi.org/10.1016/j.jcss.2009.04.001
7. Bredereck, R., Fluschnik, T., Kaczmarczyk, A.: Multistage committee election. CoRR abs/2005.02300 (2020). https://arxiv.org/abs/2005.02300
8. Chen, J., Huang, X., Kanj, I.A., Xia, G.: Strong computational lower bounds via parameterized complexity. J. Comput. Syst. Sci. **72**(8), 1346–1367 (2006). https://doi.org/10.1016/j.jcss.2006.04.007
9. Chimani, M., Troost, N., Wiedera, T.: Approximating multistage matching problems. CoRR abs/2002.06887 (2020). https://arxiv.org/abs/2002.06887
10. Chrobak, M., Dürr, C.: Reconstructing hv-convex polyominoes from orthogonal projections. Inf. Process. Lett. **69**(6), 283–289 (1999). https://doi.org/10.1016/S0020-0190(99)00025-3
11. Cook, S.A.: The complexity of theorem-proving procedures. In: Proceedings of 3rd STOC, pp. 151–158. ACM (1971). https://doi.org/10.1145/800157.805047
12. Cygan, M., et al.: Parameterized Algorithms. Springer, Heidelberg (2015). https://doi.org/10.1007/978-3-319-21275-3
13. Dell, H., van Melkebeek, D.: Satisfiability allows no nontrivial sparsification unless the polynomial-time hierarchy collapses. J. ACM **61**(4), 23:1–23:27 (2014). https://doi.org/10.1145/2629620
14. Downey, R.G., Fellows, M.R.: Parameterized complexity. In: Monographs in Computer Science. Springer, New York (1999). https://doi.org/10.1007/978-1-4612-0515-9
15. Eisenstat, D., Mathieu, C., Schabanel, N.: Facility location in evolving metrics. In: Esparza, J., Fraigniaud, P., Husfeldt, T., Koutsoupias, E. (eds.) ICALP 2014. LNCS, vol. 8573, pp. 459–470. Springer, Heidelberg (2014). https://doi.org/10.1007/978-3-662-43951-7_39

16. Even, S., Itai, A., Shamir, A.: On the complexity of timetable and multicommodity flow problems. SIAM J. Comput. **5**(4), 691–703 (1976). https://doi.org/10.1137/0205048

17. Fellows, M.R., Hermelin, D., Rosamond, F., Vialette, S.: On the parameterized complexity of multiple-interval graph problems. Theoret. Comput. Sci. **410**(1), 53–61 (2009). https://doi.org/10.1016/j.tcs.2008.09.065

18. Flum, J., Grohe, M.: Parameterized Complexity Theory. TTCSAES. Springer, Heidelberg (2006). https://doi.org/10.1007/3-540-29953-X

19. Fluschnik, T., Niedermeier, R., Rohm, V., Zschoche, P.: Multistage vertex cover. In: Proceedings of 14th IPEC. LIPIcs, vol. 148, pp. 14:1–14:14. Schloss Dagstuhl - Leibniz-Zentrum für Informatik (2019). https://doi.org/10.4230/LIPIcs.IPEC.2019.14

20. Fluschnik, T., Niedermeier, R., Schubert, C., Zschoche, P.: Multistage *s-t* path: confronting similarity with dissimilarity. In: Proceedings of 31st ISAAC. LIPIcs, Schloss Dagstuhl–Leibniz-Zentrum für Informatik (2020). Accepted for publication. https://arxiv.org/abs/2002.07569

21. Fortnow, L., Santhanam, R.: Infeasibility of instance compression and succinct PCPs for NP. J. Comput. Syst. Sci. **77**(1), 91–106 (2011). https://doi.org/10.1016/j.jcss.2010.06.007

22. Gupta, A., Talwar, K., Wieder, U.: Changing bases: multistage optimization for matroids and matchings. In: Esparza, J., Fraigniaud, P., Husfeldt, T., Koutsoupias, E. (eds.) ICALP 2014. LNCS, vol. 8572, pp. 563–575. Springer, Heidelberg (2014). https://doi.org/10.1007/978-3-662-43948-7_47

23. Hansen, P., Jaumard, B.: Minimum sum of diameters clustering. J. Classif. **4**(2), 215–226 (1987). https://doi.org/10.1007/BF01896987

24. Heeger, K., Himmel, A., Kammer, F., Niedermeier, R., Renken, M., Sajenko, A.: Multistage problems on a global budget. CoRR abs/1912.04392 (2019). http://arxiv.org/abs/1912.04392

25. Impagliazzo, R., Paturi, R.: On the complexity of *k*-sat. J. Comput. Syst. Sci. **62**(2), 367–375 (2001). https://doi.org/10.1006/jcss.2000.1727

26. Impagliazzo, R., Paturi, R., Zane, F.: Which problems have strongly exponential complexity? J. Comput. Syst. Sci. **63**(4), 512–530 (2001). https://doi.org/10.1006/jcss.2001.1774

27. Karp, R.M.: Reducibility among combinatorial problems. In: Proceedings of a symposium on the Complexity of Computer Computations, held 20–22 March 1972, at the IBM Thomas J. Watson Research Center, Yorktown Heights, New York, USA, pp. 85–103. The IBM Research Symposia Series. Plenum Press, New York (1972). https://doi.org/10.1007/978-1-4684-2001-2_9

28. Krom, M.R.: The decision problem for a class of first-order formulas in which all disjunctions are binary. Math. Logic Q. **13**(1–2), 15–20 (1967). https://doi.org/10.1002/malq.19670130104

29. Lokshtanov, D., Marx, D., Saurabh, S.: Lower bounds based on the exponential time hypothesis. Bull. EATCS **105**, 41–72 (2011). http://eatcs.org/beatcs/index.php/beatcs/article/view/92

30. Raghavan, R., Cohoon, J., Sahni, S.: Single bend wiring. J. Algorithms **7**(2), 232–257 (1986). https://doi.org/10.1016/0196-6774(86)90006-4

31. Ramnath, S.: Dynamic digraph connectivity hastens minimum sum-of-diameters clustering. SIAM J. Discrete Math. **18**(2), 272–286 (2004). https://doi.org/10.1137/S0895480102396099

The Weisfeiler-Leman Algorithm and Recognition of Graph Properties

Frank Fuhlbrück[1], Johannes Köbler[1], Ilia Ponomarenko[2,3], and Oleg Verbitsky[1(✉)]

[1] Humboldt-Universität zu Berlin, Unter den Linden 6, 10099 Berlin, Germany
{fuhlbfra,koebler,verbitsk}@informatik.hu-berlin.de
[2] Central China Normal University, Wuhan, China
[3] Steklov Institute of Mathematics at St. Petersburg, St. Petersburg, Russia
inp@pdmi.ras.ru

Abstract. The k-dimensional Weisfeiler-Leman algorithm (k-WL) is a very useful combinatorial tool in graph isomorphism testing. We address the applicability of k-WL to recognition of graph properties. Let G be an input graph with n vertices. We show that, if n is prime, then vertex-transitivity of G can be seen in a straightforward way from the output of 2-WL on G and on the vertex-individualized copies of G. This is perhaps the first non-trivial example of using the Weisfeiler-Leman algorithm for recognition of a natural graph property rather than for isomorphism testing. On the other hand, we show that, if n is divisible by 16, then k-WL is unable to distinguish between vertex-transitive and non-vertex-transitive graphs with n vertices unless $k = \Omega(\sqrt{n})$.

1 Introduction

The k-dimensional Weisfeiler-Leman algorithm (k-WL), whose original, 2-dimensional version [20] appeared in 1968, has played a prominent role in isomorphism testing already for a half century. Given a graph G with vertex set V, k-WL computes a canonical coloring $\mathrm{WL}_k(G)$ of the Cartesian power V^k. Let $\widehat{\mathrm{WL}}_k(G)$ denote the multiset of colors appearing in $\mathrm{WL}_k(G)$. The algorithm decides that two graphs G and H are isomorphic if $\widehat{\mathrm{WL}}_k(G) = \widehat{\mathrm{WL}}_k(H)$, and that they are non-isomorphic otherwise. While a negative decision is always correct, Cai, Fürer, and Immerman [5] constructed examples of non-isomorphic graphs G and H with n vertices such that $\widehat{\mathrm{WL}}_k(G) = \widehat{\mathrm{WL}}_k(H)$ unless $k = \Omega(n)$. Nevertheless, a constant dimension k suffices to correctly decide isomorphism for many special classes of graphs (when G is in the class under consideration and H is arbitrary). For example, $k = 2$ is enough if G is an interval graph [10], $k = 3$ is enough for planar graphs [15], and there is a constant $k = k(M)$ sufficient for all graphs not containing a given graph M as a minor [13]. Last but not least, k-WL is an

O. Verbitsky was supported by DFG grant KO 1053/8-1. He is on leave from the IAPMM, Lviv, Ukraine.

T. Calamoneri and F. Corò (Eds.): CIAC 2021, LNCS 12701, pp. 245–257, 2021.
https://doi.org/10.1007/978-3-030-75242-2_17

important component in Babai's quasipolynomial-time algorithm [3] for general graph isomorphism.

In the present paper, we initiate a discussion of the applicability of k-WL to recognition of graph properties rather than to testing isomorphism. That is, given a single graph G as input, we are interested in knowing which properties of G can be *easily* detected by looking at $\mathrm{WL}_k(G)$ or, in other words, for which decision problems the execution of k-WL on an input graph is a reasonable preprocessing step. Of course, some regularity properties are recognized in a trivial way. For example, G is strongly regular if and only if 2-WL splits V^2 just in the diagonal $\{(u, u) : u \in V\}$, the adjacency relation of G, and the complement.

For a graph property $\mathcal{P}$, we use the same character $\mathcal{P}$ to denote also the class of all graphs possessing this property. While the multiset of canonical colors $\widehat{\mathrm{WL}}_k(G)$ retains the isomorphism type of the original graph G only if k is sufficiently large, the coloring $\mathrm{WL}_k(G)$ of V^k does this for every k. This means that, at least implicitly, $\mathrm{WL}_k(G)$ contains the information about all properties $\mathcal{P}$ of G. It is, however, a subtle question whether any certificate of the membership of G in $\mathcal{P}$ can be extracted from $\mathrm{WL}_k(G)$ efficiently. Even when the isomorphism type of every graph in $\mathcal{P}$ is known to be identifiable by k-WL for some k, we can only be sure that k-WL distinguishes $\mathcal{P}$ from its complement, in the following sense: If $G \in \mathcal{P}$ and $H \notin \mathcal{P}$, then $\widehat{\mathrm{WL}}_k(G) \neq \widehat{\mathrm{WL}}_k(H)$. However, given the last inequality, we might never know whether $G \in \mathcal{P}$ and $H \notin \mathcal{P}$ or whether $H \in \mathcal{P}$ and $G \notin \mathcal{P}$. As a particular example, the fact that 2-WL decides isomorphism of interval graphs or that 3-WL decides isomorphism of planar graphs does not seem to imply, on its own, any efficient recognition algorithm for these classes.

We address the applicability of k-WL to recognition of properties saying that a graph is highly symmetric.

Deciding Vertex-Transitivity. A graph G is *vertex-transitive* if every vertex can be taken to any other vertex by an automorphism of G. It is unknown whether the class of vertex-transitive graphs is recognizable in polynomial time. The isomorphism problem for vertex-transitive graphs reduces to their recognition problem, and its complexity status is also open. In the case of graphs with a prime number p of vertices, a polynomial-time recognition algorithm is known due to Muzychuk and Tinhofer [18]. Their algorithm uses 2-WL as preprocessing and then involves a series of algebraic-combinatorial operations to find a Cayley presentation of the input graph. It is known [19] that, if p is prime, then every vertex-transitive graph with p vertices is *circulant*, i.e., a Cayley graph of the cyclic group of order p. Our first result, Theorem 1, shows a very simple, purely combinatorial way to recognize vertex-transitivity of a graph G with p vertices. Indeed, vertex-transitivity can immediately be detected by looking at the outdegrees of the monochromatic digraphs in $\mathrm{WL}_2(G)$ and $\mathrm{WL}_2(G_u)$ for all copies of G with an individualized vertex u. Our algorithm takes time $O(p^4 \log p)$, which is somewhat better than the running time $O(p^5 \log^2 p)$ of the algorithm presented in [18]. However, we believe that the main beneficial factor of our approach is its conceptual and technical simplicity.

Note that the research on circulant graphs has a long history; see, e.g., [2,17]. This class of graphs can be recognized in polynomial time [7], but whether or not this can be done by means of k-WL is widely open. The dimension $k = 2$ would clearly suffice if the algorithm could identify a cyclic order of the vertices in an input graph corresponding to its cyclic automorphism. However, it would be too naive to hope for this because such an order is, in general, not unique, not preserved by automorphisms and, hence, not canonical, even when the number of vertices is prime.

The analysis of our algorithm is based on the theory of coherent configurations (we provide a digest of main concepts in Sect. 3.1). In fact, our exposition, apart from the well-known facts on circulants of prime order, uses only several results about the *schurity property* of certain coherent configurations.

Lower Bounds for the WL Dimension. Since the work of Muzychuk and Tinhofer [18] the polynomial-time recognizability of vertex-transitive graphs with a prime number of vertices remains state-of-the-art in the sense that, to the best of our knowledge, no polynomial-time algorithm is currently known that recognizes vertex-transitivity on all n-vertex input graphs for infinitely many composite numbers n. Motivated by this fact, we complement our algorithmic result by exploring the limitations of the k-WL-based combinatorial approach to vertex-transitivity. We prove that, if n is divisible by 16, then k-WL is unable to distinguish between vertex-transitive and non-vertex-transitive graphs with n vertices unless $k = \Omega(\sqrt{n})$; see Theorem 7. This excludes extension of our positive result to graphs with an arbitrary number of vertices. Indeed, since the combination of 2-WL with vertex individualization is subsumed by 3-WL, such an extension would readily imply that 3-WL distinguishes any vertex-transitive graph from any non-vertex-transitive graph, contradicting our lower bound $k = \Omega(\sqrt{n})$. This bound as well excludes any other combinatorial approach to recognizing vertex-transitivity as long as it is based solely on k-WL for a fixed dimension k. It shows that, if such an algorithm succeeds on the n-vertex input graphs for n in a set S, then S can contain only finitely many multiples of 16.

Our lower bound is based on the Cai-Fürer-Immerman construction [5], which converts a template graph F into a pair of non-isomorphic graphs G and H indistinguishable by k-WL. To prove our lower bound for the WL dimension, we have to ensure that G is vertex-transitive and H is not. This is faced with two technical complications. First, the original CFI gadget [5, Fig. 3] involves vertices of different degrees and, hence, destroys vertex-transitivity even when the template graph F is vertex-transitive. This can be overcome by using a modified version of the CFI gadget with all vertex degrees equal, which apparently first appeared in [9]; see also the survey in [12]. Note that this approach has already been used to analyze vertex-transitivity of coherent configurations; see Evdokimov's thesis [8].

The second point is more subtle. The CFI construction replaces each vertex of the template graph F with a cell of new vertices, and vertices in different cells receive different colors. In many contexts the vertex coloring can be removed by using additional gadgets, but this is hardly possible without losing the vertex-transitivity. The vertex colors constrain the automorphisms of the CFI graphs G and H and ensure that these graphs are non-isomorphic. We establish rather general conditions on a template graph F under which the CFI graphs retain their functionality even without colors. This result of independent interest provides a very straight way of making the CFI graphs colorless, which can be used in any of their numerous applications.

The analysis of the regularized and discolored version of the CFI construction and the proof of our lower bound (Theorem 7) can be found in a long version of this paper [11].

2 Notation and Definitions

We denote the vertex set of a graph G by $V(G)$. The notation $\mathrm{Aut}(G)$ stands for the automorphism group of G.

Cayley Graphs. Let Γ be a group and Z be a set of non-identity elements of Γ such that $Z^{-1} = Z$, that is, any element belongs to Z only together with its inverse. The *Cayley graph* $\mathrm{Cay}(\Gamma, Z)$ has the elements of Γ as vertices, where x and y are adjacent if $x^{-1}y \in Z$. This graph is connected if and only if the *connection set* Z is a generating set of Γ. Every Cayley graph is obviously vertex-transitive.

The Weisfeiler-Leman Algorithm. The original version of the Weisfeiler-Leman algorithm, 2-WL, operates on the Cartesian square V^2 of the vertex set of an input graph G. Below it is supposed that G is undirected. We also suppose that G is endowed with a vertex coloring c, that is, each vertex $u \in V$ is assigned a color denoted by $c(u)$. The case of uncolored graphs is covered by assuming that $c(u)$ is the same for all u. 2-WL starts by assigning each pair $(u,v) \in V^2$ the initial color $\mathrm{WL}_2^0(u,v) = (type, c(u), c(v))$, where *type* takes on one of three values, namely *edge* if u and v are adjacent, *nonedge* if distinct u and v are non-adjacent, and *loop* if $u = v$. The coloring of V^2 is then modified step by step. The $(r+1)$-th coloring is computed as

$$\mathrm{WL}_2^{r+1}(u,v) = \{\!\!\{(\mathrm{WL}_2^r(u,w),\, \mathrm{WL}_2^r(w,v))\}\!\!\}_{w \in V}, \tag{1}$$

where $\{\!\!\{\}\!\!\}$ denotes the multiset. In words, the new color of a pair uv is a "superposition" of all old color pairs observable along the extensions of uv to a triangle uwv. Let $\mathcal{S}^r$ denote the partition of V^2 determined by the coloring $\mathrm{WL}_2^r(\cdot,\cdot)$. It is easy to notice that $\mathrm{WL}_2^{r+1}(u,v) = \mathrm{WL}_2^{r+1}(u',v')$ implies $\mathrm{WL}_2^r(u,v) = \mathrm{WL}_2^r(u',v')$, which means that $\mathcal{S}^{r+1}$ is finer than or equal to $\mathcal{S}^r$. It follows that the partition stabilizes starting from some step $t \leq n^2$, where $n = |V|$, that is, $\mathcal{S}^{t+1} = \mathcal{S}^t$, which implies that $\mathcal{S}^r = \mathcal{S}^t$ for all $r \geq t$. As the stabilization is reached, 2-WL terminates and outputs the coloring $\mathrm{WL}_2^t(\cdot,\cdot)$, which will be denoted by $\mathrm{WL}_2(\cdot,\cdot)$.

Note that the length of $\mathrm{WL}_2^r(u,v)$ grows exponentially as r increases. The exponential blow-up is remedied by renaming the colors after each step.

Let ϕ be an automorphism of G. A simple induction on r shows that $\mathrm{WL}_2^r(\phi(u), \phi(v)) = \mathrm{WL}_2^r(u, v)$ for all r and, hence

$$\mathrm{WL}_2(\phi(u), \phi(v)) = \mathrm{WL}_2(u, v). \tag{2}$$

In particular, if G is vertex-transitive, then the color $\mathrm{WL}_2(u, u)$ is the same for all $u \in V$. If the last condition is fulfilled, we say that 2-WL *does not split the diagonal on G*, where by the diagonal we mean the set of all loops (u, u).

In general, the automorphism group $\mathrm{Aut}(G)$ of the graph G acts on the Cartesian square $V(G)^2$, and the orbits of this action are called *2-orbits* of $\mathrm{Aut}(G)$. Thus, the partition of $V(G)^2$ into 2-orbits is finer than or equal to the stable partition $\mathcal{S} = \mathcal{S}^t$ produced by 2-WL.

3   Vertex-Transitivity on a Prime Number of Vertices

We begin with a few simple observations about the output produced by 2-WL on an input graph G. Recall that in this paper we restrict our attention to undirected graphs. Even though G is undirected, the equality $\mathrm{WL}_2(u, v) = \mathrm{WL}_2(v, u)$ need not be true in general. Thus, the output of 2-WL on G can naturally be seen as a complete colored directed graph on the vertex set $V(G)$, which we denote by $\mathrm{WL}_2(G)$. That is, $\mathrm{WL}_2(G)$ contains every pair $(u, v) \in V(G)^2$ as an arc, i.e., a directed edge, and this arc has the color $\mathrm{WL}_2(u, v)$ returned by 2-WL. We will see $\mathrm{WL}_2(G)$ as containing no loops, but instead we assign each vertex u the color $\mathrm{WL}_2(u, u)$. Any directed subgraph of $\mathrm{WL}_2(G)$ formed by all arcs of the same color is called a *constituent digraph*.

Let (u, v) and (u', v') be arcs of a constituent digraph C of $\mathrm{WL}_2(G)$. Note that the vertices u and u' must be equally colored in $\mathrm{WL}_2(G)$. Indeed, since the color partition of $\mathrm{WL}_2(G)$ is stable, there must exist w such that $(\mathrm{WL}_2(u', w), \mathrm{WL}_2(w, v')) = (\mathrm{WL}_2(u, u), \mathrm{WL}_2(u, v))$. The equality $\mathrm{WL}_2(u', w) = \mathrm{WL}_2(u, u)$ can be fulfilled only by $w = u'$ because any non-loop (u', w) is initially colored differently from the loop (u, u) and, hence, they are colored differently after all refinements.

Note also that, if u and v are equally colored in $\mathrm{WL}_2(G)$, then they have the same outdegree in every constituent digraph C; in particular, they simultaneously belong or do not belong to $V(C)$. Otherwise, contrary to the assumption that the color partition of $\mathrm{WL}_2(G)$ is stable, the loops (u, u) and (v, v) would receive different colors in another refinement round of 2-WL. It follows that for each constituent digraph C there is an integer $d \geq 1$ such that all vertices in C with non-zero outdegree have outdegree d. We call d the *outdegree of C*.

Let $u \in V(G)$. A *vertex-individualized graph G_u* is obtained from G by assigning the vertex u a special color, which does not occur in G. If G is vertex-transitive, then all vertex-individualized copies of G are obviously isomorphic.

Consider now a simple and still instructive example. Let $G = \overline{C_7}$ be the complement of the cycle graph on seven vertices $0, 1, \ldots, 6$ passed in this order. It is not hard to see that 2-WL splits $V(G)^2$ into the four 2-orbits of $\mathrm{Aut}(G)$; the diagonal $\{(u, u) : u \in V(G)\}$ is one of them. Note that the three constituent digraphs

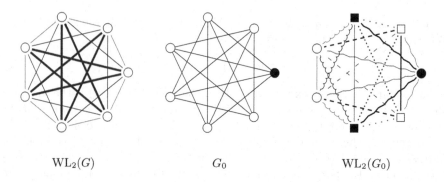

$$\mathrm{WL}_2(G) \qquad\qquad G_0 \qquad\qquad \mathrm{WL}_2(G_0)$$

Fig. 1. The output of 2-WL on input $G = \overline{C_7}$ and on its vertex-individualized copy G_0. In this example, the arcs between two equally colored vertices u and v (those with $\mathrm{WL}_2(u, u) = \mathrm{WL}_2(v, v)$) have equal colors in both directions, that is, $\mathrm{WL}_2(u, v) = \mathrm{WL}_2(v, u)$. If u and v are colored distinctly, then clearly $\mathrm{WL}_2(u, v) \neq \mathrm{WL}_2(v, u)$. As a general fact, $\mathrm{WL}_2(u, v) = \mathrm{WL}_2(u', v')$ exactly when $\mathrm{WL}_2(v, u) = \mathrm{WL}_2(v', u')$. This allows us to improve the visualization by showing only one of the two mutually reversed colors. (Color figure online)

of $\mathrm{WL}_2(G)$ are of the same degree 2; see Fig. 1. Applying 2-WL to the vertex-individualized graph G_0, it can easily be seen that 2-WL again splits $V(G)^2$ into the 2-orbits of $\mathrm{Aut}(G_0)$. Note that $\mathrm{WL}_2(G_0)$ also has exactly three constituent digraphs of outdegree 2, while all other constituent digraphs of $\mathrm{WL}_2(G_0)$ have outdegree 1. We see that the outdegrees of the constituent digraphs for G and its vertex-individualized copies are distributed similarly. This similarity proves to be a characterizing property of vertex-transitive graphs on a prime number of vertices.

Theorem 1. *Let p be a prime, and G be a graph with p vertices. Suppose that G is neither complete nor empty. Then G is vertex-transitive if and only if the following conditions are true:*

1. *If run on G, 2-WL does not split the diagonal, that is, all vertices in $\mathrm{WL}_2(G)$ are equally colored.*
2. *All constituent digraphs of $\mathrm{WL}_2(G)$ have the same outdegree $d > 1$ and, hence, there are $\frac{p-1}{d}$ constituent digraphs.*
3. *For every $u \in V(G)$, exactly $\frac{p-1}{d}$ constituent digraphs in $\mathrm{WL}_2(G_u)$ have outdegree d, and all others have outdegree 1.*

Since the color partition of $\mathrm{WL}_2(G)$ for a p-vertex graph G can be computed in time $O(p^3 \log p)$ [14], Conditions 1–3 can be verified in time $O(p^4 \log p)$, which yields an algorithm of this time complexity for recognition of vertex-transitivity of graphs with a prime number of vertices.

As it will be discussed in Remark 6 below, there are graphs G and H with a prime number of vertices such that G is vertex-transitive, H is not, and still they are indistinguishable by 2-WL. This implies that Theorem 1 is optimal in that it uses as small WL dimension as possible and, also, that the condition involving the vertex individualization cannot be dropped. Note that 1-WL, which stands for the classical degree refinement, does not suffice even when run on G and all G_u because the output of 1-WL on these inputs is subsumed by the output of 2-WL on G alone.

Theorem 1 is proved in Subsect. 3.2. The next subsection provides the necessary preliminaries.

3.1   Coherent Configurations

A detailed treatment of the material presented below can be found in [6]. The stable partition $\mathcal{S}_G$ of $V(G)^2$ produced by 2-WL on an input graph G has certain regularity properties, which are equivalent to saying that the pair $(V, \mathcal{S})$ forms a coherent configuration. This concept is defined as follows.

A *coherent configuration* $\mathcal{X} = (V, \mathcal{S})$ is formed by a set V, whose elements are called *points*, and a partition $\mathcal{S} = \{S_1, \ldots, S_m\}$ of the Cartesian square V^2, that is, $\bigcup_{i=1}^{m} S_i = V^2$ and any two S_i and S_j are disjoint. An element S_i of $\mathcal{S}$ is referred to as a *basis relation* of $\mathcal{X}$. The partition $\mathcal{S}$ has to satisfy the following three conditions:

(A) If a basis relation $S \in \mathcal{S}$ contains a loop (u, u), then all pairs in S are loops.
(B) For every $S \in \mathcal{S}$, the *transpose relation* $S^* = \{(v, u) : (u, v) \in S\}$ is also in $\mathcal{S}$.
(C) For each triple $R, S, T \in \mathcal{S}$, the number

$$p(u, v) = |\{w : (u, w) \in R, (w, v) \in S\}|$$

for a pair $(u, v) \in T$ does not depend on the choice of this pair in T.

In other words, if $\mathcal{S}$ is seen as a color partition of V^2, then such a coloring is stable under 2-WL refinement.

We describe two important sources of coherent configurations. Let $\mathcal{T}$ be an arbitrary family of subsets of the Cartesian square V^2. There exists a unique coarsest partition $\mathcal{S}$ of V^2 such that every $T \in \mathcal{T}$ is a union of elements of $\mathcal{S}$ and $\mathcal{X} = (V, \mathcal{S})$ is a coherent configuration; see [6, Section 2.6.1]. We call $\mathcal{X} = (V, \mathcal{S})$ the *coherent closure* of $\mathcal{T}$ and denote it by $Cl(\mathcal{T})$.

Given a vertex-colored undirected graph G on the vertex set V, let $\mathcal{T}$ consist of the set of the pairs $(u, v) \in V^2$ such that $\{u, v\}$ is an edge of G and the sets of loops (u, u) for all vertices u of the same color in G. Then $Cl(\mathcal{T})$ is exactly the stable partition produced by 2-WL on input G. We denote this coherent configuration by $Cl(G)$.

Given a coherent configuration $\mathcal{X} = (V, \mathcal{S})$ and a point $u \in V$, the coherent configuration $\mathcal{X}_u = Cl(\mathcal{S} \cup \{\{(u, u)\}\})$ is called a *one-point extension* of $\mathcal{X}$. This concept is naturally related to the notion of a vertex-individualized graph, in that

$$Cl(G_u) = Cl(G)_u. \tag{3}$$

Another source of coherent configurations is as follows. Let K be a permutation group on a set V. Denote the set of 2-orbits of K by $\mathcal{S}$. Then $\mathcal{X} = (V, \mathcal{S})$ is a coherent configuration, which we denote by $\mathrm{Inv}(K)$. Coherent configurations obtained in this way are said to be *schurian*.

We define an *automorphism* of a coherent configuration $\mathcal{X} = (V, \mathcal{S})$ as a bijection α from V onto itself such that, for every $S \in \mathcal{S}$ and every $(u, v) \in S$, the pair $(\alpha(u), \alpha(v))$ also belongs to S. The group of all automorphisms of $\mathcal{X}$ is denoted by $\mathrm{Aut}(\mathcal{X})$. A coherent configuration $\mathcal{X}$ is schurian if and only if

$$\mathcal{X} = \mathrm{Inv}(\mathrm{Aut}(\mathcal{X})). \tag{4}$$

Note also that the connection between the coherent closure of a graph and 2-WL implies that

$$\mathrm{Aut}(Cl(G)) = \mathrm{Aut}(G). \tag{5}$$

A set of points $X \subseteq V$ is called a *fiber* of $\mathcal{X}$ if the set of loops $\{(x, x) : x \in X\}$ is a basis relation of $\mathcal{X}$. Denote the set of all fibers of $\mathcal{X}$ by $F(\mathcal{X})$. By Property A, $F(\mathcal{X})$ is a partition of V. Property C implies that for every basis relation S of $\mathcal{X}$ there are, not necessarily distinct, fibers X and Y such that $S \subseteq X \times Y$. We use the notation $N_S(x) = \{y : (x, y) \in S\}$ for the set of all points in Y that are in relation S with x. Note that $|N_S(x)| = |N_S(x')|$ for any $x, x' \in X$. We call this number the *valency* of S. If every basis relation S of $\mathcal{X}$ has valency 1, then $\mathcal{X}$ is called *semiregular*.

Proposition 2 (see [6, Exercise 2.7.35]). *A semiregular coherent configuration is schurian.*

Given a set of points $U \subseteq V$ that is a union of fibers, let $\mathcal{S}_U$ denote the set of all basis relations $S \in \mathcal{S}$ such that $S \subseteq X \times Y$ for some, not necessarily distinct, fibers $X \subseteq U$ and $Y \subseteq U$. As easily seen, $\mathcal{X}_U = (U, \mathcal{S}_U)$ is a coherent configuration.

If a coherent configuration has a single fiber, it is called *association scheme*.

3.2   Proof of Theorem 1

Necessity. Given a vertex-transitive graph G with p vertices, where p is prime, we have to check Conditions 1–3. Condition 1 follows immediately from vertex-transitivity; see the discussion in the end of Sect. 2. For Condition 2, we use two basic results on vertex-transitive graphs with a prime number of vertices. First, every such graph is isomorphic to a *circulant graph*, i.e., a Cayley graph of a cyclic group, because every transitive group of permutations of a set of prime cardinality p contains a p-cycle (Turner [19]). Let $\mathbb{F}_p$ denote the p-element field, $\mathbb{F}_p^+$ its additive group, i.e., the cyclic group of order p, and $\mathbb{F}_p^\times$ its multiplicative group, which is isomorphic to the cyclic group of order $p-1$. Another useful fact (Alspach [1]) is that, if a set $Z \subset \mathbb{F}_p$ is non-empty and $Z \neq \mathbb{F}_p^\times$, then the automorphism group of the circulant graph $\mathrm{Cay}(\mathbb{F}_p^+, Z)$ consists of the permutations

$$x \mapsto ax + b, \ x \in \mathbb{F}_p, \text{ for all } a \in M, \ b \in \mathbb{F}_p^+, \tag{6}$$

where $M = M(Z)$ is the largest subgroup of $\mathbb{F}_p^\times$ of even order such that Z is a union of cosets of M. This subgroup is well defined because the condition $Z = -Z$ implies that Z is split into pairs $\{z, -z\}$ and, hence, is a union of cosets of the multiplicative subgroup $\{1, -1\}$. For example, $\overline{C_7} = \mathrm{Cay}(\mathbb{F}_7^+, \{2, 3, 4, 5\})$ and $M(\{2, 3, 4, 5\}) = \{1, -1\}$. Without loss of generality we assume that $G = \mathrm{Cay}(\mathbb{F}_p^+, Z)$ and denote $K = \mathrm{Aut}(G)$.

Let $\mathcal{X} = (\mathbb{F}_p^+, \mathcal{S})$ be the coherent closure of G. Recall that $\mathcal{S}$ is exactly the stable partition of $V(G)^2$ produced by 2-WL on input G. The irreflexive basis relations of $\mathcal{X}$ are exactly the constituent digraphs of $\mathrm{WL}_2(G)$, and we have to prove that all of them have the same valency.

Condition 1 says that $\mathcal{X}$ is an association scheme. In general, not all association schemes with a prime number of points are schurian (see, e.g., [6, Section 4.5]). Nevertheless, the theorem by Leung and Man on the structure of Schur rings over cyclic groups implies the following fact.

Proposition 3 (see [6, Theorem 4.5.1]). *Let $\mathcal{X} = (V, \mathcal{S})$ be an association scheme with a prime number of points. If $\mathrm{Aut}(\mathcal{X})$ acts transitively on V, then $\mathcal{X}$ is schurian.*

By Equality (5), $\mathrm{Aut}(\mathcal{X}) = K$. Since the group K is transitive, Proposition 3 implies that $\mathcal{X}$ is schurian, and we have $\mathcal{X} = \mathrm{Inv}(K)$ by Equality (4). This yields Condition 2 for $d = |M|$. Indeed, every irreflexive basis relation $S \in \mathcal{S}$ has valency $|M|$. To see this, it is enough to count the number of pairs $(0, y)$ in S. Fix an arbitrary pair $(0, y) \in S$. A pair $(0, y')$ is in the 2-orbit containing $(0, y)$ if and only if $y' = ay$ for $a \in M$, for which we have $|M|$ possibilities.

It remains to prove Condition 3. By vertex-transitivity, all vertex-individualized copies of G are isomorphic and, therefore, it is enough to consider G_0. We have to count the frequencies of valencies in $\mathcal{X}_0$. Note that $\mathcal{X}_0 = Cl(G_0)$ by Equality (3).

It is generally not true that a one-point extension of a schurian coherent configuration is schurian; see [6, Section 3.3.1]. Luckily, this is the case in our setting.

Proposition 4 (see [6, Theorem 4.4.14]). *If $\mathcal{X} = \mathrm{Inv}(K)$, where K is the group of permutations of the form (6) for a subgroup M of $\mathbb{F}_p^\times$, then the one-point extension $\mathcal{X}_0$ is schurian.*

Taking into account Equality (5), we have

$$\mathrm{Aut}(\mathcal{X}_0) = \mathrm{Aut}(Cl(G_0)) = \mathrm{Aut}(G_0) = \mathrm{Aut}(G)_0 = K_0,$$

where K_0 is the one-point stabilizer of 0 in K, that is, the subgroup of K consisting of all permutations $\alpha \in K$ such that $\alpha(0) = 0$. Obviously, $K_0 = \{x \mapsto ax, \, x \in \mathbb{F}_p\}_{a \in M}$.

Let S be a 2-orbit of K_0. If S contains a pair $(0, y)$, then it consists of all pairs $(0, y')$ for $y' \in My$ and, hence, has valency $|M|$. If S contains a pair (z, y) with $z \neq 0$ and $y \neq z$, then (z, y) is the only element of S with the first coordinate z, and S has valency 1. The proof of Condition 3 is complete.

Sufficiency. Let G be a graph satisfying Conditions 1–3 stated in the theorem. Let $\mathcal{X} = Cl(G)$. Condition 1 says that $\mathcal{X}$ is an association scheme. By Equality (5), it suffices to prove that the group $\mathrm{Aut}(\mathcal{X})$ is transitive. The proof is based on the following lemma.

Lemma 5 (see [16, **Theorem 7.1**]). *Let $\mathcal{X} = (V, \mathcal{S})$ be an association scheme. Suppose that the following two conditions are true for every point $u \in V$:*

(I) the coherent configuration $(\mathcal{X}_u)_{V\setminus\{u\}}$ is semiregular, and
(II) $F(\mathcal{X}_u) = \{N_S(u) : S \in \mathcal{S}\}$.

Then the group $\mathrm{Aut}(\mathcal{X})$ acts transitively on V.

Let u be an arbitrary vertex of G. By Equality (3), $\mathcal{X}_u = Cl(G_u)$. Now, it suffices to derive Conditions I–II in the lemma from Conditions 1–3 in the theorem.

For a fiber $X \in F(\mathcal{X}_u)$, note that $\{u\} \times X$ must be a basis relation of $\mathcal{X}_u$. Since this relation has valency $|X|$, Condition 3 implies that every fiber in $F(\mathcal{X}_u)$ is either a singleton or consists of $d \geq 2$ points. Denote the number of singletons in $F(\mathcal{X}_u)$ by a. Besides of them, $F(\mathcal{X}_u)$ contains $(p - a)/d$ fibers of size d.

For every $X, Y \in F(\mathcal{X}_u)$ with $|X| = 1$ and $|Y| = d$, $X \times Y$ is a basis relation of $\mathcal{X}_u$ of valency d. It follows from Condition 3 that

$$\frac{p - 1}{d} \geq \frac{a(p - a)}{d}.$$

Therefore, $p - 1 \geq a(p - a)$ or, equivalently, $p(a - 1) \leq (a - 1)(a + 1)$. Assume for a while that $a > 1$. It immediately follows that $a \geq p - 1$. Since the equality $a = p - 1$ is impossible, we conclude that $a = p$. However, this implies that $d = 1$, a contradiction. Thus, $a = 1$. Consequently, every fiber of the coherent configuration $\mathcal{X}' = (\mathcal{X}_u)_{V\setminus\{u\}}$ is of cardinality d, and $|F(\mathcal{X}')| = (p - 1)/d$.

Let S be a basis relation of $\mathcal{X}$. If S is reflexive, then $N_S(u) = \{u\}$. If S is irreflexive, then $N_S(u)$ must be a union of fibers in $F(\mathcal{X}')$. By Condition 2, the number of irreflexive basis relations in $\mathcal{S}$ is $(p - 1)/d$. It follows that $N_S(u)$ actually coincides with one of the fibers of $\mathcal{X}'$. This proves Condition II.

Since $\mathcal{X}_u$ contains $(p-1)/d$ basis relations of the kind $\{u\} \times X$ for $X \in F(\mathcal{X}')$, Condition 3 implies that every basis relation of $\mathcal{X}'$ is of valency 1, yielding Condition I.

The proof of Theorem 1 is complete.

Remark 6. We now argue that there is a vertex transitive graph G and a non-vertex-transitive graph H such that G and H are indistinguishable by 2-WL. Recall that a *strongly regular graph* with parameters (n, d, λ, μ) is an n-vertex d-regular graph where every two adjacent vertices have λ common neighbors, and every two non-adjacent vertices have μ common neighbors. As easily seen, two strongly regular graphs with the same parameters are indistinguishable by 2-WL, and our example will be given by G and H of this kind. Let p be a prime (or a prime power) such that $p \equiv 1 \pmod 4$. The *Paley graph* on p vertices is the Cayley graph $\mathrm{Cay}(\mathbb{F}_p^+, Y_p)$ where Y_p is the subgroup of $\mathbb{F}_p^\times$ formed by all quadratic residues modulo p. The assumption $p \equiv 1 \pmod 4$ ensures that -1 is a quadratic residue modulo p and, hence, $Y_p = -Y_p$. The Paley graph on p vertices is strongly regular with parameters $(p, \frac{p-1}{2}, \frac{p-5}{4}, \frac{p-1}{4})$.

Let G be the Paley graph on 29 vertices. It is known (Bussemaker and Spence; see, e.g., [4, Section 9.9]) that there are 40 other strongly regular graphs with parameters $(29, 14, 6, 7)$. Let H be one of them. We have only to show that H is not vertex-transitive. Otherwise, by Turner's theorem [19] this would be a circulant graph, that is, we would have $H = \mathrm{Cay}(\mathbb{F}_p^+, Z)$ for some connection set Z. In this case, the coherent closure $Cl(H)$ must be schurian by Proposition 3. Since H is strongly regular, 2-WL colors all pairs of adjacent vertices uniformly and, therefore, they form a 2-orbit of $\mathrm{Aut}(H)$. It follows that the stabilizer $\mathrm{Aut}(H)_0$ acts transitively on $N(0)$, the neighborhood of 0 in H. The aforementioned result of Alspach [1], implies that Z is the subgroup of $\mathbb{F}_p^\times$ of order $(p-1)/2$, i.e., $M = Z$ in (6). This means that $Z = Y_p$ and $H = G$, a contradiction.

4   A Lower Bound for the WL Dimension

We now state a negative result on the recognizability of vertex-transitivity by k-WL. We begin with a formal definition of the k-dimensional algorithm. Let $k \geq 2$. Given a graph G with vertex set V as input, k-WL operates on V^k. The initial coloring of $\bar{u} = (u_1, \ldots, u_k)$ encodes the equality type of this k-tuple and the ordered isomorphism type of the subgraph of G induced by the vertices $u_1, \ldots, u_k$. The color refinement is performed similarly to (1). Specifically, k-WL iteratively colors V^k by $\mathrm{WL}_k^{r+1}(\bar{u}) = \{\!\!\{(\mathrm{WL}_k^r(\bar{u}_1^w), \ldots, \mathrm{WL}_k^r(\bar{u}_k^w))\}\!\!\} w \in V(G)$, where $\bar{u}_i^w = (u_1, \ldots, u_{i-1}, w, u_{i+1}, \ldots, u_k)$. If G has n vertices, the color partition stabilizes in $t \leq n^k$ rounds, and k-WL outputs the coloring $\mathrm{WL}_k(\cdot) = \mathrm{WL}_k^t(\cdot)$.

We say that k-WL *distinguishes* graphs G and H if the final color palettes are different for G and H, that is, $\{\!\!\{\mathrm{WL}_k(\bar{u})\}\!\!\} \bar{u} \in V(G)^k \neq \{\!\!\{\mathrm{WL}_k(\bar{u})\}\!\!\} \bar{u} \in V(H)^k$ (note that color renaming in each refinement round must be performed on G and H synchronously).

Theorem 7.

1. *For every n divisible by 16 there are n-vertex graphs G and H such that G is vertex-transitive, H is not, and G and H are indistinguishable by k-WL as long as $k \leq 0.01\sqrt{n}$.*
2. *For infinitely many n there are n-vertex graphs G and H such that G is vertex-transitive, H is not, and G and H are indistinguishable by k-WL as long as $k \leq 0.001\,n$.*

5   Concluding Discussion

We have suggested a new, very simple combinatorial algorithm recognizing, in polynomial time, vertex-transitivity of graphs with a prime number of vertices.

The algorithm consists, in substance, in running 2-WL on an input graph and all its vertex-individualized copies. This is perhaps the first non-trivial example of using the Weisfeiler-Leman algorithm for recognition of a natural graph property rather than for isomorphism testing.

One can consider another, conceptually even simpler approach. If an input graph G is vertex-transitive, then k-WL colors all diagonal k-tuples $(u, \ldots, u)$, $u \in V(G)$, in the same color. Is this condition for a possibly large, but fixed k sufficient to claim vertex-transitivity? In general, a negative answer immediately follows from Theorem 7. Does there exist a fixed dimension k such that the answer is affirmative for graphs with a prime number of vertices? This is apparently a hard question; it seems that we cannot even exclude that $k = 3$ suffices.

Another interesting question is whether k-WL is able to efficiently recognize vertex-transitivity on n-vertex input graphs for n in a larger range than the set of primes. The lower bound of Theorem 7 excludes this only for n divisible by 16, in particular, for the range of n of the form $16p$ for a prime p. Can k-WL be successful on the inputs with $2p$ vertices? Conversely, can the negative result of Theorem 7 be extended to a larger range of n?[1]

Finally, we remark that the results similar to Theorems 1 and 7 can be obtained for recognition of arc-transitivity. We refer an interested reader to a long version of this paper [11].

References

1. Alspach, B.: Point-symmetric graphs and digraphs of prime order and transitive permutation groups of prime degree. J. Comb. Theory Ser. B **15**, 12–17 (1973)
2. Babai, L.: Isomorphism problem for a class of point-symmetric structures. Acta Math. Acad. Sci. Hungar. **29**(3–4), 329–336 (1977). https://doi.org/10.1007/BF01895854
3. Babai, L.: Graph isomorphism in quasipolynomial time. In: Proceedings of the 48th Annual ACM Symposium on Theory of Computing (STOC 2016), pp. 684–697 (2016). https://doi.org/10.1145/2897518.2897542
4. Brouwer, A.E., Haemers, W.H.: Spectra of Graphs. Springer, Berlin (2012). https://doi.org/10.1007/978-1-4614-1939-6
5. Cai, J., Fürer, M., Immerman, N.: An optimal lower bound on the number of variables for graph identifications. Combinatorica **12**(4), 389–410 (1992). https://doi.org/10.1007/BF01305232
6. Chen, G., Ponomarenko, I.: Coherent Configurations. Central China Normal University Press, Wuhan (2019), a draft version is available at http://www.pdmi.ras.ru/~inp/ccNOTES.pdf
7. Evdokimov, S., Ponomarenko, I.: Circulant graphs: recognizing and isomorphism testing in polynomial time. St. Petersbg. Math. J. **15**(6), 813–835 (2004)
8. Evdokimov, S.: Schurity and separability of association schemes. Ph.D. thesis, St. Petersburg University, St. Petersburg (2004)

---

[1] A simple inspection of the proof shows that Theorem 7 can be extended to the range of n divisible by $8p$ for each prime p.

9. Evdokimov, S., Ponomarenko, I.: On highly closed cellular algebras and highly closed isomorphisms. Electr. J. Comb. **6** (1999). http://www.combinatorics.org/Volume_6/Abstracts/v6i1r18.html

10. Evdokimov, S., Ponomarenko, I., Tinhofer, G.: Forestal algebras and algebraic forests (on a new class of weakly compact graphs). Discrete Math. **225**(1–3), 149–172 (2000). https://doi.org/10.1016/S0012-365X(00)00152-7

11. Fuhlbrück, F., Köbler, J., Ponomarenko, I., Verbitsky, O.: The Weisfeiler-Leman algorithm and recognition of graph properties. Tech. rep. arxiv.org/abs/2005.08887 (2020)

12. Fürer, M.: On the combinatorial power of the Weisfeiler-Lehman algorithm. In: Fotakis, D., Pagourtzis, A., Paschos, V.T. (eds.) CIAC 2017. LNCS, vol. 10236, pp. 260–271. Springer, Cham (2017). https://doi.org/10.1007/978-3-319-57586-5_22

13. Grohe, M.: Fixed-point definability and polynomial time on graphs with excluded minors. J. ACM **59**(5), 27:1-27:64 (2012). https://doi.org/10.1145/2371656.2371662

14. Immerman, N., Lander, E.: Describing graphs: a first-order approach to graph canonization. In: Selman, A.L. (eds) Complexity Theory Retrospective. Springer, New York, NY (1990). https://doi.org/10.1007/978-1-4612-4478-3_5

15. Kiefer, S., Ponomarenko, I., Schweitzer, P.: The Weisfeiler-Leman dimension of planar graphs is at most 3. J. ACM **66**(6), 44:1-44:31 (2019). https://doi.org/10.1145/3333003

16. Muzychuk, M., Ponomarenko, I.: On pseudocyclic association schemes. ARS Math. Contemp. **5**(1), 1–25 (2012)

17. Muzychuk, M.E., Klin, M.H., Pöschel, R.: The isomorphism problem for circulant graphs via Schur ring theory. In: Codes and Association Schemes. DIMACS Series in Discrete Mathematics and Theoretical Computer Science, vol. 56, pp. 241–264. DIMACS/AMS (1999). https://doi.org/10.1090/dimacs/056/19

18. Muzychuk, M.E., Tinhofer, G.: Recognizing circulant graphs of prime order in polynomial time. Electr. J. Comb. **5** (1998). http://www.combinatorics.org/Volume_5/Abstracts/v5i1r25.html

19. Turner, J.: Point-symmetric graphs with a prime number of points. J. Comb. Theory **3**, 136–145 (1967)

20. Weisfeiler, B., Leman, A.: The reduction of a graph to canonical form and the algebra which appears therein. NTI Ser. **2**(9), 12–16 (1968). https://www.iti.zcu.cz/wl2018/pdf/wl_paper_translation.pdf

The Parameterized Suffix Tray

Noriki Fujisato[1,2](✉), Yuto Nakashima[1,2], Shunsuke Inenaga[1,2,3],
Hideo Bannai[2,4], and Masayuki Takeda[1,2]

[1] Department of Informatics, Kyushu University, Fukuoka, Japan
{noriki.fujisato,yuto.nakashima,inenaga,takeda}@inf.kyushu-u.ac.jp
[2] Japan Society for the Promotion of Science, Tokyo, Japan
[3] PRESTO, Japan Science and Technology Agency, Kawaguchi, Japan
[4] M&D Data Science Center, Tokyo Medical and Dental University, Tokyo, Japan
hdbn.dsc@tmd.ac.jp

Abstract. Let Σ and Π be disjoint alphabets, respectively called the static alphabet and the parameterized alphabet. Two strings x and y over $\Sigma \cup \Pi$ of equal length are said to *parameterized match* (*p-match*) if there exists a renaming bijection f on Σ and Π which is identity on Σ and maps the characters of x to those of y so that the two strings become identical. The indexing version of the problem of finding p-matching occurrences of a given pattern in the text is a well-studied topic in string matching. In this paper, we present a state-of-the-art indexing structure for p-matching called the *parameterized suffix tray* of an input text T, denoted by $\mathsf{PSTray}(T)$. We show that $\mathsf{PSTray}(T)$ occupies $O(n)$ space and supports pattern matching queries in $O(m + \log(\sigma + \pi) + occ)$ time, where n is the length of t, m is the length of a query pattern P, π is the number of distinct symbols of $|\Pi|$ in T, σ is the number of distinct symbols of $|\Sigma|$ in T and occ is the number of p-matching occurrences of P in T. We also present how to build $\mathsf{PSTray}(T)$ in $O(n)$ time from the parameterized suffix tree of T.

1 Introduction

Parameterized Pattern Matching (PPM), first introduced by Baker [3] in 1990s, is a well-studied class of pattern matching motivated by plagiarism detection, software maintenance, and RNA structural matching [3, 20, 24].

PPM is defined as follows: Let Σ and Π be disjoint alphabets. Two equal-length strings x and y from $\Sigma \cup \Pi$ are said to parameterized match (p-match) if x can be transformed to y by applying a bijection which renames the elements of Π in x (the elements of Σ in x must remain unchanged). PPM is to report every substring in a text T that p-matches a pattern P.

In particular, the indexing version of PPM, where the task is to preprocess an input text string T so that parameterized occurrences of P in T can be reported quickly, has attracted much attention for more than two decades since the seminal paper by Baker [3].

T. Calamoneri and F. Corò (Eds.): CIAC 2021, LNCS 12701, pp. 258–270, 2021.
https://doi.org/10.1007/978-3-030-75242-2_18

Basically, the existing indexing structures for p-matching are designed upon indexing structure for exact pattern matching. Namely, *parameterized suffix trees* [3], *parameterized suffix arrays* [8], *parameterized DAWGs* [21], *parameterized CDAWGs* [21], *parameterized position heaps* [11,18], and *parameterized BWTs* [14] are based on their exact matching counterparts: suffix trees [25], suffix arrays [19], DAWGs [5], CDAWGs [4], position heaps [9,18], and BWTs [6], respectively. It should be emphasized that extending exact-matching indexing structures to parameterized matching is not straightforward and poses algorithmic challenges. Let n, m, π and σ be the lengths of a text T, a pattern P, the number of distinct symbols of $|\Pi|$ that appear in T and the number of distinct symbols of $|\Sigma|$ that appear in T, respectively. While there exist a number of algorithms which construct the suffix array for T in $O(n)$ time in the case of integer alphabets of polynomial size in n [1,10,15–17,22], the best known algorithms build the parameterized suffix array for T (denoted $\mathsf{PSA}(T)$) in $O(n \log(\sigma + \pi))$ time via the suffix tree [3,24], or directly in $O(n\pi)$ time [12]. The existence of a pure linear-time algorithm for building $\mathsf{PSA}(T)$ and the parameterized suffix tree (denoted $\mathsf{PSTree}(T)$) in the case of integer alphabets remains open.

PPM queries can be supported in $O(m + \log n + occ)$ time by $\mathsf{PSA}(T)$ coupled with the parameterized LCP array (denoted $\mathsf{PLCP}(T)$) [8], or in $O(m \log(\sigma + \pi) + occ)$ time by $\mathsf{PSTree}(T)$ [3], where occ is the number of occurrences to report.

In this paper, we propose a new indexing structure for p-matching, the *parameterized suffix tray* for T (denoted $\mathsf{PSTray}(T)$). $\mathsf{PSTray}(T)$ is a combination of $\mathsf{PSTree}(T)$ and $\mathsf{PSA}(T)$ and is an analogue to the *suffix tray* indexing structure for exact matching [7]. We show that our $\mathsf{PSTray}(T)$

(1) occupies $O(n)$ space,
(2) supports PPM queries in $O(m + \log(\sigma + \pi) + occ)$, and
(3) can be constructed in $O(n)$ time from $\mathsf{PSTree}(T)$ and $\mathsf{PSA}(T)$.

Result (3) implies that $\mathsf{PSTray}(T)$ can be constructed in $O(n \min\{\log(\sigma + \pi), \pi\})$ time using $O(n)$ working space [3,12,24]. Results (1) and (2) together with this imply that our $\mathsf{PSTray}(T)$ is the fastest linear-space indexing structure for PPM which can be built in time linear in n.

We emphasize that extending suffix trays for exact matching [7] to parameterized matching is also not straightforward. The suffix tray of a string $T \in \Sigma^*$ is a hybrid data structure of the suffix tree and suffix array of T, designed as follows: Each of the $O(\frac{n}{\sigma})$ carefully-selected nodes of the suffix tree stores an array of fixed size σ, so that pattern traversals within these selected nodes take $O(m)$ time (this also ensures a total space to be $O(\frac{n}{\sigma} \times \sigma) = O(n)$). Once the pattern traversal reaches an unselected node, then the search switches to the sub-array of the suffix array of size $O(\sigma)$. This ensures a worst-case $O(m + \log \sigma + occ)$-time pattern matching with the suffix tray.

Now, recall that the previous-encoded suffixes of T are sequences over an alphabet $\Sigma \cup \{0, \ldots, n-1\}$ of size $\Theta(\sigma + n) \subseteq O(n)$, while the alphabet size of T is $\sigma + \pi$. This means that naïve extensions of suffix trays to PPM would only result in either super-linear $O(\frac{n^2}{\sigma + \pi})$ space, or $O(m + \log n + occ)$ query time which can be achieved already with the parameterized suffix array. We overcome this difficulty by using the *smallest parameterized encoding* (*spe*) of strings which was previously proposed by the authors in the context of PPM on labeled trees [13], and this leads to our $O(n)$-space parameterized suffix trays with desired $O(m + \log(\sigma + \pi) + occ)$ query time.

2   Preliminaries

Let Σ and Π be disjoint ordered sets of characters, respectively called the static alphabet and the parameterized alphabet. We assume that any character in Π is lexicographically smaller than any character in Σ. An element of $(\Sigma \cup \Pi)^*$ is called a *p-string*. For a (p-)string $w = xyz$, x, y and z are called a *prefix*, *substring*, and *suffix* of w. The i-th character of a (p-)string w is denoted by $w[i]$ for $1 \le i \le |w|$, and the substring of a (p-)string w that begins at position i and ends at position j is denoted by $w[i : j]$ for $1 \le i \le j \le |w|$. For convenience, let $w[i : j] = \varepsilon$ if $j < i$. Also, let $w[i :] = w[i : |w|]$ for any $1 \le i \le |w|$, and $w[: j] = w[1 : j]$ for any $1 \le j \le |w|$. For any (p-)string w, let w^R denote the reversed string of w. If a p-string x is lexicographically smaller than a p-string y, then we write $x < y$.

Definition 1 (Parameterized match [2]). *Two p-strings x and y of the same length are said to* parameterized match *(p-match) iff there is a bijection f on $\Sigma \cup \Pi$ such that $f(c) = c$ for any $c \in \Sigma$ and $x[i] = f(y[i])$ for any $1 \le i \le |x|$.*

We write $x \approx y$ iff two p-strings x, y p-match. For instance, if $\Sigma = \{\text{A}, \text{B}\}$, $\Pi = \{\text{x}, \text{y}, \text{z}\}$, then $X = \text{xyzAxxxByzz}$ and $Y = \text{zxyAzzzBxyy}$ p-match since there is a bijection f such that $f(\text{A}) = \text{A}$, $f(\text{B}) = \text{B}$ and $f(\text{x}) = \text{z}, f(\text{y}) = \text{x}, f(\text{z}) = \text{y}$ and $f(\text{x})f(\text{y})f(\text{z})f(\text{A})f(\text{x})f(\text{x})f(\text{x})f(\text{B})f(\text{y})f(\text{z})f(\text{z}) = \text{zxyAzzzBxyy} = Y$

Definition 2 (Parameterized Pattern Matching problem (PPM) [2]). *Given a text p-string T and a pattern p-string P, find all positions i in T such that $T[i : i + |P| - 1] \approx p$.*

For instance, if $\Sigma = \{\text{A}\}$, $\Pi = \{\text{x}, \text{y}, \text{z}\}$, $T = \text{xyzAxxxAyyzAzx}$, and $P = \text{yAzz}$, then the out put for PPM is $\{3, 7\}$. We call the positions in the output of PPM the *p-beginning positions* for given text T and pattern P. We say that the pattern *p-appears* in the text T iff the pattern and a substring of the text p-match. In this paper, we suppose that a given text T terminates with a special end-marker $\$$ which occurs nowhere else in T. We assume that $\$$ is an element of Σ and $\$$ is lexicographically larger than any elements from Σ and Π.

Definition 3 (Previous encoding [2]). *For a p-string w, the previous encoding* prev(w) *is a string of length $|w|$ such that for each $1 \le i \le |w|$,*

$$\mathsf{prev}(w)[i] = \begin{cases} w[i] & \text{if } w[i] \in \Sigma, \\ 0 & \text{if } w[i] \in \Pi \text{ and } w[j] \ne w[i] \text{ for any } 1 \le j < i, \\ i - j & \text{otherwise, } w[i] = w[j] \text{ and } w[i] \ne w[k] \text{ for any } j < k < i. \end{cases}$$

Intuitively, when we transform w to prev(w), the first occurrence of each element of Π is replaced with 0 and any other occurrence of the element of Π is replaced by the distance to the previous occurrence of the same character, and each element of Σ remains the same.

Definition 4 (Smallest Parameterized Encoding (SPE) [13]**).** *For a p-string* w, *the smallest parameterized encoding* spe(w) *is the lexicographically smallest p-string such that* $w \approx$ spe(w).

Namely, spe(w) maps a given string w to the representative of the equivalence class of p-strings under p-matching $\approx$.

For any two p-strings w_1, w_2, prev(w_1) = prev(w_2) $\Leftrightarrow$ spe(w_1) = spe(w_2) $\Leftrightarrow$ $w_1 \approx w_2$. For instance, let $\Sigma = \{\texttt{A},\texttt{B}\}$, $\Pi = \{\texttt{x},\texttt{y},\texttt{z}\}$, $X = \texttt{yxzAyyyBxzz}$, and $Y = \texttt{zxyAzzzBxyy}$. Then prev(X) = $\texttt{000A411B771}$ = prev(Y) and spe(X) = $\texttt{xyzAxxxByzz}$ = spe(Y).

3   Parameterized Suffix Trays

In this section, we propose a new indexing structure called the *parameterized suffix tray* for PPM, and we discuss its space requirements.

Our parameterized suffix trays are a "hybrid" data structure of parameterized suffix trees and parameterized suffix arrays, which are defined as follows:

Definition 5 (Parameterized suffix trees [3]**).** *The* parameterized suffix tree *for a p-string* T, *denoted* PSTree(T), *is a compact trie that stores the set* $\{$prev($T[i:])$ $\mid$ $1 \leq i \leq |T|\}$ *of the previous encodings of all suffixes of* T.

See Fig. 1 for examples of PSTree(T). We assume that the leaves of PSTree(T) are sorted in lexicographical order, so that the sequence of the leaves corresponds to the parameterized suffix array for T, which is defined below.

Definition 6 (Parameterized suffix arrays [8]**).** *The* parameterized suffix array *of a p-string* T, *denoted* PSA(T), *is an array of integers such that* PSA(T)$[i]$ = j *if and only if* prev($T[j:])$ *is the* i*th lexicographically smallest string in* $\{$prev($T[i:])$ $\mid$ $1 \leq i \leq |T|\}$.

Definition 7 (Parameterized longest common prefix arrays [8]**).** *The parameterized longest common prefix array of a p-string* T, *denoted* PLCP(T), *is an array of integers such that* PLCP(T)$[1]$ = 0 *and* $2 \leq i \leq |T|$ PLCP(T)$[i]$ *stores the length of the longest common prefix between* prev($T[$PSA(T)$[i-1]:])$ *and* prev($T[$PSA(T)$[i]:])$.

See Fig. 2 for examples of PSA(T) and PLCP(T).

In addition to the above data structures from the literature, we introduce the following new notions and data structures. For convenience, we will sometimes identify each node of the parameterized suffix tree with the string which is represented by that node.

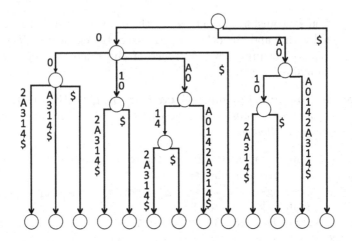

Fig. 1. PSTree(T) for a p-string $T = \mathtt{zAxAyyxyAxxy}$, where $\Sigma = \{\mathtt{A}, \$\}, \Pi = \{\mathtt{x}, \mathtt{y}, \mathtt{z}\}$.

i	PSA(T)[i]	$T[\text{PSA}(T)[i]:]$	PLCP(i)
1	6	0 0 2 A 3 1 4 $	0
2	7	0 0 A 3 1 4 $	2
3	11	0 0 $	2
4	5	0 1 0 2 A 3 1 4 $	1
5	10	0 1 0 $	3
6	3	0 A 0 1 4 2 A 3 1 4 $	1
7	8	0 A 0 1 4 $	5
8	1	0 A 0 A 0 1 4 2 A 3 1 4 $	3
9	12	0 $	1
10	4	A 0 1 0 2 A 3 1 4 $	0
11	9	A 0 1 0 $	4
12	2	A 0 A 0 1 4 2 A 3 1 4 $	2
13	13	$	0

Fig. 2. PSA(T) and PLCP(T) for a p-string $T = \mathtt{zAxAyyxyAxxy}$, where $\Sigma = \{\mathtt{A}, \$\}, \Pi = \{\mathtt{x}, \mathtt{y}, \mathtt{z}\}$.

In what follows, let $\Pi_T = \{T[i] \in \Pi \mid 1 \le i \le |T|\}$ and $\Sigma_T = \{T[i] \in \Sigma \mid 1 \le i \le |T|\}$, namely, Π_T (resp. Σ_T) is the set of distinct characters of Π (resp. Σ) that occur in T. Let $\pi = |\Pi_T|$ and $\sigma = |\Sigma_T|$.

Definition 8 (P-nodes, branching p-nodes). *Let T be a p-string over $\Sigma \cup \Pi$. A node v in PSTree(T) is called a p-node if the number of leaves in the subtree of PSTree(T) rooted at v is at least $\max\{\sigma, \pi\}$. A p-node v is called a branching p-node if at least two children of v in PSTree(T) are p-nodes.*

See Fig. 3 for examples of p-nodes and branching p-nodes.

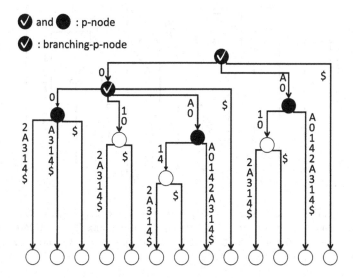

Fig. 3. PSTree(T) for a p-string $T = $ zAxAyyxyAxxy, where $\Sigma = \{$A, $\$\}$, $\Pi = \{$x, y, z$\}$. Then black nodes are p-nodes because the number of leaves in the subtree of PSTree(T) rooted at them are at least max$\{\sigma, \pi\} = 3$. Checked nodes are branching p-nodes because at least two children of them in PSTree(T) are p-nodes.

For any $x \in \Pi_T$, let rank$_T(x)$ denote the lexicographical rank of x in $\Pi_T \cup \Sigma_T$. Assuming that Π and Σ are integer alphabets of polynomial size in n, we can compute rank$_T(x)$ for every $x \in \Pi_T$ in $O(n)$ time by bucket sort. We will abbreviate rank$_T(x)$ as rank(x) when it is not confusing.

Definition 9 (P-array). *Let* prev(v) *be any branching p-node of* PSTree(T), *where* v *is some substring of* T. *The* p-array $A(\text{prev}(v))$ *for* prev(v) *is an array of length* $\sigma + \pi$ *such that for each* $x \in \Sigma \cup \Pi$, $A(\text{prev}(v))[\text{rank}(x)]$ *stores a pointer to the child* u *of* prev(v) *such that* prev$(\text{spe}(v)x)$ *is a prefix of* u *if such a child exists, and* $A(\text{prev}(v))[\text{rank}(x)]$ *stores nil otherwise.*

See Fig. 4 for an example of a p-array.

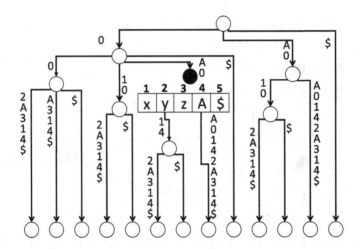

Fig. 4. PSTree(T) for a p-string $T = $ zAxAyyxyAxxy, where $\Sigma_T = \{$A$,\$\}$, $\Pi_T = \{$x, y, z$\}$.
Consider a branching p-node prev(v) = 0A0 where v is e.g. zAx. Then, $A($0A0$)[$rank(y)$] = $
$A($0A0$)[2]$ stores a pointer to node 0A014 because spe(v) = xAy and prev(spe(v)y =
xAyy) = 0A01 is a prefix of node 0A014.

Definition 10 (Parameterized suffix tray). *The parameterized suffix tray
of a p-string T, denoted* PSTray(T), *is a hybrid data structure consisting of*
PSA(T), PLCP(T), *and* PSTree(T) *where each branching p-node is augmented
with the p-array.*

We can show the following lemmas regarding the space requirements of
PSTray(T), by similar arguments to [7] for suffix trays on standard strings.

Lemma 1. *For any p-string T of length n over $\Sigma \cup \Pi$, the number of branching
p-nodes in* PSTree(T) *is* $O(\frac{n}{\pi+\sigma})$.

Lemma 2. *For any p-string T of length n,* PSTray(T) *occupies $O(n)$ space.*

4   PPM Using Parameterized Suffix Trays

In this section, we present our algorithm for parameterized pattern matching
(PPM) on PSTray(T). For any node v in PSTray(T), let $l_v = (i, j)$ denote the
range of PSA(T) that v corresponds, namely, $l_v = (i, j)$ iff the leftmost and
rightmost leaves in the subtree rooted at v correspond to the ith and jth entries
of PSA(T), respectively. For any *p-node* v, we store l_v in v. Also, for any *non-
branching* p-node u, we store a pointer to the unique child of u that is a p-node.
These can be easily computed in a total of $O(n)$ time by a standard traversal
on PSTree(T).

The basic strategy for PPM with PSTray(T) follows the (exact) pattern
matching algorithm with suffix trays on standard strings [7]. Namely, we traverse
PSTree(T) with a given pattern P from the root, and as soon as we encounter a

node that is not a p-node, then we switch to the corresponding range of $\mathsf{PSA}(T)$ and perform a binary search to locate the pattern occurrences. The details follow.

Let P be a pattern p-string of length m. We assume that Π and Σ are disjoint integer alphabets, where $\Pi = \{0, \ldots, c_1 n\}$ and $\Sigma = \{c_1 n + 1, \ldots, n^{c_2}\}$ for some positive constants c_1 and c_2. Using an array (bucket) B of size $|\Pi| = c_1 n \in O(n)$, we can compute $\mathsf{prev}(P)$ in $O(m)$ time by scanning P from left to right and keeping the last occurrence of each character $x \in \Pi$ in P in $B[x]$. We can compute $\mathsf{spe}(P)$ in $O(m)$ time in a similar manner with a bucket. These buckets are a part of our indexing structure that occupies $O(n)$ total space.

After computing $\mathsf{prev}(P)$ and $\mathsf{spe}(P)$, we traverse $\mathsf{prev}(P)$ on $\mathsf{PSTray}(T)$. If $\mathsf{prev}(P[:i])$ for prefix $P[:i]$ ($1 \leq i \leq m$) is represented by a p-node, we can find the out-going edge whose label begins with $\mathsf{prev}(P)[i+1]$ in constant time by accessing the p-array entry $A(\mathsf{prev}(P[:i]))[\mathsf{spe}(P)[i+1]]$. Therefore, we can solve PPM in $O(m + occ)$ time if $\mathsf{prev}(P)$ is a prefix of some p-node in $\mathsf{PSTray}(T)$. Otherwise (if $\mathsf{prev}(P)$ is not a prefix of any p-node), there exists integer i such that $\mathsf{prev}(P[:i])$ is not a p-node but the parent of $\mathsf{prev}(P[:i])$ is a p-node. In this case, we will use the next lemma.

Lemma 3 (PPM in PSA range (adapted from [8])). *Given a pattern p-string P of length m and a range $[j, k]$ in $\mathsf{PSA}(T)$ such that the occ occurrences of P in T lie in the range $[j, k]$ of $\mathsf{PSA}(T)$, we can find them in $O(m + \log(k-j) + occ)$ time by using $\mathsf{PSA}(T)$ and $\mathsf{PLCP}(T)$.*

Let $I_{\mathsf{prev}(P[:i])} = (j, k)$ denote the range in $\mathsf{PSA}(T)$ where $\mathsf{prev}(P)$ is a prefix of the suffixes in the range. We apply Lemma 3 to this range so we can find the parameterized occurrences of P in T in $O(m + \log(k - j) + occ)$ time. By Definition 8 we have $k - j \leq \pi + \sigma$ (recall that $\mathsf{prev}(P[:i])$ is not a p-node). Thus, $O(m + \log(k - j) + occ) \subseteq O(m + \log(\pi + \sigma) + occ)$, implying the next theorem.

Theorem 1. *Suppose $|\Pi| = O(n)$. Then, $\mathsf{PSTray}(T)$ supports PPM queries in $O(m + \log(\pi + \sigma) + occ)$ time each, where m is the length of a query pattern P and occ is the number of occurrences to report.*

5   Construction of Parameterized Suffix Trays

Let T be a p-string of length n. In this section, we show how to construct $\mathsf{PSTray}(T)$ provided that $\mathsf{PSTree}(T)$ has already been built. Throughout this section we assume that Π and Σ are disjoint integer alphabets, both being of polynomial size in n, namely, $\Pi = \{0, ..., n^{c_1}\}$ and $\Sigma = \{n^{c_1} + 1, ..., n^{c_2}\}$ for some positive constants c_1 and c_2. For convenience, we define the following two notions.

Definition 11 (P-function). *Let q, r be p-strings such that $q \approx r$. The p-function $\mathsf{f}_{q,r} : \Sigma \cup \Pi \to \Sigma \cup \Pi$ transforms q to r, namely, for every $1 \leq i \leq h$*

$$\mathsf{f}_{q,r}(q[i]) = r[i].$$

For instance, if $\Pi = \{x, y, z\}$, $q = xyxzyyxz$, $r = zxzyxxzy$ and $q \approx r$, then $f_{q,r}(x) = z$, $f_{q,r}(y) = x$, $f_{q,r}(z) = y$ since q can be transformed r by this function.

Definition 12 (F-array). *Let q be a p-string and $x \in \Pi_T$. The first (left-most) occurrence of x in q is denoted by $i_{q,x}$. The f-array of q, denoted $\mathsf{fpos}(q)$, is an array of length π such that $\mathsf{fpos}(q)[\mathsf{rank}(x)] = i_{q,x}$.*

For instance, if $\Pi_T = \{x, y, z\}$ and $q = xyxzyyxz$, then $\mathsf{fpos}(q)[\mathsf{rank}(x)] = \mathsf{fpos}(q)[1] = 1$, $\mathsf{fpos}(q)[\mathsf{rank}(y)] = \mathsf{fpos}(q)[2] = 2$, and $\mathsf{fpos}(q)[\mathsf{rank}(z)] = \mathsf{fpos}(q)[3] = 4$.

Given $\mathsf{PSTree}(T)$, we show how to construct $\mathsf{PSTray}(T)$. It is well known that $\mathsf{PSA}(T)$ and $\mathsf{PLCP}(T)$ can be constructed from $\mathsf{PSTree}(T)$ in $O(n)$ time. In the following, we consider how to compute $A(\mathsf{prev}(v))$ for every p-node $\mathsf{prev}(v)$ in $\mathsf{PSTree}(T)$.

First, we consider how to compute (branching) p-nodes in $\mathsf{PSTree}(T)$. This can be done by a similar method to the suffix tray for exact matching [7], namely:

Lemma 4 (Computing p-node). *We can compute all p-nodes and branching p-nodes in $\mathsf{PSTree}(T)$ in $O(n)$ total time.*

Our algorithm performs a bottom-up traversal on $\mathsf{PSTree}(T)$ and propagates pairs $(\mathsf{fpos}(T[i :]), i)$ from leaves to their ancestors. Each internal p-node $\mathsf{prev}(v)$ will store only a single pair $(\mathsf{fpos}(T[i :]), i)$, where i is the largest position in T such that $\mathsf{prev}(T[i :])$ is a leaf in the subtree rooted at $\mathsf{prev}(v)$[1]. See also Fig. 5. One can easily compute the pairs for all p-nodes in a total of $O(n)$ time. Then, we compute $f_{v,\mathsf{spe}(v)}$ for every p-node $\mathsf{prev}(v)$ from the pair $(\mathsf{fpos}(T[i :]), i)$ that is stored in the p-node $\mathsf{prev}(v)$. Finally, for every p-node $\mathsf{prev}(v)$ we compute $A_{\mathsf{prev}(v)}$ from $f_{v,\mathsf{spe}(v)}$ and i.

In what follows, we first show how to compute $A_{\mathsf{prev}(v)}$ from $f_{v,\mathsf{spe}(v)}$ and i in Lemmas 5 and 6. We then present how to compute $f_{v,\mathsf{spe}(v)}$ and i from $\mathsf{fpos}(T[i :])$ in Lemma 7, and how to compute $\mathsf{fpos}(T[i :])$ in Lemma 8. These lemmas will ensure the correctness and time complexity of our algorithm.

We consider how to compute $A(\mathsf{prev}(v))$ for a given p-node $\mathsf{prev}(v)$.

Lemma 5. *Let s be a p-string. If $\mathsf{prev}(s)[|s|] = k \in \{0, \ldots, |T| - 1\}$, then $\mathsf{spe}(s)[|s|] = \mathsf{spe}(s)[|s| - k]$.*

Proof. Clear from the definitions of $\mathsf{prev}(\cdot)$ and $\mathsf{spe}(\cdot)$. □

---

[1] Indeed, our $\mathsf{PSTray}(T)$ construction algorithm works with any position i in the subtree rooted at $\mathsf{prev}(v)$, and we propagate the largest leaf position i to each internal p-node for simplicity.

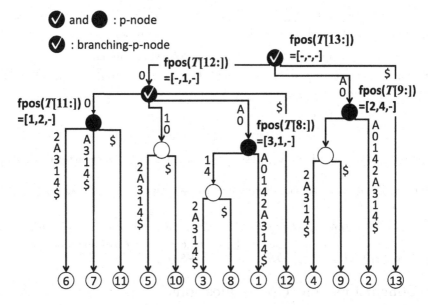

Fig. 5. PSTree(T) for a p-string $T = \mathtt{zAxAyyxyAxxy}$, where $\Sigma = \{\mathtt{A}, \$\}$ and $\Pi = \{\mathtt{x}, \mathtt{y}, \mathtt{z}\}$. For instance, we propagate fpos($T[12 :]$) (coupled with the corresponding position 12) to the p-node 0.

In the sequel, let $\mathsf{prev}(T[i :])$ be any leaf in the subtree rooted at $\mathsf{prev}(v)$, where $1 \leq i \leq |T|$. By Lemma 5, we can compute $A(\mathsf{prev}(v))$ if we know $\mathsf{spe}(v)[|v| - k + 1]$, where $k = \mathsf{prev}(T[i :])[|v| + 1]$.

Lemma 6. $\mathsf{spe}(v)[|v| - k + 1] = \mathsf{f}_{T[i:i+|v|-1],\mathsf{spe}(T[i:i+|v|-1])}T[i+|v|+k-2]$.

Proof. Clear from the definitions of $\mathsf{f}_{\cdot,\cdot}$ and $\mathsf{spe}(\cdot)$. $\quad\square$

We can compute $\mathsf{spe}(v)[|v| - k + 1]$ if we know $\mathsf{f}_{T[i:i+|v|-1],\mathsf{spe}(v)}$. In the next lemma, we show how to compute $\mathsf{f}_{T[i:i+|v|-1],\mathsf{spe}(v)}$ for all p-nodes $\mathsf{prev}(v)$.

Lemma 7. *For every p-node $\mathsf{prev}(v)$, we can compute $\mathsf{f}_{T[i:i+|v|-1],\mathsf{spe}(T[i:i+|v|-1])}$ with $\mathsf{prev}(v) = \mathsf{prev}(T[i : i + |v| - 1]))$ in amortized $O(\pi)$ time if we know pair $\mathsf{fpos}(T[i :], i)$.*

Proof. Let x_j be the jth smallest element of Π in lexicographically order. Let y_l denote the parameterized character in $S = T[i : i + |v| - 1]$ such that if j is the left-most occurrence of y_l in S (i.e. $j = \min\{h \mid S[h] = y_l\}$), then $|\Pi_{S[1..j]}| = l$. For instance, for $S = \mathtt{xAzAyA}$, then $y_1 = \mathtt{x}$, $y_2 = \mathtt{z}$, and $y_3 = \mathtt{y}$. Let $(i_{\mathsf{prev}(v)}, \mathsf{fpos}(T[i_{\mathsf{prev}(v)} :]))$ denote the pair stored in p-node $\mathsf{prev}(v)$. Consider a set $\mathbf{I} = \{(i_{\mathsf{prev}(v)}, \mathsf{fpos}(T[i_{\mathsf{prev}(v)} :])[k]) \mid \mathsf{prev}(v) \text{ is a p-node}, 1 \leq k \leq \pi\}$ of integer pairs. We sort the elements of $\mathbf{I}$ so that we can compute in $O(1)$ time $\mathsf{f}_{T[i:i+|v|-1],\mathsf{spe}(v)}(y_l) = x_l$ for all p-nodes $\mathsf{prev}(v)$, where x_l is the lth smallest parameterized character that occurs in $\mathsf{spe}(v)$. We can sort the elements of $\mathbf{I}$ in a total of $O(n)$ time by radix sort, since there are $O(\frac{n}{\sigma+\pi})$ p-nodes and each f-array $\mathsf{fpos}(T[i :])$ is of length π. This completes the proof.   □

We can easily compute $\mathsf{fpos}(T[i :])$ by the following lemma:

Lemma 8. *Let q be a p-string. Let $x \in \Pi$, $y \in \Pi \cup \Sigma$ and $x \neq y$. Then the following equations hold:*

$$\mathsf{fpos}(xq)[\mathsf{rank}(x)] = 1,$$
$$\mathsf{fpos}(xq)[\mathsf{rank}(y)] = \mathsf{fpos}(q)[\mathsf{rank}(y)] + 1.$$

Thus we can compute f-arrays $\mathsf{fpos}(T[i :])$ for all $1 \leq i \leq n$ in a total of $O(n)$ time.

Since all the afore-mentioned procedures take $O(n)$ time each, we obtain the main theorem of this section.

Theorem 2. *Given a p-string T of length n over alphabet $\Sigma \cup \Pi$ with $\Pi = \{0, ..., n^{c_1}\}$ and $\Sigma = \{n^{c_1} + 1, ..., n^{c_2}\}$ for some positive constants c_1 and c_2, we can construct $\mathsf{PSTray}(T)$ in $O(n)$ time from $\mathsf{PSTree}(T)$.*

6   Conclusions and Open Questions

In this paper, we proposed an indexing structure for parameterized pattern matching (PPM) called the parameterized suffix tray $\mathsf{PSTray}(T)$, where T is a given text string. Our $\mathsf{PSTray}(T)$ uses $O(n)$ space and supports pattern matching queries in $O(m + \log(\sigma + \pi) + occ)$ time, where $n = |T|$, m is the query pattern length, σ and π are respectively the numbers of distinct static characters and distinct parameterized characters occurring in T, and occ is the number of pattern occurrences to report. We also showed how to construct $\mathsf{PSTray}(T)$ in $O(n+s(n))$ time, where $s(n)$ denotes the time complexity to build the parameterized suffix tree $\mathsf{PSTree}(T)$ for T. It is known that $s(n) = \min\{n\pi, n(\log(\pi + \sigma)\}$ [3,12,24].

On the other hand, if we use hashing for implementing the branches of the parameterized suffix tree $\mathsf{PSTree}(T)$, one can trivially answer PPM queries in $O(m + occ)$ time with $O(n)$ space. The best linear-space deterministic hashing we are aware of is the one by Ružić [23], which can be built in $O(n(\log \log n)^2)$ time for a set of n keys in the word RAM model with machine word size $\Omega(\log n)$. By associating each node of $\mathsf{PSTree}(T)$ with a unique integer (e.g. the pre-order rank), one can regard each branch in $\mathsf{PSTree}(T)$ as an integer from the universe of polynomial size in n, each fitting in a constant number of machine words. This gives us a deterministic $O(n(\log \log n)^2 + s(n))$-time algorithm for building $\mathsf{PSTree}(T)$ with $O(m + occ)$-time PPM queries. Still, it is not known whether a similar data structure can be build in $O(n + s(n))$ time. We conjecture that our $O(m + \log(\sigma + \pi) + occ)$ PPM query time would be the best possible for any indexing structure that can be build in $O(n + s(n))$ time. Proving or disproving such a lower bound is an intriguing open problem.

Acknowledgments. This work was supported by JSPS KAKENHI Grant Numbers JP18K18002 (YN), JP17H01697 (SI), JP20H04141 (HB), JP18H04098 (MT), and JST PRESTO Grant Number JPMJPR1922 (SI).

References

1. Baier, U.: Linear-time suffix sorting - a new approach for suffix array construction. In: CPM 2016, pp. 23:1–23:12 (2016)
2. Baker, B.S.: A theory of parameterized pattern matching: algorithms and applications. STOC **1993**, 71–80 (1993)
3. Baker, B.S.: Parameterized pattern matching: algorithms and applications. J. Comput. Syst. Sci. **52**(1), 28–42 (1996)
4. Blumer, A., Blumer, J., Haussler, D., Mcconnell, R., Ehrenfeucht, A.: Complete inverted files for efficient text retrieval and analysis. J. ACM **34**(3), 578–595 (1987)
5. Blumer, A., Blumer, J., Haussler, D., Ehrenfeucht, A., Chen, M.T., Seiferas, J.: The smallest automaton recognizing the subwords of a text. Theor. Comput. Sci. **40**, 31–55 (1985)
6. Burrows, M., Wheeler, D.J.: A block sorting lossless data compression algorithm (1994)
7. Cole, R., Kopelowitz, T., Lewenstein, M.: Suffix trays and suffix trists: structures for faster text indexing. Algorithmica **72**(2), 450–466 (2015)
8. Deguchi, S., Higashijima, F., Bannai, H., Inenaga, S., Takeda, M.: Parameterized suffix arrays for binary strings. PSC **2008**, 84–94 (2008)
9. Ehrenfeucht, A., McConnell, R.M., Osheim, N., Woo, S.W.: Position heaps: a simple and dynamic text indexing data structure. J. Discrete Algorithms **9**(1), 100–121 (2011)
10. Farach-Colton, M., Ferragina, P., Muthukrishnan, S.: On the sorting-complexity of suffix tree construction. J. ACM **47**(6), 987–1011 (2000)
11. Fujisato, N., Nakashima, Y., Inenaga, S., Bannai, H., Takeda, M.: Right-to-left online construction of parameterized position heaps. PSC **2018**, 91–102 (2018)
12. Fujisato, N., Nakashima, Y., Inenaga, S., Bannai, H., Takeda, M.: Direct linear time construction of parameterized suffix and LCP arrays for constant alphabets. In: Brisaboa, N.R., Puglisi, S.J. (eds.) SPIRE 2019. LNCS, vol. 11811, pp. 382–391. Springer, Cham (2019). https://doi.org/10.1007/978-3-030-32686-9_27
13. Fujisato, N., Nakashima, Y., Inenaga, S., Bannai, H., Takeda, M.: The parameterized position heap of a Trie. In: Heggernes, P. (ed.) CIAC 2019. LNCS, vol. 11485, pp. 237–248. Springer, Cham (2019). https://doi.org/10.1007/978-3-030-17402-6_20
14. Ganguly, A., Shah, R., Thankachan, S.V.: pBWT: achieving succinct data structures for parameterized pattern matching and related problems. SODA **2017**, 397–407 (2017)
15. Kärkkäinen, J., Sanders, P., Burkhardt, S.: Linear work suffix array construction. J. ACM **53**(6), 918–936 (2006)
16. Kim, D.K., Sim, J.S., Park, H., Park, K.: Constructing suffix arrays in linear time. J. Discrete Algorithms **3**(2–4), 126–142 (2005)
17. Ko, P., Aluru, S.: Space efficient linear time construction of suffix arrays. J. Discrete Algorithms **3**(2–4), 143–156 (2005)
18. Kucherov, G.: On-line construction of position heaps. J. Discrete Algorithms **20**, 3–11 (2013)
19. Manber, U., Myers, G.: Suffix arrays: a new method for on-line string searches. SIAM J. Comput. **22**(5), 935–948 (1993)
20. Mendivelso, J., Thankachan, S.V., Pinzón, Y.J.: A brief history of parameterized matching problems. Discret. Appl. Math. **274**, 103–115 (2020)

21. Nakashima, K., et al.: DAWGs for parameterized matching: online construction and related indexing structures. In: CPM 2020, pp. 26:1–26:14 (2020)
22. Nong, G., Zhang, S., Chan, W.H.: Two efficient algorithms for linear time suffix array construction. IEEE Trans. Comput. **60**(10), 1471–1484 (2011)
23. Ružić, M.: Constructing efficient dictionaries in close to sorting time. In: Aceto, L., Damgård, I., Goldberg, L.A., Halldórsson, M.M., Ingólfsdóttir, A., Walukiewicz, I. (eds.) ICALP 2008. LNCS, vol. 5125, pp. 84–95. Springer, Heidelberg (2008). https://doi.org/10.1007/978-3-540-70575-8_8
24. Shibuya, T.: Generalization of a suffix tree for RNA structural pattern matching. Algorithmica **39**(1), 1–19 (2004)
25. Weiner, P.: Linear pattern-matching algorithms. In: Proceeding of 14th IEEE Annual Symposium on Switching and Automata Theory, pp. 1–11 (1973)

Exploring the Gap Between Treedepth and Vertex Cover Through Vertex Integrity

Tatsuya Gima[1], Tesshu Hanaka[1], Masashi Kiyomi[2], Yasuaki Kobayashi[3], and Yota Otachi[1(✉)]

[1] Nagoya University, Nagoya, Japan
{gima,hanaka,otachi}@nagoya-u.jp
[2] Seikei University, Musashino-shi, Tokyo, Japan
kiyomi@st.seikei.ac.jp
[3] Kyoto University, Kyoto, Japan
kobayashi@iip.ist.i.kyoto-u.ac.jp

Abstract. For intractable problems on graphs of bounded treewidth, two graph parameters treedepth and vertex cover number have been used to obtain fine-grained complexity results. Although the studies in this direction are successful, we still need a systematic way for further investigations because the graphs of bounded vertex cover number form a rather small subclass of the graphs of bounded treedepth. To fill this gap, we use vertex integrity, which is placed between the two parameters mentioned above. For several graph problems, we generalize fixed-parameter tractability results parameterized by vertex cover number to the ones parameterized by vertex integrity. We also show some finer complexity contrasts by showing hardness with respect to vertex integrity or treedepth.

Keywords: vertex integrity · vertex cover number · treedepth

1 Introduction

Treewidth, which measures how close a graph is to a tree, is arguably one of the most powerful tools for designing efficient algorithms for graph problems. The application of treewidth is quite wide and the general theory built there often gives a very efficient algorithm (e.g., [2,10,15]). However, still many problems are found to be intractable on graphs of bounded treewidth (e.g., [40]). To cope with such problems, one may use pathwidth, which is always larger than or equal to treewidth. Unfortunately, this approach did not quite work as no natural problem was known to change its complexity with respect to treewidth and pathwidth, until very recently [8]. Treedepth is a further restriction of pathwidth. However,

Partially supported by JSPS KAKENHI Grant Numbers JP18H04091, JP18K11168, JP18K11169, JP19K21537, JP20K19742, JP20H05793.

T. Calamoneri and F. Corò (Eds.): CIAC 2021, LNCS 12701, pp. 271–285, 2021.
https://doi.org/10.1007/978-3-030-75242-2_19

still most of the problems do not change their complexity, except for some problems with hardness depending on the existence of long paths (e.g., [21,31]). One successful approach in this direction is parameterization by the vertex cover number, which is a strong restriction of treedepth. Many problems that are intractable parameterized by treewidth have been shown to become tractable when parameterized by vertex cover number [1,14,22,24,25,33].

One drawback of the vertex-cover parameterization is its limitation to a very small class of graphs. To overcome the drawback, we propose a new approach for parameterizing graph problems by vertex integrity [5]. The *vertex integrity* of a graph G, denoted $\mathsf{vi}(G)$, is the minimum integer k satisfying that there is $S \subseteq V(G)$ such that $|S| + |V(C)| \le k$ for each component C of $G - S$. We call such S a $\mathsf{vi}(k)$-*set* of G. This parameter is bounded from above by vertex cover number + 1 and from below by treedepth. As a structural parameter in parameterized algorithms, vertex integrity (and its close variants) was used only in a couple of previous studies [12,20,26]. Our goal is to fill some gaps between treedepth and vertex cover number by presenting finer algorithmic and complexity results parameterized by vertex integrity. Note that the parameterization by vertex integrity is equivalent to the one by ℓ-component order connectivity + ℓ [19].

Short Preliminaries. For the basic terms and concepts in the parameterized complexity theory, we refer the readers to standard textbooks, e.g. [16,18].

For a graph G, we denote its treewidth by $\mathsf{tw}(G)$, pathwidth by $\mathsf{pw}(G)$, treedepth by $\mathsf{td}(G)$, and vertex cover number by $\mathsf{vc}(G)$. It is known that $\mathsf{tw}(G) \le \mathsf{pw}(G) \le \mathsf{td}(G) - 1 \le \mathsf{vi}(G) - 1 \le \mathsf{vc}(G)$ for every graph G. We say informally that a problem is fixed-parameter tractable "parameterized by vi", which means "parameterized by the vertex integrity of the input graphs." We also say "graphs of $\mathsf{vi} = c$ (or $\mathsf{vi} \le c$)".

Our Results. The main contribution of this paper is to generalize several known FPT algorithms parameterized by vc to the ones by vi. We also show some results considering parameterizations by vc, vi, or td to tighten the complexity gaps between parameterizations by vc and by td. See Table 1 for the summary of results. Due to the space limitation, we included only selected results in this version. For the omitted results and proofs, see the full version [29].

Extending FPT Results Parameterized by vc. We show that IMBALANCE, MAXIMUM COMMON (INDUCED) SUBGRAPH, CAPACITATED VERTEX COVER, CAPACITATED DOMINATING SET, PRECOLORING EXTENSION, EQUITABLE COLORING, and EQUITABLE CONNECTED PARTITION are fixed-parameter tractable parameterized by vertex integrity. We present the algorithms for IMBALANCE as a simple but still powerful example that generalizes known results (Sect. 2) and for MAXIMUM COMMON SUBGRAPH as one of the most involved examples (Sect. 3). See the full version for the other problems. A commonly used trick is to reduce the problem instance to a number of instances of integer linear programming, while each problem requires a nontrivially tailored reduction depending on its structure. It was the same for parameterizations by vc, but the reductions here are more involved because of the generality of vi. Finding the similarity among the reductions and algorithms would be a good starting point to develop a general way for handling problems parameterized by vi (or vc). Additionally, we show that BANDWIDTH is W[1]-hard parameterized by td, while we were not able to extend the algorithm parameterized by vc to the one by vi.

Table 1. Summary. The results stated without references are shown in this paper.

PROBLEM	Lower bounds	Upper bounds
IMBALANCE	NP-h [9]	FPT by tw + Δ [34]
		FPT by vi
MAX COMMON SUBGRAPH	NP-h for vi(G_2) = 3	FPT by vi(G_1) + vi(G_2)
MAX COMMON IND. SUBGRAPH	NP-h for vc(G_2) = 0	
CAPACITATED VERTEX COVER	W[1]-h by td [17]	FPT by vi
CAPACITATED DOMINATING SET	W[1]-h by td + k [17]	FPT by vi
PRECOLORING EXTENSION	W[1]-h by td [23]	FPT by vi
EQUITABLE COLORING	W[1]-h by td [23]	FPT by vi
EQUITABLE CONNECTED PART.	W[1]-h by pw [22]	FPT by vi
BANDWIDTH	W[1]-h by td	FPT by vc [24]
	NP-h for pw = 2 [38]	P for pw $\leq$ 1 [4]
GRAPH MOTIF	NP-h for vi = 4	FPT by vc [14]
		P for vi $\leq$ 3
STEINER FOREST	NP-h for vi = 5 [28]	XP by vc
UNWEIGHTED STEINER FOREST	NP-h for tw = 3 [28]	FPT by vc
UNARY MIN MAX OUTDEG. ORI.	W[1]-h by vc	XP by tw [41]
BINARY MIN MAX OUTDEG. ORI.	NP-h for vc = 3	P for vc $\leq$ 2
METRIC DIMENSION	W[1]-h by pw [13]	FPT by tw + Δ [6]
		FPT by td
DIRECTED (p, q)-EDGE DOM. SET	W[1]-h by pw [7]	FPT by tw + p + q [7]
		FPT by td
LIST HAMILTONIAN PATH	W[1]-h by pw [35]	FPT by td

Filling Some Complexity Gaps. MIN MAX OUTDEGREE ORIENTATION gives an example that a known hardness for td can be strengthened to the one for vc (Sect. 4). In the full version, we show that GRAPH MOTIF and STEINER FOREST have different complexity with respect to vc and vi . This implies that not all FPT algorithms parameterized by vc can be generalized to the ones by vi. We also observe that some W[1]-hard problems parameterized by tw become tractable parameterized by td. Such problems include METRIC DIMENSION, DIRECTED (p, q)-EDGE DOMINATING SET, and LIST HAMILTONIAN PATH.

2   Imbalance

In this section, we show that IMBALANCE is fixed-parameter tractable parameterized by vi. Let $G = (V, E)$ be a graph. Given a linear ordering σ on V, the *imbalance* $\mathrm{im}_\sigma(v)$ of $v \in V$ is the absolute difference of the numbers of the neighbors of v that appear before v and after v in σ. The *imbalance* of G, denoted $\mathrm{im}(G)$, is defined as $\min_\sigma \sum_{v \in V} \mathrm{im}(v)$, where the minimum is taken over all linear orderings on V. Given a graph G and an integer b, IMBALANCE asks whether $\mathrm{im}(G) \leq b$.

Fellows et al. [24] showed that IMBALANCE is fixed-parameter tractable parameterized by vc. Recently, Misra and Mittal [36] have extended the result by showing that IMBALANCE is fixed-parameter tractable parameterized by the sum of the twin-cover number and the maximum twin-class size. Although twin-cover number is incomparable with vertex integrity, the combined parameter in [36] is always larger than or equal to the vertex integrity of the same graph. On the other hand, the combined parameter can be arbitrarily large for some graphs of constant vertex integrity (e.g., disjoint unions of P_3's). Hence, our result here properly extends the result in [36] as well.

Key Concepts. Before proceeding to the algorithm, we need to introduce two important concepts that are common in our algorithms parameterized by vi.

1. *ILP parameterized by the number of variables.* It is known that the feasibility of an instance of integer linear programming (ILP) parameterized by the number of variables is fixed-parameter tractable [32]. Using the algorithm for the feasibility problem as a black box, one can show the same fact for the optimization version as well. This fact has been used heavily for designing FPT algorithms parameterized by vc (see e.g. [24]). We are going to see that some of these algorithms can be generalized for the parameterization by vi, and IMBALANCE is the first such example.

2. *Equivalence relation among components.* For a vertex set S of G, we define an equivalence relation $\sim_{G,S}$ among components of $G - S$ by setting $C_1 \sim_{G,S} C_2$ if and only if there is an isomorphism g from $G[S \cup V(C_1)]$ to $G[S \cup V(C_2)]$ that fixes S; that is, $g|_S$ is the identity function. When $C_1 \sim_{G,S} C_2$, we say that C_1 and C_2 have the same (G, S)-*type* (or just the same *type* if G and S are clear from the context). See Fig. 1. We say that a component C of $G - S$ is of (G, S)-*type* t (or just *type* t) by using a canonical form t of the members of the (G, S)-type equivalence class of C. We can set the canonical form t in such a way that it can be computed from S and C in time depending only on $|S \cup V(C)|$.[1] Observe that if S is a $\mathrm{vi}(k)$-set of G, then the number of $\sim_{G,S}$ classes depends only on k since $|S \cup V(C)| \leq k$ for each component C of $G - S$. Hence, we can compute for all types t the number of type-t components of $G - S$ in $O(f(k) \cdot n)$ total running time, where $n = |V|$ and $f(k)$ is a computable function depending

---

[1] For example, by fixing the ordering of vertices in S as $v_1, \ldots, v_{|S|}$, we can set t to be the adjacency matrix of $G[S \cup V(C)]$ such that the ith row and column correspond to v_i for $1 \leq i \leq |S|$ and under this condition the string $t[1, 1], \ldots, t[1, s], t[2, 1], \ldots, t[s, s]$ is lexicographically minimal, where $s = |S \cup V(C)|$.

only on k. Note that this information (the numbers of type-t components for all t) completely characterizes the graph G up to isomorphism.

Theorem 2.1. IMBALANCE *is fixed-parameter tractable parameterized by* vi.

Proof. Let S be a $\mathrm{vi}(k)$-set of G. Such a set can be found in $O(k^{k+1}n)$ time [19]. We first guess and fix the relative ordering of S in an optimal ordering. There are only $k!$ candidates for this guess. For each $v \in S$, let $\ell(v)$ and $r(v)$ be the numbers of vertices in $N(v) \cap S$ that appear before v and after v, respectively, in the guessed relative ordering of S.

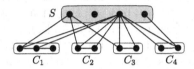

Fig. 1. The components C_2 and C_3 of $G - S$ have the same (G, S)-type.

Observe that the imbalance of a vertex v in a component C of $G-S$ depends only on the relative ordering of $S \cup V(C)$ since $N(v) \subseteq S \cup V(C)$. For each type t and for each relative ordering p of $S \cup V(C)$, where C is a type-t component of $G-S$, we denote by $\mathrm{im}(t,p)$ the sum of imbalance of the vertices in C. Similarly, the numbers of vertices in a type-t component C that appear before $v \in S$ and after v depend only on the relative ordering p of $S \cup V(C)$; we denote these numbers by $\ell(v,t,p)$ and $r(v,t,p)$, respectively. The numbers $\mathrm{im}(t,p)$, $\ell(v,t,p)$, and $r(v,t,p)$ can be computed from their arguments in time depending only on k, and thus they are treated as constants in the following ILP.

We represent by a nonnegative variable $x_{t,p}$ the number of type-t components that have relative ordering p with S. Note that the number of combinations of t and p depends only on k. For each $v \in S$, we represent (an upper bound of) the imbalance of v by an auxiliary variable y_v. This can be done by the following constraints:

$$y_v \geq (\ell(v) + \textstyle\sum_{t,p} \ell(v,t,p) \cdot x_{t,p}) - (r(v) + \textstyle\sum_{t,p} r(v,t,p) \cdot x_{t,p}),$$
$$y_v \geq (r(v) + \textstyle\sum_{t,p} r(v,t,p) \cdot x_{t,p}) - (\ell(v) + \textstyle\sum_{t,p} \ell(v,t,p) \cdot x_{t,p}).$$

Then the imbalance of the whole ordering, which is our objective function to minimize, can be expressed as

$$\textstyle\sum_{v \in S} y_v + \sum_{t,p} \mathrm{im}(t,p) \cdot x_{t,p}.$$

Now we need the following constraints to keep the total number of type-t components right:

$$\textstyle\sum_p x_{t,p} = c_t \quad \text{for each type } t,$$

where c_t is the number of components of type t in $G - S$.

By finding an optimal solution to the ILP above for each guess of the relative ordering of S, we can find an optimal ordering. Since the number of guesses and the number of variables depend only on k, the theorem follows.   □

3 Maximum Common (Induced) Subgraph

In this section, we show that MAXIMUM COMMON SUBGRAPH (MCS) and MAX-
IMUM COMMON INDUCED SUBGRAPH (MCIS) are fixed-parameter tractable
parameterized by vi of both graphs. (See the full version for the proof for MCIS.)
The results extend known results and fill some complexity gaps as described
below.

A graph Q is *subgraph-isomorphic* to G, denoted $Q \preceq G$, if there is an
injection η from $V(Q)$ to $V(G)$ such that $\{\eta(u), \eta(v)\} \in E(G)$ for every $\{u, v\} \in$
$E(Q)$. A graph Q is *induced subgraph-isomorphic* to G, denoted $Q \preceq_{\mathrm{I}} G$, if
there is an injection η from $V(Q)$ to $V(G)$ such that $\{\eta(u), \eta(v)\} \in E(G)$ if and
only if $\{u, v\} \in E(Q)$. Given two graphs G and Q, SUBGRAPH ISOMORPHISM
(SI) asks whether $Q \preceq G$, and INDUCED SUBGRAPH ISOMORPHISM (ISI) asks
whether $Q \preceq_{\mathrm{I}} G$. The results of this section are on their generalizations. Given
two graphs G_1 and G_2, MCS asks to find a graph H with maximum $|E(H)|$
such that $H \preceq G_1$ and $H \preceq G_2$. Similarly, MCIS asks to find a graph H with
maximum $|V(H)|$ such that $H \preceq_{\mathrm{I}} G_1$ and $H \preceq_{\mathrm{I}} G_2$.

If we restrict the structure of only one of the input graphs, then both problems
remain quite hard. Since PARTITION INTO TRIANGLES [27] is a special case of SI
where the graph Q is a disjoint union of triangles, MCS is NP-hard even if one of
the input graphs has vi = 3. Also, since INDEPENDENT SET [27] is a special case
of ISI where Q is an edge-less graph, MCIS is NP-hard even if one of the input
graphs has vc = 0. Furthermore, since SI and ISI generalize CLIQUE [18], MCS
and MCIS are W[1]-hard parameterized by the order of one of the input graphs.
When parameterized by vc of one graph, an XP algorithm for (a generalization
of) MCS is known [11].

For parameters restricting both input graphs, some partial results were
known. It is known that SI is fixed-parameter tractable parameterized by vi
of both graphs, while it is NP-complete when both graphs have td $\leq$ 3 [12].
The hardness proof in [12] can be easily adapted to ISI without increasing td.
It is known that MCIS is fixed-parameter tractable parameterized by vc of both
graphs [1].

Theorem 3.1. MAXIMUM COMMON SUBGRAPH *is fixed-parameter tractable
parameterized by* vi *of both input graphs.*

Proof. Let $G_1 = (V_1, E_1)$ and $G_2 = (V_2, E_2)$ be the input graphs of vertex
integrity at most k. We will find isomorphic subgraphs $\Gamma_1 = (U_1, F_1)$ of G_1
and $\Gamma_2 = (U_2, F_2)$ of G_2 with maximum number of edges, and an isomorphism
$\eta \colon U_1 \to U_2$ from Γ_1 to Γ_2.

Step 1. Guessing matched vi(2k)-*sets* R_1 *and* R_2. Let S_1 and S_2 be vi(k)-sets
of G_1 and G_2, respectively. At this point, there is no guarantee that $S_i \subseteq U_i$ or
$\eta(S_1) = S_2$. To have such assumptions, we make some guesses about η and find
vi(2k)-sets R_1 and R_2 of the graphs such that $\eta(R_1) = R_2$.

Step 1-1. Guessing subsets $X_i, Y_i \subseteq S_i$ *for* $i \in \{1, 2\}$. We guess disjoint subsets
X_1 and Y_1 of S_1 such that $X_1 = S_1 \cap \eta^{-1}(U_2 \cap S_2)$ and $Y_1 = S_1 \cap \eta^{-1}(U_2 \setminus S_2)$.

We also guess disjoint subsets X_2 and Y_2 of S_2 defined similarly as $X_2 = S_2 \cap \eta(U_1 \cap S_1)$ and $Y_2 = S_2 \cap \eta(U_1 \setminus S_1)$. Note that $\eta(X_1) = X_2$. There are $3^{|S_1|} \cdot 3^{|S_2|} \le 3^{2k}$ candidates for the combinations of X_1, Y_1, X_2, and Y_2.

Observe that the vertices in $S_i \setminus (X_i \cup Y_i)$ do not contribute to the isomorphic subgraphs and can be safely removed. We denote the resultant graphs by H_i.

Step 1-2. Guessing η on $X_1 \cup Y_1$ and η^{-1} on $X_2 \cup Y_2$. Given the guessed subsets X_1, Y_1, X_2, and Y_2, we further guess how η maps these subsets. There are $|X_1|! \le k!$ candidates for the bijection $\eta|_{X_1}$ (equivalently for $\eta^{-1}|_{X_2} = (\eta|_{X_1})^{-1}$).

Now we guess $\eta|_{Y_1}$ from at most 2^{k^3} non-isomorphic candidates as follows. Recall that $\eta(Y_1) \subseteq V_2 \setminus S_2$. Observe that each subset $A \subseteq V_2 \setminus S_2$ is completely characterized up to isomorphism by the numbers of ways A intersects type-t components for all (H_2, S_2)-types t. Since there are at most $2^{\binom{k}{2}}$ types and each component has order at most k, the total number of non-equivalent subsets of components is at most $2^{\binom{k}{2}} \cdot 2^k \le 2^{k^2}$. Since $\eta(Y_1)$ is the union of at most $|Y_1|$ such subsets, the number of non-isomorphic candidates of $\eta(Y_1)$ is at most $(2^{k^2})^{|Y_1|} \le 2^{k^3}$. In the analogous way, we can guess $\eta^{-1}|_{Y_2}$ from at most 2^{k^3} non-isomorphic candidates.

Now we set $Z_1 = \eta^{-1}(Y_2)$ and $Z_2 = \eta(Y_1)$. Let $R_1 = X_1 \cup Y_1 \cup Z_1$ and $R_2 = X_2 \cup Y_2 \cup Z_2$. Observe that each component C of $H_1 - R_1$ satisfies that $|C| \le k - |S_1| \le k$ and $|C| + |R_1| \le (k - |S_1|) + (|S_1| + |\eta^{-1}(Y_2)|) \le 2k$. Hence, R_1 is a vi($2k$)-set of H_1. Similarly, we can see that R_2 is a vi($2k$)-set of H_2. Furthermore, we know that $\eta(R_1) = R_2$.

Step 2. Extending the guessed parts of η. Assuming that the guesses we made so far are correct, we now find the entire η. Recall that we are seeking for isomorphic subgraphs $\Gamma_1 = (U_1, F_1)$ of G_1 and $\Gamma_2 = (U_2, F_2)$ of G_2 with maximum number of edges, and the isomorphism $\eta\colon U_1 \to U_2$ from Γ_1 to Γ_2. Since we already know the part $\eta|_{R_1}\colon R_1 \to R_2$, it suffices to find a bijective mapping from a subset of $V(H_1 - R_1)$ to a subset of $V(H_2 - R_1)$ that maximizes the number of matched edges where the connections to R_i are also taken into account.

As we describe below, the subproblem we consider here can be solved by formulating it as an ILP instance with $2^{O(k^3)}$ variables. The trick here is that instead of directly finding the mapping, we find which vertices and edges in $H_i - R_i$ are used in the common subgraph.

In the following, we are going to use a generalized version of *types* since the vertex set of a component of $H_i - R_i$ does not necessarily induce a connected subgraph of Γ_i. It is defined in a similar way as (H_i, R_i)-types except that it is defined for each pair (A, B) of a connected subgraph A of $H_i - R_i$ and a subset B of the edges between A and R_i. Let (A_1, B_1) and (A_2, B_2) be such pairs in $H_i - R_i$. We say that (A_1, B_1) and (A_2, B_2) have the same *g-(H_i, R_i)-type* (or just *g-type*) if there is an isomorphism from $H_i(A_1, B_1)$ to $H_i(A_2, B_2)$ that fixes R_i, where $H_i(A_j, B_j)$ is the subgraph of H_i formed by B_j and the edges in A_j. See Fig. 2. We say that a pair (A, B) is of *g-(H_i, R_i)-type* t (or just *g-type* t) by using a canonical form t of the g-(H_i, R_i)-type equivalence class of (A, B).

Observe that all possible canonical forms of g-types can be computed in time depending only on k.

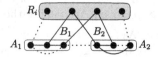

Fig. 2. The pairs (A_1, B_1) and (A_2, B_2) have the same g-(H_i, R_i)-type.

Step 2-1. Decomposing components of $H_i - R_i$ into smaller pieces. We say that an edge $\{u, v\}$ in H_1 is *used by* η if $u, v \in U_1$ and H_2 has the edge $\{\eta(u), \eta(v)\}$. Similarly, an edge $\{u, v\}$ in H_2 is *used by* η if $u, v \in U_2$ and H_1 has the edge $\{\eta^{-1}(u), \eta^{-1}(v)\}$.

Let $i \in \{1, 2\}$, t be an (H_i, R_i)-type, and T be a multiset of g-(H_i, R_i)-types. Let C be a type t component of $H_i - R_i$, C' the subgraph of C formed by the edges used by η, and E' the subset of the edges between C' and R_i used by η. If T coincides with the multiset of g-types of the pairs (A, B) such that A is a component of C' and B is the subset of E' connecting A and R_i, then we say that η *decomposes* the type-t component C into T.

We represent by a nonnegative variable $x_{t,T}^{(i)}$ the number of type-t components of $H_i - R_i$ that are decomposed into T by η. We have the following constraint:

$$\sum_T x_{t,T}^{(i)} = c_t^{(i)} \quad \text{for each } (H_i, R_i)\text{-type } t \text{ and } i \in \{1, 2\},$$

where the sum is taken over all possible multisets T of g-(H_i, R_i)-types, and $c_t^{(i)}$ is the number of components of type t in $H_i - R_i$. Additionally, if there is no way to decompose a type-t component into T, we add a constraint $x_{t,T}^{(i)} = 0$.

As each component of $H_i - R_i$ has order at most k, T contains at most k elements. Since there are at most $2^{\binom{2k}{2}}$ g-types, there are at most $(2^{\binom{2k}{2}})^k$ options for choosing T. Thus the number of variables $x_{t,T}^{(i)}$ is at most $2 \cdot 2^{\binom{2k}{2}} \cdot (2^{\binom{2k}{2}})^{k+1}$.

Now we introduce a nonnegative variable $y_t^{(i)}$ that represents the number of pairs (A, B) of g-type t obtained from the components of $H_i - R_i$ by decomposing them by η. The definition of $y_t^{(i)}$ gives the following constraint:

$$y_t^{(i)} = \sum_{t', T} \mu(T, t) \cdot x_{t', T}^{(i)} \quad \text{for each g-}(H_i, R_i)\text{-type } t \text{ and } i \in \{1, 2\},$$

where $\mu(T, t)$ is the multiplicity of g-type t in T and the sum is taken over all possible (H_i, R_i)-types t' and multisets T of g-(H_i, R_i)-types. As in the previous case, we can see that the number of variables y_t depends only on k.

Step 2-2. Matching decomposed pieces. Observe that for each g-(H_1, R_1)-type t_1, there exists a unique g-(H_2, R_2)-type t_2 such that there is an isomorphism g from $H_1(A_1, B_1)$ to $H_2(A_2, B_2)$ with $g|_{R_1} = \eta|_{R_1}$, where (A_i, B_i) is a pair of g-(H_i, R_i)-type t_i for $i \in \{1, 2\}$. We say that such g-types t_1 and t_2 *match*. Since η is an isomorphism from Γ_1 to Γ_2, η maps each g-(H_1, R_1)-type t_1 pair to a g-(H_2, R_2)-type t_2 pair, where t_1 and t_2 match. This implies that $y_{t_1}^{(1)} = y_{t_2}^{(2)}$, which we add as a constraint. Now the total number of edges used by η can be computed from $y_t^{(1)}$. Let m_t be the number of edges in $H_1(A, B)$, where (A, B) is a pair of g-(H_1, R_1)-type t. Let r be the number of matched edges in R_1; that is, $r = |\{\{u, v\} \in E(H_1[R_1]) \mid \{\eta(u), \eta(v)\} \in E(G_2[R_2])\}|$. Then, the number of matched edges is $r + \sum_t m_t \cdot y_t^{(1)}$. On the other hand, given an assignment to the variables, it is easy to find isomorphic subgraphs with that many edges. Since r is a constant here, we set $\sum_t m_t \cdot y_t^{(1)}$ to the objective function to be maximized.

Since the number of candidates in the guesses we made and the number of variables in the ILP instances depend only on k, the theorem follows.   □

4 Min Max Outdegree Orientation

Given an undirected graph $G = (V, E)$, an edge weight function $w\colon E \to \mathbb{Z}^+$, and a positive integer r, MIN MAX OUTDEGREE ORIENTATION (MMOO) asks whether there exists an orientation Λ of G such that each vertex has outdegree at most r under Λ, where the outdegree of a vertex is the sum of the weights of outgoing edges. If each edge weight is given in binary, we call the problem BINARY MMOO, and if it is given in unary, we call the problem UNARY MMOO. Note that in the binary version, the weight of an edge can be exponential in the input size, whereas the unary version does not allow such weights.

UNARY MMOO admits an $n^{O(\mathsf{tw})}$-time algorithm [41], but it is W[1]-hard parameterized by td [40].[2] In this section, we show a stronger hardness parameterized by vc. BINARY MMOO is known to be NP-complete for graphs of vi $= 4$ [3]. In the full version, we show a stronger hardness result that the binary version is NP-complete for graphs of vc $= 3$. This result is tight as we can show that the binary version is polynomial-time solvable for graphs of vc ≤ 2.

Theorem 4.1. UNARY MMOO *is W[1]-hard parameterized by* vc.

Proof. We give a parameterized reduction from UNARY BIN PACKING. Given a positive integer t and n positive integers $a_1, a_2, \ldots, a_n$ in unary, UNARY BIN PACKING asks the existence of a partition $S_1, \ldots, S_t$ of $\{1, 2, \ldots, n\}$ such that $\sum_{i \in S_j} a_i = \frac{1}{t} \sum_{1 \leq i \leq n} a_i$ for $1 \leq j \leq t$. UNARY BIN PACKING is W[1]-hard parameterized by t [30].

We assume that $t \geq 3$ since otherwise the problem can be solved in polynomial time as the integers a_i are given in unary. Let $B = \frac{1}{t} \sum_{1 \leq i \leq n} a_i$ and $W = (t-1)B = \sum_{1 \leq i \leq n} a_i - B$. The assumption $t \geq 3$ implies that $B \leq W/2$. Observe that if $a_i \geq B$ for some i, then the instance is a trivial no instance (when $a_i > B$) or the element a_i is irrelevant (when $a_i = B$). Hence, we assume that $a_i < B$ (and thus $a_i < W/2$) for every i.

---

[2] In [40], W[1]-hardness was stated for tw but the proof shows it for td as well.

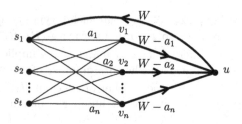

Fig. 3. Reduction from UNARY BIN PACKING to UNARY MMOO.

The reduction to UNARY MMOO is depicted in Fig. 3. From the integers $a_1, a_2, \ldots, a_n$, we construct the graph obtained from a complete bipartite graph on the vertex set $\{u, s_1, s_2, \ldots, s_t\} \cup \{v_1, \ldots, v_n\}$ by adding the edge $\{u, s_1\}$. We set $w(\{v_i, s_j\}) = a_i$ for all i, j, $w(\{v_i, u\}) = W - a_i$ for all i, and $w(\{u, s_1\}) = W$. The vertices $s_1, s_2, \ldots, s_t, u$ form a vertex cover of size $t+1$. We set the target maximum outdegree r to W. We show that this instance of UNARY MMOO is a yes instance if and only if there exists a partition $S_1, \ldots, S_t$ of $\{1, 2, \ldots, n\}$ such that $\sum_{i \in S_j} a_i = B$ for all j. Intuitively, we can translate the solutions of the problems by picking a_i into S_j if $\{v_i, s_j\}$ is oriented from v_i to s_j, and vice versa.

Assume that there exists a partition $S_1, \ldots, S_t$ of $\{1, 2, \ldots, n\}$ such that $\sum_{i \in S_j} a_i = B$ for all j. We first orient the edge $\{u, s_1\}$ from u to s_1 and each edge $\{v_i, u\}$ from v_i to u. (See the thick edges in Fig. 3.) Then, we orient $\{v_i, s_j\}$ from v_i to s_j if and only if $i \in S_j$. Under this orientation, all vertices have outdegree exactly W: $a_i + (W - a_i)$ for each v_i and $\sum_{i \notin S_j} a_i = \sum_{1 \leq i \leq n} a_i - B$ for each s_j.

Conversely, assume that there is an orientation such that each vertex has outdegree at most W. Since the sum of the edge weights is $(n + t + 1)W$ and the graph has $n + t + 1$ vertices, the outdegree of each vertex has to be exactly W. Since $a_i < W/2$ for all i, each edge $\{v_i, u\}$ has weight larger than $W/2$. Hence, for u, the only way to obtain outdegree exactly W is to orient $\{u, s_1\}$ from u to s_1 and $\{v_i, u\}$ from v_i to u for all i. Furthermore, for each i, there exists exactly one vertex s_j such that $\{v_i, s_j\}$ is oriented from v_i to s_j. Let $S_j \subseteq \{1, 2, \ldots, n\}$ be the set of indices i such that $\{v_i, s_j\}$ is oriented from v_i to s_j. The discussion above implies that $S_1, \ldots, S_t$ is a partition of $\{1, \ldots, n\}$. The outdegree of s_j is $\sum_{i \notin S_j} a_i$, which is equal to $W = \sum_{1 \leq i \leq n} a_i - B$. Thus, $\sum_{i \in S_j} a_i = \sum_{1 \leq i \leq n} a_i - W = B$. $\square$

5   Bandwidth

Let $G = (V, E)$ be a graph. Given a linear ordering σ on V, the *stretch* of $\{u, v\} \in E$, denoted $\mathsf{str}_\sigma(\{u, v\})$, is $|\sigma(u) - \sigma(v)|$. The *bandwidth* of G, denoted $\mathsf{bw}(G)$, is defined as $\min_\sigma \max_{e \in E} \mathsf{str}_\sigma(e)$, where the minimum is taken over all linear orderings on V. Given a graph G and an integer w, BANDWIDTH asks whether $\mathsf{bw}(G) \leq w$. BANDWIDTH is NP-complete on trees of $\mathsf{pw} = 3$ [37] and on graphs of $\mathsf{pw} = 2$ [38]. Fellows et al. [24] presented an FPT algorithm for BANDWIDTH parameterized by vc. Here we show that BANDWIDTH is W[1]-hard parameterized by td on trees. The proof is inspired by the one by Muradian [38].

Theorem 5.1. BANDWIDTH *is W[1]-hard parameterized by* td *on trees.*

Proof. Let $(a_1, \ldots, a_n; t)$ be an instance of UNARY BIN PACKING with $t \geq 2$. Let $B = \frac{1}{t} \sum_{1 \leq i \leq n} a_i$ be the target weight. We construct an equivalent instance $(T = (V, E), w)$ of BANDWIDTH as follows (see Fig. 4). We start with a path $(z_0, x_1, y_1, z_1, \ldots, x_t, y_t, z_t)$ of length $3t$. For $1 \leq i \leq t - 1$, we attach $12tnB$ leaves to z_i. To z_0 and z_t, we attach $12tnB + 4n + 1$ leaves. For $1 \leq i \leq n$, we take a star with $6tn \cdot a_i - 1$ leaves centered at v_i. Finally, we connect each v_i to x_1 with a path with $6t - 4$ inner vertices. We set the target width w to $6tnB + 2n + 1$. Note that $|V| = (3t + 2)w + 1$.

We can see an upper bound of $\mathsf{td}(T)$ as follows. We remove x_1 and all the leaves from T. This decreases treedepth by at most 2. The remaining graph is a disjoint union of paths and a longest path has order $6t - 3$. Since $\mathsf{td}(P_n) = \lceil \log_2(n + 1) \rceil$ [39], we have $\mathsf{td}(T) \leq 2 + \lceil \log_2(6t - 2) \rceil \leq \log_2 t + 6$.

Now we show that (T, w) is a yes instance of of BANDWIDTH if and only if $(a_1, \ldots, a_n; t)$ is a yes instance of UNARY BIN PACKING.

($\Longrightarrow$) First assume that $\mathsf{bw}(T) \leq w$ and that σ is a linear ordering on V such that $\max_{e \in E} \mathsf{str}_\sigma(e) \leq w$. Since $\deg(z_0) = 12tnB + 4n + 2 = 2w$, its closed neighborhood $N[z_0]$ has to appear in σ consecutively, where z_0 appears at the middle of this subordering. Furthermore, no edge can connect a vertex appearing before z_0 in σ and a vertex appearing after z_0 as such an edge has stretch larger than w. Since the edges not incident to z_0 form a connected subgraph, we can conclude that the vertices in $V - N[z_0]$ appear either all before $N[z_0]$ or all after $N[z_0]$ in σ. By symmetry, we can assume that those vertices appear after $N[z_0]$ in σ. This implies that $\sigma(z_0) = w + 1$. By the same argument, we can show that all vertices in $N[z_t]$ appear consecutively in the end of σ and $\sigma(z_t) = |V| - w = (3t + 1)w + 1$. Since $\sigma(z_t) - \sigma(z_0) = 3tw$ and the path $(z_0, x_1, y_1, z_1, \ldots, x_t, y_t, z_t)$ has length $3t$, each edge in this path has stretch exactly w in σ. Namely, $\sigma(x_i) = (3i - 1)w + 1$, $\sigma(y_i) = 3iw + 1$, and $\sigma(z_i) = (3i + 1)w + 1$.

For each leaf ℓ attached to z_i $(1 \leq i \leq t - 1)$, $\sigma(y_i) < \sigma(\ell) < \sigma(x_{i+1})$ holds. Other than these leaves, there are $2(w-1) - 12tnB = 4n$ vertices placed between y_i and x_{i+1}. Let V_i be the set consisting of v_i and the leaves attached to it. For $j \in \{1, \ldots, t\}$, let I_j be the set of indices i such that v_i is put between z_{j-1} and z_j. If $i \in I_j$, then all $6tn \cdot a_i$ vertices in V_i are put between y_{j-1} and x_{j+1}. (We set $y_0 := z_0$.)

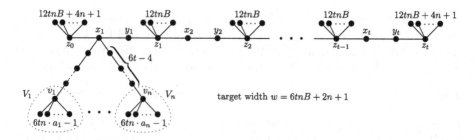

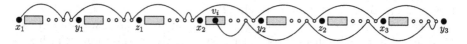

Fig. 4. Reductions from UNARY BIN PACKING to BANDWIDTH.

Fig. 5. Embedding the path from x_1 to v_i. The gray boxes are the occupied position and the white points are the vacant positions. ($n = 2$, $j = 2$, $t = 3$.)

If $\sum_{i \in I_j} a_i \geq B + 1$, then $|\bigcup_{i \in I_j} V_i| \geq 6tn(B + 1) > w + 8n - 1$ as $t \geq 2$. This number of vertices cannot be put between y_{j-1} and x_{j+1} after putting the leaves attached to z_{j-1} and z_j: we can put at most $4n$ vertices between y_{j-1} and x_j, at most $4n$ vertices between y_j and x_{j+1}, and at most $w - 1$ vertices between x_j and y_j. Since $I_1, \ldots, I_t$ form a partition of $\{1, \ldots, n\}$ and $\sum_{1 \leq i \leq n} a_i = tB$, we can conclude that $\sum_{i \in I_j} a_i = B$ for $1 \leq j \leq t$.

($\Longleftarrow$) Next assume that there exists a partition $S_1, \ldots, S_t$ of $\{1, 2, \ldots, n\}$ such that $\sum_{i \in S_j} a_i = B$ for all $1 \leq j \leq t$.

We put $N[z_0]$ at the beginning of σ and $N[z_t]$ at the end. We set $\sigma(x_i) = (3i - 1)w + 1$, $\sigma(y_i) = 3iw + 1$, and $\sigma(z_i) = (3i + 1)w + 1$. For $1 \leq i \leq t - 1$, we put the leaves attached to z_i so that a half of them have the first $6tnB$ positions between y_i and z_i and the other half have the first $6tnB$ positions between z_i and x_{i+1}. For each S_j, we put the vertices in $\bigcup_{i \in S_j} V_i$ so that they take the first $6tnB$ positions between x_j and y_j.

Now we have $2n$ vacant positions at the end of each interval between x_i and y_i for $1 \leq i \leq t$, between y_i and z_i for $1 \leq i \leq t - 1$, and between z_i and x_{i+1} for $1 \leq i \leq t - 1$. To these positions, we need to put the inner vertices of the paths connecting x_1 and $v_1, \ldots, v_n$. Let P_i be the inner part of x_1–v_i path. The path P_i uses the $(2i - 1)$st and $(2i)$th vacant positions in each interval as follows (see Fig. 5).

Let $i \in S_j$. Starting from x_1, P_i proceeds from left to right and visits the two positions in each interval consecutively until it arrives the interval between x_j and y_j. At the interval between x_j and y_j, P_i switches to the phase where it only visits the $(2i)$th vacant position in each interval and still proceeds from left to right until it reaches the interval between x_t and y_t. Then P_i changes the direction and switches to the phase where it visits the $(2i - 1)$st vacant position only in each interval until it reaches the interval between x_j and y_j.

Now all the vertices are put at distinct positions and it is easy to see that no edge has stretch more than w. This completes the proof.   □

6   Conclusion

Using vertex integrity as a structural graph parameter, we presented finer analyses of the parameterized complexity of well-studied problems. Although we needed a case-by-case analysis depending on individual problems, the results in this paper would be useful for obtaining a general method to deal with vertex integrity.

We succeeded to extend many fixed-parameter algorithms parameterized by vc to the ones parameterized by vi, but we were not so successful on graph layout problems. Fellows et al. [24] showed that IMBALANCE, BANDWIDTH, CUTWIDTH, and DISTORTION are fixed-parameter tractable parameterized by vc. Lokshtanov [33] showed that OPTIMAL LINEAR ARRANGEMENT is fixed-parameter tractable parameterized by vc. Are these problems fixed-parameter tractable parameterized by vi? We answered only for IMBALANCE in this paper.

References

1. Abu-Khzam, F.N.: Maximum common induced subgraph parameterized by vertex cover. Inf. Process. Lett. **114**(3), 99–103 (2014). https://doi.org/10.1016/j.ipl.2013.11.007
2. Arnborg, S., Lagergren, J., Seese, D.: Easy problems for tree-decomposable graphs. J. Algorithms **12**(2), 308–340 (1991). https://doi.org/10.1016/0196-6774(91)90006-K
3. Asahiro, Y., Miyano, E., Ono, H.: Graph classes and the complexity of the graph orientation minimizing the maximum weighted outdegree. Discret. Appl. Math. **159**(7), 498–508 (2011). https://doi.org/10.1016/j.dam.2010.11.003
4. Assmann, S.F., Peck, G.W., Sysło, M.M., Zak, J.: The bandwidth of caterpillars with hairs of length 1 and 2. SIAM J. Algebraic Discrete Methods **2**(4), 387–393 (1981). https://doi.org/10.1137/0602041
5. Barefoot, C.A., Entringer, R.C., Swart, H.C.: Vulnerability in graphs – a comparative survey. J. Combin. Math. Combin. Comput. **1**, 13–22 (1987)
6. Belmonte, R., Fomin, F.V., Golovach, P.A., Ramanujan, M.S.: Metric dimension of bounded tree-length graphs. SIAM J. Discret. Math. **31**(2), 1217–1243 (2017). https://doi.org/10.1137/16M1057383
7. Rémy, B., Hanaka, T., Katsikarelis, I., Kim, E.J., Lampis, M.: New results on directed edge dominating set. In: MFCS 2018, volume 117 of LIPIcs, pp. 67:1–67:16 (2018). https://doi.org/10.4230/LIPIcs.MFCS.2018.67
8. Rémy, B., Kim, E.J., Lampis, M., Mitsou, V., Otachi, Y.: Grundy distinguishes treewidth from pathwidth. In: ESA 2020, volume 173 of LIPIcs, pp. 14:1–14:19 (2020). https://doi.org/10.4230/LIPIcs.ESA.2020.14
9. Biedl, T.C., Chan, T.M., Ganjali, Y., Hajiaghayi, M.T., Wood, D.R.: Balanced vertex-orderings of graphs. Discret. Appl. Math. **148**(1), 27–48 (2005). https://doi.org/10.1016/j.dam.2004.12.001

10. Bodlaender, H.L.: Dynamic programming on graphs with bounded treewidth. In: Lepistö, T., Salomaa, A. (eds.) ICALP 1988. LNCS, vol. 317, pp. 105–118. Springer, Heidelberg (1988). https://doi.org/10.1007/3-540-19488-6_110

11. Bodlaender, H.L., Hanaka, T., Jaffke, L., Ono, H., Otachi, Y., van der Zanden, T.C.: Hedonic seat arrangement problems (extended abstract). In: AAMAS 2020, pp. 1777–1779 (2020) https://doi.org/10.5555/3398761.3398979

12. Bodlaender, H.L., Hanaka, T., Okamoto, Y., Otachi, Y., van der Zanden, T.C.: SubGraph isomorphism on graph classes that exclude a substructure. In: Heggernes, P. (ed.) CIAC 2019. LNCS, vol. 11485, pp. 87–98. Springer, Cham (2019). https://doi.org/10.1007/978-3-030-17402-6_8

13. Bonnet, É., Purohit, N.: Metric dimension parameterized by treewidth. In: IPEC 2019, volume 148 of LIPIcs, pp. 5:1–5:15 (2019). https://doi.org/10.4230/LIPIcs.IPEC.2019.5

14. Bonnet, É., Sikora, F.: The graph motif problem parameterized by the structure of the input graph. Discret. Appl. Math. **231**, 78–94 (2017). https://doi.org/10.1016/j.dam.2016.11.016

15. Courcelle, B.: The monadic second-order logic of graphs III: tree-decompositions, minor and complexity issues. RAIRO Theor. Inform. Appl. **26**, 257–286 (1992). https://doi.org/10.1051/ita/1992260302571

16. Cygan, M., et al.: Parameterized Algorithms. Springer, Cham (2015). https://doi.org/10.1007/978-3-319-21275-3

17. Dom, M., Lokshtanov, D., Saurabh, S., Villanger, Y.: Capacitated domination and covering: a parameterized perspective. In: Grohe, M., Niedermeier, R. (eds.) IWPEC 2008. LNCS, vol. 5018, pp. 78–90. Springer, Heidelberg (2008). https://doi.org/10.1007/978-3-540-79723-4_9

18. Downey, R.G., Fellows, M.R.: Parameterized Complexity. Springer, Cham (1999). https://doi.org/10.1007/978-1-4612-0515-9

19. Drange, P.G., Dregi, M.S., van 't Hof, P.: On the computational complexity of vertex integrity and component order connectivity. Algorithmica, **76**(4), 1181–1202 (2016). https://doi.org/10.1007/s00453-016-0127-x

20. Dvořák, P., Eiben, E., Ganian, R., Knop, D., Ordyniak, S.: Solving integer linear programs with a small number of global variables and constraints. In: IJCAI 2017, pp. 607–613 (2017). https://doi.org/10.24963/ijcai.2017/85

21. Dvořák, P., Knop, D.: Parameterized complexity of length-bounded cuts and multicuts. Algorithmica **80**(12), 3597–3617 (2018). https://doi.org/10.1007/s00453-018-0408-7

22. Enciso, R., Fellows, M.R., Guo, J., Kanj, I., Rosamond, F., Suchý, O.: What makes equitable connected partition easy. In: Chen, J., Fomin, F.V. (eds.) IWPEC 2009. LNCS, vol. 5917, pp. 122–133. Springer, Heidelberg (2009). https://doi.org/10.1007/978-3-642-11269-0_10

23. Fellows, M.R., et al.: On the complexity of some colorful problems parameterized by treewidth. Inf. Comput. **209**(2), 143–153 (2011). https://doi.org/10.1016/j.ic.2010.11.026

24. Fellows, M.R., Lokshtanov, D., Misra, N., Rosamond, F.A., Saurabh, S.: Graph layout problems parameterized by vertex cover. In: Hong, S.-H., Nagamochi, H., Fukunaga, T. (eds.) ISAAC 2008. LNCS, vol. 5369, pp. 294–305. Springer, Heidelberg (2008). https://doi.org/10.1007/978-3-540-92182-0_28

25. Fiala, J., Golovach, P.A., Kratochvíl, J.: Parameterized complexity of coloring problems: treewidth versus vertex cover. Theor. Comput. Sci. **412**(23), 2513–2523 (2011). https://doi.org/10.1016/j.tcs.2010.10.043

26. Ganian, R., Klute, F., Ordyniak, S.: On structural parameterizations of the bounded-degree vertex deletion problem. Algorithmica (2020). https://doi.org/10.1007/s00453-020-00758-8

27. Garey, M.R., Johnson, D.S.: Computers and Intractability: A Guide to the Theory of NP-Completeness. Freeman, W. H (1979)

28. Gassner, E.: The steiner forest problem revisited. J. Discrete Algorithms 8(2), 154–163 (2010). https://doi.org/10.1016/j.jda.2009.05.002

29. Gima, T., Hanaka, T., Kiyomi, M., Kobayashi, Y., Otachi, Y.: Exploring the gap between treedepth and vertex cover through vertex integrity. CoRR, abs/2101.09414, arXiv preprint arXiv:2101.09414 (2021)

30. Jansen, K., Kratsch, S., Marx, D., Schlotter, I.: Bin packing with fixed number of bins revisited. J. Comput. Syst. Sci. 79(1), 39–49 (2013). https://doi.org/10.1016/j.jcss.2012.04.004

31. Kellerhals, L., Koana, T.: Parameterized complexity of geodetic set. In: IPEC 2020, volume 180 of LIPIcs, pp. 20:1–20:14 (2020). https://doi.org/10.4230/LIPIcs.IPEC.2020.20

32. Lenstra Jr, H.W.: Integer programming with a fixed number of variables. Math. Oper. Res. 8, 538–548 (1983). https://doi.org/10.1287/moor.8.4.538

33. Lokshtanov, D.: Parameterized integer quadratic programming: variables and coefficients. CoRR, abs/1511.00310 (2015). arXiv preprint arXiv:1511.00310

34. Lokshtanov, D., Misra, N., Saurabh, S.: Imbalance is fixed parameter tractable. Inf. Process. Lett. 113(19–21), 714–718 (2013). https://doi.org/10.1016/j.ipl.2013.06.010

35. Meeks, K., Alexander, S.: The parameterised complexity of list problems on graphs of bounded treewidth. Inf. Comput. 251, 91–103 (2016). https://doi.org/10.1016/j.ic.2016.08.001

36. Misra, N., Mittal, H.: Imbalance parameterized by twin cover revisited. In: Kim, D., Uma, R.N., Cai, Z., Lee, D.H. (eds.) COCOON 2020. LNCS, vol. 12273, pp. 162–173. Springer, Cham (2020). https://doi.org/10.1007/978-3-030-58150-3_13

37. Monien, B.: The bandwidth minimization problem for caterpillars with hair length 3 is NP-complete. SIAM J. Algebraic Discrete Methods 7(4), 505–512 (1986). https://doi.org/10.1137/0607057

38. Muradian, D.: The bandwidth minimization problem for cyclic caterpillars with hair length 1 is NP-complete. Theor. Comput. Sci. 307(3), 567–572 (2003). https://doi.org/10.1016/S0304-3975(03)00238-X

39. Nešetřil, J., de Mendez, P.O.: Sparsity: Graphs, Structures, and Algorithms. Algorithms and Combinatorics. Springer, Cham (2012). https://doi.org/10.1007/978-3-642-27875-4

40. Szeider, S.: Not so easy problems for tree decomposable graphs. Ramanujan Mathematical Society, Lecture Notes Series, No. 13, pp. 179–190 (2010) arXiv preprint arXiv:1107.1177

41. Szeider, S.: Monadic second order logic on graphs with local cardinality constraints. ACM Trans. Comput. Log. 12(2), 12:1–12:21 (2011). https://doi.org/10.1145/1877714.1877718

Covering a Set of Line Segments
with a Few Squares

Joachim Gudmundsson[1], Mees van de Kerkhof[2], André van Renssen[1],
Frank Staals[2], Lionov Wiratma[3], and Sampson Wong[1(✉)]

[1] University of Sydney, Sydney, Australia
{joachim.gudmundsson,andre.vanrenssen}@sydney.edu.au,
swon7907@uni.sydney.edu.au
[2] Utrecht University, Utrecht, Netherlands
{m.a.vandekerkhof,f.staals}@uu.nl
[3] Parahyangan Catholic University, Bandung, Indonesia
lionov@unpar.ac.id

Abstract. We study three covering problems in the plane. Our original
motivation for these problems come from trajectory analysis. The first is
to decide whether a given set of line segments can be covered by up to four
unit-sized, axis-parallel squares. The second is to build a data structure
on a trajectory to efficiently answer whether any query subtrajectory
is coverable by up to three unit-sized axis-parallel squares. The third
problem is to compute a longest subtrajectory of a given trajectory that
can be covered by up to two unit-sized axis-parallel squares.

Keywords: Computational geometry · Geometric coverings ·
Trajectory analysis

1 Introduction

Geometric covering problems are a classic area of research in computational
geometry. The traditional *geometric set cover problem* is to decide whether one
can place k axis-parallel unit-sized squares (or disks) to cover n given points in
the plane. If k is part of the input, the problem is known to be NP-hard [5,11].
Thus, efficient algorithms are known only for small values of k. For $k = 2$ or 3,
there are linear time algorithms [4,17], and for $k = 4$ or 5, there are $O(n \log n)$
time algorithms [12,15]. For general k, the $O(n^{\sqrt{k}})$ time algorithm for unit-sized
disks [10] most likely generalises to unit-sized axis-parallel squares [1].

Motivated by trajectory analysis, we study a line segment variant of the
geometric set cover problem where the input is a set of n line segments. Given
a set of line segments, we say it is *k-coverable* if there exist k unit-sized axis-
parallel squares in the plane so that every line segment is in the union of the k
squares (we may write coverable to mean k-coverable when k is clear from the
context). The first problem we study in this paper is:

The full version of this paper can be found at [7].

© Springer Nature Switzerland AG 2021
T. Calamoneri and F. Corò (Eds.): CIAC 2021, LNCS 12701, pp. 286–299, 2021.
https://doi.org/10.1007/978-3-030-75242-2_20

Problem 1. Decide if a set of line segments is k-coverable, for $k \in O(1)$.

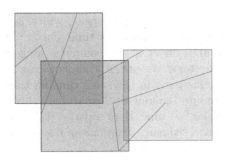

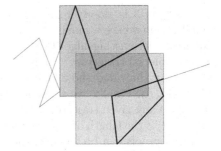

Fig. 1. A set of 3-coverable segments.   **Fig. 2.** A 2-coverable subtrajectory.

A key difference in the line segment variant and the point variant is that each segment need not be covered by a single square, as long as each segment is covered by the union of the k squares. See Fig. 1.

Hoffmann [9] provides a linear time algorithm for $k = 2$ and 3, however, a proof was not included in his extended abstract. Sadhu et al. [14] provide a linear time algorithm for $k = 2$ using constant space. In Sect. 2, we provide a proof for a $k = 3$ algorithm and a new $O(n \log n)$ time algorithm for $k = 4$.

Next, we study trajectory coverings. A trajectory T is a polygonal curve in the plane parametrised by time. A subtrajectory $T[s, t]$ is the trajectory T restricted to a contiguous time interval $[s, t] \subseteq [0, 1]$, see Fig. 2 for an example. Trajectories are commonly used to model the movement of an object (e.g. a bird, a vehicle, etc.) through time and space. The analysis of trajectories have applications in animal ecology [3], meteorology [18], and sports analytics [6].

To the best of our knowledge, this paper is the first to study k-coverable trajectories for $k \geq 2$. A k-coverable trajectory may, for example, model a commonly travelled route, and the squares could model a method of displaying the route (i.e. over multiple pages, or multiple screens), or alternatively, the location of several facilities. We build a data structure that can efficiently decide whether a subtrajectory is k-coverable.

Problem 2. Construct a data structure on a trajectory, so that given any query subtrajectory, it can efficiently answer whether the subtrajectory is k-coverable, for $k \in O(1)$.

For $k = 2$ and $k = 3$ we preprocess a trajectory T with n vertices in $O(n \log n)$ time, and store it in a data structure of size $O(n \log n)$, so that we can test if an arbitrary subtrajectory (not necessarily restricted to vertices) $T[s, t]$ can be k-covered.

Finally, we consider a natural extension of Problem 2, that is, to calculate the *longest* k-coverable subtrajectory of any given trajectory. This problem is similar in spirit to the problem of covering the maximum number of points by k unit-sized axis-parallel squares [2,13].

Problem 3. Given a trajectory, compute its longest k-coverable subtrajectory, for $k \in O(1)$.

Problem 3 is closely related to computing a trajectory *hotspot*, which is a small region where a moving object spends a large amount of time. For $k = 1$ squares, the existing algorithm by Gudmundsson et al. [8] computes longest 1-coverable subtrajectory of any given trajectory. We notice a missing case in their algorithm, and show how to resolve this issue in the same running time of $O(n \log n)$. Finally, we show how to compute the longest 2-coverable subtrajectory of any given trajectory in $O(n\beta_4(n) \log^2 n)$ time. Here, $\beta_4(n) = \lambda_4(n)/n$, and $\lambda_s(n)$ is the length of a Davenport-Schinzel sequence of order s on n symbols. Omitted proofs can be found in the full version [7].

2   Problem 1: The Decision Problem

2.1   Is a Set of Line Segments 2-Coverable?

This section restates known results that will be useful for the recursive step in Sect. 2.2 and for the data structure in Sect. 3.1. The first result relates the bounding box, which is the smallest axis-aligned rectangle that contains all the segments, to a covering, which is a set of squares that covers all line segments.

Observation 1. *Every covering must touch all four sides of the bounding box.*

The reasoning behind Observation 1 is simple: if the covering does not touch one of the four sides, say the left side, then the covering could not have covered the leftmost vertex of the set of segments. An intuitive way for two squares to satisfy Observation 1 is to place the two squares in opposite corners of the bounding box. This intuition is formalised in Observation 1.

Lemma 1 (Sadhu et al. [14]). *A set of segments is 2-coverable if and only if there is a covering with squares in opposite corners of the bounding box of the set of segments.*

Lemma 1 is useful in that it narrows down our search for a 2-covering. It suffices to check the two configurations where squares are in opposite corners of the bounding box. For each of these two configurations, we simply check if each segment is in the union of the two squares, which takes linear time in total, leading to the following theorem:

Theorem 1. *One can decide if a set of n segments is 2-coverable in $O(n)$ time.*

2.2   Is a Set of Line Segments 3-Coverable?

We notice that for a covering consisting of three squares, Lemma 1 and the pigeon-hole principle imply that there must be one square that touches at least two sides of the bounding box. An intuitive way to achieve this is if one of the squares in the 3-covering is in a corner of the bounding box. We formalise this intuition in Lemma 2.

Lemma 2. *A set of segments is 3-coverable if and only if there is a covering with a square in a corner of the bounding box of the set of segments.*

Again, this lemma allows us to narrow down our search for a 3-covering. We consider four cases, one for each corner of the bounding box. After placing the first square in one of the four corners, we would like to check whether two additional squares can be placed to cover the remaining segments. We start by computing the remaining segments that are not yet covered. We subdivide each segment into at most one subsegment that is covered by the corner square, and up to two subsegments that are not yet covered. Then we can use Theorem 1 to (recursively) check whether two additional squares can be placed to cover all the uncovered subsegments.

The running time for subdividing each segment takes linear time in total. There are at most a linear number of remaining segments. Checking if the remaining segments are 2-coverable takes linear time by Theorem 1. Hence, we have the following theorem:

Theorem 2. *One can decide if a set of n segments is 3-coverable in $O(n)$ time.*

2.3   Is a Set of Line Segments 4-Coverable?

For a 4-covering, it remains true that any covering must touch all four sides of the bounding box. Unlike the three squares case, we cannot use the pigeon-hole principle to deduce that there is a square touching at least two sides of the bounding box. Fortunately, we have only two cases: either there exists a square which touches at least two sides of the bounding box, or each square touches exactly one side of the bounding box. This implies:

Lemma 3. *A set of segments is 4-coverable if and only if: (i) there is a covering with a square in a corner of the bounding box, or (ii) there is a covering with each square touching exactly one side of the bounding box.*

In the first case we can use the same strategy as in the three squares case by placing the first square in a corner and then (recursively) checking if three additional squares can cover the remaining subsegments. This gives a linear time algorithm for the first case.

For the remainder of this section, we focus on solving the second case.

Definition 1. *Define L, B, T and R to be the square that touches the left, bottom, top and right sides of the bounding box respectively. See Fig. 3.*

Without loss of generality, suppose that T is to the left of B. This implies that the left to right order of the squares is L, T, B, R. Suppose for now there were a way to compute the initial placement of L. Then we can deduce the position of T in the following way.

Lemma 4. *Given the position of L, if three additional squares can be placed to cover the remaining subsegments, then it can be done with T in the top-left corner of the bounding box of the remaining subsegments.*

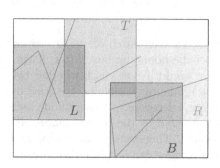

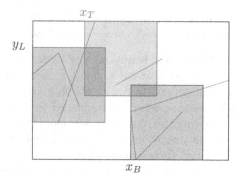

Fig. 3. The squares L, T, B and R.   **Fig. 4.** The variables y_L, x_T and x_B.

The intuition behind this lemma is that after placing the first square, T is the topmost and leftmost of the remaining squares. A formal proof for Lemma 4 is given in the full version [7]. For an analogous reason, after placing the first two squares, we can place B in the bottom-left corner of the bounding box of the remaining segments. Finally, we cover the remaining segments with R, if possible.

We have therefore shown that the position of L along the left boundary uniquely determines the positions of the squares T, B and R along their respective boundaries. Unfortunately, we do not know the position of L in advance, so instead we consider all possible initial positions of L via parametrisation. Let y_L be the y-coordinate of the top side of L, and similarly let x_T, x_B be the x-coordinates of the left side of T and B respectively. See Fig. 4.

Finally, we will try to cover all remaining subsegments with the square R. Define x_{R_1} and x_{R_2} to be the x-coordinates of the leftmost and rightmost uncovered points after the first three squares have been placed. Similarly, define y_{R_1} and y_{R_2} to be the y-coordinates of the topmost and bottommost uncovered points. Then it is possible to cover the remaining segments with R if and only if $x_{R_1} - x_{R_2} \leq 1$ and $y_{R_1} - y_{R_2} \leq 1$.

Since the position of L uniquely determines T, B and R, we can deduce that the variables x_T, x_B, x_{R_1}, x_{R_2}, y_{R_1} and y_{R_2} are all functions of y_L. We will show that each of these functions is piecewise linear and can be computed in $O(n \log n)$ time. We begin by computing x_T as a function of variable y_L.

Lemma 5. *The variable x_T as a function of variable y_L is a piecewise linear function and can be computed in $O(n \log n)$ time.*

Next, we show that x_B is a piecewise linear function of y_L, with complexity $O(n)$, and can be computed in $O(n \log n)$ time.

Lemma 6. *The variable x_B as a function of variable y_L is a piecewise linear function and can be computed in $O(n \log n)$ time.*

Then we compute x_{R_1}, x_{R_2}, y_{R_1} and y_{R_2} in a similar fashion.

Lemma 7. *The variables* x_{R_1}, x_{R_2}, y_{R_1}, y_{R_2} *as functions of variable* y_L *are piecewise linear functions and can be computed in* $O(n \log n)$ *time.*

Finally, we check if there exists a value of y_L so that $x_{R_1} - x_{R_2} \leq 1$ and $y_{R_1} - y_{R_2} \leq 1$. If so, there exist positions for L, B, T and R that covers all the segments, otherwise, there is no such position for L, T, B and R. This yields the following result:

Theorem 3. *One can decide if a set of* n *segments is 4-coverable in* $O(n \log n)$ *time.*

3 Problem 2: The Subtrajectory Data Structure Problem

In this section, we briefly describe some of the main ideas for building the data structures that can answer whether a subtrajectory is either 2-coverable or 3-coverable. Details of the data structures can be found in the full version of this paper [7].

We begin by building three preliminary data structures. Given a piecewise linear trajectory of complexity n, our preliminary data structures are:

Tool 1. *A bounding box data structure that preprocesses a trajectory in* $O(n)$ *time, so that given a query subtrajectory, it returns the subtrajectory's bounding box in* $O(\log n)$ *time.*

Tool 2. *An upper envelope data structure that preprocesses a trajectory in* $O(n \log n)$ *time, so that given a query subtrajectory and a query vertical line, it returns the highest intersection between the subtrajectory and the vertical line (if one exists) in* $O(\log n)$ *time. See Fig. 5.*

Tool 3. *A highest vertex data structure that preprocesses a trajectory in* $O(n \log n)$ *time, so that given a query subtrajectory and a query axis-parallel rectangle, it returns the highest vertex of the subtrajectory inside the rectangle (if one exists) in* $O(\log^2 n)$ *time. See Fig. 6.*

3.1 Query If a Subtrajectory Is 2-Coverable

Our construction procedure is to build Tool 1 and Tool 2. Our query procedure consists of two steps. The first step is to narrow down the covering to one of two configurations using Lemma 1 and Tool 1. The second step is to check whether one of these configurations indeed covers the subtrajectory. The key idea in the second step is to use Tool 2 along the boundary of the configuration to see if the subtrajectory passes through the boundary. Putting this together yields:

Theorem 4. *Let* $\mathcal{T}$ *be a trajectory with* n *vertices. After* $O(n \log n)$ *preprocessing time,* $\mathcal{T}$ *can be stored using* $O(n \log n)$ *space, so that deciding if a query subtrajectory* $\mathcal{T}[a, b]$ *is 2-coverable takes* $O(\log n)$ *time.*

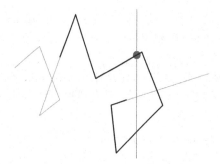

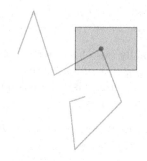

Fig. 5. Tool 2 returns the highest intersection of a subtraj. and a vertical line.

Fig. 6. Tool 3 returns the highest subtrajectory vertex in a query rectangle.

3.2 Query If a Subtrajectory Is 3-Coverable

Our construction procedure is to build Tools 1, 2, and 3. Our query procedure consists of three steps. The first step is to place the first square in a constant number of configurations using Lemma 2 and Tool 1. For each placement of the first square, the second step generates two configurations by placing the remaining two squares. The key idea in the second step is to compute the bounding box of the uncovered subsegments by using a combination of Tools 2 and 3. The third step is to check if a configuration indeed covers the subtrajectory. The key idea in the third step is to use Tool 2 along the boundary of the configuration to see if the subtrajectory passes through the boundary. We require an additional check using Tool 3 in one of the configurations. Putting this together yields:

Theorem 5. *Let T be a trajectory with n vertices. After $O(n \log n)$ preprocessing time, T can be stored using $O(n \log n)$ space, so that deciding if a query subtrajectory $T[a, b]$ is 3-coverable takes $O(\log^2 n)$ time.*

4   Problem 3: The Longest Coverable Subtrajectory

In this section we compute a longest k-coverable subtrajectory $T[p^*, q^*]$ of a given trajectory T. Note that the start and end points p^* and q^* of such a subtrajectory need not be vertices of the original trajectory. Gudmundsson, van Kreveld, and Staals [8] presented an $O(n \log n)$ time algorithm for the case $k = 1$. However, we note that there is a mistake in one of their proofs, and hence their algorithm misses one of the possible scenarios. We correct this mistake, and using the insight gained, also solve the problem for $k = 2$.

4.1   A Longest 1-Coverable Subtrajectory

Gudmundsson, van Kreveld, and Staals state that there exists an optimal placement of a unit square, i.e. one such that the square covers a longest 1-coverable subtrajectory of T, and has a vertex of T on its boundary [8, Lemma 7]. However, that is incorrect, as illustrated in Fig. 7. Let $p(t)$ be a parametrisation of the trajectory. Fix a corner c of the square and shift the square so that c follows $p(t)$. Let $q(t)$ be the point so that $T[p(t), q(t)]$ is the maximal subtrajectory contained in the square, and let $\phi(t)$ be the length of this subtrajectory. This function ϕ is piecewise linear, with inflection points

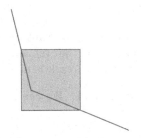

Fig. 7. An optimal placement that has no vertex on the boundary of the square.

not only when a vertex of T lies on the boundary of the square, but also when $p(t)$ or $q(t)$ hits a corner of the square. The argument in [8] misses this last case. Instead, the correct characterization is:

Lemma 8. *Given a trajectory T with vertices $v_1, .., v_n$, there exists a square H covering a longest 1-coverable subtrajectory so that either:*

- *there is a vertex v_i of T on the boundary of H, or*
- *there are two trajectory edges passing through opposite corners of H.*

We give the full proof of this lemma in the full version of this paper [7]. To compute a longest 1-coverable subtrajectory we also have to consider this scenario. We use the existing algorithm to test all placements of the first type from Lemma 8 in $O(n \log n)$ time. Next, we briefly describe how we can also test all placements of the second type in $O(n \log n)$ time.

Lemma 9. *Given a pair of non-parallel edges e_i and e_j of T, there is at most one unit square H such that the top left corner of H lies on e_i, and the bottom right corner of H lies on e_j.*

It follows that any pair of edges e_i, e_j of T generates at most a constant number of additional candidate placements that we have to consider. Let $\mathcal{H}_{ij}$ denote this set. Next, we argue that there are only $O(n)$ relevant pairs of edges that we have to consider.

We define the *reach* of a vertex v_i, denoted $r(v_i)$, as the vertex v_j such that $T[v_i, v_j]$ can be 1-covered, but $T[v_i, v_{j+1}]$ cannot. Let $\mathcal{H}_i = \mathcal{H}_{(i-1)j}$ denote the set of candidate placements corresponding to v_i and $v_j = r(v_i)$. Analogously, we define the *reverse reach* $rr(v_j)$ of v_j as the vertex v_i such that $T[v_i, v_j]$ can be 1-covered, but $T[v_{i-1}, v_j]$ cannot, and the set $\mathcal{H}'_j = \mathcal{H}_{(i-1)j}$. Finally, let $\mathcal{H} = \bigcup_{i=1}^n \mathcal{H}_i \cup \mathcal{H}'_i$ be the set of placements contributed by all reach and reverse reach pairs. Observe that this set consists of $O(n)$ placements, as all individual sets $\mathcal{H}_i$ and $\mathcal{H}'_i$ have at most one element.

Lemma 10. *Let $p^* \in e_i$ and $q^* \in e_j$ lie on edges of T, and let H be a unit square with p^* in one corner, and q^* in the opposite corner. We have that $H \in \mathcal{H}$.*

Once we have the reach $r(v_i)$ and the reverse reach $rr(v_i)$ for every vertex v_i we can easily construct $\mathcal{H}$ in linear time (given a pair of edges e_i, e_j we can construct the unit squares for which one corner lies on e_i and the opposite corner lies on e_j in constant time). We can use Tool 1 to test each candidate in $O(\log n)$ time. So all that remains is to compute the reach of every vertex of $\mathcal{T}$; computing the reverse reach is analogous.

Lemma 11. *We can compute $r(v_i)$, for each vertex $v_i \in \mathcal{T}$, in $O(n \log n)$ time in total.*

Theorem 6. *Given a trajectory $\mathcal{T}$ with n vertices, there is an $O(n \log n)$ time algorithm to compute a longest 1-coverable subtrajectory of $\mathcal{T}$.*

4.2 A Longest 2-Coverable Subtrajectory

In this section we reuse some of the observations from Sect. 4.1 to develop an $O(n\beta_4(n) \log^2 n)$ time algorithm for the $k = 2$ case. Here, $\beta_4(n) = \lambda_4(n)/n$, and $\lambda_s(n)$ is the length of a Davenport-Schinzel sequence of order s on n symbols.

Our algorithm to compute a longest 2-coverable subtrajectory $\mathcal{T}[p^*, q^*]$ of $\mathcal{T}$ consists of two steps. In the first step we compute a set S of candidate starting points on $\mathcal{T}$, so that $p^* \in S$. In the second step, we compute the longest 2-coverable subtrajectory $\mathcal{T}[p, q]$ for each starting point $p \in S$, and report a longest such subtrajectory. With slight abuse/reuse of notation, for any point $p \in S$, we denote the endpoint q of this longest 2-coverable subtrajectory $\mathcal{T}[p, q]$ by $r(p)$. This generalizes our notion of *reach* from Sect. 4.1 to arbitrary points on $\mathcal{T}$.

Computing the Reach of a Point. We modify the data structure in Theorem 4, i.e. the data structure for answering whether a given subtrajectory is 2-coverable, to answer the reach queries. We do so by applying parametric search to the query procedure. Note that applying a simple binary search will give us only the edge containing $r(p)$. Furthermore, even given this edge it is unclear how to find $r(p)$ itself, as the squares may still shift, depending on the exact position of $r(p)$.

Lemma 12. *Let $\mathcal{T}$ be a trajectory with n vertices. After $O(n \log n)$ preprocessing time, $\mathcal{T}$ can be stored using $O(n \log n)$ space, so that given a query point p on $\mathcal{T}$ it can compute the reach $r(p)$ of p in $O(\log^2 n)$ time.*

Corollary 1. *Given a trajectory $\mathcal{T}$, and a set of m candidate starting points on $\mathcal{T}$, we can compute the longest 2-coverable subtrajectory that starts at one of those points in $O(n \log n + m \log^2 n)$ time.*

Computing the Set of Starting Points. It remains only to construct a set S of candidate starting points with the property that the starting point of a longest 2-coverable subtrajectory is guaranteed to be in the set. Our construction consists of six types of starting points, which when grouped up into their respective types, we will call *events*. The six types of events are vertex events, reach events, bounding box events, bridge events, upper envelope events, and special configuration events. Figures 8, 9, and 10 illustrate these events, and show how a longest 2-coverable subtrajectory may start at such an event. We then prove that it suffices to consider *only* these six types of candidate starting points. Finally, we bound the number of events, and thus candidate starting points, and describe how to compute them. Combining this with our result from Corollary 1 gives us an efficient algorithm to compute a longest 2-coverable subtrajectory. Note that in Definitions 2–7, for simplicity we define the events only in one of the four cardinal directions. However, in our construction in Definition 8 we require all six events for all four cardinal directions.

Definition 2. *Given a trajectory T, p is a vertex event if p is a vertex of T.*

Definition 3. *Given a trajectory T, p is a reach event if $r(p)$ is a vertex of T, and no point $q < p$ satisfies $r(q) = r(p)$.*

Definition 4. *Given a trajectory T, p is a bounding box event if the topmost vertex of T within the subtrajectory $T[p, r(p)]$ has the same y-coordinate as p.*

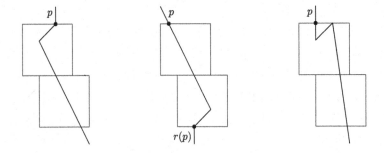

Fig. 8. A vertex event (left), a reach event (middle), and a bounding box event (right).

Definition 5. *Given a trajectory T, p is a bridge event if:*

- *the point p is the leftmost point on $T[p, r(p)]$, and*
- *the point p is one unit to the left of a point $u \in T[p, r(p)]$, and*
- *the point u is one unit above the lowest vertex of $T[p, r(p)]$.*

Definition 6. *Given a trajectory T, p is an upper envelope event if:*

- *the point p is the leftmost point on $T[p, r(p)]$, and*
- *the point p is one unit to the left of a point $u \in T[p, r(p)]$, and*
- *the point u is an intersection or vertex on the upper envelope of $T[p, r(p)]$.*

Definition 7. *Given a trajectory T, p is a special configuration event if there is a covering of squares H_1 and H_2 so that H_1 contains the top-left corner of H_2, and either:*

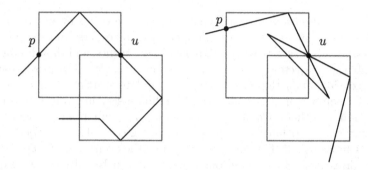

Fig. 9. Examples of a bridge event (left), and an upper envelope event box (right).

1. *point p is in the top-right corner of H_1 and $r(p)$ is in the bottom-left corner of H_1, or*
2. *point p is in the top-left corner of H_1 and the trajectory $\mathcal{T}$ passes through the bottom-right corner of H_1, or*
3. *point p is in the top-left corner of H_1, $r(p)$ is in the bottom-right corner of H_2, and the trajectory $\mathcal{T}$ passes through the two intersections of H_1 and H_2.*

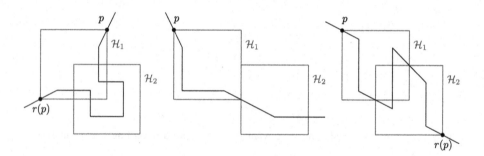

Fig. 10. Examples of the three types of special configuration events.

Using these definitions and their analogous versions in all four cardinal directions, we subdivide the trajectory $\mathcal{T}$ to obtain a trajectory $\mathcal{T}_3$ that forms our set of candidate starting points.

Definition 8. *Given a trajectory $\mathcal{T}$, let $\mathcal{T}_1$ be a copy of $\mathcal{T}$ with these additional points added to the set of vertices of $\mathcal{T}_1$:*

- *all the vertex, reach, bounding box, and bridge events of $\mathcal{T}$ for all four cardinal directions.*

Next, let $\mathcal{T}_2$ be a copy of $\mathcal{T}_1$ with these additional points added to the set of vertices of $\mathcal{T}_2$:

- all the upper envelope events of T_1 for all four cardinal directions.

Finally, let T_3 be a copy of T_2 with these additional points added to the set of vertices of T_3:

- all the special configuration events of T_2 for all configurations of $\mathcal{H}_1$ and $\mathcal{H}_2$.

Next, we argue that the set of vertices of this trajectory T_3 is a suitable set of candidate starting points.

Lemma 13. *The set T_3 is guaranteed to contain the starting point of a longest coverable subtrajectory of T.*

We now bound the number of candidate starting points.

Lemma 14. *Trajectory T_3 has $O(n\beta_4(n))$ vertices, and can be constructed in time $O(n\beta_4(n)\log^2 n)$. More specifically, for each type of event, the number of such events and the time in which we can compute them is*

	#events	computation time
Vertex events	$O(n)$	$O(n)$
Reach events	$O(n)$	$O(n\log^2 n)$
Bounding box events	$O(n)$	$O(n\log^2 n)$
Bridge events	$O(n)$	$O(n\log^2 n)$
Upper envelope events	$O(n\beta_4(n))$	$O(n\beta_3(n)\log^2 n)$
Special configuration events	$O(n\beta_4(n))$	$O(n\beta_4(n)\log^2 n)$

Proof. We briefly sketch the idea for only the upper envelope events. Refer to full version for the details, the proofs for the other events, and the description of the algorithms that compute these events [7].

Consider the set V of vertices of T and intersections of T with itself, and observe that every point $u \in V$ generates at most $O(1)$ candidate starting points p that are upper-envelope events. However, there may be $\Theta(n^2)$ such intersection points. We argue that not all of these intersection points generate valid starting points.

Let $p_1, .., p_k$ be the upper envelope events of the trajectory, and let $U = u_1, .., u_k$ be the edge of T that contains the point u on $T[p, r(p)]$ one unit to the right of p. We argue that U is a Davenport-Schinzel sequence of order 4 on $n-1$ symbols [16], and thus has complexity $k = O(n\beta_4(n))$. It follows that there are $O(n\beta_4(n))$ upper envelope events.

Since U is a Davenport-Schinzel sequence of order $s = 4$, it has no alternating (not necessarily contiguous) subsequences of length $s+2 = 6$. Our first step is to show that if the sequence of edges a, b, a occurs (not necessarily consecutively) then the starting points corresponding to the first two segments of the sequence must be x-monotone. Hence, an alternating sequence of edges of length six yields alternating, x-monotone sequence of starting points of length five. We then argue that this leads to a contradiction. □

By Lemma 14 we can compute $m = O(n\beta_4(n))$ candidate starting times for a longest 2-coverable subtrajectory of T in $O(n\beta_4(n)\log^2 n)$ time. Using Corollary 1 we can thus compute this subtrajectory in $O(n\log n + m\log^2 n) = O(n\beta_4(n)\log^2 n)$ time.

Theorem 7. *Given a trajectory T with n vertices, there is an $O(n\beta_4(n)\log^2 n)$ time algorithm to compute a longest 2-coverable subtrajectory of T.*

References

1. Agarwal, P.K., Procopiuc, C.M.: Exact and approximation algorithms for clustering. Algorithmica **33**(2), 201–226 (2002)
2. Bereg, S., et al.: Optimizing squares covering a set of points. Theor. Comput. Sci. **729**, 68–83 (2018)
3. Damiani, M.L., Issa, H., Cagnacci, F.: Extracting stay regions with uncertain boundaries from GPS trajectories: a case study in animal ecology. In: Proceedings of the 22nd ACM SIGSPATIAL International Conference on Advances in Geographic Information Systems, pp. 253–262 (2014)
4. Drezner, Z.: On the rectangular p-center problem. Naval Res. Logistics (NRL) **34**(2), 229–234 (1987)
5. Fowler, R.J., Paterson, M., Tanimoto, S.L.: Optimal packing and covering in the plane are NP-complete. Inf. Process. Lett. **12**(3), 133–137 (1981)
6. Gudmundsson, J., Horton, M.: Spatio-temporal analysis of team sports. ACM Comput. Surv. (CSUR) **50**(2), 22 (2017)
7. Gudmundsson, J., van de Kerkhof, M., Renssen, A., Staals, F., Wiratma, L., Wong, S.: Covering a set of line segments with a few squares. CoRR, abs/2101.09913 (2021)
8. Gudmundsson, J., van Kreveld, M., Staals, F.: Algorithms for hotspot computation on trajectory data. In: Proceedings of the 21st ACM SIGSPATIAL International Conference on Advances in Geographic Information Systems, pp. 134–143 (2013)
9. Hoffmann, M.: Covering polygons with few rectangles. In: Abstracts 17th European Workshop Computational Geometry, pp. 39–42 (2001)
10. Hwang, R.Z., Lee, R.C.T., Chang, R.C.: The slab dividing approach to solve the Euclidean p-center problem. Algorithmica **9**(1), 1–22 (1993)
11. Megiddo, N., Supowit, K.J.: On the complexity of some common geometric location problems. SIAM J. Comput. **13**(1), 182–196 (1984)
12. Nussbaum, D.: Rectilinear p-piercing problems. In: Proceedings of the 1997 International Symposium on Symbolic and Algebraic Computation, ISSAC, pp. 316–323 (1997)
13. Mahapatra, P.R.S., Goswami, P.P., Das, S.: Maximal covering by two isothetic unit squares. In: Canadian Conference on Computational Geometry, pp. 103–106 (2008)
14. Sadhu, S., Roy, S., Nandy, S.C., Roy, S.: Linear time algorithm to cover and hit a set of line segments optimally by two axis-parallel squares. Theor. Comput. Sci. **769**, 63–74 (2019)
15. Segal, M.: On piercing sets of axis-parallel rectangles and rings. Int. J. Comput. Geometry Appl. **9**(3), 219–234 (1999)
16. Sharir, M., Agarwal, P.K.: Davenport-Schinzel Sequences and their Geometric Applications. Cambridge University Press, Cambridge (1995)

17. Sharir, M., Welzl, E.: Rectilinear and polygonal p-piercing and p-center problems. In: Proceedings of the 12th Annual Symposium on Computational Geometry, pp. 122–132 (1996)
18. Stohl, A.: Computation, accuracy and applications of trajectories–a review and bibliography. Dev. Environ. Sci. **1**, 615–654 (2002)

Circumventing Connectivity
for Kernelization

Pallavi Jain[1], Lawqueen Kanesh[2(✉)], Shivesh Kumar Roy[3], Saket Saurabh[4,5],
and Roohani Sharma[6]

[1] Indian Institute of Technology Jodhpur, Jodhpur, India
pallavi@iitj.ac.in
[2] School of Computing, National University of Singapore, Singapore, Singapore
lawqueen@comp.nus.edu.sg
[3] Eindhoven University of Technology, Eindhoven, The Netherlands
s.k.roy@tue.nl
[4] Institute of Mathematical Sciences, HBNI, Chennai, India
saket@imsc.res.in
[5] Department of Informatics, University of Bergen, Bergen, Norway
[6] Max Planck Institute for Informatics, Saarland Informatics Campus,
Saarbrücken, Germany
rsharma@mpi-inf.mpg.de

Abstract. Classical vertex subset problems demanding connectivity are of the following form: given an input graph G on n vertices and an integer k, find a set S of at most k vertices that satisfies a property and $G[S]$ is connected. In this paper, we initiate a systematic study of such problems under a specific connectivity constraint, from the viewpoint of Kernelization (Parameterized) Complexity. The specific form that we study does not demand that $G[S]$ is connected but rather $G[S]$ has a *closed walk* containing all the vertices in S. In particular, we study CLOSED WALK-SUBGRAPH VERTEX COVER (CW-SVC, in short), where given a graph G, a set $X \subseteq E(G)$, and an integer k; the goal is to find a set of vertices S that hits all the edges in X and can be traversed by a closed walk of length at most k in G. When X is $E(G)$, this corresponds to CLOSED WALK-VERTEX COVER (CW-VC, in short). One can similarly define these variants for FEEDBACK VERTEX SET, namely CLOSED WALK-SUBGRAPH FEEDBACK VERTEX SET (CW-SFVS, in short) and CLOSED WALK-FEEDBACK VERTEX SET (CW-FVS, in short). Our results are as follows:

- CW-VC and CW-SVC both admit a polynomial kernel, in contrast to CONNECTED VERTEX COVER that does not admit a polynomial kernel unless NP $\subseteq$ coNP/poly.
- CW-FVS admits a polynomial kernel. On the other hand CW-SFVS does not admit even a polynomial Turing kernel unless the polynomial-time hierarchy collapses.

We complement our kernelization algorithms by designing single-exponential FPT algorithms – $2^{\mathcal{O}(k)} n^{\mathcal{O}(1)}$ – for all the problems mentioned above.

© Springer Nature Switzerland AG 2021
T. Calamoneri and F. Corò (Eds.): CIAC 2021, LNCS 12701, pp. 300–313, 2021.
https://doi.org/10.1007/978-3-030-75242-2_21

1   Introduction

In last few years, classical vertex subset problems demanding connectivity - given an input graph G and an integer k, find a set S of size at most k that satisfies a property and $G[S]$ is connected - has gained a lot of attention in the paradigm of parameterized complexity [7,13,15,17,19,20,22]. Two of the well-studied problems in this stream are CONNECTED VERTEX COVER (CVC, in short) and CONNECTED FEEDBACK VERTEX SET (CFVS, in short), in which apart from $G[S]$ being connected, we demand that $G-S$ is edgeless and acyclic, respectively. Both the problems are *fixed-parameter tractable* (FPT) [7,15], however, they are "hard" from the kernelization point-of-view, that is, both the problems do not admit a polynomial kernel unless PH= Σ_p^3 [11,19]. Here, a natural question that arises is the following. Is there a "specific" connectivity constraint that brings us back to the world of polynomial kernels? We answer this question affirmatively. In this paper, instead of demanding that $G[S]$ is connected, we demand that $G[S]$ has a closed walk that visits every vertex of the graph. In [2], authors have studied such a constraint for the VERTEX COVER problem, under the name of TOUR COVER, and proposed a constant factor approximation algorithm. However, it remains unstudied from the viewpoint of parameterized complexity.

This constraint is also motivated from a routing problem, apparently posed by a real newspaper company in Buenos Aires. The problem is described as follows. Consider the job of delivering newspapers to the subscribers by a truck that stops at street crossings and the driver of the truck manually delivers the newspapers to all the customers in the lanes of the street crossings. The goal now is to minimize the number of streets (counted as the path between two crossings) covered by the truck. The decision version of the problem can be formally described as being given a graph G (representing the topology of the city), a set of edges $X \subseteq E(G)$ (representing the lanes where the subscribers live), and an integer k; the goal is to find a closed walk (an alternating sequence of vertices and edges which starts and ends at the same vertex such that an edge is incident on the vertices immediately before and after this edge in the sequence) of length at most k in G whose vertices hit all the edges of X, that is, the vertex set of the closed walk is a vertex cover of G_X, where G_X is a graph on the vertex set $V(G)$ and the edge set X. We call this problem as CLOSED WALK-SUBGRAPH VERTEX COVER (CW-SVC, in short)[1], which is known to be NP-hard [9]. Note that a variant of the VERTEX COVER problem that demands a closed walk whose vertex set is a vertex cover of the given graph is a special case of CW-SVC when $X = E(G)$. We call this problem as CLOSED WALK-VERTEX COVER (CW-VC, in short).

We next generalize the definition of CW-SVC to $\mathcal{Q}$-VERTEX DELETION problem, where $\mathcal{Q}$ is a graph property (a class of graphs), as follows. Let $V(W)$

---

[1] This problem was studied in [9] under the name of STAR ROUTING. We changed the name of this problem here to fit the theme of names of problems that we describe ahead in this article.

denote the vertex set of the walk W. We define CLOSED WALK Q-SUBGRAPH VERTEX DELETION as follows.

CLOSED WALK Q-SUBGRAPH VERTEX DELETION   **Parameter:** k
(CW-Q-SVD)
Input: A graph G, a set $X \subseteq E(G)$, and an integer k
Question: Does there exist a closed walk W in G of length at most k such that $G_X - V(W)$ is in Q?

When $X = E(G)$, we call this problem as CLOSED WALK Q-VERTEX DELETION (CW-Q-VD, in short). Note that when Q is the family of all graphs with no edges, the problems CW-Q-SVD and CW-Q-VD are the same as the problems CW-SVC and CW-VC, respectively. When Q is the family of all acyclic graphs (that is, the family of all forests), we call these problems as CLOSED WALK-SUBGRAPH FEEDBACK VERTEX SET (CW-SFVS, in short) and CLOSED WALK-FEEDBACK VERTEX SET (CW-FVS, in short), respectively.

Our Results and Methods: For our study, we consider a natural parameter, the *solution size*, in the parameterized complexity. Our contribution to this article begins with the study of kernelization complexity of CW-SVC, CW-VC, CW-SFVS, and CW-FVS. We show that unlike CVC and CFVS, CW-SVC, CW-VC, and CW-FVS admit a polynomial kernel, while CW-SFVS is WK[1]-hard , that is, CW-SFVS does not admit a polynomial Turing kernel unless the polynomial-time hierarchy collapses. In particular, we prove the following results.

Theorem 1. (♣)[2] CW-SVC *admits an* $\mathcal{O}(k^5)$ *vertex kernel.*

Theorem 2. (♣) CW-VC *admits an* $\mathcal{O}(k^2)$ *vertex kernel.*

Theorem 3. CW-SFVS *is* WK[1]-*hard.*

Theorem 4. CW-FVS *admits an* $\mathcal{O}(k^{17})$ *vertex kernel.*

At the first sight, it might seem a little weird as to what gives rise to such a contrast in the kernelization complexity of the problems with a closed walk constraint versus a general connectivity constraint. For instance, it is easy to see that if there is a solution to CVC/CFVS of size k, then there is a solution to CW-VC/CW-FVS of size at most $2k$; and if there is solution to CW-VC/CW-FVS of size k, then there is a solution to CVC/CFVS of size at most k. Regardless of this, the problems behave differently primarily because for the CVC/CFVS problems a minimal solution to the VC/FVS problems has to be connected via a steiner tree (which is a hard task), whereas in the CW-VC/CW-FVS problems, a solution can be obtained from a minimal solution of VC/FVS by connecting a pair of vertices greedily by any available shortest path. We exploit this greedy choice of shortest paths to construct a closed walk containing the vertices of a

---

[2] Proofs of results marked with ♣ have been removed from this short version due to space constraints.

vertex cover/ feedback vertex set to get the desired kernels. More precisely, our main strategy to give polynomial kernels for CW-VC, CW-SVC (CW-FVS) is to first find a set that preserves minimal solutions of VC (FVS) problem. Then, to ensure the closed walk property, it is enough for us to preserve the shortest path between every pair of vertices in this set. We use this observation and the constructed set to design reduction rules to reduce the instance. To show that CW-SFVS is WK[1]-hard, we give a parameter preserving reduction from MULTICOLORED CYCLE which is known to be WK[1]-complete [16]. The construction of this reduction is the same as the one used to show WK[1]-hardness of RURAL POSTMAN PROBLEM in [5].

We next move towards designing FPT algorithms for CW-SVC and CW-SFVS. In particular, we prove the following results.

Theorem 5. (♣) CW-SVC *admits an* FPT *algorithm that runs in time* $\mathcal{O}(3.2362^k n^{\mathcal{O}(1)})$, *where n is the number of vertices in the input graph.*

Theorem 6. (♣) CW-SFVS *admits an* FPT *algorithm that runs in time* $\mathcal{O}(34^k n^{\mathcal{O}(1)})$, *where n is the number of vertices in the input graph.*

We next give an informal description of our algorithms. Full details have been removed due to space constraints. We first need to define the GROUP CLOSED WALK problem and the notion of the compact representations.

GROUP CLOSED WALK (GCW)   **Parameter:** k
Input: A simple graph G, a family $\mathcal{C}$ of pairwise vertex disjoint subsets of $V(G)$, and a positive integer k
Question: Does there exist a closed walk W in G of length at most k such that $|V(W) \cap C| \geq 1$ for each $C \in \mathcal{C}$?

Observe that GCW is NP-hard as it generalizes the classical HAMILTONIAN CYCLE problem.

Theorem 7. (♣) GCW *admits an* FPT *algorithm that runs in time* $\mathcal{O}(2^k n^{\mathcal{O}(1)})$, *where n is the number of vertices in the input graph.*

To design an FPT algorithm for GCW, we use dynamic programming over subsets of $\mathcal{C}$. Next, we define a compact representation for an instance of a $\mathcal{Q}$-VERTEX DELETION problem, in which given a graph G and an integer k, we shall decide the existence of a set, $S \subseteq V(G)$, of size at most k such that $G - S \in \mathcal{Q}$. Recall that $\mathcal{Q}$ is a graph class.

Let (G, k) be an instance of $\mathcal{Q}$-VD. Let $\mathcal{C}^*$ be a k-sized family of vertex disjoint subsets of $V(G)$. We say that $\mathcal{C}^*$ *respects* a set S^*, if the set S^* can be obtained by selecting exactly one vertex from each set in $\mathcal{C}^*$, that is, $|S^*| = |\mathcal{C}^*|$ and $|S^* \cap C| = 1$ for each $C \in \mathcal{C}^*$. If every set that $\mathcal{C}^*$ respects is a minimal solution to (G, k) of $\mathcal{Q}$-VD, then we say that $\mathcal{C}^*$ is a *valid* family of (G, k) for $\mathcal{Q}$-VD.

Definition 1 (Q-VD-compact representation(Q-VD-cr)). Given an instance (G, k) of Q-VD, a family $\mathcal{F}$ is a Q-VD-compact representation (Q-VD-cr, in short) of (G, k), if $\mathcal{F}$ is a collection of at most k-sized families of vertex disjoint subsets of $V(G)$ such that each family in $\mathcal{F}$ is valid and for every minimal solution S to (G, k) of Q-VD, there exists at least one family $\mathcal{C}$ in $\mathcal{F}$ such that $\mathcal{C}$ respects S.

Assuming an algorithm for computing a Q-VD family (if it exists), we state the following result about CW-Q-SVD.

Theorem 8. (♣) *Given an instance (G, k) of Q-VD, suppose that there is an algorithm, which runs in time $\tau(k)n^{\mathcal{O}(1)}$, that either outputs a Q-VD-cr $\mathcal{F}$ of (G, k), or outputs that (G, k) is a No-instance of Q-VD. Then, CW-Q-SVD admits an algorithm that runs in time $\tau(k)n^{\mathcal{O}(1)} + |\mathcal{F}| \cdot 2^k n^{\mathcal{O}(1)}$, where n is the number of vertices in the input graph.*

The main idea to prove the above theorem is to run the algorithm in Theorem 7 for GCW on each family in Q-VD-cr. When Q is the family of all graphs with no edges (resp. all acyclic graphs), we denote Q-VD-cr by VC-cr (resp. FVS-cr). The following theorem gives an FPT algorithm to compute VC-cr, if it exists.

Theorem 9. (♣) *Given a graph G and a positive integer k, there is an algorithm that either correctly outputs that (G, k) is a No-instance of VC or outputs a VC-cr $\mathcal{F}$ of (G, k). Moreover, $|\mathcal{F}|$ is bounded by $\mathcal{O}(1.6181^k)$ and the algorithm runs in time $\mathcal{O}(1.6181^k n^{\mathcal{O}(1)})$, where n is the number of vertices in the graph G.*

Theorem 10 improves the bound given by Guo et al. [14] for FVS-cr.

Theorem 10. (♣) *Given a graph G and a positive integer k, there is an algorithm that either correctly outputs that (G, k) is a No-instance of FVS or outputs an FVS-cr $\mathcal{F}$ of (G, k). Moreover, $|\mathcal{F}|$ is bounded by $\mathcal{O}(17^k)$ and the algorithm runs in time $\mathcal{O}(17^k n^{\mathcal{O}(1)})$, where n is the number of vertices in the graph G.*

To design FPT algorithms to compute VC-cr and FVS-cr, we use branching technique. Using Theorem 8 and 9, we obtain Theorem 5 and using Theorem 8 and 10, we obtain Theorem 6.

Before moving to the technical details, we give some formal definitions. A *walk* in a graph G is a sequence $W = v_0 e_1 v_1 \ldots v_{k-1} e_k v_k$ whose terms alternate between vertices and edges (not necessary distinct) such that $e_i = v_{i-1} v_i$ for $1 \leq i \leq k$. Here, k denotes the length of the walk. If $v_0 = v_k$, then W is said to be a *closed walk*. By $V(W)$, we denote the set $\{v_0, \ldots, v_k\}$. We say that a path P is a *degree two induced path* in G if each internal vertex in P has degree exactly two in G (there is no restriction on degree of endpoints). For standard graph theoretic notations, we refer the reader to [10]. For more introduction to FPT algorithms and kernelization, we refer the reader to [6,12]. For WK[1]-hardness, we refer the reader to [16].

2 Kernel Lower Bound for Subset Closed Walk FVS

In this section, we will prove Theorem 3, that is, we will show that CW-SFVS does not admit even a polynomial Turing kernel. Towards this, we give a polynomial-time parameter preserving reduction from the MULTICOLORED CYCLE (MCC) problem, which is defined as follows. Given a graph G, and a partition of $V(G)$ into k sets $\{V_1, \ldots, V_k\}$, the question is does there exist a cycle C in G such that $|V(C) \cap V_i| = 1$, for all $i \in [k]$? It is known that MCC is WK[1]-hard [16], when parameterized by the solution size. In the following, we present a polynomial-time parameter preserving reduction from MCC to CW-SFVS.

Given an instance $(G, V_1, \ldots, V_k)$ of MCC, we construct an instance (G', X, k) of CW-SFVS as follows. Initially, we set $V(G') = V(G)$ and $E(G') = E(G)$. Let $V_j = u_j^1, \ldots, u_j^\ell$, $j \in [k]$. For every $j \in [k]$, we add a cycle on the vertices of V_j. Formally, for every $j \in [k]$, we add edges $E_j = \cup_{i \in [\ell-1]}\{u_j^i u_j^{i+1}\} \cup \{u_j^1 u_j^\ell\}$ in G'. This completes the construction of the graph G'. Let $X = \cup_{j \in [k]} E_j$. We claim that $(G, V_1, \ldots, V_k)$ is a YES-instances of MCC if and only if (G', X, k) is a YES-instance of CW-SFVS, which together with the fact that MCC is WK[1]-hard gives Theorem 3.

3 A Kernelization Algorithm for CLOSED WALK-FEEDBACK VERTEX SET

In this section, we will prove Theorem 4. Note that we allow instances with self-loops and parallel edges. Given an instance (G, k) of CW-FVS, intuitively the algorithm works as follows. The algorithm first computes an approximate solution, R, of FVS for the graph G and either concludes that (G, k) is a No-instance of CW-FVS or proceeds to construct a set $F \subseteq V(G)$ such that for every vertex $v \in F$, v is contained in every solution to (G, k) of FVS. Using set F, the algorithm either concludes that (G, k) is a No-instance of CW-FVS or constructs a subgraph $G^\star$ of G, which has the following properties: (i) the graph $G^\star$ together with the set F preserves all minimal solution to (G, k) of FVS; (ii) the minimum degree of $G^\star$ is at least two; (iii) the number of vertices of degree at least three is bounded by $\mathcal{O}(k^3)$; and (iv) the number of maximal degree two paths in $G^\star$ is bounded by $\mathcal{O}(k^3)$. Observe that the following reduction rules which are a common practice in FVS and its variants: Deleting vertices in F and reducing the budget, deleting an edge which is redundant with respect to FVS, short-circuiting degree two vertices, cannot be applied here; as a vertex can be useful for the purpose of hitting cycles as well as to provide connectivity and by deleting a vertex or an edge we might lose a closed walk. Considering this observation, the algorithm constructs a set M, using the sets F, R and the graph $G^\star$, which in some sense preserves a solution to (G, k) of CW-FVS, if exists. Then, by using the set M, the algorithm applies few reduction rules exhaustively in the order in which they are stated to get the desired kernel. Due to space constraints we have removed some proofs from the short version of the paper.

Tools and Notations: For a graph H, we refer vertices of degree at least 3 in H as *high degree* vertices and denote this set of vertices by $V_{\geq 3}(H)$. Let $W = v_1 e_1 \ldots e_{\ell-1} v_\ell$, be a closed walk. By v_i to v_j subwalk of W, we mean a walk $v_i e_i \ldots e_{j-1} v_j$ which is contained in W. We define a W_{ij} *subwalk of* W as follows. If $j > i$, then W_{ij} represents v_i to v_j subwalk of W; otherwise, $W_{ij} = W_{i\ell} \cdot W_{1j}$. By $V(W_{ij})$, we denote the set of vertices in the walk W_{ij}. By *replace* subwalk W_{ij} by a v_i to v_j path P in walk W operation, we mean the following. If $j > i$, then replace v_i to v_j subwalk in W by the path P. If $j \leq i$, then replace v_i to v_ℓ subwalk in W by path P and v_1 to v_j subwalk in W by v_j. Note that if the length of P is at most the length of W_{ij}, then replace operation does not increase the length of the walk W. Let H be a graph and v be a vertex in H. Then, for a positive integer t, we define a t-*flower* at v as a set of t distinct cycles in H, such that their pairwise intersection is exactly at $\{v\}$. The following results will help us streamline later arguments.

Proposition 1 [8,18,21]. *For a graph H, a vertex $v \in V(H)$ without a self-loop, and an integer k, there is a polynomial-time algorithm, which either outputs a $(k+1)$-flower at v, or correctly concludes that no such $(k+1)$-flower exists. Moreover, if there is no $(k+1)$-flower at v, it outputs a set $Z_v \subseteq V(H) \setminus \{v\}$ of size at most $2k$, such that Z_v intersects every cycle containing v in G.*

Next, we state a lemma which will be used in designing kernelization algorithm. The following lemma basically reduces the number of high degree vertices and the number of maximal degree two paths in the given graph, and preserves all minimal feedback vertex sets of size at most k in the given graph. Thus, it can be used as a tool to design kernel of several other variants of FVS, for example, CONFLICT FREE FEEDBACK VERTEX SET for specific classes of graphs (as in general, this problem is W[1]-hard) [1], and many other, as after this step, main effort is in bounding the degree two vertices and preserving "other" required properties of the solution.

Lemma 1. *Let (H, k) be an instance of FVS such that there is no vertex in H with a $(k+1)$-flower. Then, there is a polynomial-time algorithm that either correctly outputs that (H, k) is a NO-instance of FVS or outputs a subgraph $H^\star$ of H, which satisfies the following properties.*

1. *A set S is a minimal solution to (H, k) of FVS if and only if S is a minimal solution to $(H^\star, k)$ of FVS.*
2. *The minimum degree of $H^\star$ is at least 2.*
3. *$|V_{\geq 3}(H^\star)|$ is bounded by $\mathcal{O}(k^3)$.*
4. *The number of maximal degree two paths in $H^\star$ is bounded by $\mathcal{O}(k^3)$.*

Kernelization Algorithm: Let (G, k) be an instance of CW-FVS. We assume that the size of a minimum feedback vertex set of G is at least four, otherwise we can solve the problem in polynomial time using the brute-force algorithm, and accordingly return a trivial instance of CW-FVS of constant size. We will define a sequence of reduction rules, which are applied exhaustively in the order in which they are stated. For the ease of notation, throughout the section, we will use (G, k) for the reduced instance after an application of a reduction rule.

We begin with some simple reduction rules, that is, if the size of feedback vertex set for G is large, then (G, k) is a No-instance of CW-FVS. Towards this, we will use a well-known result that there is a factor 2-approximation algorithm for FVS [3,4].

Reduction Rule 1. *Given an instance (G, k) of* CW-FVS, *let R be a factor 2-approximate feedback vertex set for the graph G. If $|R| > 2k$, then return a trivial No-instance of* CW-FVS *of constant size.*

Reduction Rule 2. *Given an instance (G, k) of* CW-FVS, *let C be an isolated cycle in G. Then, if $G - C$ is not a forest, then return a trivial No-instance of* CW-FVS *of constant size, otherwise return a trivial Yes-instance of* CW-FVS *of constant size.*

From now onwards, we will assume that the graph G does not have an isolated cycle. Next, we use the algorithm in Proposition 1 and construct a set $F \subseteq V(G)$ such that for every vertex $v \in F$, there is a $(k+1)$-flower at v in G. Observe that a vertex with $(k+1)$-flower belongs to every solution to (G, k) of FVS, and hence to every solution of CW-FVS. Therefore, we apply the following reduction rule.

Reduction Rule 3. *Given an instance (G, k) of* CW-FVS, *let $F \subseteq V(G)$ be a set of all vertices with $(k+1)$-flower. If $|F| > k$, then return a trivial No-instance of* CW-FVS *of constant size.*

When Reduction Rule 3 is not applicable, then in polynomial time, we can find the set of vertices with $(k + 1)$-flower, F, in G of size at most k.

Reduction Rule 4. *Given an instance (G, k) of* CW-FVS, *let $F \subseteq V(G)$ be a set of all vertices with $(k + 1)$-flower. If there exists a vertex $v \in F$ in G such that there is no self-loop at v, then add a self-loop vv in G and return (G', k), where $G' = G + vv$.*

Next, we run the algorithm in Lemma 1 on the instance $(G - F, k - |F|)$ of FVS, and apply the following reduction rule.

Reduction Rule 5. *Given an instance (G, k) of* CW-FVS, *let $F \subseteq V(G)$ be a set of all vertices with $(k + 1)$-flower. If the algorithm in Lemma 1 outputs that $(G - F, k - |F|)$ is a No-instance of FVS, then return a trivial No-instance of* CW-FVS *of constant size.*

Suppose that (G, k) is such an instance of CW-FVS for which Reduction Rule 5 does not return a trivial No-instance. This implies that Lemma 1 returned a graph, say $G^\star$. Observe that every minimal solution to $(G^\star, k)$ of FVS contains at most one vertex of degree two from a degree two path, P, in $G^\star$, and this could be any vertex of P, as any cycle that contains a vertex from P contains all the vertices in P, consecutively. Using this observation and by item 1 of Lemma 1, we have the following properties about a minimal solution to (G, k) of FVS.

Lemma 2 *(1). If a set S is a minimal solution to (G, k) of FVS, then $S \setminus F$ is a minimal solution to $(G^\star, k)$ of FVS. Furthermore, if a set S is a minimal solution to $(G^\star, k)$ of FVS, then $S \cup F$ is a minimal solution to (G, k) of FVS. (2) Every minimal solution to (G, k) of FVS contains at most one vertex of degree two from a degree two path, P, in $G^\star$, and this could be any degree two vertex of P.*

Now, before moving further in the algorithm, we need to define some notations. Let $Z = F \cup R \cup V_{\geq 3}(G^\star)$. Let P and P' be two maximal degree two paths in $G^\star$, and u, v be degree two vertices in P and P', respectively. Suppose that Q is a u to v path in $G - Z$ and the internal vertices of Q are not contained in paths P and P'. Then, we say that Q is a $\{P, P'\}$-*connecting path* in $G - Z$. Since $G - Z$ is a forest, as R is an approximate solution of FVS for the graph G, we have that such connecting paths are *unique* for a pair of maximal degree two paths in $G^\star$. Let $\mathcal{B}$ be the set of $\{P, P'\}$-connecting paths, for all maximal degree two paths P, P' in $G^\star$, of length at most k in $G - Z$ and Y be the set of endpoints of paths in $\mathcal{B}$. Next, we construct a set $M \subseteq V(G)$. Eventually, we will show that we can delete all the vertices in G which are neither in M nor in $G^\star$.

Construction 1 (Construction of M)

Step 1. $M = Z \cup Y$.

Step 2. For every connecting path $P \in \mathcal{B}$, add vertices of P to M.

Step 3. For every pair of vertices $u, v \in Y \cup Z$, if length of a shortest u to v path in G is at most k, then add vertices of an arbitrary shortest u to v path in G to M.

Step 4. For every pair of vertices $u, v \in Z$, and for every maximal degree two path P in $G^\star$, if there exists a vertex w in P such that the sum of the length of a shortest u to w path and the length of a shortest v to w path in G is smallest among all vertices in P and at most k, then add vertices of an arbitrary shortest u to w path and vertices of an arbitrary shortest v to w path in G to M.

Lemma 3. *Let M be the set constructed in Construction 1. Then, $|M| = \mathcal{O}(k^{13})$.*

Reduction Rule 6. *Let (G, k) be an instance of CW-FVS, $G^\star$ be the graph returned by the algorithm in Lemma 1, and M be the set constructed by Construction 1. If there exists a vertex v in G, such that $v \notin M \cup V(G^\star)$, then delete v from G and return (G', k), where $G' = G - v$.*

Lemma 4. *Reduction Rule 6 is safe.*

Proof. In the forward direction, let $W = v_1 e_1 \ldots e_{\ell-1} v_\ell$ be a solution to (G, k) of CW-FVS. Let $S \subseteq V(W)$ be a minimal solution to (G, k) of FVS. Since $S \subseteq F \cup V(G^\star)$ (by Lemma 2), we have that $v \notin S$. Thus, due to the subgraph property, S is also a solution to (G', k) of FVS. If W does not contain vertex v, then observe that W is also a closed walk in G', and hence a solution to (G', k) of CW-FVS. Next, we consider the case when W contains v. Let $i, j \in [\ell]$ such that $v_i, v_j \in S$ (v_i and v_j are in the walk W as $S \subseteq V(W)$) such that v is contained in W_{ij} subwalk in W and no vertex from $V(W_{ij}) \setminus \{v_i, v_j\}$ is contained in S. Note that v_i, v_j exist as the size of a minimum feedback vertex set of G is at least four, due to our assumption. We consider the following cases:

1. Both the vertices v_i and v_j are in Z. In Step 3 of Construction 1, vertices of an arbitrary v_i to v_j shortest path, Q, in G are added to M. Then, by replacing W_{ij} subwalk in W by path Q, we obtain another walk $W^\star$ in G as well as in G'. Since Q is a shortest v_i to v_j path in G, the length of $W^\star$ is at most the length of W. Since S does not contain any vertex from $V(W_{ij}) \setminus \{v_i, v_j\}$, we can infer that $S \subseteq V(W^\star)$. Thus, $W^\star$ is a solution to (G', k) of CW-FVS.

2. Only one of v_i and v_j is in Z. Suppose that $v_i \in Z$ and $v_j \notin Z$ ($v_j \in Z$ and $v_i \notin Z$ can be argued similarly). Recall that $Z = F \cup R \cup V_{\geq 3}(G^\star)$ and $S \subseteq F \cup V(G^\star)$. Since $v_j \in S \setminus Z$ and minimum degree of $G^\star$ is at least two (by Property 1 of Lemma 1), we have that v_j is a vertex of degree exactly two in $G^\star$. Clearly, v_j is contained in a maximal degree two path in $G^\star$. Let P be such a path. Recall that the size of a minimum feedback vertex set of G is at least four. Therefore, there exists $i' \in [\ell]$ such that $i' \notin \{i, j\}$, $v_{i'} \in S$ and no vertex from the subwalk $W_{ji'}$ in W is contained in S except v_j and $v_{i'}$. In particular, $V(W_{ji'}) \cap S = \{v_j, v_{i'}\}$. We further consider the following cases.

2(i). A vertex from $V(W_{ji'})$, say $v_{j'}$, is in Z. Note that $v_{j'} \neq v_j$ as $v_j \notin Z$. In Step 4 of Construction 1, for the vertices $v_i, v_{j'} \in Z$ and the path P, let w be the vertex of degree two in P such that the vertices of a shortest v_i to w path, say Q_1, in G and the vertices of a shortest w to $v_{j'}$ path, say Q_2, in G are added to M. Since v_j and w both are degree two vertices in the path P which is a degree two path in $G^\star$, due to Lemma 2, we have that $S' = (S \setminus \{v_j\}) \cup \{w\}$ is also a minimal solution to (G, k) of FVS. By replacing $W_{ij'}$ subwalk in W by path $Q_1 \cdot Q_2$, we obtain another walk $W^\star$. Observe that the length of $W^\star$ is at most the length of W, since sum of the length of paths Q_1 and Q_2 is smallest among all v_i to w' and w' to $v_{j'}$ paths in G for each vertex w' in P. Next, we argue that $v \notin V(W^\star)$. Recall that v is in the subwalk W_{ij} in W. Further, due to the choice of $v_{j'}$, W_{ij} is also a subwalk of $W_{ij'}$. Since we replaced $W_{ij'}$ subwalk by path $Q_1 \cdot Q_2$ to create a new walk $W^\star$, it follows that v is not in $V(W^\star)$. Since $(V(W_{ij}) \cup V(W_{ji'})) \cap S = \{v_i, v_j, v_{i'}\}$, we can infer that $S' \subseteq V(W^\star)$, and hence a solution to (G', k) of FVS (by subgraph property). Hence, $W^\star$ is a solution to (G', k) of CW-FVS.

2(ii). No vertex from $V(W_{ji'})$ is in Z. Since $v_{i'} \in S \setminus Z$, we have that $v_{i'}$ is a vertex of degree exactly two in $G^\star$. Therefore, $v_{i'}$ is contained in a maximal degree two path, say P', in $G^\star$. Since v_j and $v_{i'}$ both are in a minimal solution S and are degree two vertices in maximal degree two path in $G^\star$, due to Lemma 2, we have that P and P' are distinct. Since no vertex of $V(W_{ji'})$ is in Z, the subwalk $W_{ji'}$ in W is a walk in $G - Z$. Note that the $\{P, P'\}$-connecting path is contained in the subwalk $W_{ji'}$ in W, otherwise we will obtain a cycle in $G - Z$. Let $v_{j'}, v_{j''} \in Y$ be the endpoints of $\{P, P'\}$-connecting path. Without loss of generality, let us assume that the vertex $v_{j'}$ is in P. Since both v_j and $v_{j'}$ are degree two vertices in P, which is a degree two path in $G^\star$, due to Lemma 2, we have that $S' = (S \setminus \{v_j\}) \cup \{v_{j'}\}$ is also a minimal solution to (G, k) of FVS. In Step 3 of Construction 1, for $v_i \in Z$ and $v_{j'} \in Y$, vertices of an arbitrary v_i to $v_{j'}$ shortest path, say Q, of length at most k in G are added to M. Note that such a path exist as $v_i, v_{j'} \in V(W)$ and W is a walk of length at most k. Then, by replacing $W_{ij'}$ subwalk in W by path Q, we obtain another walk $W^\star$. Since Q is a shortest v_i to $v_{j'}$ path in G, the length of $W^\star$ is at most the length of W. Next, we argue that $W^\star$ is also a walk in G'. Towards this, clearly, it is sufficient to prove that $v \notin V(W^\star)$. Recall that v is in W_{ij} subwalk in W. Further, due to the choice of j', W_{ij} is also a subwalk in $W_{ij'}$. Since the vertices of Q are in M, v is not in Q. Therefore, $v \notin V(W^\star)$. Next, we argue that S' is a solution to (G', k) of FVS. Since $(V(W_{ij}) \cup V(W_{ji'})) \cap S = \{v_i, v_j, v_{i'}\}$, due to the construction of S', we can infer that $S' \subseteq V(W^\star)$, and hence a solution to (G', k) of FVS (by subgraph property). Hence, $W^\star$ is a solution to (G', k) of CW-FVS.

3. Neither v_i nor v_j is in Z. As argued in previous cases, let v_i, v_j be contained in two distinct maximal degree two paths, say P and P', respectively in $G^\star$. **3(i).** A vertex $v_t \in V(W_{ij})$ is in Z. Note that v_i and v_j are not in Z. Let $v \in W_{(t+1)(j-1)}$ ($v \in W_{(i+1)(t-1)}$ can be argued similarly). This case can be handled similar to the case 2. In the proof, v_t plays the role of v_i and v_j is same as in case 2. **3(ii).** No vertex from $V(W_{ij})$ is in Z. Before dwelling to the proof, we wish to mention here that this case is not same as case 2(ii) because in case 2(ii), v is not in $W_{ji'}$ subwalk in W. Now, we return to our proof. As argued in case 2(ii), since vertices in $V(W_{ij})$ are not in Z, the subwalk W_{ij} in W is a walk in $G - Z$, which implies that the $\{P, P'\}$-connecting path is contained in the subwalk W_{ij} in W. Let $v_t, v_{t'} \in Y$ be endpoints of $\{P, P'\}$-connecting path. Without loss of generality, let us assume that the vertex v_t and $v_{t'}$ are in P and P', respectively. We first note that $\{v_i, v_j\} \neq \{v_t, v_{t'}\}$ as in 2, we have added vertices of all connecting paths to M, and $v \in V(W_{ij})$ but $v \notin M$. Suppose that $v_j \neq v_{t'}$ ($v_i \neq v_t$ case can be argued similarly). If $v_i = v_t$, then $v \in W_{(t'+1)(j-1)}$ as $v \notin M$, otherwise either $v \in W_{(i+1)(t-1)}$ or $v \in W_{(t'+1)(j-1)}$. Let $v \in W_{(t'+1)(j-1)}$ ($v \in W_{(i+1)(t-1)}$ can be argued similarly). Recall that the size of minimum feedback vertex set of G is at least 4 and no vertex from $V(W_{ij} \setminus \{v_i, v_j\})$ is in S. Therefore, there exists $i' \in [\ell]$ such that $v_{i'} \in S$ and no vertex from the subwalk $W_{ji'}$ in W is contained in S except v_j and $v_{i'}$. In particular, $V(W_{ji'}) \cap S = \{v_j, v_{i'}\}$. Since both v_j and $v_{t'}$ are degree two vertices in P', which is a degree two path in $G^\star$, due to Lemma 2, we have that $S' = (S \setminus \{v_j\}) \cup \{v_{t'}\}$ is also a minimal solution to (G, k) of FVS, and hence a solution to (G', k) of FVS (by subgraph property). Next, we consider the following cases. **(a)** A vertex $v_r \in V(W_{ji'})$ is in Z. Recall that the vertex v_j is not in Z. In Step 3 of Construction 1, for $v_{t'} \in Y$ and $v_r \in Z$, vertices of an arbitrary $v_{t'}$ to v_r shortest path, say Q, in G are added to M. Then, by replacing the subwalk $W_{t'r}$ in W by the path Q, we obtain another walk, say $W^\star$. Note that $W_{t'r}$ contains v and Q does not contain v as $v \notin M$. Thus, $v \notin V(W^\star)$. Since Q is a shortest v_t to v_r path in G, the length of $W^\star$ is at most the length of W. Since $(V(W_{ij}) \cup V(W_{ji'})) \cap S = \{v_i, v_j, v_{i'}\}$, due to the construction of S', we can infer that $S' \subseteq V(W^\star)$, and hence a solution to (G', k) of FVS (by subgraph property). Therefore, $W^\star$ is a solution to (G', k) of CW-FVS. (b) No vertex from $V(W_{ji'})$ is in Z. This case is same as case 2(ii), where v'_t plays the role of v_i.

This completes the proof in the forward direction.

In the backward direction, let $W^\star$ be a solution to (G', k) of CW-FVS. Let $S^* \subseteq V(W^*)$ be a minimal solution to (G', k) of FVS. Since Reduction Rule 4 is no longer applicable, there exists a self-loop at every vertex of the set F in G'. Hence, $F \subseteq S^*$. Note that $G^\star$ is a subgraph of G', and $V(G^\star) \cap F = \emptyset$ as we called Lemma 1 on the graph $G - F$. Therefore, $S' = S^* \setminus F$ is a solution to $(G^\star, k - |F|)$ of FVS. Let $S'' \subseteq S'$ be a minimal solution to $(G^\star, k - |F|)$ of FVS. Then, by Lemma 2, $S'' \cup F$ is a minimal solution to (G, k) of FVS. By subgraph property $W^\star$ is also a closed walk in G, and hence a solution to (G, k) of CW-FVS. □

Observe that, when Reduction Rule 6 is no longer applicable, then vertices in G are either contained in M or are of degree exactly two in $G^\star$, i.e. $V(G) = V_{=2}(G^\star) \cup M$. By Lemma 3, M is bounded. Now, we are only remaining to bound the vertices in G which are of degree exactly two in $G^\star$ and are not in M. Towards that, we first apply the following reduction rule that ensures that if a vertex in G, which is not in M, is a degree two vertex in $G^\star$, then it is also a degree two vertex in G.

Reduction Rule 7. *Let (G, k) be an instance of CW-FVS, $G^\star$ be the graph returned by the algorithm in Lemma 1, and M be the set constructed by Construction 1. If there exists a vertex $v \in G$ such that $v \in V(G^\star) \setminus M$ and there exists a vertex $u \in G$ such that uv is an edge in G but uv is not an edge in $G^\star$, then delete edge uv from G and return (G', k), where $G' = G - uv$.*

When Reduction Rules 4-7 are no longer applicable, then a degree two vertex in $G^\star$ (vertex in a maximal degree two path in $G^\star$), which is not contained in M, is also a degree two vertex in G. Then, observe that if P is a degree two path in $G^\star$ such that the internal vertices of P are not in M, then each internal vertex in P is of degree exactly two in G. Next, we apply the following reduction rule to bound the degree two vertices in G.

Reduction Rule 8. *Let (G, k) be an instance of CW-FVS. Suppose that $P = v_1, \ldots, v_\ell$ is a degree two path in G such that $v_2, \ldots, v_{\ell-1} \notin M$. If $\ell \geq k + 4$, then delete $v_{\ell-1}$ and add edge $v_{\ell-2}v_\ell$ to the graph G. Return (G', k), where $G' = G - v_{\ell-1} + v_{\ell-2}v_\ell$.*

Lemma 5. *The number of vertices in $G - M$ is bounded by $\mathcal{O}(k^{17})$.*

The bound in Theorem 4 follows from Lemmas 3 and 5. This completes the proof of Theorem 4.

4   Conclusion

In this article, we studied variants of two classical NP-complete problems, viz.
VERTEX COVER (VC) and FEEDBACK VERTEX SET (FVS), in the realm of
parameterized complexity. The studied variants can be more closely contrasted to
the popular versions CONNECTED VERTEX COVER and CONNECTED FEEDBACK
VERTEX SET. On one end, where adding the connectivity constraint on the
solutions to the VC and FVS problems, deprives them from the existence of a
polynomial kernel, we show that adding a specific connectivity constraint, the
"closed walk constraint", brings them back into the world that allows them to
exhibit polynomial kernels.

The study leads to interesting open problems on various fronts. First, what
happens to the kernelization complexity of these classical vertex deletion problems, when other "specific connectivity" constraints are demanded from the solution. For example, one generic question in this setting would be when additionally
a connected graph H on k vertices is given and the input graph induced on the
solution vertices is required to have the graph H as a spanning subgraph. For
which graphs H, the classical vertex deletion problems admit polynomial kernel
and for which they do not? Can we hope to get such kind of a dichotomy result
for some classical problems? Another interesting question would be to give a
dichotomy of the vertex deletion problems that admit polynomial kernels with
the closed walk constraint.

Acknowledgments. This project has received funding from the European Research
Council (ERC) under the European Union's Horizon 2020 research and innovation

programme (grant no. 819416).

References

1. Agrawal, A., Jain, P., Kanesh, L., Misra, P., Saurabh, S.: Exploring the kernelization borders for hitting cycles. In: 13th International Symposium on Parameterized
   and Exact Computation, IPEC, pp. 14:1–14:14 (2018)
2. Arkin, E.M., Halldórsson, M.M., Hassin, R.: Approximating the tree and tour
   covers of a graph. IPL **47**(6), 275–282 (1993)
3. Bafna, V., Berman, P., Fujito, T.: A 2-approximation algorithm for the undirected
   feedback vertex set problem. SIAM J. Discrete Math. **12**(3), 289–297 (1999)
4. Becker, A., Geiger, D.: Optimization of pearl's method of conditioning and greedy-
   like approximation algorithms for the vertex feedback set problem. Artif. Intell.
   83(1), 167–188 (1996)
5. van Bevern, R., Fluschnik, T., Tsidulko, O.Y.: On approximate data reduction for the rural postman problem: theory and experiments. arXiv preprint
   arXiv:1812.10131 (2018)
6. Cygan, M., et al.: Parameterized Algorithms, vol. 4. Springer, Cham (2015).
   https://doi.org/10.1007/978-3-319-21275-3

7. Cygan, M.: Deterministic parameterized connected vertex cover. In: Fomin, F.V., Kaski, P. (eds.) SWAT 2012. LNCS, vol. 7357, pp. 95–106. Springer, Heidelberg (2012). https://doi.org/10.1007/978-3-642-31155-0_9

8. Cygan, M., et al.: Parameterized Algorithms. Springer, Cham (2015). https://doi.org/10.1007/978-3-319-21275-3

9. Delle Donne, D., Tagliavini, G.: Star routing: between vehicle routing and vertex cover. In: Kim, D., Uma, R.N., Zelikovsky, A. (eds.) COCOA 2018. LNCS, vol. 11346, pp. 522–536. Springer, Cham (2018). https://doi.org/10.1007/978-3-030-04651-4_35

10. Diestel, R.: Graph theory, volume 173 of. Graduate texts in mathematics, p. 7 (2012)

11. Dom, M., Lokshtanov, D., Saurabh, S.: Kernelization lower bounds through colors and ids. ACM Trans. Algorithms (TALG) 11(2), 1–20 (2014)

12. Fomin, F.V., Lokshtanov, D., Saurabh, S., Zehavi, M.: Kernelization: Theory of Parameterized Preprocessing. Cambridge University Press, Cambridge (2019)

13. Fomin, F.V., Lokshtanov, D., Saurabh, S., Thilikos, D.M.: Linear kernels for (connected) dominating set on h-minor-free graphs. In: Proceedings of the twenty-third annual ACM-SIAM symposium on Discrete Algorithms, pp. 82–93 (2012)

14. Guo, J., Gramm, J., Hüffner, F., Niedermeier, R., Wernicke, S.: Compression-based fixed-parameter algorithms for feedback vertex set and edge bipartization. JCSS 72(8), 1386–1396 (2006)

15. Guo, J., Niedermeier, R., Wernicke, S.: Parameterized complexity of generalized vertex cover problems. In: Dehne, F., López-Ortiz, A., Sack, J.-R. (eds.) WADS 2005. LNCS, vol. 3608, pp. 36–48. Springer, Heidelberg (2005). https://doi.org/10.1007/11534273_5

16. Hermelin, D., Kratsch, S., Soltys, K., Wahlström, M., Wu, X.: A completeness theory for polynomial (turing) kernelization. Algorithmica 71(3), 702–730 (2015)

17. Krithika, R., Majumdar, D., Raman, V.: Revisiting connected vertex cover: FPT algorithms and lossy kernels. Theory Comput. Syst. 62(8), 1690–1714 (2018)

18. Misra, N., Philip, G., Raman, V., Saurabh, S.: On parameterized independent feedback vertex set. Theor. Comput. Sci. 461, 65–75 (2012)

19. Misra, N., Philip, G., Raman, V., Saurabh, S., Sikdar, S.: FPT algorithms for connected feedback vertex set. J. Comb. Optim. 24(2), 131–146 (2012)

20. Ramanujan, M.: An approximate kernel for connected feedback vertex set. In: 27th Annual European Symposium on Algorithms (ESA 2019) (2019)

21. Thomassé, S.: A $4k^2$ kernel for feedback vertex set. ACM Trans. Algorithms 6(2), 32:1-32:8 (2010)

22. Wang, J., Yang, Y., Guo, J., Chen, J.: Planar graph vertex partition for linear problem kernels. J. Comput. Syst. Sci. 79(5), 609–621 (2013)

Online and Approximate Network Construction from Bounded Connectivity Constraints

Jesper Jansson[1]([✉]), Christos Levcopoulos[2], and Andrzej Lingas[2]

[1] The Hong Kong Polytechnic University, Hung Hom, Kowloon, Hong Kong
jesper.jansson@polyu.edu.hk
[2] Department of Computer Science, Lund University, 22100 Lund, Sweden
{Christos.Levcopoulos,Andrzej.Lingas}@cs.lth.se

Abstract. The *Network Construction problem*, studied by Angluin et al., Hodosa et al., and others, asks for a minimum-cost network satisfying a set of *connectivity constraints* which specify subsets of the vertices in the network that have to form connected subgraphs. More formally, given a set V of vertices, construction costs for all possible edges between pairs of vertices from V, and a sequence $S_1, S_2, \ldots \subseteq V$ of connectivity constraints, the objective is to find a set E of edges such that each S_i induces a connected subgraph of the graph (V, E) and the total cost of E is minimized. First, we study the online version where every constraint must be satisfied immediately after its arrival and edges that have already been added can never be removed. We give an $O(B^2 \log n)$-competitive and $O((B + \log r) \log n)$-competitive polynomial-time algorithms along with an $\Omega(B)$-competitive lower bound, where B is an upper bound on the size of constraints, while r, n denote the number of constraints and the number of vertices, respectively. In the cost-uniform case, we provide an $\Omega(\sqrt{B})$-competitive lower bound and an $O(\sqrt{n}(\log n + \log r))$-competitive upper bound with high probability, when constraints are unbounded. All our randomized competitive bounds are against an adaptive adversary, except for the last one which is against an oblivious adversary. Next, we discuss a hybrid approximation method for the (offline) Network Construction problem combining an approximation algorithm of Hosoda et al. with one of Angluin et al. and an application of the hybrid method to bioinformatics. Finally, we consider a natural strengthening of the connectivity requirements in the Network Construction problem, where each constraint is supposed to induce a subgraph (of the constructed graph) of diameter at most d. Among other things, we provide a polynomial-time $(\binom{B}{2} - B + 2)\binom{B}{2}$-approximation algorithm for the Network Construction problem with the d-diameter requirements.

Keywords: Network optimization · Induced subgraph · Connectivity · Approximation algorithm · Online algorithm

Research supported in part by VR grant 2017-03750 (Swedish Research Council).

© Springer Nature Switzerland AG 2021
T. Calamoneri and F. Corò (Eds.): CIAC 2021, LNCS 12701, pp. 314–325, 2021.
https://doi.org/10.1007/978-3-030-75242-2_22

1   Introduction

Korach and Stern introduced the problem of interconnecting possibly overlapping groups of users by a network such that the users in the same group do not need to use connections outside the group [9]. The optimization objective is to minimize the total cost of the pairwise connections. Angluin et al. and Chockler et al. studied this problem in [2] and [5], respectively.

Angluin et al. showed in [2] that if $P \neq NP$ and n is the number of vertices in the network then the problem cannot be approximated within a factor that is sublogarithmic in n, even in the uniform edge cost case. On the other hand, they proved that a greedy heuristic can approximate the optimal solution within a factor of $O(\log r)$, where r is the number of constraints. As observed in [2], the lower bound matches the upper bound in case r is polynomial in n.

Angluin et al. also studied the online version of this problem where each constraint has to be satisfied directly after its arrival [2]. Their motivation for this problem variant was to help infer the structure of a social network describing the spread of diseases in a community and to decide where to allocate resources to fight an epidemic efficiently. They assumed that the individuals affected by each outbreak of a disease are specified by a connectivity constraint, that the outbreaks occur over time, and that resources that have been committed cannot be released. They provided an $O(n \log n)$-competitive online algorithm for the online version along with an $\Omega(n)$-competitive lower bound. They also considered the uniform cost case of this online version, providing an $O(n^{2/3} \log^{2/3} n)$-competitive algorithm against an oblivious adversary and an $\Omega(\sqrt{n})$-competitive lower bound against an adaptive adversary.

Hosoda et al. studied a *B-constraint-bounded* variant of the Network Construction problem, where the cardinality of each connectivity constraint S_i does not exceed B [8]. This corresponds to constructing a minimum overlay network for a topic-based peer-to-peer pub/sub system where users (represented by vertices) who are interested in a common topic (represented by connectivity constraints) form connected subgraphs, and moreover, the number of users following each topic is bounded by a constant due to the publisher of that topic having a limited number of available slots for users. Hosoda et al. provided a polynomial-time approximation algorithm for this variant and proved its APX-completeness in [8].

A natural generalization of the Network Construction problem, where some pairwise connections are given a priori has applications in bioinformatics [11]. The purpose is to infer protein-protein interactions that are missing from a database based on a collection of known, overlapping protein complexes (see Sect. 4.1 for details).

1.1   The Structure of the Paper and Our New Results

The next section defines the Network Construction problem and its B-constraint-bounded variant, where each constraint includes at most B vertices. We also recall the Minimum Weight Set Cover problem and some facts about its approximability. In Sect. 3, we study the online version of the B-constraint-bounded Network Construction problem. We present $O(B^2 \log n)$-competitive and $O((B + \log r) \log n)$-competitive polynomial-time algorithms for the online B-constraint-bounded Network Construction problem, where r, n stand for the number of constraints and the number of vertices in the network, respectively. In the cost-uniform case when constraints are unbounded, we provide an $O(\sqrt{n}(\log n + \log r))$-competitive upper bound with high probability. All our randomized competitive bounds are against an adaptive adversary but for the last one which is against an oblivious adversary. We also provide a $(B-1)$-competitive lower bound in case of arbitrary edge costs and a $\sqrt{B}$-competitive lower bound in case of uniform edge costs. In Sect. 4, we study approximation algorithms for the offline Network Construction problem and its extensions. First, we discuss a hybrid approximation method combining the approximation algorithm of Hosoda et al. from [8] with that of Angluin et al. from [2] in the context of the application to bioinformatics. Next, we consider a natural strengthening of the connectivity requirements in the Network Construction problem. Each constraint is supposed to induce a subgraph (of the constructed graph) of diameter at most d, where d is given a priori. We provide a polynomial-time $\left(\binom{B}{2} - B + 2\right)\binom{B}{2}$-approximation algorithm for the aforementioned problems with the d-diameter requirements, when each constraint has at most B vertices. Also, we present a polynomial-time algorithm achieving a non-trivial approximation ratio in the general case of the d-diameter variant, where the size of constraints is unbounded. We conclude with final remarks.

Our approximate or online solutions to the aforementioned variants with bounded constraints can be used to solve approximately or online the corresponding variants with unbounded constraints by splitting the constraints into small and large ones (Sects. 3, 4).

2 Preliminaries

For a positive integer r, the term $[r]$ will denote $\{1, ..., r\}$, and for sets S, V, $|S|$ will stand for the cardinality of S while V^2 for $\{\{v, u\} | v, u \in V\}$.

A *subgraph* of a graph (V, E) is a graph (V', E') such that $V' \subseteq V$ and $E' \subseteq E$. The subgraph of a graph (V, E) *induced* by a subset S of V is the graph $(S, E \cap S^2)$. A *perfect cut* of a graph (V, E) is a partition of V into subsets V' and V'' such that $E \cap \{\{v, u\} | v \in V' \ \& \ u \in V''\} = \emptyset$. The *diameter* of a graph (V, E) is the minimum number ℓ such that any pair of vertices in V can be connected by a path composed of at most ℓ edges in E. If the graph is disconnected, its diameter is undefined.

The *Network Construction* problem is as follows [2]. We are given a set V of vertices and for each possible edge $e = \{v_i, v_j\}$, the cost $c(e)$ of its construction. We are also given a collection of connectivity constraints $S = \{S_1, ..., S_r\}$, where each S_i is a subset of V. The objective is to construct a set E of edges in V^2 such that for $i = 1, ..., r$, the subgraph of the graph (V, E) induced by S_i is connected and the total cost of the edges in E is minimal. In the *uniform-cost* case of the problem, we have $c(e) = 1$ for all edges in V^2. We can naturally generalize the problem to include the *Network Extension* problem, where some subset E' of edges is already given (constructed) a priori. Note that when zero construction costs of edges are allowed the Network Construction problem is equivalent to that of Network Extension. Simply it is sufficient to set the construction costs of the edges given a priori to zero in order to obtain an equivalent version of the Network Construction problem. In order to avoid duplications in our statements, in the aforementioned situation we shall mention only the Network Construction problem.

Among other things, the following fact was established by Angluin et al. in [2].

Fact 1 *(Theorem 2 in [2]). There is a polynomial-time $O(\log r)$-approximation algorithm for the Network Construction problem on r constraints.*

We shall also consider a *B-constraint-bounded* variant of the Network Construction problem, where the cardinality of each connectivity constraint S_i does not exceed B. It was studied by Hosoda et al. in [8]. They provided a polynomial-time approximation algorithm for this variant and showed its APX-completeness.

Fact 2 *(Theorem 4 in [8]). There is a polynomial-time $\lfloor B/2 \rfloor \lceil B/2 \rceil$-approximation (i.e., $\approx B^2/4$-approximation) algorithm for the B-constraint-bounded Network Construction problem.*

In the context of the Network Construction and Extension problems, we refer to two types of edges: those already constructed and the remaining ones that potentially could be constructed. For instance, when referring to a perfect cut, we consider the edges of the first type while when we refer to edges crossing a perfect cut we mean the edges of the second type.

Recall the definition of *Minimum Weight Set Cover* problem. The input to this problem is a universal set U on n elements and a family F of m subsets of U. Each subset in F is assigned a non-negative weight. A *set cover* is a sub-family of F whose union is equal to U. The objective is to find a set cover of minimum total weight. The decision version of this optimization problem is already NP-hard in the uniform-weight case [6]. The so-called *Minimum Weight Hitting Set* problem is an equivalent formulation of the Minimum Weight Set Cover problem with the roles of elements and subsets exchanged. Here, the input is a finite set S of weighted elements and a family C of subsets of S. The objective is to find a minimum weight subset of S that hits all the subsets in C, i.e., that has a non-empty intersection with each of the subsets in C. This problem is known to be equivalent to Minimum Weight Set Cover [3]. Consequently, approximation algorithms and inapproximability results for each of them carry over to the other one.

Hochbaum [7] used a relaxation of an integer linear programming formulation to obtain an approximation of the Minimum Weight Set Cover in cubic time. The same approximation ratio was obtained by Bar-Yehuda and Even [4] with a more direct, linear-time method. We summarize their results as follows.

Fact 3. *The Minimum Weight Set Cover problem (U, F), where each element of the universal set U occurs in at most B subsets of U in F, can be approximated within multiplicative factor B in linear time. Consequently, the Minimum Weight Hitting set problem, where each subset in the given family has cardinality at most B, can be approximated within B in linear time.*

3 Online B-Constraint-Bounded Network Construction

In this section, we consider the online version of the Network Construction problem studied in [2]. It arises naturally in the situation when the knowledge about the relationships between the entities represented by vertices changes over time. In the online version, the collection of connectivity constraints is given one at a time. When a constraint S_i is presented, the online algorithm is in round i. The algorithm is now supposed to satisfy this constraint during this round by constructing, if necessary, additional edges before the start of the next round (no previously constructed edges may be removed). The next constraint is then presented in round $i+1$. To study the worst-case performance of our online algorithms, we shall use an *adaptive adversary* that can wait with setting the next constraint until the online algorithm satisfied the previous one. We shall use competitive analysis of our online algorithms. An online algorithm is *c-competitive* if the cost of its solution does not exceed c times the cost of an optimal offline solution.

3.1 Upper Bounds

First consider the following online Fractional Network Construction problem: For a set V of vertices and edge costs $c(e)$ for $e \in V^2$, and sequence of connectivity constraints $S_1, ..., S_r$, assign fractional capacities $w(e)$ to the edges e such that for each $i \in [r]$, for each pair of vertices in S_i, the maximum flow between them is at least 1. The optimization objective is to minimize $\sum_e w(e)c(e)$.

Fact 4 *(Lemma 2 in [2]).* *There is an $O(\log n)$-competitive polynomial-time algorithm for the online Fractional Network Construction problem on n vertices.*

By using this fact, we obtain the following theorem.

Theorem 1. *There is an $O(B^2 \log n)$-competitive polynomial-time algorithm for the online B-constraint-bounded Network Construction problem on n vertices.*

Proof. Run the online $O(\log n)$-competitive algorithm for the online Fractional Network Construction problem from Fact 4. Disregard all edges that are assigned capacity smaller than B^{-2} by the online solution to the fractional problem and construct all the remaining edges. Note that after the edges of capacity smaller than B^{-2} are removed, for any pair of vertices in any B bounded constraint the maximum flow is still at least $1 - \binom{B}{2}B^{-2} \geq \frac{1}{2}$. Hence, there is a path composed of the constructed edges between such a pair. The cost of the constructed edges is at most B^2 times larger than the cost of the fractional solution, i.e., the sum of products of edge cost and edge capacity over all edges. □

To derive another competitive upper bound for the online B-constraint-bounded Network Construction problem, we shall consider the online version of the Minimum Weight Set Cover problem. In this version, a family of subsets of the universal set is given a priori while the elements of a subset of the universal are presented online one at a time [1]. A new element has to be covered before the arrival of the next one. Analogously, in the online version of the equivalent Minimum Weight Hitting set problem, the set of hitting elements is given a priori, and the sets to be hit arrive online one at a time. A new set has to be hit before the arrival of the next one. Alon et al. established the following fact in [1].

Fact 5. *There is an $O(\log n \log r)$-competitive polynomial-time algorithm for the online Minimum Weight Set Cover problem, where n is the cardinality of the universal set and r is the cardinality of the given family of subsets of the universal set. Consequently, there is an $O(\log n \log r)$-competitive polynomial-time algorithm for the online Minimum Weight Hitting set, where n is the cardinality of the family of sets to hit and r is the cardinality of the set of all possible hitting elements.*

By combining Fact 5 with the reduction of the Network Construction problem to the Minimum Weight Set Cover problem given by Angluin et al. in [2], we obtain another competitive upper bound for the online B-constraint-bounded Network Construction problem.

Theorem 2. *There is an $O((B + \log r) \log n)$-competitive polynomial-time algorithms for the online B-constraint-bounded Network Construction problem with n vertices and r constraints.*

Proof. We shall reduce the online Network Construction problem to the online Minimum Weight Hitting Set problem, following the reduction of the former problem to the online Minimum Weight Set Cover problem from [2]. The set of the possible hitting elements given a priori is just the set of all possible edges. Each edge has weight equal to the cost of its construction. Next, each constraint upon its arrival online, for each perfect cut of the subgraph induced by the constraint, yields the set of all (additional potential) edges crossing the perfect cut, i.e., having endpoints in the two different parts of the bipartition. Note that the constraint is satisfied if and only if each perfect cut in the subgraph induced by it is crossed by some edge accounted to the online formed hitting set. It follows that the cost of an optimal solution to the resulting online Minimum Hitting Set problem is the same as that to the original Network Construction problem. Now it is sufficient to observe that the former problem has $O(n^2)$ possible hitting elements and at most $r2^B$ sets to hit, and then to apply Fact 5.   □

We can use Theorem 2 to derive a competitive upper bound for the uniform-cost variant of the Network Construction problem with unbounded constraints. Angluin et al. considered also the uniform cost variant of the Network Construction problem in [2], providing an $O(n^{2/3} \log^{2/3} n)$-competitive algorithm against an oblivious adversary and an $\Omega(\sqrt{n})$-competitive lower bound against an adaptive adversary. Our upper bound in the uniform case is as that of Angluin et al. against an oblivious adversary, i.e., an adversary not knowing the randomized results of the algorithm. The key idea is to split the constraints into small and large ones, and use Theorem 2 to process the former ones.

Theorem 3. *The uniform cost online Network Construction problem on n vertices and r constraints admit an $O(kn^{0.5}(\log n + \log r))$-competitive polynomial-time solution, for every positive k, with probability at least $1 - O((nr)^{-1}) - \frac{1}{k}$ provided that r is known in advance and the adversary is oblivious.*

Proof. Split the set of constraints into two sets, one consisting of all constraints of size $\leq n^{0.5}$ and one consisting of the rest. We can apply the $O((B + \log r) \log n)$ competitive algorithm from Theorem 2 to the small constraints obtaining an $O((n^{0.5} + \log r) \log n)$ competitive solution. To satisfy the large constraints with more than $n^{0.5}$ vertices we proceed as follows.

Let Q be the set of vertices involved in the large constraints, and let q stand for the cardinality of Q. We initialize an empty vertex set S. Upon an arrival of a new large constraint, each vertex v in the constraint that is outside of S is added to S with probability $q^{-0.5}(\ln n + 2 \ln r)$. We may assume w.l.o.g. that $\ln n + 2 \ln r < q^{0.5}$. Furthermore, if v is added to S then all missing edges incident to v are constructed. It follows that the expected total number, and hence, the expected total cost of the so constructed edges amounts to $q^{0.5}(q-1)(\ln n + 2 \ln r)$. Thus, the total cost is at most $kq^{0.5}(q - 1)(\ln n + 2 \ln r)$ with probability at least $1 - \frac{1}{k}$ by Markov's inequality. For each large constraint, the probability that it does not contain any vertex from S is at most

$$(1 - \frac{q^{0.5}(\ln n + 2\ln r)}{q})^{n^{0.5}} \leq (1 - \frac{1}{n^{0.5}})^{n^{0.5}(\ln n + 2\ln r)} \leq O(\frac{1}{nr^2})$$

Since there are at most r large constraints, the cost of an optimal solution is at least $q - 1$ and $q \leq n$, we obtain an $kn^{0.5}(\ln n + 2 \ln r)$ competitive upper bound for the large constraints with probability at least $1 - O((nr)^{-1}) - \frac{1}{k}$.   □

3.2   Lower Bounds

We present two lower bounds on the competitiveness of algorithms for the online B-constraint-bounded Network Construction problem.

When the edge costs can be arbitrary, it is not possible to achieve a competitive ratio smaller than $B - 1$.

Theorem 4. *For any $c < 1$, there is no $c(B-1)$-competitive algorithm for the online B constraint-bounded Network Construction problem.*

Proof. We modify the proof of Theorem 6 in [2] for the competitive ratio in the general case of online Network Construction, in our case the optimal offline solution is not necessarily a path. Following [2], we set the cost of edges among the first $n - 1$ vertices to zero, and the cost of all edges incident to the last vertex to 1. The adversary divides the first $n - 1$ vertices into blocks of $B - 1$ vertices. Assume first that $n - 1$ is divisible by $B - 1$. Then for $i = 1, ..., (n - 1)/(B - 1)$, the adversary repetitively picks an l-tuple of vertices, $l \in [2, B]$, that includes the last vertex and all vertices from the i-th block that are not endpoints of already constructed edges incident to the last vertex. In this way, the algorithm is forced to construct all $B - 1$ edges connecting the vertices in the i-th block with the last vertex while in the optimal offline solution only such a last edge is needed while the vertices in the i-th block are connected in the order of their removal by a constructed path. Thus, the algorithm constructs $(B - 1) \times \frac{n-1}{B-1}$ edges of cost 1 while the optimal offline solution uses only $\frac{n-1}{B-1}$ edges of cost 1. It follows that the algorithm cannot be $c(B - 1)$ competitive.

If $n - 1$ is not divisible by $B - 1$ then the algorithm constructs at least $(B - 1) \times \lfloor \frac{n-1}{B-1} \rfloor + 1$ edges of cost 1 while the optimal offline solution uses only $\lceil \frac{n-1}{B-1} \rceil$ edges of cost 1. Hence, for enough large n, the algorithm cannot be $c(B - 1)$-competitive.   □

For the uniform cost case, we can present a weaker lower bound. The proof of the following theorem can be found in the full version of this paper.

Theorem 5. *The online uniform cost B constraint-bounded Network Construction problem has an $\Omega(\sqrt{B})$ competitive lower bound.*

4   Offline Approximation Algorithms

In this section, we discuss first a hybrid approximation method for the offline Network Construction problem and its application to bioinformatics. It combines the approximation algorithm of Hosoda et al. from [8] with that of Angluin et al. from [2]. Next, we present approximation algorithms for a strengthened version of the Network Construction problem, where each constraint is supposed to induce a subgraph (of the constructed graph) of diameter at most d for a d given a priori.

4.1   A Hybrid Method with Biological Applications

An application of the Network Extension problem to bioinformatics was given in [11]. There, the goal was to infer protein-protein interactions (PPIs) that were missing from a database based on a collection of known, overlapping protein complexes. More precisely, the vertices V in the input graph were used to represent proteins, the set E' of a priori given edges represented PPIs already in the database, and each input connectivity constraint S_i consisted of the proteins belonging to a single protein complex. Using the assumption that each protein complex must induce a connected subgraph, solving instances of the Network Extension problem gave lower bounds on the number of missing PPIs in various widely used PPI databases. The overwhelming majority of complexes in the existing PPI databases seem to be of small size, containing at most 10 proteins each, but a few larger ones with up to 100 proteins also occur (for details, see Table A3 in the Supporting Information file for [11]).

The aforementioned statistics suggest a hybrid method consisting of applying the approximation algorithm of Hosoda et al. from [8] to the constraints corresponding to small complexes and that of Angluin et al. from [2] to the constraints corresponding to larger complexes. We can express it in terms of the Network Construction problem by the equivalence observed in Sect. 2. The output is the union of the output of each of the two algorithms applied separately. Hence, by combining Fact 1 with Fact 2, we obtain the following theorem.

Theorem 6. *Consider an instance of the Network Construction problem. For $B \in [n]\backslash\{1\}$, let r_B be the number of constraints with more than B vertices in the instance. A solution to the instance (for the respective problem) of total cost not exceeding $\min_{B \in [n]\backslash\{1\}} \lfloor B/2 \rfloor \lceil B/2 \rceil + O(\log r_B)$ times the minimum can be found in polynomial time.*

The hybrid method will be useful when there is a relatively small $B \in [n]\backslash\{1\}$ such that the number r_B of large constraints including more than B vertices, i.e., the number of large complexes in the biological application, is small. More details about the hybrid method can be found in the full version of this paper.

4.2 Bounded Diameter Requirements

One can naturally strengthen the connectivity requirements in the Network Construction or Extension problems by demanding that each constraint should induce a subgraph of the constructed network of diameter at most d, where $d \in [n-1]$ is given a priori (cf. [5]).

For instance, Chockler et al. studied the Network Construction problem in [5] using a different terminology. They considered the problem of constructing an optimal overlay (network) that for each topic (constraint) includes a dissemination tree composed of nodes interested in the topic (i.e., belonging to the constraint). One of the measures of the quality of such an overlay suggested on p. 116 of [5] is the diameter. Intuitively, having a low diameter is good because it means that two users interested in the same topic do not need to rely on many intermediate parties, which leads to more efficient communication and better performance.

We shall term the strengthened version of the Network Construction problem as the d-diameter Network Construction problem. In fact, the latter problem restricted to instances with a single constraint is already hard. The restriction can be simply rephrased as follows: given a vertex set V, edge costs $c(e)$ for potential edges in V^2, find a cheapest graph spanning V with diameter not exceeding d.

The d-diameter Network Construction problem restricted to single constraint instances is known to be NP-hard already for $d = 2$ [10]. In contrast, when restricted to instances with uniform edge costs, this problem variant becomes trivial as any spanning star graph provides an optimal solution.

Analogously to the preceding sections, we can consider the d-diameter Network Construction problem with constraints of cardinality not exceeding B. By using an *auxiliary problem*, we can obtain a $\left(\binom{B}{2} - B + 2\right)\binom{B}{2}$ approximation in polynomial time for the B-constraint-bounded d-diameter Network Construction problem. The auxiliary problem is as follows.

For an instance of the B-constraint-bounded d-diameter Network Construction problem with a vertex set V, edge construction costs $c(e)$, a set E' of edges e with $c(e) = 0$, and connectivity constraints $S_1, ..., S_r$ find a minimum cost edge set $E'' \subseteq V^2 \backslash E'$ such that for $i = 1, ..., r$, if the diameter of the subgraph of $G' = (V, E)$ induced by S_i is larger than d then $E'' \cap S_i^2 \neq \emptyset$.

The following lemma provides an approximation algorithm for the auxiliary problem.

Lemma 1. *The auxiliary problem can be approximated within* $\binom{B}{2}$ *in polynomial time.*

Proof. Consider an instance of the auxiliary problem with a vertex set V, edge construction costs $c(e)$, a set E' of edges with zero construction cost, and connectivity constraints $S_1, ..., S_r$. We may assume w.l.o.g. that for $i = 1, ..., r$, the diameter of the subgraph of the graph $G' = (V, E')$ induced by S_i is larger than d since otherwise the constraint S_i can be disregarded. To solve the auxiliary problem, for $i = 1, ..., r$, form the set E_i of all edges in $S_i^2 \backslash E'$. The auxiliary problem is equivalent to finding a minimum weight subset of the set of all potential edges that hits all the sets $E_1, ..., E_r$, where the weights of the edges are equal to their construction costs. By our assumptions, for $i = 1, ..., r$, $|E_i| \leq \binom{B}{2}$ hold. Now it is sufficient to apply Fact 3 in order to obtain a $\binom{B}{2}$ approximation for the auxiliary problem in time linear in the total size of the family $\{E_1, ..., E_r\}$ and $|V|^2$. The latter size is in turn polynomial in the size of the input instance of the auxiliary problem. $\square$

Now, in order to provide an approximate solution to an instance of the B-constraint-bounded d-diameter Network Construction problem, we iterate the method of Lemma 1 as shown in Fig. 1.

Theorem 7. *The B-constraint-bounded d-diameter Network Construction problem can be approximated within* $\left(\binom{B}{2} - B + 2\right)\binom{B}{2}$ *in polynomial time.*

Proof. We shall analyze the iterative method based on Lemma 1. Since for $i = 1, ..., r$, $|S_i| \leq B$, the subgraph of the original graph $G' = (V, E')$ induced by S_i can be completed by at most $\binom{B}{2}$ edges. Hence, at most $\binom{B}{2}$ iterations of the while block are sufficient. In fact, already $\binom{B}{2} - B + 2$ iterations are sufficient since in a graph with B vertices and at least $\binom{B}{2} - (B - 2)$ edges each pair of non-adjacent vertices has a common neighbor. Note that the cost of an optimal solution to any of the at most $\binom{B}{2} - B + 2$ auxiliary problems approximately solved in consecutive iterations of the while block cannot be greater than that of an optimal solution to the original B-constraint-bounded d-diameter Network Construction problem. Hence, the upper bound $\left(\binom{B}{2} - B + 2\right)\binom{B}{2}$ on the approximation factor of the iterative method follows from Lemma 1. $\square$

Input: a vertex set V, edge construction costs $c(e)$, a set E' of edges e with $c(e) = 0$, and d-diameter constraints $S_1, ..., S_r$ of cardinality $\leq B$.

Output: A set of edges yielding an $(\binom{B}{2} - B + 2)\binom{B}{2}$ approximation to the d-diameter Network Construction problem for the input instance.

1: $E'' \leftarrow E'$
2: **while** there is a not yet satisfied constraint S_i **do**
3: Use the method of Lemma 1 to solve the auxiliary problem approximately by an edge set $E''' \subseteq V^2 \setminus E''$
4: $E'' \leftarrow E'' \cup E'''$ (E'' can be interpreted as the set of already constructed edges)
5: Remove all connectivity constraints S_i such that the subgraph of $G' = (V, E'')$ induced by S_i has diameter $\leq d$.
6: Set the construction costs of the edges in E''' to zero.
7: **end while**
8: **return** $E'' \setminus E'$

Fig. 1. The $(\binom{B}{2} - B + 2)\binom{B}{2}$ approximation algorithm for the B-constraint-bounded-diameter Network Construction problem.

In the general case with unbounded constraints, straightforward greedy approaches do not seem to work. However, if the edge costs are uniform, we can obtain a large but still a nontrivial approximation factor in polynomial time by splitting the constraints into small and large ones, and using Lemma 1 to obtain an approximation for the former. The proof is analogous to that of Theorem 3. It can be found in the full version of this paper.

Theorem 8. *The uniform cost d-diameter Network Construction problem with n vertices and r constraints admits an $O(n^{0.8}(\ln n + \ln r))$ approximation with probability at least $1 - (nr)^{-1}$ in polynomial time.*

5 Final Remarks

It would be useful to tighten the upper and lower competitiveness bounds on the online version of the B-constraint-bounded Network Construction problem. It would be especially interesting to know if the factor that is logarithmic in n can be removed from the upper bounds.

As mentioned in Sect. 4.2, straightforward greedy approaches do not seem to work for the d-diameter Network Construction problem with unbounded constraints. One reason for this is that natural candidates for potential functions in greedy methods seem to lack the submodularity property. It is an interesting question if it is possible to achieve a reasonable approximation factor for this problem in the general case, at least when edge costs are uniform.

Acknowledgments. We would like to thank Tatsuya Akutsu and Natsu Nakajima for introducing us to the problem studied in this paper.

References

1. Alon, N., Awerbuch, B., Azar, Y., Buchbinder, N., Naor, J.: The online set cover problem. SIAM J. Comput. **39**(2), 361–370 (2009)
2. Angluin, D., Aspnes, J., Reyzin, L.: Network construction with subgraph connectivity constraints. J. Comb. Optim. **29**(2), 418–432 (2013). https://doi.org/10.1007/s10878-013-9603-2
3. Ausiello, G., D'Atri, A., Protasi, M.: Structure preserving reductions among convex optimization problems. J. Comput. Syst. Sci. **21**, 136–153 (1980)
4. Bar-Yehuda, R., Even, S.: A linear-time approximation algorithm for the weighted vertex cover problem. J. Algorithms **2**(2), 198–203 (1981)
5. Chockler, G.V., Melamed, R., Tock, Y., Vitenberg, R.: Constructing scalable overlays for pub-sub with many topics. In: Proceedings of the Twenty-Sixth Annual ACM Symposium on Principles of Distributed Computing (PODC 2007), pp. 109–118 (2007)
6. Garey, M.R., Johnson, D.S.: Computers and Intractability - A Guide to the Theory of NP-Completeness. W. H. Freeman and Company, New York (1979)
7. Hochbaum, D.S.: Approximation algorithms for the set covering and vertex cover problems. SIAM J. Comput. **11**(3), 555–556 (1982)
8. Hosoda, J., Hromkovič, J., Izumi, T., Ono, H., Steinová, M., Wada, K.: On the approximability and hardness of minimum topic connected overlay and its special instances. Theor. Comput. Sci. **429**, 144–154 (2012)
9. Korach, E., Stern, M.: The clustering matroid and the optimal clustering tree. Math. Program. **98**, 385–414 (2003). https://doi.org/10.1007/s10107-003-0410-x
10. Li, C., McCormick, S.T., Simchi-Levi, D.: On the minimum-cardinality-bounded-diameter and the bounded-cardinality-minimum-diameter edge addition problems. Oper. Res. Lett. **11**, 303–308 (1992)
11. Nakajima, N., Hayashida, M., Jansson, J., Maruyama, O., Akutsu, T.: Determining the minimum number of protein-protein interactions required to support known protein complexes. PLOS ONE **13**(4) (2018). Article e0195545

Globally Rigid Augmentation of Minimally Rigid Graphs in $\mathbb{R}^2$

Csaba Király[1,2](✉) (iD) and András Mihálykó[2] (iD)

[1] MTA-ELTE Egerváry Research Group, Eötvös Loránd Research Network (ELKH),
Budapest, Hungary
[2] Department of Operations Research, ELTE Eötvös Loránd University,
Budapest, Hungary
{cskiraly,mihalyko}@cs.elte.hu

Abstract. The two main concepts of Rigidity Theory are rigidity, where the framework has no continuous deformation, and global rigidity, where the given distance set determines the locations of the points up to isometry. We consider the following augmentation problem. Given a minimally rigid graph $G = (V, E)$ in $\mathbb{R}^2$, find a minimum cardinality edge set F such that the graph $G' = (V, E + F)$ is globally rigid in $\mathbb{R}^2$. We provide a min-max theorem and an $O(|V|^2)$ time algorithm for this problem.

Keywords: Global Rigidity · Augmentation · Rigidity · Combinatorial Algorithm

1 Introduction

Let us consider the following motivating question: Given some sensors in the plane and the distances between some pairs of them, at least how many of them need to be localized so that we could reconstruct the exact sensor-locations? This is the so-called *global rigidity pinning* (or anchoring) problem. Sometimes measuring the exact locations is too expensive or even impossible. Instead, one may ask at least how many new distances need to be measured so that the distances uniquely determine the positions of the sensors (up to isometry). This problem is called the *global rigidity augmentation* problem. The concept of global rigidity, which appears in the previous network localization problems, plays an important role in rigidity theory [3,5,12].

Let us consider the aforementioned problems by the means of Rigidity Theory. A d-dimensional **framework** is a pair (G, p), where $G = (V, E)$ is a graph and $p : V \to \mathbb{R}^d$ is a map of the vertices to the d-dimensional space. We call p a **realization** of G in $\mathbb{R}^d$. Two frameworks (G, p) and (G, q) are **equivalent** if $||p(u) - p(v)|| = ||q(u) - q(v)||$ for every $uv \in E$. (G, p) and (G, q) are **congruent** if $||p(u) - p(v)|| = ||q(u) - q(v)||$ holds for every vertex pair $u, v \in V$, or in other words, when (G, q) can be obtained from (G, p) by an isometry of $\mathbb{R}^d$. We say that the framework (G, p) is **globally rigid** if each framework (G, q) which is equivalent to (G, p) is also congruent to (G, p), that is, the length of the edges in

T. Calamoneri and F. Corò (Eds.): CIAC 2021, LNCS 12701, pp. 326–339, 2021.
https://doi.org/10.1007/978-3-030-75242-2_23

(G, p) uniquely determines the realization up to isometry of $\mathbb{R}^d$. (For example, Fig. 1(c) is a globally rigid framework in $\mathbb{R}^2$.) A framework (G, p) is called **rigid** if there exists an $\varepsilon > 0$ such that each framework (G, q), which is equivalent to (G, p) and for which $\|p(v) - q(v)\| < \varepsilon$ holds for each $v \in V$, is also congruent to (G, p), that is, if every edge-length preserving continuous motion of the framework results in a framework which is congruent to (G, p). (See Fig. 1(a) for an example of a non-rigid and Fig. 1(b) for a rigid framework in $\mathbb{R}^2$.)

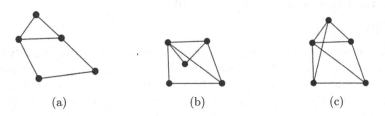

(a)   (b)   (c)

Fig. 1. Frameworks of various rigidity in $\mathbb{R}^2$. (a) A non-rigid framework. (b) A rigid framework which is not globally rigid. (c) A globally rigid framework.

Deciding whether a given framework is rigid (globally rigid, respectively) in $\mathbb{R}^d$ is NP-hard for $d \geq 2$ ($d \geq 1$, respectively) [1,21]. The analysis gets more tractable if we consider *generic frameworks* where the set of coordinates of the points is algebraically independent over the rationals. In this case, the rigidity and the global rigidity of the framework depends only on the underlying graph G [5,8,23]. (We note that reconstructing the position of the points is a challenging task, even if they are uniquely determined by the framework, see [2,16,22]. In this paper we do not address this problem.)

A graph G is called **rigid** (or **globally rigid**) in $\mathbb{R}^d$ if each (or equivalently some) of its generic realizations as a framework is rigid (or globally rigid, respectively). The combinatorial characterization of rigid and globally rigid graphs is known for $d = 1, 2$ [11,20] while it is a major open problem of rigidity theory for $d \geq 3$. We shall use these combinatorial characterizations in our work.

For generic frameworks, the global rigidity augmentation problem can be modelled as follows:

Problem 1. *Given a graph $G = (V, E)$, find an edge set F of minimum cardinality on the same vertex set, such that $G + F = (V, E \cup F)$ is globally rigid in $\mathbb{R}^2$.*

The complexity of Problem 1 is open. There are some partial results in connection with it, for example, Fekete and Jordán [6] gave a constant factor approximation for the global rigidity pinning problem in $\mathbb{R}^2$ for generic frameworks, however, the complexity of that problem is also open. In Sect. 5 we show how the result of [6] can be applied to give a constant factor approximation for Problem 1.

In this paper we shall solve Problem 1 optimally for a special case. A graph $G = (V, E)$ is called **minimally rigid**, if G is rigid but $G - e$ is not rigid for any $e \in E$. We show that, if G is minimally rigid in Problem 1, then we can give a min-max theorem and also an $O(|V|^2)$ time algorithm that solves the problem optimally. Moreover, it follows from this result that the globally rigid pinning problem also can be solved optimally for minimally rigid graphs (see Sect. 5). The most of the proofs are left for the full version of this extended abstract [17].

2   Preliminaries and Definitions

2.1   Rigidity in $\mathbb{R}^2$

In this subsection we collect the basic definitions and results from rigidity theory that we shall use. There are several equivalent approaches to graph rigidity, for our purpose, a combinatorial one is the most practical. For a detailed introduction to rigidity theory including the equivalence of our approach, the reader is referred to [14].

A graph $G = (V, E)$ is called **sparse** if $i(X) \leq 2|X| - 3$ for all $X \subseteq V$ with $|X| \geq 2$, where $i(X)$ denotes the number of edges induced by X. A graph $G = (V, E)$ is called **tight** (or sometimes Laman) if it is sparse and $|E| = 2|V| - 3$. This definition can be used for the characterization of the rigid graphs in $\mathbb{R}^2$ by the fundamental results of Pollaczek-Geiringer and Laman.

Theorem 1 ([19,20]). *A graph G is minimally rigid in $\mathbb{R}^2$ if and only if G is tight. Thus, a graph G is rigid in $\mathbb{R}^2$ if it contains a spanning tight subgraph.*

As we work in $\mathbb{R}^2$ we omit this indication from the rest of this paper. A graph $G = (V, E)$ is called **k-connected** if $|V| > k$ and $G - X$ is connected for any vertex set $X \subset V$ of cardinality at most $k - 1$. Connectivity has several connections to rigidity. An often used folklore result is the following (see [14]).

Lemma 1. *If $G = (V, E)$ is a tight graph for which $|V| \geq 3$, then G is 2-connected.*

The most important result related to our problem is the following characterization of global rigidity in $\mathbb{R}^2$ due to Jackson and Jordán. An edge e of a rigid graph G is called **redundant** if $G - e$ is rigid. A graph is **redundantly rigid** if all of its edges are redundant.

Theorem 2 ([11]). *A graph $G = (V, E)$ with $|V| > 3$ is globally rigid in $\mathbb{R}^2$ if and only if it is redundantly rigid and 3-connected.*

Based on the above results, the problem we shall solve in this paper is equivalent to the following.

Problem 2. *Given a tight graph $G = (V, E)$, find a graph $H = (V, F)$ with a minimum cardinality edge set F, such that $G \cup H$ is redundantly rigid and 3-connected.*

If G has at most 3 vertices then G is tight if and only if it is globally rigid [11], hence the solution of Problem 2 is obvious. Thus we may suppose in what follows that G contains at least 4 vertices.

2.2 The Redundant Rigidity Augmentation Problem and Co-tight Sets

Let us first investigate the problem of augmenting a tight graph $G = (V, E)$ to a redundantly rigid graph by a minimum number of edges. This problem was considered and solved before by García and Tejel [7]. A generalization of this augmentation problem to (k, ℓ)-tight graphs appears in a work by the authors of this paper [18]. We use some ideas from both of these works.

Tight graphs have some well known properties. By definition, any subgraph of a sparse graph is also sparse and any tight subgraph of a sparse graph is an induced subgraph. With standard submodular techniques one can prove the well-known fact that the intersection and the union of two tight subgraphs of a sparse graph is also tight if they have at least two common vertices (see [14]). Given two vertices $u, v \in V$ of a tight graph $G = (V, E)$, this fact implies that the intersection of all the tight subgraphs of G which contain both of u and v is also tight, and hence it is the unique minimal tight subgraph of G containing both of u and v. Let us denote this unique minimal tight subgraph of G containing both of u and v by $\boldsymbol{T(uv)}$ (or simply by $T(e)$ when e is an edge between u and v). It is easy to see that the edge set of $T(e)$ is exactly the set of those edges of G which become redundant if we add the edge e to G (see [7]). Similarly, if we add the edges $e_1, \ldots, e_k$ to G, (the edges of) some subgraph of G will become redundant, which we denote by $\boldsymbol{R(e_1, \ldots, e_k)}$. For the sake of convenience, we will not distinguish a graph from its edge set, that is, we denote the edge set of $T(e)$ and $R(e_1, \ldots, e_k)$ by $T(e)$ and $R(e_1, \ldots, e_k)$, respectively. The following statement generalizes the fact that $R(e_1) = T(e_1)$.

Lemma 2 ([7, Lemma 4]). *Let $G = (V, E)$ be a tight graph. Then $R(e_1, \ldots, e_k) = T(e_1) \cup \cdots \cup T(e_k)$ for arbitrary edges $e_1, \ldots, e_k$.*

Lemma 2 is the base of our method hence we will use it throughout the paper without explicitly referring to it.

Given a tight graph $G = (V, E)$, a non-empty set $C \subsetneq V$ is called **co-tight** if $V - C$ induces a tight subgraph. This is equivalent to the following: C is co-tight in G if $0 < |C| \leq |V| - 2$ and $2|C| = i(C) + d(C, V - C)$, where $\boldsymbol{d(X, Y)}$ denotes the number of edges between two disjoint sets $X, Y \subsetneq V$. For the sake of brevity, let us abbreviate the name of minimal co-tight sets by **MCT** sets. See Fig. 2 for an example. Observe that every tight graph G on at least 4 vertices contains at least two co-tight sets that do not contain each other, as any edge forms a tight subgraph of G.

Let C be a co-tight set of a tight graph G. If $\{u, v\} \cap C = \emptyset$, then $V(T(uv)) \cap C = \emptyset$ by the definition of $T(uv)$. Thus the next lemma follows easily by Lemma 2.

Lemma 3 ([18, Observation 5.3]). *The vertex set of any edge set that augments a tight graph G to a redundantly rigid graph must intersect every co-tight set.*

Let $\boldsymbol{C^*}$ denote the family of all MCT sets of G. We shall use the following key result on MCT sets (which are called minimal co-rigid sets in [14]).

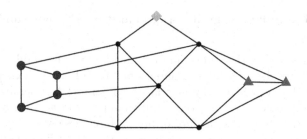

Fig. 2. A tight graph with two MCT sets, the set formed by the big (blue) circles and the set formed by the (gray) square. Adding an edge between any (blue) circle vertex and the (gray) square vertex augments G to a redundantly rigid graph, which is not globally rigid, as it is not 3-connected. Adding an edge between a (red) triangle vertex and the (gray) square vertex augments G to a 3-connected but not redundantly rigid graph. (Color figure online)

Lemma 4 ([14, **Theorem 3.9.13**]). *Let G be a tight graph. Then the members of $\mathcal{C}^*$ are pairwise disjoint or there are two vertices $v, w \in V$ such that $\{v, w\} \cap C \neq \emptyset$ for all $C \in \mathcal{C}^*$.*

If there are at least two intersecting MCT sets, then it is easy to deduce from Lemma 4 that the edge $e = vw$ (for the pair $v, w \in V$ provided by the lemma) is an optimal solution of the redundant augmentation problem, that is, $R(e) = T(e) = G$. In the general case, the following theorem determines the cardinality of the optimal augmentation.

Theorem 3 ([18, **Theorem 1.1**]). *Let G be a tight graph on at least 4 vertices. Then $\min\{|F| : F$ is an edge set on V for which $G + F$ is a redundantly rigid graph$\} = \max\left\{\left\lceil\frac{|\mathcal{C}|}{2}\right\rceil : \mathcal{C}$ is a family of disjoint co-tight sets in $G\right\}$.*

2.3   The 3-Connectivity Augmentation Problem

By Lemma 1, every tight graph is 2-connected and thus we need to augment a 2-connected graph to a 3-connected graph. There exists several methods to deal with this particular problem, even linear time algorithms [10]. However, we also need to augment G to a redundantly rigid graph hence we stick to a simpler approach following the ideas of [13].

Let us call $u, v \in V$ a **cut-pair** of G, if $G - \{u, v\}$ is not connected. If u, v is a cut-pair in G, then let $b_{(u,v)}(G)$ denote the number of components of $G - \{u, v\}$. Let $b(G)$ denote the maximum value of $b_{(u,v)}(G)$ over all cut-pairs u, v of G. If there are no cut-pairs in G, let $b(G) := 1$. Let $N(X)$ denote the neighbor set of $X \subseteq V$, that is, $N(X) := \{v \in V - X :$ there exists an edge uv such that $u \in X\}$. A set $P \subset V$ is called a **3-fragment** if $|N(P)| = 2$ and $P \cup N(P) \neq V$. The maximum number of pairwise disjoint 3-fragments is denoted by $t(G)$.

To augment a 2-connected graph G to a 3-connected graph, we need to increase the number of neighbors of each 3-fragment of G, and hence the vertex set of any edge set that augments G to a 3-connected graph must intersect all 3-ends. Moreover, any edge set F that augments G to a 3-connected graph needs to span a connected graph on the components of $G - \{u, v\}$ for every cut-pair u, v. Thus $|F| \geq b(G) - 1$. These imply the following well-known statement.

Lemma 5. *Given a 2-connected graph G, the minimum number of edges that augments G to a 3-connected graph is at least* $\max\left\{b(G) - 1, \left\lceil \frac{t(G)}{2} \right\rceil\right\}$.

In fact, any 2-connected graph can be augmented to a 3-connected graph by a set of $\max\left\{b(G) - 1, \left\lceil \frac{t(G)}{2} \right\rceil\right\}$ edges (see [10, 13]).

Let us call an inclusion-wise minimal 3-fragment a **3-end**. As every 3-fragment contains at least one 3-end, $t(G)$ is equal to the number of pairwise disjoint 3-ends. In a rigid graph, this latter value is equal to the number of 3-ends since their disjointness follows by the following result of Jackson and Jordán [11].

Lemma 6 ([11]). *Let G be a rigid graph in $\mathbb{R}^2$. Then, for any two disjoint cut-pairs v_1, v_2 and u_1, u_2 of G, u_1 and u_2 are in the same component of $G - \{v_1, v_2\}$.*

3 Min-Max Theorem

In this section we shall merge the results on the redundant rigidity and 3-connectivity augmentation problems to a new min-max theorem for the global rigidity augmentation problem by mixing the statements of Theorem 3 and Lemma 5, as follows.

Theorem 4. *Let $G = (V, E)$ be a tight graph on at least 4 vertices. Then*

$$\min\{|F| : F \text{ is an edge set on } V \text{ for which } G + F \text{ is globally rigid}\} = \max\left\{b(G) - \right.$$

$$\left. 1, \max\left\{\left\lceil \frac{|\mathcal{A}|}{2} \right\rceil : \mathcal{A} \text{ is a family of disjoint co-tight sets and 3-fragments}\right\}\right\}.$$

Proof (Sketch). Recall that a graph on at least 4 vertices is globally rigid if and only if it is 3-connected and redundantly rigid by Theorem 2. The min $\geq$ max implication in Theorem 4 is obvious since the set of endvertices of the optimal augmenting edge set must intersect all co-tight sets and 3-fragments by Lemmas 3 and 5. Notice that, if G is 3-connected, then Theorem 4 follows directly by Theorem 3. Hence from now on, we may assume that G is not 3-connected. In this case we shall extend the proof of Theorem 3 given in [18] with the ideas of the 3-connectivity augmentation method given by Jordán [13]. Hence to prove the min $\leq$ max part, let us consider the family of all MCT sets and 3-ends of a tight graph G. Let us call the inclusion-wise minimal elements of this family the **atoms** of G. (In Fig. 2 these are the three sets formed by the highlighted vertices: the big (blue) circles form an MCT set, the (gray) square vertex form an MCT set which is also a 3-end, and the (red) triangle vertices form a 3-end.) Let us denote the family of atoms by $\mathcal{A}^*$. We shall show that the atoms are pairwise disjoint and there exists a set of $\max\left\{b(G) - 1, \left\lceil \frac{|\mathcal{A}^*|}{2} \right\rceil \right\}$ edges that augments G to a globally rigid graph. Hence we first need the following counterpart of Lemma 4 for atoms.

Lemma 7. Let $G = (V, E)$ be a tight graph which is not 3-connected. Then the atoms of G are pairwise disjoint.

Note that if G is 3-connected, Lemma 7 does not always hold (see Lemma 4). As we have seen before in Sect. 2.3, the 3-ends of G are pairwise disjoint and Lemma 4 implies that two MCT sets can only intersect each other in special circumstances. Beside these facts, the proof of Lemma 7 uses the following inter-mediate result. The proofs of both lemmas can be found in the full version [17] of this extended abstract.

Lemma 8. Suppose that $G = (V, E)$ is a tight graph. Let $a \in A$ be a vertex from an atom $A \in \mathcal{A}^*$ of G. Then there is no $v \in V$ such that a, v forms a cut-pair.

Now, we turn to prove that there exists a set of $\max\left\{b(G) - 1, \left\lceil \frac{|\mathcal{A}^*|}{2} \right\rceil \right\}$ edges that augments G to a globally rigid graph. A set X is called a **transversal** of a family $\mathcal{S}$ if $|X \cap S| = 1$ for each $S \in \mathcal{S}$ and $|X| = |\mathcal{S}|$. As the members of $\mathcal{A}^*$ are pairwise disjoint if G is not 3-connected by Lemma 7, choosing one arbitrary vertex from every member of $\mathcal{A}^*$ leads to a transversal of $\mathcal{A}^*$.

Let P be a transversal of $\mathcal{A}^*$. Observe that P is a minimum cardinality vertex set that intersects all MCT sets and 3-ends, and consequently all co-tight sets and 3-fragments. Hence $|\mathcal{A}| \leq |P|$ holds for an arbitrary family $\mathcal{A}$ of disjoint co-tight sets and 3-fragments. We shall show now that a connected graph on P of $\mathcal{A}^*$ augments G to a globally rigid graph, that is, 3-connected and redundantly rigid. Later, we will reduce the number of edges needed for this augmentation to the optimum value. First it is easy to observe that any connected graph on P augments G to a 3-connected graph since P covers all 3-ends (by the definition of the atoms and Lemma 7) and contains no vertex from any cut-pair by Lemma 8.

Lemma 9. Suppose that G is a tight graph which is not 3-connected. Let P be a transversal of $\mathcal{A}^*$. Then, for any connected graph $H = (P, F)$ on P, $G \cup H$ is 3-connected.

To show that the above augmentation gives a redundantly rigid graph, one can extend the ideas of the proof of Theorem 3 from [18] for atoms by using Lemma 7 instead of Lemma 4. (Again, see [17] for the full proofs.) Recall that $R(F)$ denotes the set of redundant edges of G in $G + F$.

Lemma 10 (Extension of [18, Lemma 5.8]). Suppose that G is a tight graph which is not 3-connected. Let P be a transversal of $\mathcal{A}^*$ and let F be the edge *set of a connected graph on $P' \subseteq P$. Then $R(F)$ is the minimal tight subgraph containing all elements of P'. In particular, if F is the edge set of a star $K_{1,|P|-1}$ on the vertex set P, then $G + F$ is redundantly rigid.*

Observation 1. *Lemmas 9 and 10 imply that $G + F$ is globally rigid if F is an edge set of an arbitrary connected graph (in particular, a tree) on a transversal P of $\mathcal{A}^*$.*

The idea of Observation 1 can be found in [15], where the authors got to this fact from a different approach, with the so-called extreme vertices. The connection between these two approaches is presented in [18, Lemma 5.10].

By the min $\geq$ max part of Theorem 4, $\left\lceil \frac{|\mathcal{A}^*|}{2} \right\rceil$ edges are always needed to augment G to a globally rigid graph. However, if $|\mathcal{A}^*| \leq 3$ then it is indeed enough to do so by Observation 1. On the other hand, if $|\mathcal{A}^*| > 3$, then we need to reduce the number of edges used by the augmentation provided by Observation 1. To this end, we shall use the following straightforward adaptation of [18, Lemma 5.9] (see [17] for the proof).

Lemma 11. *Let $G = (V, E)$ be a tight graph which is not 3-connected and let P be a transversal of $\mathcal{A}^*$. Suppose that $x_1, x_2, x_3, y \in P$ are distinct vertices. Let $T^* = T(x_1 y) \cup T(x_2 y) \cup T(x_3 y)$. Then $T^* = T(x_1 y) \cup T(x_2 x_3)$ or $T^* = T(x_2 y) \cup T(x_1 x_3)$ holds.*

Observe that the operation in Lemma 11 allows us to reduce the cardinality of the edge set used for the augmentation by maintaining the property that it augments G to a redundantly rigid graph. However, we also need to maintain the 3-connectivity of the augmentation to complete the proof of Theorem 4.

To reduce the number of edges needed for the augmentation in such a way that the global rigidity of the augmented graph is maintained, we do the following procedure. Initially, let $F := \emptyset$ and $N := P$. During the procedure, the set $N \subseteq P$ stands for "not fixed" vertices while vertices in $P - N$ are the "fixed" vertices. We can **fix** an edge $f_1 f_2$ by removing f_1 and f_2 from N and adding $f_1 f_2$ to F. In each step of the procedure we carefully choose two vertices from N and fix the edges between them (decreasing the number of vertices in N by two and increasing the number of edges in F). Hence the edge set F always covers the vertices of $P - N$. We shall keep the following properties during the whole procedure:

1. For an arbitrary star S_N on the vertex set N, $G + F + S_N$ is a redundantly rigid graph.
2. In every 3-end of $G + F$, there is at least one vertex from N
3. $\max\left\{ b(G + F) - 1, \left\lceil \frac{|N|}{2} \right\rceil \right\} + |F| = \max\left\{ b(G) - 1, \left\lceil \frac{|P|}{2} \right\rceil \right\}$.

Notice that Properties 1–3 hold for $N = P$ and $F = \emptyset$ by Lemmas 9 and 10.

Remark 1. *Properties 1 and 2 ensure that $G + F + S_N$ is redundantly rigid and 3-connected, and thus globally rigid by Theorem 2. Property 3 ensures the optimality.*

Remark 2. *If $|N| \geq 4$, then from any two edges chosen on $x_1, x_2, x_3 \in N$ fixing one of them maintains Property 1 by Lemma 11.*

By Remark 2 we always aim to find at least two possibilities to fix such that Property 2 holds. Also, if it can be done so that $\max\left\{b(G+F) - 1, \left\lceil \frac{|N|}{2} \right\rceil\right\}$ decreases by one, then we can maintain Properties 1–3. Roughly, we distinguish 4 different possibilities in each of which we find 3 vertices from N such that we can apply Remark 2 and hence we can fix one edge while maintaining Properties 1–3.

Lemma 12. *Let G be a tight graph which is not 3-connected such that $|\mathcal{A}^*| \geq 4$. Let P be a transversal on $\mathcal{A}^*$. Let $N \subseteq P$ be a vertex set and F be an edge set on P such that they satisfy Properties 1–3. If $|N| \geq \max\{4, b(G+F)+1\}$, then we can choose $f_1, f_2 \in N$, such that for $N - \{f_1, f_2\}$ and $F + \{f_1 f_2\}$ (that is, for fixing $f_1 f_2$) Properties 1–3 also hold.*

Proof. We use the following method for the proof. This is the core of our algorithm which we will describe in Sect. 4.

1 If $b(G+F) - 1 \geq \left\lceil \frac{|N|}{2} \right\rceil$, **then**

2   If there is only one cut-pair (u, v) such that $b_{(u,v)}(G+F) = b(G+F)$, **then**

        Choose x_1, x_2 from a component of $G + F - \{u, v\}$ that contains at least two vertices from N. Let $x_3 \in N$ be a vertex from a component of $G + F - \{u, v\}$ that does not contain x_1 and x_2.

3   **else**

        Let (u_1, v_1) and (u_2, v_2) be two cut-pairs for which $b_{(u_1,v_1)}(G + F) = b(G+F) = b_{(u_2,v_2)}(G+F)$. Choose $x_1, x_2 \in N$ from two different components of $G+F-\{u_1, v_1\}$ that do not contain $\{u_2, v_2\}$. Choose $x_3 \in N$ from a component of $G + F - \{u_2, v_2\}$ that does not contain $\{u_1, v_1\}$.

4 else

5   If there is a cut-pair $\{u, v\}$ such that for one component of $G + F - \{u, v\}$, say K, $|N \cap K| \geq 2$ and $|(V - K) \cap N| \geq 2$, **then**

        Choose x_1, x_2 from $N \cap K$ and choose $x_3 \in N$ from $(V - K) \cap N$.

6   **else** (Notice that if $b(G + F) = 1$, then this is the only possible case.)

        Choose $x_1, x_2, x_3 \in N$ arbitrarily.

7 If $G + F + S(N - \{x_1, x_3\}) + x_1 x_3$ is redundantly rigid, **then**
        $f_1 := x_1, f_2 := x_3$.

    else

        $f_1 := x_2, f_2 := x_3$.

First we prove that the above method is consistent, that is, we can execute each of its steps. As $|N| \geq b(G+F)+1$ and P contains no vertex from a cut-pair of G by Lemma 8, $|N| > b_{(u,v)}(G+F)$ for an arbitrary cut-pair $\{u,v\}$. Hence, there exists a component of $G+F-\{u,v\}$ that contains at least two vertices from N. This shows that we can choose vertices in STEPS 2 and 5 consistently. Meanwhile, in STEP 3 there are at least two components of $G+F-\{u_1,v_1\}$ that do not contain $\{u_2,v_2\}$ since $|N| \geq 4$ and thus $b_{(u_1,v_1)}(G+F) \geq 3$.

Now let us turn to show that the choice of f_1 and f_2 maintains Property 2.

Claim. Suppose that there is a cut-pair $\{u,v\}$ such that for one component of $G+F-\{u,v\}$, say K, $x_1, x_2 \in N \cap K$ and $x_3, y \in (V-K) \cap N$. Then fixing either $x_1 x_3$ or $x_2 x_3$ maintains Property 2.

Proof. Notice that the role of x_1 and x_2 is symmetric thus we might suppose that we fixed the edge $x_1 x_3$. Suppose that we form a new 3-end L in $G+F$ with it. Then necessarily $x_1, x_3 \in L$. If $x_2 \in L$ or $y \in L$, then Property 2 holds automatically. On the other hand, if none of them is in L, then there is a cut-pair in $K \cup \{u\}$ or in $K \cup \{v\}$ which separates x_1 from x_2. There is another cut-pair in $V-K$ (other than $\{u,v\}$, say $\{u',v'\}$) which separates x_3 from y. Both remain cut-pairs after fixing the edge $x_1 x_3$. However, this contradicts the assumption that L is 3-end in $G+F$, as $|N(L)| = 2$ must hold for a 3-end. $\square$

Notice that the conditions of the above Claim hold in STEPS 2, 3 and 5 thus with our choice of x_1, x_2, and x_3 Property 2 is maintained. If $G+F$ is already 3-connected, then Property 2 is obvious. Otherwise, in STEP 6, every cut-pair cuts $G+F$ into two component, one of which contains exactly one element from N by the condition of STEP 5. For the sake of a contradiction, assume that $G+F+f_1 f_2$ contains a 3-end L which contains no element of $N-\{f_1, f_2\}$. Let $N(L) = \{u,v\}$. Then $N \cap L = \{f_1.f_2\}$, $V-L-\{u,v\} \neq \emptyset$, and u,v is a cut pair of $G+F$. By our above condition, (u,v) cuts $G+F$ into two component one of which contains exactly one element from N. Hence exactly L and $V-L-\{u,v\}$ are these two components. Moreover, as $|L \cap N| = 2$, this implies $|N \cap (V-L-\{u,v\})| = 1$, contradicting $|N| \geq 4$.

Now we show that our method maintains Property 3. Fixing any edge decreases $\left\lceil \frac{|N|}{2} \right\rceil$ by one while increases F by one. By STEPS 5 and 6 it is enough to keep Property 3 true as in this case $\max\left\{ b(G+F)-1, \left\lceil \frac{|N|}{2} \right\rceil \right\} > b(G+F)-1$. We need to show that if the condition in STEP 1 is true, then we also decrease $b(G+F)$. If $b(G+F)-1 \geq \left\lceil \frac{|N|}{2} \right\rceil$, then there can be at most two cut-pairs of $G+F$ satisfying $b_{(u,v)}(G+F) = b(G+F)$ by a simple calculation on the number of 3-ends (see [13]). If there is only one, the pair (u,v) chosen in STEP 2, then we only need to decrease $b_{(u,v)}(G+F)$. Since $x_1 x_3$ and $x_2 x_3$ both connect two different components of $G+F-\{u,v\}$, $b_{(u,v)}(G+F)$ decreases by one after fixing any of them. If there are exactly two such cut-pairs, (u_1, v_1) and (u_2, v_2) chosen in STEP 3, then we need to decrease $b_{(u_1,v_1)}(G+F)$ and $b_{(u_2,v_2)}(G+F)$ simultaneously. Again our choice of $x_1 x_3$ and $x_2 x_3$ guarantees this.

Therefore, by Remark 2 applied to STEP 7, fixing $f_1 f_2$ maintains Properties 1–3. This completes the proof of Lemma 12.   □

We apply Lemma 12 recursively until $|N| < \max\{4, b(G+F)+1\}$. To complete the proof of Theorem 4, we need to show the following.

Claim. If $|N| \leq \max\{3, b(G+F)\}$, then, for an arbitrary star S_N on N, $G + F + S_N$ forms a globally rigid graph for which $|F| + |S_N| = \max\left\{b(G) - 1, \left\lceil \frac{|P|}{2} \right\rceil\right\}$.

Proof. $G + F + S_N$ is globally rigid by Remark 1. By Property 3 it is enough to show that $\max\left\{b(G+F) - 1, \left\lceil \frac{|N|}{2} \right\rceil\right\} = |S_N| = |N| - 1$. If $|N| = b(G+F)$, then $\max\left\{b(G+F) - 1, \left\lceil \frac{|N|}{2} \right\rceil\right\} = |N| - 1$ as $\left\lceil \frac{|N|}{2} \right\rceil \leq |N| - 1$. On the other hand, if $|N| < b(G+F)$, then $2 \leq |N| \leq 3$ thus $\left\lceil \frac{|N|}{2} \right\rceil = |N| - 1$.   □

Recall that $\mathcal{A}^*$ consists of pairwise disjoint MCT sets and 3-ends of G and hence the maximum in Theorem 4 is at least $\max\left\{b(G) - 1, \left\lceil \frac{|\mathcal{A}^*|}{2} \right\rceil\right\}$. On the other hand, the above claim implies that G can be augmented to a globally rigid graph by an addition of an edge set of cardinality $\max\left\{b(G) - 1, \left\lceil \frac{|P|}{2} \right\rceil\right\} = \max\left\{b(G) - 1, \left\lceil \frac{|\mathcal{A}^*|}{2} \right\rceil\right\}$. This completes the proof of Theorem 4.   □

4 Algorithmic Aspects

It is easy to see that the proof of Theorem 4 provides an algorithm for Problem 2 when the input tight graph $G = (V, E)$ is not 3-connected. On the other hand, the algorithm of García and Tejel [7] or that by the authors of this paper in [18] provides an algorithm for the case where G is 3-connected since in this case we only need a redundantly rigid augmentation of G. In this section we sketch how one can provide an $O(|V|^2)$ time algorithm for Theorem 4.

Theorem 5. *Let $G = (V, E)$ be a tight graph. There exists an $O(|V|^2)$ time algorithm that finds a graph $H = (V, F)$ with a minimum cardinality edge set F for which $G + H$ is a globally rigid graph.*

Proof (sketch). Note that the tightness of G implies that $|E| = 2|V| - 3$. Hence the 3-connectivity of G and all cut-pairs and 3-ends of G can be found in $O(|V|)$ time by the algorithm of Hopcroft and Tarjan [9].

The algorithm of Berg and Jordán [4] checks the tightness of G in $O(|V|^2)$ time, moreover, after this it can be used to calculate $T(ij)$ for each pair of vertices $i, j \in V$ in linear time. This fact was used to show that the algorithms in [7] and [18] both provide an optimal redundantly rigid augmentation of G in $O(|V|^2)$ time which completes the proof when the input is 3-connected.

To start the algorithm of Lemma 12, we first need a transversal of $\mathcal{A}^*$. (And this is also needed to solve the case where $|\mathcal{A}^*| \leq 3$.) This can be calculated in $O(|V|^2)$ time by using [18, Algorithm 6.1] and [18, Algorithm 6.9] with some slight modifications. We leave the details to the full version of this paper [17].

Since F is a matching throughout the algorithm of Lemma 12, we need to run the algorithm recursively $O(|V|)$ times, and $G + F$ has $O(|V|)$ edges in each recursive call of the algorithm. To execute the steps of the algorithm, we need to know every cut-pair (u, v) of the graph $G + F$ along with the value of $b_{(u,v)}(G + F)$, and we need to check whether the condition of STEP 5 holds. These all can be checked in $O(|V|)$ time based on the structure provided by the algorithm of Hopcropft and Tarjan [9], see again the full version [17] for more details. Finally, STEP 7 of the algorithm of Lemma 12 can also be executed in $O(|V|)$ time since we only need to calculate the subgraphs $T(xx_1)$, $T(xx_2)$, $T(xx_3)$, and $T(x_1x_3)$ (which needs $O(|V|)$ running time by [4]) for an arbitrarily chosen $x \in N - \{x_1, x_2, x_3\}$ and check whether $T(xx_1) \cup T(xx_2) \cup T(xx_3) = T(xx_2) \cup T(x_1x_3)$.   □

5   Concluding Remarks

In this paper, we solved Problem 1 in the case where the input is a tight graph. For general inputs, a constant factor approximation can be given, as follows.

Let us recall the global rigidity pinning problem. In this problem, the goal is to anchor a minimum set of points of a framework such that the resulting framework is globally rigid. In the generic case, pinning can be modelled by adding a complete graph on the anchored vertices to the graph (see [6]). Moreover, instead of a complete graph we can add any globally rigid graph on the anchored vertex set, for example the square graph of a cycle. (A square of a graph arises by connecting all pairs of vertices which has distance at most 2 in the original graph). Notice that the square graph of the cycle on the vertex set V consists of $2|V|$ edges. This way one can see that a constant approximation to the global rigidity pinning problem gives a constant approximation to the global rigidity augmentation problem and *vice versa*. Fekete and Jordán [6] investigated the global rigidity pinning problem and gave a constant approximation algorithm to it. This implies that there exists a polynomial time constant approximation algorithm to Problem 1 (and it has an approximation ratio at most 4 times more than that of the pinning problem).

For tight input graphs, we can solve the global rigidity pinning problem optimally as follows. It can be shown easily that we must pin at least one vertex from each atom. On the other hand, a complete graph on a transversal of $\mathcal{A}^*$ indeed augments G to a globally rigid graph as it contains also the optimal edge set given by Theorem 4. Thus one vertex from each atom pins the graph optimally. (When G is 3-connected, we may apply the method of [18, Sect. 8] directly.)

Acknowledgements. Project no. NKFI-128673 has been implemented with the support provided from the National Research, Development and Innovation Fund of Hungary, financed under the FK_18 funding scheme. The first author was supported by the János Bolyai Research Scholarship of the Hungarian Academy of Sciences and by the ÚNKP-19-4 and ÚNKP-20-5 New National Excellence Program of the Ministry for Innovation and Technology. The second author was supported by the European Union, co-financed by the European Social Fund (EFOP-3.6.3-VEKOP-16-2017-00002). The authors are grateful to Tibor Jordán for his help, the inspiring discussions and his comments.

References

1. Abbot, T.G.: Generalizations of Kempe's universality theorem. Master's thesis, MIT (2008). http://web.mit.edu/tabbott/www/papers/mthesis.pdf
2. Anderson, B.D.O., Shames, I., Mao, G., Fidan, B.: Formal theory of noisy sensor network localization. SIAM J. Discrete Math. **24**, 684–698 (2010)
3. Aspnes, J., et al.: A theory of network localization. IEEE Trans. Mob. Comput. **5**(12), 1663–1678 (2006)
4. Berg, A.R., Jordán, T.: Algorithms for graph rigidity and scene analysis. In: Di Battista, G., Zwick, U. (eds.) ESA 2003. LNCS, vol. 2832, pp. 78–89. Springer, Heidelberg (2003). https://doi.org/10.1007/978-3-540-39658-1_10
5. Connelly, R.: Generic global rigidity. Discrete Comput. Geom. **33**(4), 549–563 (2005). https://doi.org/10.1007/s00454-004-1124-4
6. Fekete, Z., Jordán, T.: Uniquely localizable networks with few anchors. In: Niko-letseas, S.E., Rolim, J.D.P. (eds.) ALGOSENSORS 2006. LNCS, vol. 4240, pp. 176–183. Springer, Heidelberg (2006). https://doi.org/10.1007/11963271_16
7. García, A., Tejel, J.: Augmenting the rigidity of a graph in $\mathbb{R}^2$. Algorithmica **59**(2), 145–168 (2011). https://doi.org/10.1007/s00453-009-9300-9
8. Gortler, S.J., Healy, A.D., Thurston, D.P.: Characterizing generic global rigidity. Am. J. Math. **132**(4), 897–939 (2010)
9. Hopcroft, J., Tarjan, R.: Dividing a graph into triconnected components. SIAM J. Comput. **2**, 135–158 (1973)
10. Hsu, T.S., Ramachandran, V.: A linear time algorithm for triconnectivity augmentation. In: Proceedings of the Annual Symposium on Foundations of Computer Science, pp. 548–559 (1991)
11. Jackson, B., Jordán, T.: Connected rigidity matroids and unique realizations of graphs. J. Comb. Theory Ser. B **94**, 1–29 (2005)
12. Jackson, B., Jordán, T.: Graph theoretic techniques in the analysis of uniquely localizable sensor networks. In: Mao, G., Fidan, B. (eds.) Localization Algorithms and Strategies for Wireless Sensor Networks, pp. 146–173. IGI Global (2009)
13. Jordán, T.: On the optimal vertex-connectivity augmentation. J. Comb. Theory Ser. B **63**, 8–20 (1995)
14. Jordán, T.: Combinatorial rigidity: graphs and matroids in the theory of rigid frameworks. In: Discrete Geometric Analysis, Volume 34 of MSJ Memoirs, pp. 33–112. Mathematical Society of Japan (2016)
15. Jordán, T., Mihálykó, A.: Minimum cost globally rigid subgraphs. In: Bárány, I., Katona, G.O.H., Sali, A. (eds.) Building Bridges II. BSMS, vol. 28, pp. 257–278. Springer, Heidelberg (2019). https://doi.org/10.1007/978-3-662-59204-5_8
16. Kaewprapha, P., Li, J., Puttarak, N.: Network localization on unit disk graphs. In: 2011 IEEE Global Telecommunications Conference - GLOBECOM 2011, pp. 1–5 (2011)

17. Király, Cs., Mihálykó, A.: Globally rigid augmentation of minimally rigid graphs in $\mathbb{R}^2$. Technical report TR-2020-07, Egerváry Research Group, Budapest (2020). www.cs.elte.hu/egres

18. Király, Cs., Mihálykó, A.: Sparse graphs and an augmentation problem. Technical report TR-2020-06, Egerváry Research Group, Budapest (2020). www.cs.elte.hu/egres. An extended abstract appeared in Bienstock, D., Zambelli, G. (eds.) Integer Programming and Combinatorial Optimization, IPCO 2020. Lecture Notes in Computer Science, vol. 12125, pp. 238–251. Springer, Cham (2020)

19. Laman, G.: On graphs and rigidity of plane skeletal structures. J. Eng. Math. **4**, 331–340 (1970). https://doi.org/10.1007/BF01534980

20. Pollaczek-Geiringer, H.: Über die Gliederung ebener Fachwerke. ZAMM-J. Appl. Math. Mech. **7**(1), 58–72 (1927)

21. Saxe, J.B.: Embeddability of weighted graphs in k-space is strongly NP-hard. Technical report, Computer Science Department, Carnegie-Mellon University, Pittsburgh, PA (1979)

22. So, A., Ye, Y.: Theory of semidefinite programming for sensor network localization. Math. Program. **109**, 405–414 (2005). https://doi.org/10.1007/s10107-006-0040-1

23. Whiteley, W.: Some matroids from discrete applied geometry. In: Bonin, J.E., Oxley, J.G., Servatius, B. (eds.) Matroid Theory, Volume 197 of Contemporary Mathematics, pp. 171–311. AMS (1996)

Extending Partial Representations of Rectangular Duals with Given Contact Orientations

Steven Chaplick[1] , Philipp Kindermann[2] , Jonathan Klawitter[2(✉)] ,
Ignaz Rutter[3] , and Alexander Wolff[2]

[1] Maastricht University, Maastricht, The Netherlands
[2] Universität Würzburg, Würzburg, Germany
[3] Universität Passau, Passau, Germany

Abstract. A rectangular dual of a graph G is a contact representation of G by axis-aligned rectangles such that (i) no four rectangles share a point and (ii) the union of all rectangles is a rectangle. The partial representation extension problem for rectangular duals asks whether a given partial rectangular dual can be extended to a rectangular dual, that is, whether there exists a rectangular dual where some vertices are represented by prescribed rectangles. Combinatorially, a rectangular dual can be described by a regular edge labeling (REL), which determines the orientations of the rectangle contacts. We characterize the RELs that admit an extension, which leads to a linear-time testing algorithm. In the affirmative, we can construct an extension in linear time.

Keywords: rectangular dual · partial representation extension

1 Introduction

A *geometric intersection representation* of a graph G is a mapping $\mathcal{R}$ that assigns to each vertex w of G a geometric object $\mathcal{R}(w)$ such that two vertices u and v are adjacent in G if and only if $\mathcal{R}(u)$ and $\mathcal{R}(v)$ intersect. In a *contact representation* we further require that, for any two vertices u and v, the objects $\mathcal{R}(u)$ and $\mathcal{R}(v)$ have disjoint interiors. The *recognition problem* asks whether a given graph admits an intersection or contact representation whose sets have a specific geometric shape. Classic examples are interval graphs [1], where the objects are intervals of $\mathbb{R}$, or coin graphs [18], where the objects are interior-disjoint disks in the plane. The *partial representation extension problem* is a natural generalization of this question where, for each vertex u of a given subset of the vertex set, the geometric object is already prescribed, and the question is whether this partial representation can be extended to a full representation of the input graph. In the last decade the partial representation extension problem has been intensely

Partially supported by DFG grants Ru 1903/3-1 and Wo 758/11-1.

T. Calamoneri and F. Corò (Eds.): CIAC 2021, LNCS 12701, pp. 340–353, 2021.
https://doi.org/10.1007/978-3-030-75242-2_24

studied for various classes of intersection graphs, such as (unit or proper) interval graphs [15,16], circle graphs [6], trapezoid graphs [20], as well as for contact representations [5] and bar-visibility representations [7].

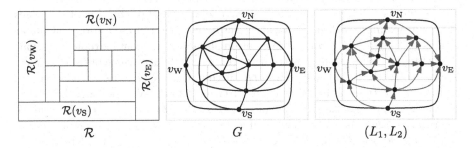

Fig. 1. A rectangular dual $\mathcal{R}$ for the graph G; the REL (L_1, L_2) induced by $\mathcal{R}$.

Rectangular Duals. In this paper we consider the partial representation extension problem for the following type of representation. A *rectangular dual* of a graph G is a contact representation $\mathcal{R}$ of G by axis-aligned rectangles such that (i) no four rectangles share a point and (ii) the union of all rectangles is a rectangle; see Fig. 1. We observe that G may admit a rectangular dual only if it is planar and internally triangulated. Furthermore, a rectangular dual can always be augmented with four additional vertices (one on each side) so that only four rectangles touch the outer face of the representation. It is customary that the four vertices on the outer face are denoted by v_S, v_W, v_N, and v_E corresponding to the geographic directions, and to require that $\mathcal{R}(v_W)$ is the leftmost rectangle, $\mathcal{R}(v_E)$ is rightmost, $\mathcal{R}(v_S)$ is bottommost, and $\mathcal{R}(v_N)$ is topmost; see Fig. 1. We call these vertices the *outer vertices* and the remaining ones the *inner vertices*. It is known that a plane internally-triangulated graph has a representation with only four rectangles touching the outer face if and only if its outer face is a 4-cycle and it has no separating triangles, that is, a triangle whose removal disconnects the graph [19]. Such a graph is called a *properly-triangulated planar (PTP)* graph. Kant and He [14] have shown that a rectangular dual of a given PTP graph G can be computed in linear time.

Historically, rectangular duals have been studied due to their applications in architecture [26], VLSI floor-planning [23,27], and cartography [12]. Besides the question of an efficient construction algorithm [14], other problems concerning rectangular duals are area minimization [4], sliceability [22], and area-universality, that is, rectangular duals where the rectangles can have any given areas [9]. The latter question highlights the close relation between rectangular duals and rectangular cartograms. Rectangular cartograms were introduced in 1934 by Raisz [25] and combine statistical and geographical information in thematic maps, where geographic regions are represented as rectangles and scaled in proportion to some statistic. There has been lots of work on efficiently computing rectangular cartograms [3,13,21]; Nusrat and Kobourov [24] recently surveyed this topic. As a dissection of a rectangle into smaller rectangles, a rectangular dual is also related to other types of dissections, for example with squares [2] or hexagons [8]; see also Felsner's survey [10].

Regular Edge Labelings. The combinatorial aspects of a contact representation of a graph G can often be described with a coloring and orientation of the edges of G. For example, Schnyder woods describe contact representations of planar graphs by triangles [11]. Such a description also exists for contact representations by rectangles, for example for triangle-free rectangle arrangements [17] or rectangular duals [14]. More precisely, a rectangular dual $\mathcal{R}$ gives rise to a 2-coloring and an orientation of the inner edges of G as follows. We color an edge $\{u,v\}$ blue if the contact between $\mathcal{R}(u)$ and $\mathcal{R}(v)$ is a horizontal line segment, and we color it red otherwise. We orient a blue edge $\{u,v\}$ as (u,v) if $\mathcal{R}(u)$ lies below $\mathcal{R}(v)$, and we orient a red edge $\{u,v\}$ as (u,v) if $\mathcal{R}(u)$ lies to the left of $\mathcal{R}(v)$; see Fig. 1. The resulting coloring and orientation has the following properties:

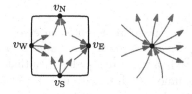

1. All inner edges incident to v_W, v_S, v_E, and v_N are red outgoing, blue outgoing, red incoming, and blue incoming, respectively.
2. The edges incident to each inner vertex form four counterclockwise ordered non-empty blocks of red incoming, blue incoming, red outgoing, and blue outgoing, respectively.

Fig. 2. Edge order at the four outer vertices and an inner vertex. (Color figure online)

A coloring and orientation with these properties is called a *regular edge labeling (REL)* or *transversal structure.* We let (L_1, L_2) denote a REL, where L_1 is the set of blue edges and L_2 is the set of red edges. Let $L_1(G)$ and $L_2(G)$ denote the two subgraphs of G induced by L_1 and L_2, respectively. Note that both $L_1(G)$ and $L_2(G)$ are st-graphs, that is, directed acyclic graphs with exactly one source and exactly one sink. It is well known that a PTP graph has a rectangular dual if and only if it admits a REL [14]. A rectangular dual $\mathcal{R}$ *realizes* a REL (L_1, L_2) if the REL induced by $\mathcal{R}$ is (L_1, L_2). Note that while a rectangular dual uniquely defines a REL, there exist different rectangular duals that realize any given REL.

Partial Rectangular Duals. For a graph G, let $E(G)$ denote the set of edges and $V(G)$ the set of vertices of G. Let U be a subset of $V(G)$. Let $G[U]$ be the subgraph of G induced by U. A *partial rectangular dual* of $G[U]$ is a contact representation $\mathcal{P}$ that maps each $u \in U$ to an axis-aligned rectangle $\mathcal{P}(u)$. For each $u \in U$, we call $\mathcal{P}(u)$ a *fixed* rectangle. For a given graph G, a subset U of $V(G)$, and a partial rectangular dual $\mathcal{P}$ of $G[U]$, the *partial rectangular dual extension problem* asks whether $\mathcal{P}$ can be extended to a rectangular dual $\mathcal{R}$ of G. In particular, for such an extension $\mathcal{R}$ and each $u \in U$, we require that $\mathcal{P}(u) = \mathcal{R}(u)$. In this paper, we study the variant of this problem where we are not only given G, U, and $\mathcal{P}$, but also a REL (L_1, L_2) of G and ask whether there is an extension $\mathcal{R}$ of $\mathcal{P}$ that realizes (L_1, L_2).

Closely related work includes partial representation extension of segment contact graphs [5] and bar-visibility representations [7]. Both problems are NP-complete. However, the hardness reductions crucially rely on low connectivity for choices in the planar embedding. Since PTP graphs are triconnected, they have a unique planar embedding and hence these results cannot be easily transferred.

Contribution and Outline. Our first contribution is a characterization of RELs that admit an extension of a given partial rectangular dual via the existence of what we will call a *boundary path set*; see Sect. 2. Next, we provide an algorithm that constructs a boundary path set (if possible) as well as an algorithm that computes a representation extension from a boundary path set. Both algorithms run in $\mathcal{O}(nh)$ time, where $n = |V(G)|$ and $h = |U|$, and are detailed in Sect. 3. Finally, we show that by checking only for the existence of a boundary path set, but not explicitly constructing one, we can solve the partial representation extension problem in linear time; see Sect. 4. We summarize our contribution as follows.

Theorem 1. *The partial representation extension problem for rectangular duals with a fixed regular edge labeling can be solved in linear time. For yes-instances, an explicit rectangular dual can be constructed within the same time bound.*

2   Characterization

In this section, we characterize when a given PTP graph G with REL (L_1, L_2), $U \subset V(G)$, and partial representation $\mathcal{P}$ of $G[U]$ admits an extension $\mathcal{R}$ that realizes (L_1, L_2). Before we can explain our main idea, we require an observation and a few definitions.

We may assume that v_W, v_S, v_E, and v_N are in U. (Otherwise, we simply place the outer rectangles appropriately around $\mathcal{P}$ such that they touch potential neighbours in $\mathcal{P}$.) The rectangles $\mathcal{P}(v_W)$, $\mathcal{P}(v_S)$, $\mathcal{P}(v_E)$, and $\mathcal{P}(v_N)$ thus form a *frame* with the area inside partially covered and partially uncovered. To make the question of whether this uncovered area can be filled with the rectangles for $V(G) \setminus U$ more accessible, we subdivide the uncovered area into smaller parts and then try to fill them one by one. More precisely and as illustrated in Fig. 3, we draw a vertical

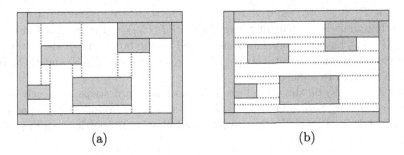

(a)   (b)

Fig. 3. Dissection of the interior of the frame into (a) vertical and (b) horizontal strips.

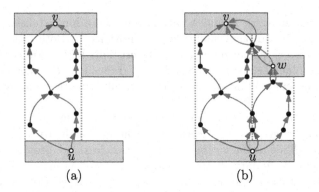

(a)   (b)

Fig. 4. (a) A boundary path pair for the strip with start rectangle $\mathcal{R}(u)$ and end rectangle $\mathcal{R}(v)$. (b) Neighboring strips can have overlapping boundary paths.

segment through the vertical sides of each fixed rectangle of an inner vertex until another fixed rectangle is hit. This divides the uncovered area inside the frame into *vertical strips*. We call the fixed rectangles bounding a vertical strip S from below and above the *start* and *end* rectangles of S, respectively. We define *horizontal strips* symmetrically. The start and end rectangle of a horizontal strip are to its left and right, respectively.

The idea for our characterization is as follows. Consider an extension $\mathcal{R}$ of $\mathcal{P}$, where the strips are now filled with rectangles. The vertical strips thus naturally induce subgraphs in $L_1(G)$ containing the vertices that lie inside them plus their start and end rectangle. Together, these subgraphs cover the whole of $L_1(G)$. In particular, for a vertical strip S with start rectangle $\mathcal{P}(u)$ and end rectangle $\mathcal{P}(v)$, the outer face of its induced subgraph consists of a path containing the rectangles along the left side of S and a path along the right side of S. The idea is that, even with $\mathcal{R}$ not known, we have to be able to cover $L_1(G)$ (and $L_2(G)$) with subgraphs defined by pairs of boundary paths. We now make this precise.

For two paths P and P' in $L_1(G)$, we write $P \preceq P'$ if no vertex of P lies to the right of P', i.e., there is no path from a vertex in P' to a vertex in $P \setminus P'$ in $L_2(G)$. Let S be a vertical strip with start rectangle $\mathcal{P}(u)$ and end rectangle $\mathcal{P}(v)$. A *boundary path pair* of S is a pair of paths $\langle P_l(S), P_r(S) \rangle$ from u to v in $L_1(G)$ such that $P_l(S) \preceq P_r(S)$ and $V(P_l(S) \cup P_r(S)) \cap U = \{u, v\}$; see Fig. 4(a). Based on the boundary path pair of S, we define $S(L_1(G))$ as the maximal subgraph of $L_1(G)$ that has precisely $P_l(S)$ and $P_r(S)$ as the boundary of the outer face. The definitions for horizontal strips, where we order paths $P_b(S)$ and $P_t(S)$ from bottom to top, are analogous.

Let $\mathcal{S}_1$ and $\mathcal{S}_2$ be the sets of vertical and horizontal strips, respectively. We define a *boundary path set* of a REL (L_1, L_2) as a set of boundary path pairs, one for each strip in $\mathcal{S}_1$ and $\mathcal{S}_2$, that satisfy the following properties (see Fig. 4(b)):

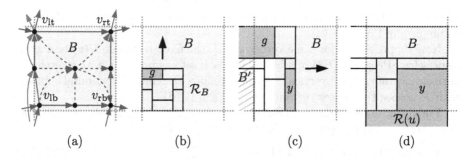

Fig. 5. (a) Graph G_B for a box B; (b) representation $\mathcal{R}_B$ for G_B; (c) adjusting the left boundary of $\mathcal{R}_B$ to $\mathcal{R}_{B'}$; and (d) adjusting the bottom boundary to $\mathcal{R}(u)$.

(B1) For strips S and S' in $\mathcal{S}_1$ with S left of S', it holds that $P_r(S) \preceq P_l(S')$.
(B2) For strips S and S' in $\mathcal{S}_2$ with S below S', it holds that $P_t(S) \preceq P_b(S')$.
(B3) The vertical strips cover $L_1(G)$, and the horizontal strips cover $L_2(G)$, that is, $\bigcup_{S \in \mathcal{S}_1} S(L_1(G)) = L_1(G)$ and $\bigcup_{S \in \mathcal{S}_2} S(L_2(G)) = L_2(G)$.

An extension $\mathcal{R}$ of $\mathcal{P}$ directly induces a boundary path set. In the following, we show that the converse is also true.

Theorem 2. *Let G be a PTP graph, let $U \subset V(G)$, and let $\mathcal{P}$ be a partial rectangular dual of $G[U]$. A REL (L_1, L_2) of G admits an extension of $\mathcal{P}$ if and only if (L_1, L_2) admits a boundary path set.*

Proof. Suppose that (L_1, L_2) admits a boundary path set. We show how to use this set to construct an extension of $\mathcal{P}$.

Let S be a vertical strip, let S' be a horizontal strip, and assume that $B = S \cap S'$ is nonempty. We call B a *box*. All such boxes together with $\mathcal{P}(U)$ form a rectangle. We now fill the boxes from the bottom-left to the top-right.

The paths in the pairs $\langle P_l(S), P_r(S) \rangle$ and $\langle P_b(S'), P_t(S') \rangle$ pairwise intersect in single vertices $v_{lb}, v_{lt}, v_{rb}, v_{rt}$. Note that some or even all of these four vertices may coincide. Let G_B be the subgraph of G whose outer cycle is formed by the boundary path pairs between these vertices; see Fig. 5(a). If we enclose G_B appropriately with a 4-cycle, we can apply the algorithm of Kant and He [14] to compute a rectangular dual $\mathcal{R}_B$ of G_B.

By the order in which we fill boxes, we have already treated those immediately to the left and below B; either of them may also be a fixed rectangle. Without loss of generality, we assume that there is a box B' that touches B from the left and a fixed rectangle that touches B from below.

First, we modify $\mathcal{R}_B$ such that it fits to the rectangular dual $\mathcal{R}_{B'}$ that is drawn inside of B'. Property (B1) of a boundary path set ensures that the rectangles in $\mathcal{R}_{B'}$ that are adjacent to the right side of $\mathcal{R}_{B'}$ are "compatible" to the rectangles in $\mathcal{R}_B$ that are adjacent to the left side of $\mathcal{R}_B$. Hence, starting with a tiny version of $\mathcal{R}_B$ placed in the lower left corner of B, we can stretch $\mathcal{R}_B$ vertically along suitable horizontal cuts such that, for every vertex u in $V(G_{B'}) \cap V(G_B)$, the left piece of $\mathcal{R}(u)$ (in B') and the right piece of $\mathcal{R}(u)$ (in B) fit together; see the green rectangle g in Fig. 5(b)–(c).

Now suppose that, for some vertex $u \in U$, the fixed rectangle $\mathcal{P}(u)$ bounds B from below. Property (B3) of a boundary path set ensures that if we stretch $\mathcal{R}_B$ horizontally along some vertical cut, then we have the correct horizontal contacts with $\mathcal{P}(u)$; see the yellow rectangle y in Fig. 5(c)–(d).

Finally, note that property (B3) ensures that, at the end of this construction, every vertex of G is represented by a rectangle in $\mathcal{R}$.   □

We close this section with an observation about the potential size of boundary path sets. As we have noted above, a vertex may lie on multiple boundary paths; in fact, it may even lie on all of them as the example in Fig. 6 shows. Hence, the size of a boundary path set can be in $\Omega(nh)$, where $n = |V(G)|$ and $h = |U|$.

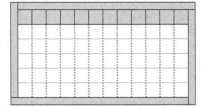

Fig. 6. A rectangular dual with a boundary path set of size $\mathcal{O}(nh)$.

3   Finding a Boundary Path Set

We now show how to compute a boundary path set for a given REL (L_1, L_2) and a partial representation $\mathcal{P}$. The idea is as follows. As we did for the boxes in the proof of Theorem 2, we handle the vertical strips in $\mathcal{S}_1$ from bottom-left to top-right. When computing the boundary path pair for a vertical strip $S \in \mathcal{S}_1$, we want the resulting graph $S(L_1(G))$ to include all necessary vertices but otherwise as few vertices as possible. In particular, there may be rectangles that by (L_1, L_2) need to have their left boundary align with the left boundary of S and thus need to be in $P_l(S)$. To make this more precise, let $\mathcal{P}(v_1), \mathcal{P}(v_2), \ldots, \mathcal{P}(v_k)$ be the fixed rectangles whose right sides touch the left side of S. Let x be a vertex that lies on a path from u to v in $L_1(G)$. Then we say x is *left-bounded* in S if and only if one of the following conditions applies (see Fig. 7(a)):

(L1) $x = u$ or $x = v$ and the left side of $\mathcal{P}(x)$ aligns with the left side of S;
(L2) (v_i, x), for some $i \in \{1, \ldots, k\}$, is an edge in $L_2(G)$;
(L3) (y, x) is the leftmost outgoing edge of y and the leftmost incoming edge of x in $L_1(G)$, and y is left-bounded;
(L4) (x, y) is the leftmost outgoing edge of x and the leftmost incoming edge of y in $L_1(G)$, and y is left-bounded.

Condition (L2) applies if $\mathcal{R}(x)$ has to be directly to the right of a fixed rectangle left of S. Condition (L3) and (L4) apply if the left side of $\mathcal{R}(x)$ has to align with the left side of a left-bounded rectangle $R(y)$ directly above or below, respectively. Note that in this case there exists also a vertex y' that is right-bounded in a strip S' left of S and $(y', x) \in E(L_2(G))$.

Next, let $\mathcal{P}(v_1'), \mathcal{P}(v_2'), \ldots, \mathcal{P}(v_k')$ be the fixed rectangles whose left sides touch the right side of S. Then x is *right-bounded* in S if and only if one of the following conditions applies (see Fig. 7(b)):

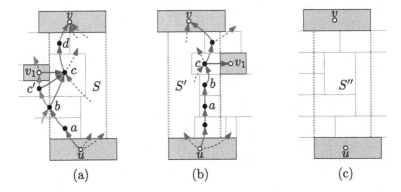

Fig. 7. (a) Vertex u is left-bounded in S on Condition (L1), c on Condition (L2), a on Condition (L3), and b is not left-bounded; (b) vertex c is right-bounded in S' on Condition (R2), and a and b on Condition (R3); (c) S'' has neither left- nor right-bounded vertices except for u and v.

(R1) $x = u$ or $x = v$ and the right side of $\mathcal{P}(x)$ aligns with the right side of S;

(R2) (x, v_i'), for some $i \in \{1, \ldots, k\}$, is an edge in G_2;

(R3) (y, x) is the rightmost outgoing edge of y and the rightmost incoming edge of x in $L_1(G)$, and y is right-bounded;

(R4) (x, y) is the rightmost outgoing edge of x and the rightmost incoming edge of y in $L_1(G)$, and y is right-bounded.

Note that x can be both left- and right-bounded. Furthermore, starting from $u, v_1', \ldots, v_k', v$, these conditions can easily be checked for each strip. Overall, we can thus find all left- and right-bounded vertices of all strips in $\mathcal{O}(n)$ time.

Theorem 3. *Let G be a PTP graph with n vertices and REL (L_1, L_2), let $U \subset V(G)$, let $h = |U|$, and let $\mathcal{P}$ be a partial rectangular dual of $G[U]$.*
In $\mathcal{O}(nh)$ time, we can decide whether (L_1, L_2) admits a boundary path set with respect to $\mathcal{P}$ and, in the affirmative, compute it.

Proof. We show how to compute the boundary path pairs for vertical strips; horizontal strips can be treated analogously. Let $\mathcal{S}_1^{\checkmark}$ be the strips in $\mathcal{S}_1$ that have already been processed. Let S be a strip with start rectangle $\mathcal{P}(u)$ and end rectangle $\mathcal{P}(v)$ such that every strip left of S is in $\mathcal{S}_1^{\checkmark}$.

An edge (x, y) of $L_1(G)$ is *suitable* if one of the following conditions applies:

(E1) $y = v$;

(E2) $y \in P_{\mathrm{r}}(S') \backslash U$, where $S' \in \mathcal{S}_1^{\checkmark}$ is directly left of S and y is not right-bounded in S';

(E3) $y \notin U$ and (x, y) is not an edge of $\mathcal{S}_1^{\checkmark}(L_1(G))$.

Condition (E2) means that $\mathcal{R}(y)$ can span from S' into S since it is not right-bounded in S'. Thus, in Fig. 7(a) (w, x) is suitable but (x, y') is not. Furthermore, (z, v) is suitable by Condition (E1), and (u, w), (w, x), and (y, z) are suitable by

348   S. Chaplick et al.

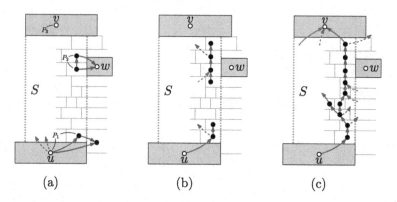

(a)   (b)   (c)

Fig. 8. Computation of $P_r(S)$ (a) starting with subpaths induced by u, right-bounded vertices and v, (b) extending these subpaths with their rightmost predecessor and successor, and (c) leftwards until they meet. (Extensions downwards are shown green.) (Color figure online)

Condition (E3). Note that $P_l(S)$ may only use suitable edges. Hence, to compute $P_l(S)$, we can start at u and always add the leftmost suitable outgoing edge until we reach v. It follows that if, at some point, there is no suitable edge available, then (L_1, L_2) does not admit a boundary path set. Taking the leftmost suitable outgoing edge ensures that $P_l(S)$ passes through all left-bounded vertices in S.

We now show how to construct $P_r(S)$, enforcing that all right-bounded vertices lie on $P_r(S)$. We thus start with the set of disjoint subpaths $P_1, P_2, \ldots, P_k$ induced by u, the right-bounded vertices, and v ordered from bottom to top; see Fig. 8(a). Note that for a right-bounded vertex x its rightmost outgoing edge also has to be in $P_r(S)$, unless $x = v$, and its rightmost incoming edge also has to be in $P_r(S)$, unless $x = u$. Therefore, we extend each subpath with these rightmost outgoing and incoming edges; see Fig. 8(b). For $i \in \{1, \ldots, k-1\}$, we then simultaneously extend P_i and P_{i+1} by always taking the leftmost suitable outgoing and incoming edge, respectively, but without crossing $P_l(S)$. If the extensions of P_i and P_{i+1} meet, we join them; see Fig. 8(c). Otherwise, both extensions will stop (due to a lack of suitable edges). In this case there is no path $P_r(S)$, and then the REL (L_1, L_2) does not admit a boundary path set.

Once $P_l(S)$ and $P_r(S)$ have been computed successfully, we update the edge set of $\mathcal{S}_1'(L_1(G))$ before processing the next strip.

The runtime is linear in the size of the boundary path set, that is, $\mathcal{O}(nh)$. □

Next, we show how to obtain an extension of $\mathcal{P}$ from a boundary path set.

Theorem 4. *Let G be a PTP graph with n vertices and REL (L_1, L_2), let $U \subset V(G)$, let $h = |U|$, and let $\mathcal{P}$ be a partial rectangular dual of $G[U]$. Given a boundary path set of (L_1, L_2), we can find an extension of $\mathcal{P}$ in $\mathcal{O}(nh)$ time.*

Proof. In the proof of Theorem 2, we gave an algorithm that finds for every box B the graph G_B of vertices whose rectangles (partially) lie inside or on the boundary of B. The algorithm computes a rectangular dual of G_B, which requires $\mathcal{O}(|V(G_B)|)$ time per box [14], and then fits each dual into the extension built so far, which can also be done in $\mathcal{O}(|V(G_B)|)$ time per box.

We now argue that $\sum_B |V(G_B)| = \mathcal{O}(nh)$. Namely, a box B either lies completely inside a rectangle, in which case $|V(G_B)| = 1$, or it contains part of the boundary of every rectangle that corresponds to a vertex in $V(G_B)$. For any vertex $v \in V(G) \setminus U$, each of the four boundary sides of $\mathcal{R}(v)$ lies either inside a single strip or on the boundary between two strips, so the boundary of $\mathcal{R}(v)$ can lie in only $\mathcal{O}(h)$ boxes in total. As there are $\mathcal{O}(h^2)$ boxes, we have $\sum_B |V(G_B)| \in \mathcal{O}(h^2 + nh) = \mathcal{O}(nh)$. $\qquad\square$

4   Linear-Time Algorithm

Explicitly constructing a boundary path set, as in Theorem 3, requires time proportional in the size of the set, which can however be in $\Omega(nh)$. In this section, we show that even without an explicit construction, we can decide if a boundary path set exists, and if so, compute an extension. Both the decision and the computation can be done in linear time.

Our approach relies on the following observations. Suppose a boundary path set exists. Let v be a vertex in $V(G) \setminus U$ that lies on a boundary path of vertical strips $S_1, \ldots, S_k$, ordered from left to right. Then the left boundary of $\mathcal{R}(v)$ lies in S_1 and the right boundary in S_k. Thus, to compute the x-coordinates of $\mathcal{R}(v)$, it suffices to know the leftmost and the rightmost boundary path on which v lies. Instead of constructing all boundary path pairs of vertical strips, we only construct the subgraph H_1 of $L_1(G)$ induced by U and the vertices on the boundary path pairs. We call H_1 the *vertical boundary graph* of (L_1, L_2). Furthermore, for each edge e in H_1, we store the leftmost and the rightmost strip for which e lies on a boundary path. Note that as H_1 is a subgraph of $L_1(G)$, the size of H_1 is in $\mathcal{O}(n)$. Analogously, we define H_2 for the horizontal strips.

Before we show how to construct the boundary graphs H_1 and H_2, we prove that they suffice to compute an extension of $\mathcal{P}$.

Lemma 5. *Let G be a PTP graph with n vertices, let $U \subset V(G)$, let $\mathcal{P}$ be a partial representation of $G[U]$ and (L_1, L_2) a REL of G. If boundary graphs H_1 and H_2 of (L_1, L_2) are given, then an extension of $\mathcal{P}$ that realizes (L_1, L_2) can be computed in $\mathcal{O}(n)$ time.*

Proof. We show how to compute the x-coordinates of rectangles using H_1; the y-coordinates can be computed analogously with H_2. The idea is to compute a rectangular dual for each inner face of H_1, which in total will yield a full rectangular dual. Note that the boundary of each face of H_1 consists of two directed paths between a start and an end vertex. Therefore, each face has a single source, a single sink, a left path, and a right path.

We distinguish two types of inner faces of H_1, namely, whether they describe a region inside a strip or a part of the boundary of a strip. A face f of the latter type can be identified by the occurrence of a right-bounded vertex v on the left path of f where v is not the source or sink of f; see Fig. 9(a). Note that in this

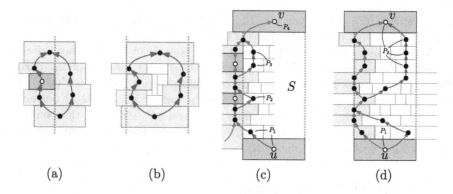

Fig. 9. Correspondence between a face of H_1 and (a) a part of the boundary of a strip or (b) a region inside a strip; extending H_1 with (c) $P_l(S)$ and (d) $P_r(S)$.

case all inner vertices of the left path of f are right-bounded and all inner vertices of the right path of f are left-bounded. We then set the right x-coordinate of every inner vertex on the left path and the left x-coordinate of every inner vertex on the right path to the x-coordinate of the respective boundary of the strip.

Otherwise, an inner face f of H_1 describes a region inside a strip of S; see Fig. 9(b). We define the graph G_f as the subgraph of G with all vertices that lie on or inside the cycle that is defined by f. By adding an outer four-cycle appropriately, we obtain a rectangular dual $\mathcal{R}_f$ of G_f with the algorithm by Kant and He [14]. We then scale $\mathcal{R}_f$ to the width of S and set the x-coordinates for the vertices in G_f inside S accordingly, that is, the right x-coordinate for the vertices on the left path of f, the left x-coordinate for vertices on the right path of f, and both x-coordinates for interior vertices of G_f.

After we have processed all faces, both x-coordinates of all vertices are set since each vertex is either fixed, has a face to the left and a face to the right, or lies inside a face. Since the faces are ordered from left to right in accordance with their respective strips, the computed x-coordinates of the rectangles also form the correct horizontal adjacencies. We repeat this process with H_2 to compute the y-coordinates and thus to obtain also the correct vertical adjacencies.

Processing a face f takes time linear in the size of G_f. Hence, the total running time is linear in the size of $L_1(G)$ and $L_2(G)$, and thus in $\mathcal{O}(n)$.   □

Lemma 6. *Let G be a PTP graph with n vertices, let $U \subset V(G)$, let $\mathcal{P}$ be a partial representation of $G[U]$ and (L_1, L_2) a REL of G. In $\mathcal{O}(n)$ time, we can decide whether (L_1, L_2) admits a boundary path set with respect to $\mathcal{P}$ and, in the affirmative, compute boundary graphs H_1 and H_2.*

Proof. As in the proof of Theorem 3, we focus again on vertical strips and process them from bottom-left to top-right. Let $\mathcal{S}_1^{\checkmark}$ be the strips in $\mathcal{S}_1$ that have already been processed. Let S be a strip with start rectangle $\mathcal{P}(u)$ and end rectangle $\mathcal{P}(v)$ such that every strip left of S is in $\mathcal{S}_1^{\checkmark}$. The idea is to only compute the parts of $P_l(S)$ that do not coincide with a boundary path of a strip $S' \in \mathcal{S}_1^{\backslash}$ and the parts of $P_r(S)$ that do not coincide with $P_l(S)$. Initially, set H_1 as $L_1(G)[U]$.

We start with $P_l(S)$; see Fig. 9(c). Observe that $P_l(S)$ should only consist of u, vertices left-bounded in S, v, and vertices on paths $P_r(S')$ for $S' \in \mathcal{S}_1^{\backslash}$. Let $P_1, \ldots, P_k$ be the subpaths induced by u, vertices left-bounded in S, and v. Let x_i be the first vertex of P_i and y_i the last. Further let (w_i, x_i) be the leftmost incoming edge of x_i and let (y_i, z_i) be the leftmost outgoing edge of y_i. For $i \in \{2, \ldots, n\}$, w_i already has to be in H_1 but not in U; otherwise $P_l(S)$ does not exist. The analogous condition needs to hold for (y_i, z_i). If this holds for each $i \in \{1, \ldots, k\}$, we add the vertices and edges in each P_i as well as the edges (w_i, x_i) and (y_i, z_i) to H_1. Finally, we can test that the P_i's are in the correct order in H_1 with an st-ordering of $L_1(G)$.

Checking the existence of $P_r(S)$ works like the construction in Theorem 3; see Fig. 9(d). Recall that we extended subpaths induced by u, right-bounded vertices, and v, first with right-most incoming and outgoing edges appropriately, and then tried to join subsequent subpaths by taking the left-most outgoing and incoming edges, respectively. Observe that reaching $P_l(S)$ during such an extension, now means that we encounter a vertex in $H_1 \setminus U$; see Fig. 9. Hence, here we stop extensions when encounter a vertex v that is already in H_1 but not in U. If connecting the subpaths to each other or H_1 is successful, we test their order again with an st-ordering of $L_1(G)$ and finally add them to H_1.

During the construction of H_1 we also need to label the faces with the strips they belong to. Therefore when a subpath P_i, for $i \in \{1, \ldots, k-1\}$, is added to H_1 as part of a left boundary path $P_l(S)$, we tell its left face that is lies on the left boundary of S. For any subpath added to H_1 as part of a right boundary path $P_r(S)$, we tell its left face that it lies inside S.

Lastly, note that the running time is linear in the size of H_1, H_2 and G. $\square$

As a result of Lemmas 5 and 6 we get our main result, Theorem 1.

Theorem 1. *The partial representation extension problem for rectangular duals with a fixed regular edge labeling can be solved in linear time. For yes-instances, an explicit rectangular dual can be constructed within the same time bound.*

5   Concluding Remarks

We have characterized the partial rectangular duals that admit an extension realizing a given REL in terms of boundary path sets. Based on this, we have given an algorithm that computes an extension, if it exists, in time proportional to the size of the boundary path set. We have sped up this algorithm by considering only the underlying simple graph of a boundary path set – the boundary graph.

The partial rectangular dual extension problem remains open when no REL is specified. Eppstein et al. [9] gave algorithms that compute constrained area-universal rectangular duals and solved the extension problem for RELs. A partial rectangular dual induces a partial REL. Hence an extension of a partial rectangular dual $\mathcal{P}$ can be found by computing every extension of this partial REL and by testing for each whether it admits an extension of $\mathcal{P}$, using our linear-time algorithm. There can, however, be exponentially many extensions of a partial REL. We are interested in a faster approach.

References

1. Booth, K., Lueker, G.: Testing for the consecutive ones property, interval graphs, and graph planarity using PQ-tree algorithms. J. Comput. Syst. Sci. **13**(3), 335–379 (1976). https://doi.org/10.1016/S0022-0000(76)80045-1
2. Brooks, R.L., Smith, C.A.B., Stone, A.H., Tutte, W.T.: The dissection of rectangles into squares. Duke Math. J. **7**(1), 312–340 (1940). https://doi.org/10.1215/S0012-7094-40-00718-9
3. Buchin, K., Speckmann, B., Verdonschot, S.: Evolution strategies for optimizing rectangular cartograms. In: Xiao, N., Kwan, M.-P., Goodchild, M.F., Shekhar, S. (eds.) GIScience 2012. LNCS, vol. 7478, pp. 29–42. Springer, Heidelberg (2012). https://doi.org/10.1007/978-3-642-33024-7_3
4. Buchsbaum, A.L., Gansner, E.R., Procopiuc, C.M., Venkatasubramanian, S.: Rectangular layouts and contact graphs. ACM Trans. Algorithms **4**(1) (2008). https://doi.org/10.1145/1328911.1328919
5. Chaplick, S., Dorbec, P., Kratochvíl, J., Montassier, M., Stacho, J.: Contact representations of planar graphs: extending a partial representation is hard. In: Kratsch, D., Todinca, I. (eds.) WG 2014. LNCS, vol. 8747, pp. 139–151. Springer, Cham (2014). https://doi.org/10.1007/978-3-319-12340-0_12
6. Chaplick, S., Fulek, R., Klavík, P.: Extending partial representations of circle graphs. J. Graph Theory **91**(4), 365–394 (2019). https://doi.org/10.1002/jgt.22436
7. Chaplick, S., Guśpiel, G., Gutowski, G., Krawczyk, T., Liotta, G.: The partial visibility representation extension problem. Algorithmica **80**(8), 2286–2323 (2017). https://doi.org/10.1007/s00453-017-0322-4
8. Duncan, C.A., Gansner, E.R., Hu, Y., Kaufmann, M., Kobourov, S.G.: Optimal polygonal representation of planar graphs. Algorithmica **63**(3), 672–691 (2012). https://doi.org/10.1007/s00453-011-9525-2
9. Eppstein, D., Mumford, E., Speckmann, B., Verbeek, K.: Area-universal and constrained rectangular layouts. SIAM J. Comput. **41**(3), 537–564 (2012). https://doi.org/10.1137/110834032
10. Felsner, S.: Rectangle and square representations of planar graphs. In: Pach, J. (ed.) Thirty Essays on Geometric Graph Theory, pp. 213–248. Springer, New York (2013). https://doi.org/10.1007/978-1-4614-0110-0_12
11. de Fraysseix, H., de Mendez, P.O., Rosenstiehl, P.: On triangle contact graphs. Comb. Probab. Comput. **3**(2), 233–246 (1994). https://doi.org/10.1017/S0963548300001139
12. Gabriel, K.R., Sokal, R.R.: A new statistical approach to geographic variation analysis. Syst. Biol. **18**(3), 259–278 (1969). https://doi.org/10.2307/2412323
13. Heilmann, R., Keim, D.A., Panse, C., Sips, M.: RecMap: rectangular map approximations. In: Ward, M.O., Munzner, T. (eds.) IEEE Symposium on Information Visualization. pp. 33–40. IEEE Computer Society (2004). https://doi.org/10.1109/INFVIS.2004.57
14. Kant, G., He, X.: Regular edge labeling of 4-connected plane graphs and its applications in graph drawing problems. Theoret. Comput. Sci. **172**(1), 175–193 (1997). https://doi.org/10.1016/S0304-3975(95)00257-X

15. Klavík, P., et al.: Extending partial representations of proper and unit interval graphs. Algorithmica **77**(4), 1071–1104 (2016). https://doi.org/10.1007/s00453-016-0133-z

16. Klavík, P., Kratochvíl, J., Otachi, Y., Saitoh, T., Vyskočil, T.: Extending partial representations of interval graphs. Algorithmica **78**(3), 945–967 (2016). https://doi.org/10.1007/s00453-016-0186-z

17. Klawitter, J., Nöllenburg, M., Ueckerdt, T.: Combinatorial properties of triangle-free rectangle arrangements and the squarability problem. In: Di Giacomo, E., Lubiw, A. (eds.) GD 2015. LNCS, vol. 9411, pp. 231–244. Springer, Cham (2015). https://doi.org/10.1007/978-3-319-27261-0_20

18. Koebe, P.: Kontaktprobleme der konformen Abbildung. Berichte über die Verhandlungen der Sächsischen Akademie der Wissenschaften zu Leipzig. Math. Phys. Klasse **88**, 141–164 (1936)

19. Koźmiński, K., Kinnen, E.: Rectangular duals of planar graphs. Networks **15**(2), 145–157 (1985). https://doi.org/10.1002/net.3230150202

20. Krawczyk, T., Walczak, B.: Extending partial representations of trapezoid graphs. In: Bodlaender, H.L., Woeginger, G.J. (eds.) WG 2017. LNCS, vol. 10520, pp. 358–371. Springer, Cham (2017). https://doi.org/10.1007/978-3-319-68705-6_27

21. van Kreveld, M.J., Speckmann, B.: On rectangular cartograms. Comput. Geom. **37**(3), 175–187 (2007). https://doi.org/10.1016/j.comgeo.2006.06.002

22. Kusters, V., Speckmann, B.: Towards characterizing graphs with a sliceable rectangular dual. In: Di Giacomo, E., Lubiw, A. (eds.) GD 2015. LNCS, vol. 9411, pp. 460–471. Springer, Cham (2015). https://doi.org/10.1007/978-3-319-27261-0_38

23. Leinwand, S.M., Lai, Y.-T.: An algorithm for building rectangular floor-plans. In: 21st Design Automation Conference Proceedings, pp. 663–664 (1984). https://doi.org/10.1109/DAC.1984.1585874

24. Nusrat, S., Kobourov, S.G.: The state of the art in cartograms. Comput. Graph. Forum **35**(3), 619–642 (2016). https://doi.org/10.1111/cgf.12932

25. Raisz, E.: The rectangular statistical cartogram. Geogr. Rev. **24**(2), 292–296 (1934). https://doi.org/10.2307/208794

26. Steadman, P.: Graph theoretic representation of architectural arrangement. In: Architectural Research and Teaching, pp. 161–172 (1973)

27. Yeap, G.K.H., Sarrafzadeh, M.: Sliceable floorplanning by graph dualization. SIAM J. Discrete Math. **8**(2), 258–280 (1995). https://doi.org/10.1137/S0895480191266700

Can Local Optimality Be Used for Efficient Data Reduction?

Christian Komusiewicz[ORCID] and Nils Morawietz[✉]

Fachbereich Mathematik und Informatik, Philipps-Universität Marburg,
Marburg, Germany
{komusiewicz,morawietz}@informatik.uni-marburg.de

Abstract. An independent set S in a graph G is k-swap optimal if there is no independent set S' such that $|S'| > |S|$ and $|(S' \setminus S) \cup (S \setminus S')| \leq k$. Motivated by applications in data reduction, we study whether we can determine efficiently if a given vertex v is contained in some k-swap optimal independent set or in all k-swap optimal independent sets. We show that these problems are NP-hard for constant values of k even on graphs with constant maximum degree. Moreover, we show that the problems are Σ_2^P-hard when k is not constant, even on graphs of constant maximum degree. We obtain similar hardness results for determining whether an edge is contained in a k-swap optimal max cut. Finally, we consider a certain type of edge-swap neighborhood for the LONGEST PATH problem. We show that for a given edge we can decide in $f(\Delta + k) \cdot n^{\mathcal{O}(1)}$ time whether it is in some k-optimal path.

Keywords: local search · independent set · max-cut · longest path · NP-hardness

1 Introduction

Local search and data reduction are two widely successful strategies for coping with hard computational problems. Local search, which applies most naturally to optimization problems, aims at computing good heuristic solutions by using the following generic approach: Define a *local neighborhood relation* on the set of feasible solutions. Then, compute some feasible solution S. Now check whether there is a better solution S' in the local neighborhood of S. If this is the case, then replace S by S'. Otherwise, output the locally optimal solution S and stop.

Local search algorithms have been explored thoroughly from a practical and theoretical point of view [2,8–10,14]. The theoretical framework most closely related to our investigations is *parameterized local search*. Here, the local search neighborhood comes equipped with an operational parameter k that bounds the radius of the local search neighborhood. The size of the neighborhood is assumed to be polynomial in the input size for every fixed value of k. For example,

N. Morawietz—Supported by the Deutsche Forschungsgemeinschaft (DFG), project OPERAH, KO 3669/5-1.

T. Calamoneri and F. Corò (Eds.): CIAC 2021, LNCS 12701, pp. 354–366, 2021.
https://doi.org/10.1007/978-3-030-75242-2_25

in INDEPENDENT SET one is given a graph G and asks for a largest vertex set S such that no two vertices in S are adjacent. The feasible solutions are the independent sets of G. A natural neighborhood with a radius k is the k-swap neighborhood: Two vertex sets S and S' are in their respective k-swap neighborhoods if $|S \oplus S'| \leq k$, where $\oplus$ denotes the symmetric difference of two sets. In LS-INDEPENDENT SET, one is then given a graph G, an independent set S, and an integer k and asks whether the k-swap neighborhood of S contains a larger independent set S'. LS-INDEPENDENT SET can be solved in $\Delta^{\mathcal{O}(k)} \cdot n$ time [6,10]; the currently best fixed-parameter algorithm for LS-INDEPENDENT SET is efficient in practice, solving the parameterized local search problem for $k \approx$ 20 even on large real-world graphs [10]. Summarizing, for moderate values of k, a k-swap optimal independent set can be computed faster than an optimal one.

In data reduction, the idea is to preprocess any instance of a hard problem by identifying those parts of the instance that are easy to solve. Usually, data reduction algorithms are stated as a collection of data reduction rules which can be applied if a certain precondition is fulfilled and reduce the instance size whenever they apply. Two classic trivial reduction rules for INDEPENDENT SET are as follows: First, remove any vertex v that has no neighbors in G and add v to the independent set that is computed for the remaining instance. Second, remove any vertex v that has two degree-one neighbors.

In this work, we aim to explore the usefulness of local search or, more precisely, of local optimality, in the context of data reduction for hard problems. The correctness of the first reduction rule above is rooted in the observation that v is contained in every maximum independent set. Similarly, the correctness of the second rule is rooted in the fact that v is contained in no maximum independent set. Most of the known data reduction rules employ this principle and the crux of proving the correctness of a data reduction rule lies in proving that v is contained in all or no optimal solution. In general, given a vertex v and computing whether v is contained in a largest independent set is just as hard as computing an optimal solution in the first place. This is why data reduction rules use specific preconditions that allow proving optimality of including or excluding v without computing an optimal solution.

One may avoid such specific preconditions by relying on *locally* optimal solutions instead of optimal solutions: If we could compute efficiently that a vertex v is *not* contained in a locally optimal solution for *some* local search neighborhood, then we could remove v from the graph. Conversely, if we could compute efficiently that some vertex v is contained in *every* locally optimal solution, then we could remove v and its neighborhood from the graph as described above. The hope is that, since computing locally optimal solutions is easier for suitable local search neighborhoods, this approach could help in circumventing the previous dilemma: to say something about the optimal solution, one more or less needs to compute it. Going back to the INDEPENDENT SET problem, we would like to determine whether a given vertex is in some k-swap optimal independent set.

SOME LOCALLY OPTIMAL INDEPENDENT SET (∃-LO-IS)
Input: An undirected graph $G = (V, E)$, a vertex $v \in V$, and $k \in \mathbb{N}$.
Question: Is v contained in some k-swap optimal independent set in G?

If the answer is no, then v can be removed from G without destroying any optimal solution. We may also ask whether v is in every locally optimal independent set.

EVERY LOCALLY OPTIMAL INDEPENDENT SET ($\forall$-LO-IS)
Input: An undirected graph $G = (V, E)$, a vertex $v \in V$, and $k \in \mathbb{N}$.
Question: Is v contained in every k-swap optimal independent set in G?

If the answer is yes, then v belongs to every optimal independent set and we can apply a data reduction rule that removes v and its neighbors and adds v to the independent set that is computed for the remaining graph. Motivated by the usefulness of efficient algorithms for these two problems in data reduction for INDEPENDENT SET, we study their complexity. We consider related problems for two further classic NP-hard problems: MAX CUT and LONGEST PATH.

Our Results. For INDEPENDENT SET our results are decidedly negative. $\exists$-LO-IS and $\forall$-LO-IS are NP-complete and coNP-complete, respectively, even if $k = 3$ and $\Delta = 4$ and also if $k = 5$ and $\Delta = 3$. These results are tight[1] in the following sense: when $k = 1$ or when $\Delta = 2$, then both problems can be solved in polynomial time. Moreover, when k is not constant, we show that the problems are even Σ_2^P-complete and Π_2^P-complete, respectively. Thus, both problems are substantially harder than LS-INDEPENDENT SET. For MAX CUT the situation is similar: Deciding whether some edge is contained in a k-swap optimal cut or in every k-swap optimal cut is NP-complete and coNP-complete even if $k = 1$ and $\Delta = 5$. Moreover, if k is not constant then the problems are, again, Σ_2^P-complete and Π_2^P-complete, respectively.

Finally, we consider LONGEST PATH with a certain edge-swap neighborhood. We show that for this neighborhood we can determine in $f(\Delta + k) \cdot n^{\mathcal{O}(1)}$ time whether some edge is contained in a k-optimal path. If the answer is no, then this edge can be safely removed from the input graph. Since LONGEST PATH is NP-hard on cubic graphs, our results indicate that there are scenarios in which testing for the containment of edges in locally optimal solutions is a viable approach to obtain data reduction rules for NP-hard problems.

Related Work. A related problem is to determine if there is a maximal (and, thus, 1-swap optimal) independent set S containing only vertices of a specific subset of vertices $U \subseteq V$ [3,12]. In other words, this problem asks for a 1-swap optimal independent set containing no vertex of $V \setminus U$, a generalization of the complement problem of $\exists$-LO-IS for $k = 1$. This problem is NP-hard even on graphs where INDEPENDENT SET can be solved in polynomial time, like bipartite graphs [3]. In the scope of 1-swaps, the INDEPENDENT SET RECONFIGURATION problem was analyzed extensively [1,4,5,13]. Here, we are given two independent sets X and Y of a graph G and an integer k and we want to determine if there is a sequence $S_1, \ldots, S_r$ of independent sets such that $S_1 = X$, $S_r = Y$, $|S_j| \geq k$ for

---

[1] For even k, an independent set is k-swap optimal if and only if it is $(k-1)$-swap optimal [10]. Thus, only odd values of k are interesting.

all $j \in [1, r]$, and $|S_j \oplus S_{j+1}| = 1$ for all $j \in [1, r-1]$. Hence, one wants to add or remove a single vertex at a time and transform X into Y without decreasing the size of the current independent set below k. This problem is PSPACE-complete even on bipartite graphs [13].

The proofs of statements marked with a (*) are deferred to a full version.

2 Preliminaries

Sets and Graphs. For a set A, we denote with $\binom{A}{2} := \{\{a, b\} \mid a \in A, b \in A\}$ the collection of all size-two subsets of A. For two sets A and B, we denote with $A \oplus B := (A \setminus B) \cup (B \setminus A)$ the *symmetric difference* of A and B.

An (undirected) graph $G = (V, E)$ consists of a set of vertices V and a set of edges $E \subseteq \binom{V}{2}$. For vertex sets $S \subseteq V$ and $T \subseteq V$ we denote with $E_G(S, T) := \{\{s, t\} \in E \mid s \in S, t \in T\}$ the edges between S and T. Moreover, we define $E_G(S) := E_G(S, S)$. For a vertex $v \in V$, we denote with $N_G(v) := \{w \in V \mid \{v, w\} \in E\}$ the *open neighborhood* of v in G and with $N_G[v] := N_G(v) \cup \{v\}$ the *closed neighborhood* of v in G. Analogously, for a vertex set $S \subseteq V$, we define $N_G[S] := \bigcup_{v \in S} N_G[v]$ and $N_G(S) := N_G[S] \setminus S$. If G is clear from the context, we may omit the subscript. Moreover, we denote with $\Delta(G) := \max\{|N_G(v)| \mid v \in V\}$ the *maximum degree* of G.

A sequence of distinct vertices $P = (v_0, \ldots, v_k)$ is a *path* or (v_0, v_k)-*path* of length k in G if $\{v_{i-1}, v_i\} \in E(G)$ for all $i \in [1, k]$. We denote with $V(P)$ the vertices of P and with $E(P)$ the edges of P.

Satisfiability. For *variable set* Z, we define the set of *literals* $\mathcal{L}(Z) := Z \cup \{\neg z \mid z \in Z\}$. A literal set $\tilde{Z} \subseteq \mathcal{L}(Z)$ is an *assignment* of Z if $|\{z, \neg z\} \cap \tilde{Z}| = 1$ for all $z \in Z$. For a subset $X \subseteq Z$ of the variables we denote with $\tau_Z(X) := X \cup \{\neg z \mid z \in Z \setminus X\}$, the assignment of Z where all variables of X occur positively and all variables of $Z \setminus X$ occur negatively. A *clause* $\phi \subseteq \mathcal{L}(Z)$ is *satisfied* by an assignment τ of Z if $\phi \cap \tau \neq \emptyset$, and we write $\tau \models \phi$. Analogously, a set $\Phi \subseteq \mathbb{P}(\mathcal{L}(Z))$ of clauses is satisfied by τ if $\tau \models \phi$ for all $\phi \in \Phi$, and we write $\tau \models \Phi$.

3 Independent Set

In this section, we analyze the complexity of ∃-LO-IS and ∀-LO-IS. First, let us set the following notation. Let $G = (V, E)$ be a graph and let k be an integer. We call a subset $W \subseteq V$ a *k-swap* in G if $|W| \leq k$. A set $S \subseteq V$ is an *independent set* if $\{v, w\} \notin E$ for all $v, w \in S$. Further, an independent set S is *k-swap optimal* in G if there is no k-swap W in G such that $S \oplus W$ is an independent set and $|S| < |S \oplus W|$. We will make use of the following observation on improving k-swaps.

Observation 1 ([10, Lemma 2]). *Let S be an independent set for a graph $G = (V, E)$ and let k be an integer. Then, S is k-swap optimal if and only if there is no swap $C \subseteq V$ of size at most k such that $G[C]$ is connected and $|S \oplus C| = |S| + 1$.*

Observation 2. *An instance* (G, v, k) *of* $\forall$-LO-IS *is a yes-instance if and only if* (G, w, k) *is a no-instance of* $\exists$-LO-IS *for every* $w \in N(v)$.

First, we observe that we can solve $\exists$-LO-IS in polynomial time for the following almost trivial cases. Note that for $k = 1$, we ask whether a vertex is contained in some maximal independent set.

Proposition 1 (*). $\exists$-LO-IS *and* $\forall$-LO-IS *can be solved in polynomial time if* $k = 1$ *or* $\Delta \leq 2$.

We now show that we cannot improve upon Proposition 1, neither in terms of k nor in terms of Δ.

Theorem 3 (*). $\exists$-LO-IS *is* NP-*complete and* $\forall$-LO-IS *is* coNP-*complete even if* $k = 3$ *and* $\Delta = 4$.

Theorem 4. $\exists$-LO-IS *is* NP-*complete and* $\forall$-LO-IS *is* coNP-*complete even if* $k = 5$ *and* $\Delta = 3$.

Proof. First, we show the statement for $\exists$-LO-IS via reducing from SAT. Given an instance $I = (Z, \Phi)$ of SAT, we construct in polynomial time an equivalent instance $I' = (G = (V, E), v_{\exists}, k)$ of $\exists$-LO-IS where $k = 5$ and where G has maximum degree three. We may assume that every clause has size three, every variable of Z occurs twice positively and twice negatively in Φ, and every variable occurs at most once per clause since SAT remains NP-hard in this case [15].

Let ψ denote the number of clauses and let $\Phi = \{\phi_1, \ldots, \phi_\psi\}$. We start with an empty graph G and add for every variable $z \in Z$ a cycle $(z^1, \neg z^1, z^2, \neg z^2, z^1)$. We add for every clause $\phi_i = \{\ell_1, \ell_2, \ell_3\} \in \Phi$ the subgraph $G_{\phi_i} = (V_{\phi_i}, E_{\phi_i})$ shown in Fig. 1a. The graph G_{ϕ_i} contains the vertex u_i, for each $j \in \{1, 2, 3, \vee\}$, the path $(a_j^i, b_j^i, c_j^i, d_j^i, e_j^i)$, and the edges $\{b_1^i, a_\vee^i\}$, $\{b_2^i, a_\vee^i\}$, $\{b_\vee^i, u_i\}$, and $\{b_3^i, u_i\}$. We connect a gadget G_ϕ with the cycles of the variables of ϕ as follows: for every $j \in [1, 3]$ we add the edge $\{\ell_j^1, a_j^i\}$ if ℓ_j^1 is not connected to any clause gadget already. Otherwise, we add the edge $\{\ell_j^2, a_j^i\}$. Since every variable occurs twice positively and twice negatively in Φ, the vertices ℓ_j^1 and ℓ_j^2 are each connected to exactly one subgraph G_{ϕ_i} at the end of the construction. The idea behind G_{ϕ_i} is that every 5-swap optimal independent set S for G containing the vertices a_1^i, a_2^i, and a_3^i, also contains the vertex u_i. Hence, if no vertex representing any literal of ϕ_i is contained in S (that is, if ϕ_i is not satisfied), then u_i is contained in S which will imply that $v_{\exists}$ is not contained in S by the remaining gadgets.

Next, we add a binary tree with leaf vertices $\{u_i \mid \phi_i \in \Phi\}$, the set of inner vertices T, and root r to G. Afterwards, we remove all edges of this tree and connect every parent vertex p with his two child vertices c_1 and c_2 by adding the subgraph $G_p = (V_p, E_p)$ shown in Fig. 1b. This subgraph contains the vertices of the binary tree p, c_1 and c_2, the vertices p', q, q_3^1, q_3^2, and a cycle $(p_0, p_1^1, p_2^1, p_3^1, p_4^1, p_5, p_4^2, p_3^2, p_2^2, p_1^2, p_0)$. Further, the set E_p contains the edges $\{p_3^1, q_3^1\}$, $\{p_3^2, q_3^2\}$, $\{p_1^1, c_1\}$, $\{p_1^2, c_2\}$, $\{p, p'\}$, $\{p', q\}$, and $\{p', p_0\}$.

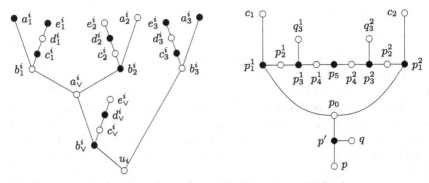

(a) The subgraph G_{ϕ_i} for a clause $\phi_i = \{\ell_1, \ell_2, \ell_3\} \in \Phi$. Black vertices belong to $V_{\phi_i}(\tau)$ with $\tau \cap \phi_i = \{\ell_2\}$.

(b) The subgraph G_p for some parent vertex $p \in T$ with the child vertices c_1 and c_2. The black vertices are contained in every 5-swap optimal independent set containing $v_\exists$.

Fig. 1. The gadgets of the reduction of Theorem 4.

Finally, we add a path (v_0, v_1, v_2, v_3, r) to G and set $v_\exists := v_1$. This completes the construction of I'. Note that the constructed graph has a maximum degree of three. We show that I is a yes-instance of SAT if and only if I' is a yes-instance of $\exists$-LO-IS.

($\Rightarrow$) Suppose that I is a yes-instance of SAT. Thus, there is an assignment τ for Z such that $\tau \models \Phi$. Let $\phi_i = \{\ell_1, \ell_2, \ell_3\}$ be a clause of Φ and let τ be an assignment of Z. We set

$$V_{\phi_i}(\tau) := \{b_j^i, d_j^i \mid \ell_j \in \tau\} \cup \{a_j^i, c_j^i, e_j^i \mid \ell_j \notin \tau\} \cup \begin{cases} \{b_\vee^i, d_\vee^i\} & \tau \models \{\ell_1, \ell_2\} \\ \{a_\vee^i, c_\vee^i, e_\vee^i\} & \text{otherwise} \end{cases}.$$

Note that $V_{\phi_i}(\tau)$ is an independent set. We set $S := \{\ell^1, \ell^2 \mid \ell \in \tau\} \cup \bigcup_{\phi \in \Phi} V_\phi(\tau) \cup \{p', p_1^1, p_3^1, p_5, p_3^2, p_1^2 \mid p \in T\} \cup \{v_1, v_3\}$. That is, S contains the vertices representing the literals of τ, the vertices of the clause gadgets according to τ, the black vertices of V_p shown in Fig. 1b for every $p \in T$, and the vertices v_3 and $v_1 = v_\exists$. By construction, S is an independent set. It remains to show that S is 5-swap optimal. To this end, we observe the following.

Claim 1 (*). For each vertex $w \in V \setminus S$ with $|N(w) \cap S| \leq 1$, we have $|N(w)| = 1$.

Since a 5-swap W for S with $|S \oplus W| > |S|$ has to contain two distinct vertices $w_1, w_2 \in V \setminus S$ with $|N(w_1) \cap S| = |N(w_2) \cap S| = 1$ we obtain by Claim 1 and the fact that degree-one vertices in G have pairwise distance at least six, that there is no such W. Consequently, S is a 5-swap optimal independent set in G with $v_\exists = v_1 \in S$.

($\Leftarrow$) Suppose that I' is a yes-instance of $\exists$-LO-IS. Thus, there is a 5-swap optimal independent set S in G with $v_\exists = v_1 \in S$. We first observe the following.

Claim 2 (*). Let $p \in T$ with the child vertices c_1 and c_2. If $p \notin S$ and $p' \in S$, then $|N(c_1) \cap S| \geq 2$ and $|N(c_2) \cap S| \geq 2$.

Note that this also implies, that $c_1 \notin S$ and $c_2 \notin S$.

Recall that r is the root of the binary tree and that r' is the unique neighbor of r in G_r. Now, observe that $\{v_1, v_3, r'\} = \{v_\exists, v_3, r'\} \subseteq S$ and $\{v_0, v_2, r\} \subseteq V \setminus S$ as, otherwise, S would not be a 5-swap optimal independent set in G. Thus, by Claim 2 one can show inductively, that for all vertices c of the binary tree it holds that $|N(c) \cap S| \geq 2$. Consequently, this also holds for the leaves of the tree, namely the vertices $\{u_i \mid i \in [1, \psi]\}$. Hence, for every $\phi_i \in \Phi$ it holds that $u_i \notin S$ and that $b_\lor^i \in S$ or $b_3^i \in S$.

We set $\tau = \{\ell \in \mathcal{L}(Z) \mid \{\ell^1, \ell^2\} \cap S \neq \emptyset\}$ and show that $\tau \models \Phi$. Note that τ contains at most one of z and $\neg z$ since S is an independent set. Let $\phi_i = \{\ell_1, \ell_2, \ell_3\}$ be a clause of Φ, we show that $\tau \models \phi_i$. First, we show that if there is some $j \in \{1, 2, 3, \lor\}$ with $N(a_j^i) \cap S = \{b_j^i\}$, then S is not 5-swap optimal. Since S is an independent set, it holds that $c_j^i \notin S$. If $d_j^i \notin S$, then $S \oplus \{a_j^i, b_j^i, c_j^i\}$ is an independent set and, thus, S is not 5-swap optimal. Otherwise, $d_j^i \in S$ and, thus, $e_j^i \notin S$. Hence, $S \oplus \{a_j^i, b_j^i, c_j^i, d_j^i, e_j^i\}$ is an independent set and, thus, S is not 5-swap optimal.

We may thus assume that if $a_j^i \notin S$, then $N(a_j^i)$ is a subset of S containing ℓ^1 or ℓ^2. We now use this fact to argue that at least one literal vertex adjacent to any vertex a_j^i is contained in S and, thus, ϕ_i is satisfied by τ. Recall that $b_\lor^i \in S$ or $b_3^i \in S$. Consequently, if $b_\lor^i \in S$, then $|N(a_\lor^i) \cap S| \geq 2$ and, therefore, $\{b_1^i, b_2^i\} \cap S \neq \emptyset$. Hence, there is some $j \in \{1, 2, 3\}$ such that $b_j^i \in S$ and, thus, $N(a_j^i) \subseteq S$. As a consequence, $\{\ell_j^1, \ell_j^2\} \cap S \neq \emptyset$ and, therefore, $\ell_j \in \tau$. Hence, ϕ_i is satisfied by τ and I is a yes-instance of SAT.

Due to Observation 2 and the fact that $N(v_0) = \{v_\exists\}$ it follows that (G, v_0, k) is a yes-instance of ∀-LO-IS if and only if I' is a no-instance of ∃-LO-IS. Consequently, ∀-LO-IS is coNP-hard even if $k = 5$ and where the input graph has a maximum degree of three.   □

Corollary 1. *For every fixed odd $k \geq 5$, ∀-LO-IS is coNP-complete and ∃-LO-IS is NP-complete even if $\Delta = 3$.*

Proof. Let $k \geq 7$ and let $I = (G = (V, E), v_\exists, 5)$ be an instance of ∃-LO-IS constructed as in the proof of Theorem 4. By adding for every degree-one vertex w in G a path P_w with $k - 5$ vertices to G and connecting w with one endpoint of P_w, we obtain an equivalent instance $I' = (G', v_\exists, k)$ of ∃-LO-IS. □

Next, we analyze the case where the swap distance k is unbounded.

Theorem 5 (*). *∀-LO-IS is Π_2^P-complete (∃-LO-IS is Σ_2^P-complete) if $\Delta = 3$.*

4   Max Cut

We now analyze the complexity of deciding whether an edge is a cut edge of some locally optimal cut for the MAX CUT problem. We formally define cuts

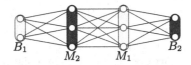

Fig. 2. A $(2,3)$-enforcer $F = (B_1 \cup B_2 \cup M_1 \cup M_2, F_E)$.

and their local neighborhoods as follows. Let $S, T \subseteq V$. The pair (S, T) is a *cut* in G if $S \cup T = V$ and $S \cap T = \emptyset$. A cut (S, T) is a *k-swap optimal cut* in G if there is no k-swap W in G such that $|E(S', T')| > |E(S, T)|$ where $S' = S \oplus W$ and $T' = T \oplus W = V \setminus S'$. Let $A, B \subseteq V$. We say that A and B are *in the same part of the cut* if $A \cup B \subseteq S$ or $A \cup B \subseteq T$. Moreover, A and B are *in opposite parts of the cut* if $A \subseteq S$ and $B \subseteq T$ or vice versa.

EVERY LOCALLY OPTIMAL MAX-CUT ($\forall$-LO-MC)
Input: An undirected graph $G = (V, E)$, an edge $e \in E$, and $k \in \mathbb{N}$.
Question: Is e contained in every k-swap optimal cut in G?

Analogously, we ask in SOME LOCALLY OPTIMAL MAX-CUT ($\exists$-LO-MC) if e is contained in some k-swap optimal cut in G.

For every fixed value of k, we can check in polynomial time if a given cut (S, T) is k-swap optimal in G. Consequently, we obtain the following.

Lemma 1. *For every fixed value of k, $\forall$-LO-MC is contained in* coNP *and $\exists$-LO-MC is contained in* NP.

We now show that both problems are hard, even for constant k and Δ. In the reduction, we use a graph called $(2,3)$-*enforcer* which is shown in Fig. 2. For a $(2,3)$-enforcer F, we denote $F(i) := B_i \cup M_i$ for $i \in \{1, 2\}$.

Proposition 2 (*). *Let G be a graph with $\Delta(G) = 5$. If G contains a $(2,3)$-enforcer $F = (B_1 \cup B_2 \cup M_1 \cup M_2, E')$ as an induced subgraph, then for every 1-swap optimal cut (S, T) in G it holds that $F(1)$ and $F(2)$ are in opposite parts of the cut.*

Theorem 6. *$\exists$-LO-MC is* NP-*complete and $\forall$-LO-MC is* coNP-*complete even if $k = 1$ and $\Delta = 5$.*

Proof. We reduce SAT to $\exists$-LO-MC. Given an instance $I = (Z, \Phi)$ of SAT, we construct in polynomial time an equivalent instance $I' = (G = (V, E), e_\exists, k)$ of $\exists$-LO-MC where $k = 1$ and where G has a maximum degree of five. We can assume without loss of generality that every clause has size three, every variable of Z occurs twice positively and twice negatively in Φ, and every variable occurs at most once per clause since SAT remains NP-hard in this case [15]. Further, we let ψ denote the number of clauses and we assume that $\psi = 2^r$ for some even r.

Let $\Phi = \{\phi_1, \ldots, \phi_\psi\}$. We start with a balanced binary tree with $r + 1$ levels. We denote with $L_p := \{u_q^p \mid q \in [1, 2^p]\}$ the vertices of the pth level of the tree for all $p \in [0, r]$. The leaf u_q^r represents the clause ϕ_q for all $q \in [1, \psi]$. Further, we add a (2,3)-enforcer $F_\exists$ with $B_1 = \{w_\exists^1, w_\exists^2\}$ and $B_2 = \{w_\forall^1, w_\forall^2\}$ and, for each variable $v \in Z$, a (2,3)-enforcer F_v with $B_1 = \{v, v'\}$ and $B_2 = \{\neg v, \neg v'\}$.

The idea is that $F_v(1)$ corresponds to the true-assignment of v and that $F_v(2)$ corresponds to the false-assignment of v since $v \in F_v(1)$ and $\neg v \in F_v(2)$. By Proposition 2, $F_v(1)$ and $F_v(2)$ are in opposite parts of every 1-swap optimal cut for G. Thus, for every 1-swap optimal cut (S, T) for G, both $S \cap \mathcal{L}(Z)$ and $T \cap \mathcal{L}(Z)$ are assignments for the variables of Z.

For each clause $\phi_q \in \Phi$ and each literal $\ell \in \phi_q$ we add the edge $\{\ell, u_q^r\}$. Furthermore, we connect the vertices of $L_1 \cup \{w_\exists^2\}$ to a cycle of length three and for $p \in [2, r-1]$ we connect the vertices of L_p to a cycle of length 2^p. Finally, we add the edges between u_1^0 and each of the vertices $w_\forall^1, w_\forall^2$, and $w_\exists^1$ and set $e_\exists = \{w_\exists^1, u_1^0\}$. This completes the construction of I'. We now show that I is a yes-instance of SAT if and only if I' is a yes-instance of $\exists$-LO-MC.

($\Rightarrow$) Suppose that I is a yes-instance of SAT. Thus, there is $\tilde{Z} \subseteq Z$ such that $\tau_Z(\tilde{Z})$ satisfies Φ. We set

$$S := F_\exists(1) \cup \bigcup_{\text{odd } p \in [1, r-1]} L_p \cup \bigcup_{v \in \tilde{Z}} F_v(1) \cup \bigcup_{v \in Z \setminus \tilde{Z}} F_v(2)$$

and $T = V \setminus S$ and show that (S, T) is a 1-swap optimal cut in G and contains $e_\exists$. Note that for every $\ell \in \mathcal{L}(Z)$ it holds that $\ell \in S$ if and only if $\ell \in \tau_Z(\tilde{Z})$.

By construction, G has a maximum degree of five. Thus, for every vertex v in any enforcer it follows directly that at least half of the neighbors of v are in the opposite part of the cut of v. Further, since for every $v \in L_p, p \in [1, r-1]$, it holds that $|N_G(v) \cap (L_{p-1} \cup L_{p+1})| = 3$ and, thus, by construction of (S, T) that at least half of the neighbors of v are in the opposite part of the cut of v. The same also holds for the root u_1^0 since $u_1^0 \in T$ and $N_G(u_1^0) \cap S = L_1 \cup \{w_\exists^1\}$. Since $L_r \subseteq T$, it remains to show that for every $q \in [1, \psi]$, $|N_G(u_q^r) \cap S| \geq 2$. By definition of (S, T) it follows that $|N_G(u_q^r) \cap L_{r-1} \cap S| = 1$. The fact that $\tau_Z(\tilde{Z}) \models \Phi$ implies that $\phi_q \cap S \neq \emptyset$. Consequently, for every $v \in V$, at least half of the neighbors of v are in the opposite part of the cut of v. Therefore, (S, T) is a 1-swap optimal cut in G and contains $e_\exists$.

($\Leftarrow$) Let (S, T) be a 1-swap optimal cut in G which contains $e_\exists$. By Proposition 2, we may assume without loss of generality that $F_\exists(1) \subseteq S$ and $F_\exists(2) \subseteq T$. Further, for every $v \in Z$, either $F_v(1) \subseteq S$ or $F_v(2) \subseteq S$. We set $\tilde{Z} := \{v \in Z \mid F_v(1) \subseteq S\}$ and show that $\tau_Z(\tilde{Z}) \models \Phi$. Note that for every literal $\ell \in \mathcal{L}(Z)$ it holds that $\ell \in \tau_Z(\tilde{Z})$ if and only if $\ell \in S$.

Claim 3 (*). For every $p \in [0, r]$, it holds that $L_p \subseteq S$ if p is odd and $L_p \subseteq T$ if p is even.

As a consequence, $u_q^r \in T$ for all $q \in [1, \psi]$. Since (S, T) is 1-swap optimal in G, it holds that u_q^r has at least two neighbors in S: the single neighbor in L_{q-1} and at least one neighbor in $N_G(u_q^r) \setminus L_{q-1} = \phi_q$ and, thus, $S \cap \phi_q \neq \emptyset$. Consequently, $\tau_Z(\tilde{Z}) \models \Phi$.

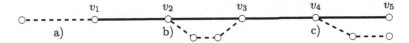

Fig. 3. The three possible kinds of minimal improving $(1, k)$-swaps: a) appending an edge to one of the endpoints, b) replacing an edge $\{u, v\}$ with a (u, v)-path of length at most k, and c) removing one endpoint and appending a path of length two to its neighbor in the path.

Thus, $\exists$-LO-MC is NP-complete if $k = 1$ and $\Delta = 5$. Due to Proposition 2 and the fact that the $(2, 3)$-enforcer $F_{\exists}$ contains both $w_{\exists}^1$ and $w_{\forall}^1$, we know that for every 1-swap optimal cut (S, T) in G it holds that $w_{\exists}^1 \in S$ if and only if $w_{\forall}^1 \in T$. Hence, $I' = (G, e_{\exists}, 1)$ is a no-instance of $\exists$-LO-MC if and only if $I'' = (G, e_{\forall}, 1)$ is a yes-instance of $\forall$-LO-MC, where $e_{\forall} := \{w_{\forall}^1, u_1^0\}$. Consequently, $\forall$-LO-MC is coNP-complete if $k = 1$ and $\Delta = 5$. □

Theorem 7 (*). $\forall$-LO-MC *is* Π_2^P-complete *and* $\exists$-LO-MC *is* Σ_2^P-complete *even if* $\Delta = 3$.

5 Longest Path

Finally, we consider LONGEST PATH which is NP-hard even on cubic graphs [7]. Again, we want to find out whether some small part of the graph is contained in a locally optimal solution. We consider the following neighborhood.

Definition 1. *Let k be an integer. Two paths P_1 and P_2 are $(1, k)$-swap neighbors, if 1) the relative ordering of the common vertices of P_1 and P_2 is equal in both paths and 2) $|E(P_1) \setminus E(P_2)| \le k$ and $|E(P_2) \setminus E(P_1)| \le 1$ or $|E(P_1) \setminus E(P_2)| \le 1$ and $|E(P_2) \setminus E(P_1)| \le k$.*

All three kinds of minimal improving $(1, k)$-swaps are shown in Fig. 3.

SOME LOCALLY OPTIMAL PATH ($\exists$-LO-PATH)
Input: An undirected graph $G = (V, E)$, an edge $e^* \in E$, and $k \in \mathbb{N}$.
Question: Is e^* contained in some $(1, k)$-swap optimal path P^* in G?

First, we observe that, unfortunately, the problem is hard already for fixed k.

Theorem 8 (*). $\exists$-LO-PATH *is contained in* NP *for every fixed value of k and* NP-*hard for every $k \ge 2$.*

We now show that on bounded-degree graphs, we can obtain an efficient algorithm for small k. Our algorithm is based on a sufficient and necessary condition for the existence of a $(1, k)$-swap optimal path in G containing a specific edge. To formulate the condition, we define two collections of vertex sets. We denote for every integer k and every pair of distinct vertices $v, w \in V$ with $\mathcal{V}_k(v, w) := \{V(P) \setminus \{v, w\} \mid P \text{ is an } (v, w)\text{-path with } 2 < |V(P)| \le k + 1\}$ the collection of sets of inner vertices of (v, w)-path with length at most k. Moreover, for every vertex $v \in V$ we denote by $\mathcal{V}_2(v) := \{\{x, y\} \mid x, y \in V, (v, x, y) \text{ is a path in } G\}$ the collection of subsets of $\tilde{V} \subseteq V \setminus \{v\}$, such that G contains a path P of length two starting in v with $V(P) = \tilde{V} \cup \{v\}$.

Lemma 2. *Let $e^* = \{v^*, w^*\}$ be an edge which is not isolated and let $k \geq 2$. There is a $(1, k)$-swap optimal path P^* containing e^* if and only if there are (not necessarily distinct) vertices s, s', t' and t such that there is an (s, t)-path P in G containing the edges $e^*, \{s, s'\}, \{t', t\}$, the vertices $N(s) \cup N(t)$, and where $V(P)$ is a hitting set for $\mathcal{V}_k(s, s') \cup \mathcal{V}_2(s') \cup \mathcal{V}_k(v^*, w^*) \cup \mathcal{V}_2(t') \cup \mathcal{V}_k(t', t)$.*

Proof. ($\Rightarrow$) Let P^* be a $(1, k)$-swap optimal path containing e^*. Then, $V(P^*)$ contains at least two vertices and, thus, there are vertices $s, s', t', t \in V$ such that P^* contains the edges $e^*, \{s, s'\}$, and $\{t', t\}$. Moreover, the set $V(P^*)$ contains $N(s) \cup N(t)$ as, otherwise, P^* is not $(1, k)$-swap optimal. Assume towards a contradiction, that $V(P^*)$ is not a hitting set for $\mathcal{V}_k(s, s') \cup \mathcal{V}_2(s') \cup \mathcal{V}_k(v^*, w^*) \cup \mathcal{V}_r(t', t) \cup \mathcal{V}_2(t')$.

Case 1: $V(P^*)$ **is not a hitting set for** $\mathcal{V}_k(x, y)$ **for some** $\{x, y\} \in \{\{s, s'\}, \{v^*, w^*\}, \{t', t\}\}$. Then, there is some $\tilde{V} \in \mathcal{V}_k(x, y)$ such that there is some (x, y)-path P' in G of length at most k and $V(P') = \tilde{V} \cup \{x, y\}$, such that $V(P') \cap V(P^*) = \{x, y\}$. Replacing the edge $\{x, y\}$ by the (x, y)-path P' is an improving $(1, k)$-swap, since $\tilde{V} \neq \emptyset$. This contradicts the fact that P^* is $(1, k)$-swap optimal in G.

Case 2: $V(P^*)$ **is not a hitting set for** $\mathcal{V}_2(x)$ **for some** $x \in \{s', t'\}$. Then, there is some $\tilde{V} \in \mathcal{V}_2(x)$ such that there is some path P' of length two in G starting in x with $V(P') = \tilde{V} \cup \{x\}$ and $V(P') \cap V(P^*) = \{x\}$. Hence, replacing the edge $\{x, x'\}$ by the path P' is an improving $(1, k)$-swap, since P' contains two edges. This contradicts the fact that P^* is $(1, k)$-swap optimal in G.

($\Leftarrow$) Suppose that there are vertices $s, s', t', t \in V$ such that there is an (s, t)-path P in G containing the edges $e^*, \{s, s'\}, \{t', t\}$, the vertices $N(s) \cup N(t)$, and where $V(P)$ is a hitting set for $\mathcal{V}_k(s, s') \cup \mathcal{V}_2(s') \cup \mathcal{V}_k(v^*, w^*) \cup \mathcal{V}_k(t', t) \cup \mathcal{V}_2(t')$.

Since P contains 1) at least one inner vertex of every (s, s')-path in G of length at least two and at most k, and 2) at least one inner vertex besides s' of every path of length two starting in s', there is no improving $(1, k)$-swap that removes $\{s, s'\}$ from P. This also holds for the edge $\{t', t\}$. Since $V(P)$ is a hitting set for $\mathcal{V}_k(v^*, w^*)$, P contains at least one inner vertex of every (v^*, w^*)-path in G of length at least two and at most k.

In the following, we will show, that every path P^* which can be reached by an arbitrary number of improving $(1, k)$-swaps, also fulfills all properties of P. Note that this implies that there is a $(1, k)$-swap optimal path P^* in G containing e^*.

Since e^* is not an isolated edge and P contains the vertices $N(s) \cup N(t)$, and the edges $e^*, \{s, s'\}, \{t', t\}$, we obtain that P contains at least three vertices. Hence, not all edges of P can be removed by a single improving $(1, k)$-swap. Note that an improving $(1, k)$-swap can only remove a vertex from the path if this vertex is one of the endpoints.

By the above, none of the edges $e^*, \{s, s'\}$, or $\{t', t\}$ can be removed by an improving $(1, k)$-swap. Hence, for every improving $(1, k)$-swap neighbor P' of P

it holds that $V(P) \subseteq V(P')$, P' contains the edges $e^*, \{s, s'\}$, and $\{t', t\}$. Moreover, since $N(s) \cup N(t) \subseteq V(P)$, we obtain that P' is also an (s, t)-path. Consequently, one can show via induction, that for every path P^* which can be reached by an arbitrary number of improving $(1, k)$-swap starting from P, that P^* is an (s, t)-path containing the edges $e^*, \{s, s'\}$, and $\{t', t\}$, the vertices $N(s) \cup N(t)$, and $V(P^*)$ is a hitting set for $\mathcal{V}_k(s, s') \cup \mathcal{V}_2(s') \cup \mathcal{V}_k(v^*, w^*) \cup \mathcal{V}_2(t') \cup \mathcal{V}_k(t', t)$. □

Theorem 9. $\exists$-LO-PATH *can be solved in time* $\mathcal{O}(f(\Delta + k) \cdot n^4)$ *for some computable function* f.

Proof. Let $I = (G = (V, E), e^* = \{v^*, w^*\}, k)$ be an instance of SOME LOCALLY OPTIMAL PATH. First, if e^* is an isolated edge, then, there is exactly one path $P^* = (v^*, w^*)$ in G that contains the edge e^*. This path is $(1, k)$-swap optimal in G if and only if G does not contain a path with two edges. Hence, to determine if I is a yes-instance of SOME LOCALLY OPTIMAL PATH, we only have to check if G contains a path with two edges, which can be done in $\mathcal{O}(n^2)$ time. Second, if e^* is not an isolated edge, then, due to Lemma 2, it is sufficient to check if there are vertices $s, s', t', t \in V$ such that there is an (s, t)-path P in G containing the edges $e^*, \{s, s'\}$, and $\{t', t\}$, the vertices $N(s) \cup N(t)$, and where $V(P)$ is a hitting set for $\mathcal{V}_k(s, s') \cup \mathcal{V}_2(s') \cup \mathcal{V}_k(v^*, w^*) \cup \mathcal{V}_k(t', t) \cup \mathcal{V}_2(t')$. In the following, we describe how we can check in $\mathcal{O}(f(\Delta + r) \cdot n^4)$ time, if such a path exists.

For every combination of vertices $s \in V, s' \in N(s), t \in V$, and $t' \in N(t)$, we compute the collections $\mathcal{V}_k(s, s'), \mathcal{V}_2(s'), \mathcal{V}_k(v^*, w^*), \mathcal{V}_k(t', t)$, and $\mathcal{V}_2(t')$. Let $\mathcal{V} := \mathcal{V}_k(s, s') \cup \mathcal{V}_2(s') \cup \mathcal{V}_k(v^*, w^*) \cup \mathcal{V}_k(t', t) \cup \mathcal{V}_2(t')$. Each of these collections contains at most Δ^{k-1} sets of size at most $k - 1$ and each can be computed in $\mathcal{O}(\Delta^k)$ time. Moreover, for each set $V' \in \mathcal{V}_k(x, y)$ it holds that $V' \subseteq N^{k/2}[x] \cup N^{k/2}[y]$, where $N^{k/2}[u]$ denotes the set of vertices having distance at most $k/2$ to u. Hence, $V^* := \bigcup_{V' \in \mathcal{V}} V'$ is a subset of $\bigcup_{x \in \{s, s', v^*, w^*, t', t\}} N^{k/2}[x] \cup N^2[s'] \cup N^2[t']$. Consequently, there is some $\lambda \in \mathcal{O}(\Delta^{\max(k/2, 2)})$ such that $|V^*| \leq \lambda$.

Next, we check for each $V' \subseteq V^*$ if V' is a hitting set for $\mathcal{V}$. If this is the case, then we check if there is an (s, t)-path P in G containing the edges $e^*, \{s, s'\}$, and $\{t', t\}$ such that $V' \cup N(s) \cup N(t) \subseteq V(P)$. This can be done by checking if there is an ordering $\pi = (x_1, \ldots, x_\ell)$ of the vertices of $\tilde{V} := V' \cup N(s) \cup N(t) \cup \{s, t, v^*, w^*\}$ such that there are pairwise vertex-disjoint (x_i, x_{i+1})-paths for all $i \in [1, \ell - 1]$ where $x_1 = s, x_2 = s', x_{\ell-1} = t', x_\ell = t$, and v^* and w^* are consecutive in π. This can be done in $g(|\tilde{V}|) \cdot n^2$ time [11] for some computable function g. Since we check all combinations of s, s', t', and t as well as every possible hitting set for $\mathcal{V}$, the algorithm is correct and has overall running time $\mathcal{O}(n^2 \cdot \Delta^2 \cdot \lambda \cdot 2^\lambda \cdot (\lambda + 4 + 2\Delta)! \cdot g(\lambda + 4 + 2\Delta) \cdot n^2) \subseteq \mathcal{O}(f(\Delta + k) \cdot n^4)$. □

6   Conclusion

We proposed a generic approach to the design of data reduction rules via local optimality and examined its viability for well-known NP-hard problems. It seems interesting to find further positive applications of this approach along the lines

of the LONGEST PATH problem. One might, for example, consider extensions of
the $(1,k)$-swap neighborhood for paths. Finally, regardless of the connection to
data reduction, it seems interesting in its own right to study which properties of
locally optimal solutions can be computed efficiently.

References

1. Belmonte, R., Hanaka, T., Lampis, M., Ono, H., Otachi, Y.: Independent set reconfiguration parameterized by modular-width. Algorithmica **82**(9): 2586–2605 (2020). https://doi.org/10.1007/s00453-020-00700-y
2. Cai, S., Su, K., Luo, C., Sattar, A.: NuMVC: an efficient local search algorithm for minimum vertex cover. J. Artif. Intell. Res. **46**, 687–716 (2013)
3. Casel, K., Fernau, H., Ghadikoalei, M.K., Monnot, J., Sikora, F.: Extension of vertex cover and independent set in some classes of graphs. In: Heggernes, P. (ed.) CIAC 2019. LNCS, vol. 11485, pp. 124–136. Springer, Cham (2019). https://doi.org/10.1007/978-3-030-17402-6_11
4. Censor-Hillel, K., Rabie, M.: Distributed reconfiguration of maximal independent sets. J. Comput. Syst. Sci. **112**, 85–96 (2020)
5. de Berg, M., Jansen, B.M.P., Mukherjee, D.: Independent-set reconfiguration thresholds of hereditary graph classes. Discret. Appl. Math. **250**, 165–182 (2018)
6. Fellows, M.R., Fedor, F.V., Lokshtanov, D., Rosamond, F.A., Saurabh, S., Villanger, Y.: Local search: is brute-force avoidable? J. Comput. Syst. Sci. **78**(3), 707–719 (2012)
7. Garey, M.R., Johnson, D.S., Tarjan, R.E.: The planar Hamiltonian circuit problem is NP-complete. SIAM J. Comput. **5**(4), 704–714 (1976)
8. Guo, J., Hartung, S., Niedermeier, R., Suchý, O.: The parameterized complexity of local search for TSP, more refined. Algorithmica **67**(1), 89–110 (2013). https://doi.org/10.1007/s00453-012-9685-8
9. Johnson, D.S., Papadimitriou, C.H., Yannakakis, M.: How easy is local search? J. Comput. Syst. Sci. **37**(1), 79–100 (1988)
10. Katzmann, M., Komusiewicz, C.: Systematic exploration of larger local search neighborhoods for the minimum vertex cover problem. In: Proceedings of the Thirty-First AAAI Conference on Artificial Intelligence, (AAAI 2017), pp. 846–852. AAAI Press (2017)
11. Kawarabayashi, K., Kobayashi, Y., Reed, B.A.: The disjoint paths problem in quadratic time. J. Comb. Theory Ser. B **102**(2), 424–435 (2012)
12. Khosravian Ghadikoalei, M., Melissinos, N., Monnot, J., Pagourtzis, A.: Extension and its price for the connected vertex cover problem. In: Colbourn, C.J., Grossi, R., Pisanti, N. (eds.) IWOCA 2019. LNCS, vol. 11638, pp. 315–326. Springer, Cham (2019). https://doi.org/10.1007/978-3-030-25005-8_26
13. Lokshtanov, D., Mouawad, A.E.: The complexity of independent set reconfiguration on bipartite graphs. ACM Trans. Algorithms **15**(1), 7:1–7:19 (2019)
14. Marx, D.: Searching the k-change neighborhood for TSP is W[1]-hard. Oper. Res. Lett. **36**(1), 31–36 (2008)
15. Tovey, C.A.: A simplified NP-complete satisfiability problem. Discret. Appl. Math. **8**(1), 85–89 (1984)

Colouring Graphs of Bounded Diameter in the Absence of Small Cycles

Barnaby Martin, Daniël Paulusma[ORCID], and Siani Smith[✉]

Department of Computer Science, Durham University, Durham, UK
{barnaby.d.martin,daniel.paulusma,siani.smith}@durham.ac.uk

Abstract. For $k \geq 1$, a k-colouring c of G is a mapping from $V(G)$ to $\{1, 2, \ldots, k\}$ such that $c(u) \neq c(v)$ for any two non-adjacent vertices u and v. The k-COLOURING problem is to decide if a graph G has a k-colouring. For a family of graphs $\mathcal{H}$, a graph G is $\mathcal{H}$-free if G does not contain any graph from $\mathcal{H}$ as an induced subgraph. Let C_s be the s-vertex cycle. In previous work (MFCS 2019) we examined the effect of bounding the diameter on the complexity of 3-COLOURING for $(C_3, \ldots, C_s)$-free graphs and H-free graphs where H is some polyad. Here, we prove for certain small values of s that 3-COLOURING is polynomial-time solvable for C_s-free graphs of diameter 2 and (C_4, C_s)-free graphs of diameter 2. In fact, our results hold for the more general problem LIST 3-COLOURING. We complement these results with some hardness result for diameter 4.

1 Introduction

Graph colouring is a well-studied topic in Computer Science due to its wide range of applications. A k-*colouring* of a graph G is a mapping $c : V(G) \to \{1, \ldots, k\}$ that assigns each vertex u a *colour* $c(u)$ in such a way that $c(u) \neq c(v)$ for any two adjacent vertices u and v of G. The aim is to find the smallest value of k (also called the *chromatic number*) such that G has a k-colouring. The corresponding decision problem is called COLOURING, or k-COLOURING if k is fixed, that is, not part of the input. As even 3-COLOURING is NP-complete [16], k-COLOURING and COLOURING have been studied for many special graph classes, as surveyed in, for example, [1,5,9,13,15,21,23,26]. This holds in particular for *hereditary* classes of graphs, which are the classes of graphs closed under vertex deletion.

It is well known and not difficult to see that a class of graphs is hereditary if and only if it can be characterized by a unique set $\mathcal{F}_\mathcal{G}$ of minimal forbidden induced subgraphs. In particular, a graph G is H-*free* for some graph H if G does not contain H as an *induced* subgraph. The latter means that we cannot modify G into H by a sequence of vertex deletions. For a set of graphs $\{H_1, \ldots, H_p\}$, a graph G is $(H_1, \ldots, H_p)$-*free* if G is H_i-free for every $i \in \{1, \ldots, p\}$.

We continue a long-term study on the complexity of 3-COLOURING for special graph classes. Let C_t and P_t be the cycle and path, respectively, on t vertices.

Research supported by the Leverhulme Trust (RPG-2016-258).

T. Calamoneri and F. Corò (Eds.): CIAC 2021, LNCS 12701, pp. 367–380, 2021.
https://doi.org/10.1007/978-3-030-75242-2_26

The complexity of 3-COLOURING for H-free graphs has not yet been classified; in particular this is still is open for P_t-free graphs for every $t \geq 8$, whereas the case $t = 7$ is polynomial [3]. For $t \geq 3$, let $C_{>t} = \{C_{t+1}, C_{t+2}, \ldots\}$. Note that for $t \geq 2$, the class of P_t-free graphs is a subclass of $C_{>t}$-free graphs. Recently, Pilipczuk, Pilipczuk and Rzążewski [22] gave for every $t \geq 3$, a quasi-polynomial-time algorithm for 3-COLOURING on $C_{>t}$-free graphs. Rojas and Stein [24] proved in another recent paper that for every odd integer $t \geq 9$, 3-COLOURING is polynomial-time solvable for $(\mathcal{C}^{odd}_{<t-3}, P_t)$-free graphs, where $\mathcal{C}^{odd}_{<t}$ is the set of all odd cycles on less than t vertices. This complements a result from [10], which implies that for every $t \geq 1$, 3-COLOURING, or more general LIST 3-COLOURING (defined later), is polynomial-time solvable for (C_4, P_t)-free graphs (see also [18]).

The graph classes in this paper are only partially characterized by forbidden induced subgraphs: we also restrict the diameter. The *distance* $\text{dist}(u,v)$ between two vertices u and v in a graph G is the length (number of edges) of a shortest path between them. The *diameter* of a graph G is the maximum distance over all pairs of vertices in G. Note that the n-vertex path P_n has diameter $n-1$, but by removing an internal vertex the diameter becomes infinite. Hence, for every integer $d \geq 2$, the class of graphs of diameter at most d is not hereditary.

For every $d \geq 3$, the 3-COLOURING problem for graphs of diameter at most d is NP-complete, as shown by Mertzios and Spirakis [20] who gave a highly non-trivial NP-hardness construction for the case where $d = 3$. In fact they proved that 3-COLOURING is NP-complete even for C_3-free graphs of diameter 3 and radius 2. The complexity of 3-COLOURING for the class of all graphs of diameter 2 has been posed as an open problem in several papers [2,4,19–21].

On the positive side, Mertzios and Spirakis [20] gave a subexponential-time algorithm for 3-COLOURING on graphs of diameter 2. Moreover, as we discuss below, 3-COLOURING is polynomial-time solvable for several subclasses of diameter 2. A graph G has an *articulation neighbourhood* if $G - (N(v) \cup \{v\})$ is disconnected for some $v \in V(G)$. The neighbourhoods $N(u)$ and $N(v)$ of two distinct (and non-adjacent) vertices u and v are *nested* if $N(u) \subseteq N(v)$. We let $K_{1,r}$ be the star on $r + 1$ vertices. The *subdivision* of an edge uw in a graph removes uw and replaces it with a new vertex v and edges uv, vw. We let $K^\ell_{1,r}$ be the ℓ-*subdivided star*, which is obtained from $K_{1,r}$ by subdividing *one* edge exactly ℓ times. The graph $S_{h,i,j}$, for $1 \leq h \leq i \leq j$, is the tree with one vertex x of degree 3 and exactly three leaves, which are of distance h, i and j from x, respectively. Note that $S_{1,1,1} = K_{1,3}$. The *diamond* is obtained from the 4-vertex complete graph by deleting an edge. The 3-COLOURING problem is polynomial-time solvable for:

- diamond-free graphs of diameter 2 with an articulation neighbourhood but without nested neighbourhoods [20];
- (C_3, C_4)-free graphs of diameter 2 [19];
- $K^2_{1,r}$-free graphs of diameter 2, for every $r \geq 1$ [19]; and
- $S_{1,2,2}$-free graphs of diameter 2 [19].

It follows from results in [8,12,17] that without the diameter-2 condition, 3-Colouring is NP-complete again in each of the above cases; in particular 3-Colouring is NP-complete for $\mathcal{C}$-free graphs for any finite set $\mathcal{C}$ of cycles.

Our Results. We aim to increase our understanding of the complexity of 3-Colouring for graphs of diameter 2. In [19] we mainly considered 3-Colouring for graphs of diameter 2 with some forbidden induced subdivided star. In this paper, we continue this study by focusing on 3-Colouring for C_s-free or (C_s, C_t)-free graphs of diameter 2 for small values of s and t; in particular for the case where $s = 4$ (cf. the aforementioned result for (C_4, P_t)-free graphs). In fact we prove our results for a more general problem, namely List 3-Colouring, whose complexity for diameter 2 is also still open. A *list assignment* of a graph $G = (V, E)$ is a function L that prescribes a *list of admissible colours* $L(u) \subseteq \{1, 2, \ldots\}$ to each $u \in V$. A colouring c *respects* L if $c(u) \in L(u)$ for every $u \in V$. For an integer $k \geq 1$, if $L(u) \subseteq \{1, \ldots, k\}$ for each $u \in V$, then L is a *list k-assignment*. The List k-Colouring problem is to decide if a graph G with an list k-assignment L has a colouring that respects L. If every list is $\{1, \ldots, k\}$, we obtain k-Colouring.

The following two theorems summarize our main results.

Theorem 1. *For $s \in \{5, 6\}$, List 3-Colouring is polynomial-time solvable for C_s-free graphs of diameter 2.*

Theorem 2. *For $t \in \{3, 5, 6, 7, 8, 9\}$, List 3-Colouring is polynomial-time solvable for (C_4, C_t)-free graphs of diameter 2.*

The case $t = 3$ in Theorem 2 directly follows from the Hoffman-Singleton Theorem [11], which states that there are only four (C_3, C_4)-free graphs of diameter 2. The cases $t \in \{5, 6\}$ immediately follows from Theorem 1. Hence, apart from proving Theorem 1, we only need to prove Theorem 2 for $t \in \{7, 8, 9\}$.

We prove Theorem 1 and the case $t = 7$ of Theorem 2 in Sect. 3. As we explain in the same section, all these results follow from the same technique, which is based on a number of (known) propagation rules. We first colour a small number of vertices and then start to apply the propagation rules exhaustively. This will reduce the sizes of the lists of the vertices. The novelty of our approach is the following: we can prove that the diameter-2 property ensures such a widespread reduction that each precolouring changes our instance into an instance of 2-List Colouring: the polynomial-solvable variant of List Colouring where each list has size at most 2 [7] (see also Sect. 2).

We prove the cases $t = 8$ and $t = 9$ of Theorem 2 in Sect. 4 using a refinement of the technique from Sect. 3. We explain this refinement in detail at the start of Sect. 4. In short, in our branching, we exploit information from earlier obtained no-answers to reduced instances of our original instance (G, L).

We complement Theorems 1 and 2 by the following result for diameter 4, whose proof we omit.

Theorem 3. *For every even integer $t \geq 6$, 3-Colouring is NP-complete on the class of $(C_4, C_6, \ldots, C_t)$-free graphs of diameter 4.*

Results of Damerell [6] imply that 3-COLOURING is polynomial-time solvable for (C_3, C_4, C_5, C_6)-free graphs of diameter 3 and for $(C_3, \ldots, C_8)$-free graphs of diameter 4 [19]. We were not able to reduce the diameter in Theorem 3 from 4 to 3; see Sect. 5 for a further discussion, including other open problems.

2   Preliminaries

Let $G = (V, E)$ be a graph. A vertex $u \in V$ is *dominating* if u is adjacent to every other vertex of G. For $S \subseteq V$, the graph $G[S] = (S, \{uv \mid u, v \in S \text{ and } uv \in E\})$ denotes the subgraph of G induced by S. The *neighbourhood* of a vertex $u \in V$ is the set $N(u) = \{v \mid uv \in E\}$ and the *degree* of u is the size of $N(u)$. For a set $U \subseteq V$, we write $N(U) = \bigcup_{u \in U} N(u) \backslash U$.

The *bull* is the graph obtained from a triangle on vertices x, y, z after adding two new vertices u and v and edges xu and yv. A *clique* is a set of pairwise adjacent vertices, and an *independent set* is a set of pairwise non-adjacent vertices.

Let G be a graph with a list assignment L. If $|L(u)| \leq \ell$ for each $u \in V$, then L is a ℓ-*list assignment*. A list k-assignment is a k-list assignment, but the reverse is not necessarily true. The ℓ-LIST COLOURING problem is to decide if a graph G with an ℓ-list assignment L has a colouring that respects L. We use a known general strategy for obtaining a polynomial-time algorithm for LIST 3-COLOURING on some class $\mathcal{G}$. That is, we will reduce the input to a polynomial number of instances of 2-LIST COLOURING and use a well-known result:

Theorem 4 ([7]). *The* 2-LIST COLOURING *problem is linear-time solvable.*

We also need an observation (proof omitted).

Lemma 1. *Let G be a non-bipartite graph of diameter 2. Then G contains a C_3 or induced C_5.*

3   The Propagation Algorithm and Three Results

We present our initial propagation algorithm, which is based on a number of (well-known) propagation rules; we illustrate Rules 4 and 5 in Figs. 1 and 2.

Rule 1 (no empty lists). If $L(u) = \emptyset$ for some $u \in V$, then return **no**.
Rule 2 (not only lists of size 2). If $|L(u)| \leq 2$ for every $u \in V$, then apply Theorem 4.
Rule 3 (single colour propagation). If u and v are adjacent, $|L(u)| = 1$, and $L(u) \subseteq L(v)$, then set $L(v) := L(v) \backslash L(u)$.
Rule 4 (diamond colour propagation). If u and v are adjacent and share two common non-adjacent neighbours x and y with $|L(x)| = |L(y)| = 2$ and $L(x) \neq L(y)$, then set $L(x) := L(x) \cap L(y)$ and $L(y) := L(x) \cap L(y)$ (so $L(x)$ and $L(y)$ get size 1).

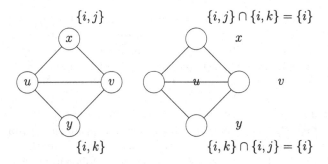

Fig. 1. Left: A diamond graph before applying Rule 4. Right: After applying Rule 4.

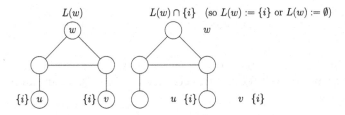

Fig. 2. Left: A bull graph before applying Rule 5. Right: After applying Rule 5.

Rule 5 (bull colour propagation). If u and v are the two degree-1 vertices of an induced bull B of G and $L(u) = L(v) = \{i\}$ for some $i \in \{1, 2, 3\}$ and moreover $L(w) \neq \{i\}$ for the degree-2 vertex w of B, then set $L(w) := L(w) \cap \{i\}$.

We say that a propagation rule is *safe* if the new instance is a yes-instance of LIST 3-COLOURING if and only if the original instance is so. We make the following observation, which is straightforward (see also [14]).

Lemma 2. *Each of the Rules 1–5 is safe and can be applied in polynomial time.*

Consider again an instance (G, L). Let N_0 be a subset of $V(G)$ that has size at most some constant. Assume that $G[N_0]$ has a colouring c that respects the restriction of L to N_0. We say that c is an L-*promising* N_0-*precolouring* of G.

In our algorithms we first determine a set N_0 of constant size and consider every L-promising N_0-precolouring of G. That is, we modify L into a list assignment L_c with $L_c(u) = \{c(u)\}$ (where $c(u) \in L(u)$) for every $u \in N_0$ and $L_c(u) = L(u)$ for every $u \in V(G) \backslash N_0$. We then apply Rules 1–5 on (G, L_c) *exhaustively*, that is, until none of the rules can be applied anymore. This is the *propagation algorithm* and we say that it did a *full c-propagation*. The propagation algorithm may output **yes** and **no** (when applying Rules 1 or 2); else it will output **unknown**.

If the algorithm returns **yes**, then (G, L) is a yes-instance of LIST 3-COLOURING by Lemma 2. If it returns **no**, then (G, L) has no L-respecting colouring coinciding with c on N_0, again by Lemma 2. If the algorithm returns unknown, then (G, L) may still have an L-respecting colouring that coincides with c on N_0. In that case the propagation algorithm did not apply Rule 1 or 2. Hence, it modified L_c into a list assignment L'_c of G such that $L'_c(u) \neq \emptyset$ for every $u \in V(G)$ and at least one vertex v of G still has a list $L'_c(v)$ of size 3, that is, $L'_c(v) = \{1, 2, 3\}$. We say that L'_c (if it exists) is the c-propagated list assignment of G.

After performing a full c-propagation for every L-promising N_0-precolouring c of G we say that we performed a *full N_0-propagation*. We say that (G, L) is N_0-*terminal* if after the full N_0-propagation one of the following cases hold:

1. for some L-promising N_0-precolouring, the propagation algorithm returned yes;
2. for every L-promising N_0-precolouring, the propagation algorithm returned no.

Note that if (G, L) is N_0-terminal for some set N_0, then we have solved LIST 3-COLOURING on instance (G, L). The next lemma formalizes our approach (proof omitted).

Lemma 3. *Let (G, L) be an instance of* LIST 3-COLOURING. *Let N_0 be a subset of $V(G)$ of constant size. Performing a full N_0-propagation takes polynomial time. Moreover, if (G, L) is N_0-terminal, then we have solved* LIST 3-COLOURING *on instance (G, L).*

We now prove our first three results on LIST 3-COLOURING for diameter-2 graphs. The first result, whose proof we omit, generalizes a corresponding result for 3-COLOURING in [19].

Theorem 5. LIST 3-COLOURING *can be solved in polynomial time for C_5-free graphs of diameter at most 2.*

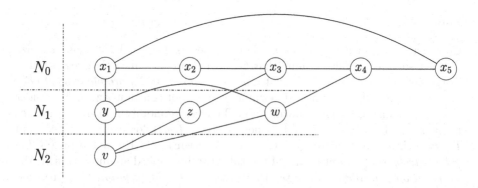

Fig. 3. The situation in the proof of Theorem 6, which is similar to the situation in the proof of Theorem 7.

Theorem 6. LIST 3-COLOURING *can be solved in polynomial time for C_6-free graphs of diameter at most 2.*

Proof. Let $G = (V, E)$ be a C_6-free graph of diameter 2 with a list 3-assignment L. If G is C_5-free, then we apply Theorem 5. If G contains a K_4, then G is not 3-colourable and hence, (G, L) is a no-instance of LIST 3-COLOURING. We check these properties in polynomial time. So, from now on, we assume that G is a K_4-free graph that contains an induced 5-vertex cycle C, say with vertex set $N_0 = \{x_1, \ldots, x_5\}$ in this order. Let N_1 be the set of vertices that do not belong to C but that are adjacent to at least one vertex of C. Let $N_2 = V \backslash (N_0 \cup N_1)$ be the set of remaining vertices.

As N_0 has size 5, we can apply a full N_0-propagation in polynomial time by Lemma 3. By the same lemma we are done if we can prove that (G, L) is N_0-terminal. We prove this claim below.

For contradiction, assume that (G, L) is not N_0-terminal. Then there must exist an L-promising N_0-precolouring c for which we obtain the c-propagated list assignment L'_c. By definition of L'_c we find that G contains a vertex v with $L'_c(v) = \{1, 2, 3\}$. Then $v \notin N_0$, as every $u \in N_0$ has $L'_c(u) = \{c(u)\}$. Moreover, $v \notin N_1$, as vertices in N_1 have a list of size at most 2 after applying Rule 3. Hence, we find that $v \in N_2$.

We first note that some colour of $\{1, 2, 3\}$ appears exactly once on N_0, as $|N_0| = 5$. Hence, we may assume without loss of generality that $c(x_1) = 1$ and that $c(x_i) \in \{2, 3\}$ for every $i \in \{2, 3, 4, 5\}$.

As G has diameter 2, there exists a vertex $y \in N_1$ that is adjacent to x_1 and v. As $L'_c(v) = \{1, 2, 3\}$ and $c(x_1) = 1$, we find that $L'_c(y) = \{2, 3\}$. As $c(x_i) \in \{2, 3\}$ for every $i \in \{2, 3, 4, 5\}$, the latter means that y is not adjacent to any x_i with $i \in \{2, 3, 4, 5\}$. Hence, as G has diameter 2, there exists a vertex $z \in N_1$ with $z \neq y$, such that z is adjacent to x_3 and v. We assume without loss of generality that $c(x_3) = 3$ and thus $c(x_2) = c(x_4) = 2$ and thus $c(x_5) = 3$. As $L'_c(v) = \{1, 2, 3\}$ and $c(x_3) = 3$, we find that $L'_c(z) = \{1, 2\}$. Hence, z is not adjacent to any vertex of $\{x_1, x_2, x_4\}$. Now the set $\{x_1, x_2, x_3, z, v, y\}$ forms a cycle on six vertices. As G is C_6-free, this cycle cannot be induced. Hence, the above implies that y and z must be adjacent; see also Fig. 3.

As G has diameter 2, there exists a vertex $w \in N_1$ that is adjacent to x_4 and v. As both y and z are not adjacent to x_4, we find that $w \notin \{y, z\}$. As $L'_c(v) = \{1, 2, 3\}$ and $c(x_4) = 2$, we find that $L'_c(w) = \{1, 3\}$. As $c(x_1) = 1$ and $c(x_3) = c(x_5) = 3$, the latter implies that w is not adjacent to any vertex of $\{x_1, x_3, x_5\}$. Consequently, w must be adjacent to y, as otherwise the 6-vertex cycle with vertex set $\{x_1, x_5, x_4, w, v, y\}$ would be induced, contradicting the C_6-freeness of G. We refer again to Fig. 3 for a display of the situation.

If w and z are adjacent, then $\{v, w, y, z\}$ induces a K_4, contradicting the K_4-freeness of G. Hence, w and z are not adjacent. Then $\{v, w, y, z\}$ induces a diamond, in which w and z are the two non-adjacent vertices. However, as $L'_c(w) = \{1, 3\}$ and $L'_c(z) = \{1, 2\}$, our algorithm would have applied Rule 4. This would have resulted in lists of w and z that are both equal to $\{1, 3\} \cap \{1, 2\} = \{1\}$. Hence, we obtained a contradiction and conclude that (G, L) is N_0-terminal. □

Theorem 7 is proven in a similar way as Theorem 6 and we omit its proof.

Theorem 7. LIST 3-COLOURING *can be solved in polynomial time for* (C_4, C_7)-*free graphs of diameter* 2.

4   The Extended Propagation Algorithm and Two Results

For our next two results, we need a more sophisticated method. Let (G, L) be an instance of LIST 3-COLOURING. Let p be some positive constant. We consider each set $N_0 \subseteq V(G)$ of size at most p and perform a full N_0-propagation. Afterwards we say that we performed a *full p-propagation*. We say that (G, L) is *p-terminal* if after the full p-propagation one of the following cases hold:

1. for some $N_0 \subseteq V(G)$ with $|N_0| \leq c$, there is an L-promising N_0-precolouring c, such that the propagation algorithm returns **yes**; or
2. for every set $N_0 \subseteq V(G)$ with $|N_0| \leq c$ and every L-promising N_0-precolouring c, the propagation algorithm returns **no**.

We can now prove the following lemma.

Lemma 4. *Let (G, L) be an instance of* LIST 3-COLOURING *and $p \geq 1$ be some constant. Performing a full p-propagation takes polynomial time. Moreover, if (G, L) is p-terminal, then we have solved* LIST 3-COLOURING *on instance (G, L).*

Proof. For every set $N_0 \subseteq V(G)$, a full N_0-propagation takes polynomial time by Lemma 3. Then the first statement of the lemma follows from this observation and the fact that we need to perform $O(n^p)$ full N_0-propagations, which is a polynomial number, as p is a constant.

Now suppose that (G, L) is p-terminal. First assume that for some $N_0 \subseteq V(G)$ with $|N_0| \leq c$, there exists an L-promising N_0-precolouring c, such that the propagation algorithm returns **yes**. Then (G, L) is a yes-instance due to Lemma 2. Now assume that for every set $N_0 \subseteq V(G)$ with $|N_0| \leq c$ and every L-promising N_0-precolouring c, the propagation algorithm returns **no**. Then (G, L) is a no-instance. This follows from Lemma 2 combined with the observation that if (G, L) was a yes-instance, the restriction of a colouring c that respects L to any set N_0 of size at most p would be an L-promising N_0-precolouring of G. □

In our next two algorithms, we perform a full p-propagation for some appropriate constant p. If we find that an instance (G, L) is p-terminal, then we are done by Lemma 4. In the other case, we exploit the new information on the structure of G that we obtain from the fact that (G, L) is not p-terminal. We omit the proof of the first theorem.

Theorem 8. LIST 3-COLOURING *can be solved in polynomial time for* (C_4, C_8)-*free graphs of diameter* 2.

Theorem 9. LIST 3-COLOURING *can be solved in polynomial time for* (C_4, C_9)-*free graphs of diameter* 2.

Proof. Let $G = (V, E)$ be a (C_4, C_9)-free graph of diameter 2 with a list 3-assignment L. If G is C_7-free, then we apply Theorem 7. If G contains a K_4, then G is not 3-colourable and hence, (G, L) is a no-instance of LIST 3-COLOURING. We check these properties in polynomial time. So, from now on, we assume that G is a K_4-free graph that contains at least one induced cycle on seven vertices.

We set $p = 7$ and perform a full p-propagation. This takes polynomial time by Lemma 2. By the same lemma, we have solved LIST 3-COLOURING on (G, L) if (G, L) is p-terminal. Suppose we find that (G, L) is not p-terminal.

We first prove the following claim.

Claim 1. For each induced 7-vertex cycle C, the propagation algorithm returned **no** *for every L-promising $V(C)$-colouring c that assigns the same colour i on two vertices of C that have a common neighbour on C and that gives every other vertex of C a colour different from i.*

We prove Claim 1 as follows. Consider an induced 7-vertex cycle C, say with vertex set $N_0 = \{x_1, \ldots, x_7\}$ in this order. Let N_1 be the set of vertices that do not belong to C but that are adjacent to at least one vertex of C. Let $N_2 = V \backslash (N_0 \cup N_1)$ be the set of remaining vertices. Let c be an L-promising $V(C)$-colouring that assigns two vertices of C with a common neighbour on C the same colour, say $c(x_1) = 1$ and $c(x_3) = 1$, and moreover, that assigns every vertex x_i with $i \in \{2, 4, 5, 6, 7\}$ colour $c(x_i) \neq 1$.

For contradiction, suppose that a full c-propagation does not yield a **no** output. As (G, L) is not p-terminal, this means that we obtained the c-propagated list assignment L'_c. By definition of L'_c we find that G contains a vertex v with $L'_c(v) = \{1, 2, 3\}$. Then $v \notin N_0$, as every $u \in N_0$ has $L'_c(u) = \{c(u)\}$. Moreover, $v \notin N_1$, as vertices in N_1 have a list of size at most 2 after applying Rule 3. Hence, we find that $v \in N_2$.

As G has diameter 2, there exist a vertex $y \in N_1$ that is adjacent to both v and x_1. Then y is not adjacent to any x_i with $i \in \{2, 4, 5, 6, 7\}$; in that case y would have a list of size 1 (as each x_i other than x_1 and x_3 is coloured 2 or 3) meaning that $L'_c(v)$ would have size at most 2. Hence, y is not adjacent to x_3 either, as otherwise $\{y, x_1, x_2, x_3\}$ would induce a C_4. As G has diameter 2, this means that there exists a vertex $y' \in N_1$ with $y' \neq y$ such that y' is adjacent to both v and x_3. By the same arguments we used for y', we find that x_3 is the only neighbour of y' on C.

If yy' is an edge then, by Rule 5, v would have had list $\{1\}$ instead of $\{1, 2, 3\}$. Hence, y and y' are not adjacent. However, now $\{y, v, y', x_3, x_4, x_5, x_6, x_7, x_1\}$ induces a C_9, a contradiction; see also Fig. 4. This proves Claim 1.

Claim 1 tells us that if G has a colouring c respecting L, then c only gives the same colour to two vertices x and x' that are of distance 2 on some induced 7-vertex cycle C if there is a third vertex x'' that is of distance 2 from either x or x' on C with $c(x'') = c(x') = c(x)$. Hence, we can safely use the following new

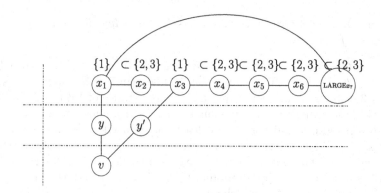

Fig. 4. The situation that is described in Claim 1 in the proof of Theorem 9. The set $\{x_1, y, v, y', x_3, x_4, x_5, x_6, x_7\}$ induces C_9, which is not possible.

rule, whose execution takes polynomial time (in this rule, $c(x_1) = c(x_6)$ is not possible: view x_1 as x and x_6 as x' and note that x'' can neither be x_3 or x_4).

Rule-C7 (C_7 colour propagation). Let C be an induced cycle on seven vertices $x_1, x_2, \ldots, x_7$ in that order. If $|L(x_i)| = 1$ for $i \in \{1, 2, 3, 4\}$, $L(\{x_1, x_2, x_3\}) = \{1, 2, 3\}$, $L(x_4) = L(x_2)$, and $L(x_1) \subseteq L(x_6)$, then set $L(x_6) := \{1, 2, 3\} \setminus L(x_1)$ (so $L(x_6)$ gets size at most 2).

We now consider an induced 7-vertex cycle C in G, say on vertices $x_1, \ldots, x_7$ in that order. Then either one colour appear once on C, or two colours appear exactly twice on C, with distance 3 from each other on C. Hence, we may assume without loss of generality that if G has a colouring c that respects L, then one of the following holds for such a colouring c (see also Figs. 5 and 6):

(1) $c(x_1) = 1$, $c(x_2) = 2$, $c(x_3) = 3$, $c(x_4) = 2$, $c(x_5) = 3$, $c(x_6) = 2$, $c(x_7) = 3$;

or

(2) $c(x_1) = 1$, $c(x_2) = 2$, $c(x_3) = 3$, $c(x_4) = 1$, $c(x_5) = 3$, $c(x_6) = 2$, $c(x_7) = 3$.

We let again $N_0 = \{x_1, \ldots, x_7\}$, N_1 be the set of vertices that do not belong to C but that are adjacent to at least one vertex of C, and $N_2 = V \setminus (N_0 \cup N_1)$ be the set of remaining vertices. We do a full c-propagation but now we also include the exhaustive use of Rule-C7. By combining Lemma 2 with the observation that Rule-C7 runs in polynomial time and reduces the list size of at least one vertex, this takes polynomial time. By combining the same lemma with the fact that Rule-C7 is safe (due to Claim 1) and the above observation that every L-respecting colouring of G coincides with c on N_0 (subject to colour permutation), we are done if we can prove that the propagation algorithm either outputs **yes** or **no**. We show that this is the case for each of the two possibilities (1) and (2) of c.

For contradiction, assume that the propagation algorithm returns **unknown**. Then we obtained the c-propagated list assignment L'_c. By definition of L'_c we find that G contains a vertex v with $L'_c(v) = \{1, 2, 3\}$. Then $v \notin N_0$, as every

$u \in N_0$ has $L'_c(u) = \{c(u)\}$. Moreover, $v \notin N_1$, as vertices in N_1 have a list of size at most 2 after applying Rule 3. Hence, we find that $v \in N_2$. We now need to distinguish between the two possibilities of c.

Case 1 $c(x_1) = 1, c(x_2) = 2, c(x_3) = 3, c(x_4) = 2, c(x_5) = 3, c(x_6) = 2, c(x_7) = 3$. As G has diameter 2, there exists a vertex $y \in N_1$ that is adjacent to x_1 and v. Hence, y is not adjacent to any vertex in $\{x_2, \ldots, x_7\}$; otherwise y would have a list of size 1 due to Rule 3, and by the same rule, v would have a list of size 2. As G has diameter 2, there exists a vertex $y' \in N_1$ that is adjacent to x_4 and v. By the same arguments as above, y' is not adjacent to any vertex of $\{x_1, x_3, x_5, x_7\}$. The latter, together with the C_4-freeness of G, implies that y' is not adjacent to x_2 and x_6 either.

First suppose that $yy' \in E$. Then $\{x_1, x_7, x_6, x_5, x_4, y', y\}$ induces a C_7; see also Fig. 5. As $c(x_1) = 1$, $c(x_7) = 3$, $c(x_6) = 2$ and $c(x_5) = 3$, we find that $L_c(\{x_1, x_7, x_6\}) = \{1, 2, 3\}$ and $L_c(x_5) = L_c(x_7)$. Then $1 \notin L_c(y')$, as otherwise the propagation algorithm would have applied Rule-C7. Moreover, $2 \notin L_c(y')$, as otherwise the propagation algorithm would have applied Rule 3. Hence, $L_c(y') = \{3\}$. However, then $|L_c(v)| \leq 2$, again due to Rule 3, a contradiction.

Now suppose that $yy' \notin E$. Then $\{x_1, x_2, x_3, x_4, y', v, y\}$ induces a C_7. As $c(x_1) = 1$, $c(x_2) = 2$, $c(x_3) = 3$, $c(x_4) = 2$, we find that $L_c(\{x_1, x_2, x_3\}) = \{1, 2, 3\}$ and $L_c(x_4) = L_c(x_2)$. Then $1 \notin L_c(v)$ due to Rule-C7. This is a contradiction, as we assumed $L_c(v) = \{1, 2, 3\}$. We conclude that the propagation algorithm returned either **yes** or **no**.

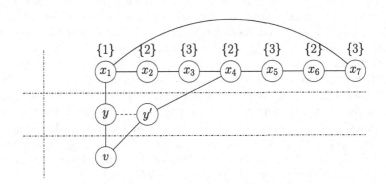

Fig. 5. The situation that is described in Case 1 in the proof of Theorem 9. If the edge yy' exists, then $\{x_1, x_7, x_6, x_5, x_4, y', y\}$ induces a C_7 to which Rule-C7 should have been applied. Otherwise the vertices $\{x_1, x_2, x_3, x_4, y', v, y\}$ induce such a C_7.

Case 2 $c(x_1) = 1, c(x_2) = 2, c(x_3) = 3, c(x_4) = 1, c(x_5) = 3, c(x_6) = 2, c(x_7) = 3$. As G has diameter 2, there is a vertex $y \in N_1$ adjacent to x_3 and v. Hence, y is not adjacent to any vertex in $\{x_1, x_2, x_4, x_6\}$; otherwise y would have a list of size 1 due to Rule 3, and by the same rule, v would have a list of size 2.

As $yx_4 \notin E$, we find that $yx_5 \notin E$ either; otherwise $\{y, x_3, x_4, x_5\}$ induces a C_4. As G has diameter 2, this means there is a vertex $y' \in N_1 \backslash \{y\}$ adjacent to x_5 and v. By the same arguments as above, y' is not adjacent to any vertex of $\{x_1, x_2, x_4, x_6\}$. As G is C_4-free, the latter implies that $y'x_3 \notin E$ and $y'x_7 \notin E$.

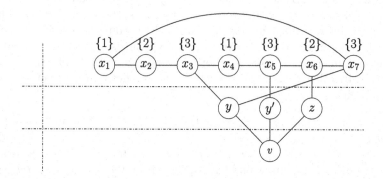

Fig. 6. The situation that is described in Case 2 in the proof of Theorem 9. The set $\{x_6, x_5, x_4, x_3, y, v, z\}$ induces a C_7 to which Rule-C7 should have been applied.

If $yy' \in E$, then v would have a list of size at most 2 due to Rule 5. Hence $yy' \notin E$. If $yx_7 \notin E$, this means that $\{x_1, x_2, x_3, y, v, y', x_5, x_6, x_7\}$ induces a C_9, which is not possible. Hence, $yx_7 \in E$.

To summarize, we found that v has two distinct neighbours y and y', where y has exactly two neighbours on C, namely x_3 and x_7, and y' has exactly one neighbour on C, namely x_5. As G has diameter 2, this means that there exists a vertex $z \in N_1$ with $z \notin \{y, y'\}$ that is adjacent to x_6 and v. Then z is not adjacent to any vertex of $\{x_1, x_3, x_4, x_5, x_7\}$, as otherwise z would have a list of size 1 due to Rule 3, and by the same rule, v would have a list of size 2. If $zy \in E$, then $\{y, z, x_6, x_7\}$ induces a C_4, which is not possible. Hence $zy \notin E$.

From the above, we find that $\{x_6, x_5, x_4, x_3, y, v, z\}$ induces a C_7; see also Fig. 6. As $c(x_6) = 2$, $c(x_5) = 3$, $c(x_4) = 1$ and $c(x_3) = 3$, we find that $L_c(\{x_6, x_5, x_4\}) = \{1, 2, 3\}$ and $L_c(x_3) = L_c(x_5)$. Then $2 \notin L_c(v)$, due to Rule-C7. Hence, $|L_c(v)| \le 2$, a contradiction. We conclude that the propagation algorithm returned either **yes** or **no** in Case 2 as well.   □

5   Conclusions

We proved that 3-COLOURABILITY is polynomial-time solvable for several subclasses of diameter 2 that are characterized by forbidding one or two small induced cycles. In order to do this we used a unified framework of propagation rules, which allowed us to exploit the diameter-2 property of the input graph. Our current techniques need to be extended to obtain further results (in particular, we cannot currently handle the increasing number of different 3-colourings of induced cycles of length larger than 9).

As open problems we pose: determine the complexity of 3-COLOURING and LIST 3-COLOURING for graphs of diameter 2; C_t-free graphs of diameter 2 for $s \in \{3, 4, 7, 8, \ldots\}$; and (C_4, C_t)-free graphs of diameter 2 for $t \geq 10$. We also note that the complexity of k-COLOURING for $k \geq 4$ and COLOURING is still open for C_3-free graphs of diameter 2 (see also [19]).

Finally, we turn to the class of graphs of diameter 3. The construction of Mertzios and Spirakis [20] for proving that 3-COLOURING is NP-complete for C_3-free graphs of diameter 3 appears to contain not only induced subdivided stars of arbitrary diameter and with an arbitrary number of leaves but also induced cycles of arbitrarily length $s \geq 4$. Hence, we pose as open problems: determine the complexity of 3-COLOURING and LIST 3-COLOURING for C_t-free graphs of diameter 3 for $t \geq 4$ and (C_4, C_t)-free graphs for $t \in \{3, 5, 6, \ldots\}$.

References

1. Alon, N.: Restricted colorings of graphs. Surveys in combinatorics, London Mathematical Society Lecture Note Series **187**, 1–33 (1993)
2. Bodirsky, M., Kára, J., Martin, B.: The complexity of surjective homomorphism problems - a survey. Discrete Appl. Math. **160**(12), 1680–1690 (2012)
3. Bonomo, F., Chudnovsky, M., Maceli, P., Schaudt, O., Stein, M., Zhong, M.: Three-coloring and list three-coloring of graphs without induced paths on seven vertices. Combinatorica **38**(4), 779–801 (2018)
4. Broersma, H., Fomin, F.V., Golovach, P.A., Paulusma, D.: Three complexity results on coloring P_k-free graphs. Eur. J. Comb. **34**(3), 609–619 (2013)
5. Chudnovsky, M.: Coloring graphs with forbidden induced subgraphs. In: Proceedings ICM 2014, vol. 4, pp. 291–302 (2014)
6. Damerell, R.M.: On Moore graphs. Proc. Camb. Philos. Soc. **74**, 227–236 (1973)
7. Edwards, K.: The complexity of colouring problems on dense graphs. Theoret. Comput. Sci. **43**, 337–343 (1986)
8. Emden-Weinert, T., Hougardy, S., Kreuter, B.: Uniquely colourable graphs and the hardness of colouring graphs of large girth. Comb. Probab. Comput. **7**(04), 375–386 (1998)
9. Golovach, P.A., Johnson, M., Paulusma, D., Song, J.: A survey on the computational complexity of colouring graphs with forbidden subgraphs. J. Graph Theor. **84**(4), 331–363 (2017)
10. Golovach, P.A., Paulusma, D., Song, J.: Coloring graphs without short cycles and long induced paths. Discrete Appl. Math. **167**, 107–120 (2014)
11. Hoffman, A.J., Singleton, R.R.: On Moore graphs with diameter 2 and 3. IBM J. Res. Devel. **5**, 497–504 (1960)
12. Holyer, I.: The NP-completeness of edge-coloring. SIAM J. Comput. **10**(4), 718–720 (1981)
13. Jensen, T.R., Toft, B.: Graph Coloring Problems. Wiley, New York (1995)
14. Klimošová, T., Malík, J., Masařík, T., Novotná, J., Paulusma, D., Slívová, V.: Colouring $(P_r + P_s)$-free graphs. In: Proceedings ISAAC 2018, LIPIcs, vol. 123, pp. 5:1–5:13 (2018)
15. Kratochvíl, J., Tuza, Zs., Voigt, M.: New trends in the theory of graph colorings: choosability and list coloring. In: Proceedings DIMATIA-DIMACS Conference, vol. 49, pp. 183–197 (1999)

16. Lovász, L.: Coverings and coloring of hypergraphs. In: Proceedings of 4th South-eastern Conference on Combinatorics, Graph Theory, and Computing, Utilitas Mathematicae, pp. 3–12 (1973)
17. Lozin, V.V., Kaminski, M.: Coloring edges and vertices of graphs without short or long cycles. Contrib. Discrete Math. **2**(1), 61–66 (2007)
18. Lozin, V.V., Malyshev, D.S.: Vertex coloring of graphs with few obstructions. Discrete Appl. Math. **216**, 273–280 (2017)
19. Martin, B., Paulusma, D., Smith, S.: Colouring H-free graphs of bounded diameter. In: Proceedings MFCS 2019, LIPIcs, vol. 138, pp. 14:1–14:14 (2019)
20. Mertzios, G.B., Spirakis, P.G.: Algorithms and almost tight results for 3-colorability of small diameter graphs. Algorithmica **74**(1), 385–414 (2016)
21. Paulusma, D.: Open problems on graph coloring for special graph classes. In: Mayr, E.W. (ed.) WG 2015. LNCS, vol. 9224, pp. 16–30. Springer, Heidelberg (2016). https://doi.org/10.1007/978-3-662-53174-7_2
22. Pilipczuk, M., Pilipczuk, M., Rzążewski, P.: Quasi-polynomial-time algorithm for independent set in P_t-free and $C_{\geq t}$-free graphs via shrinking the space of connecting subgraphs. CoRR, abs/2009.13494 (2020)
23. Randerath, B., Schiermeyer, I.: Vertex colouring and forbidden subgraphs - a survey. Graphs Comb. **20**(1), 1–40 (2004)
24. Rojas, A., Stein, M.: 3-colouring P_t-free graphs without short odd cycles. CoRR, abs/2008.04845 (2020)
25. Schaefer, T.J.: The complexity of satisfiability problems. In: Proceedings STOC, vol. 1978, pp. 216–226 (1978)
26. Tuza, Z.: Graph colorings with local constraints - a survey. Discuss. Math. Graph Theory **17**(2), 161–228 (1997)

Online Two-Dimensional Vector Packing With Advice

Bengt J. Nilsson[1(✉)] [iD] and Gordana Vujovic[2(✉)]

[1] Malmö University, Malmö, Sweden
bengt.nilsson.TS@mau.se
[2] University of Ljubljana, Ljubljana, Slovenia

Abstract. We consider the online two-dimensional vector packing problem, showing a lower bound of 11/5 on the competitive ratio of any ANYFIT strategy for the problem. We provide a strategy with competitive ratio $\max\{2, 6/(1 + 3\tan(\pi/4 - \gamma/2)) + \epsilon\}$ and logarithmic advice, for any instance where all the input vectors are restricted to have angles in the range $[\pi/4 - \gamma/2, \pi/4 + \gamma/2]$, for $0 \leq \gamma < \pi/3$. In addition, we give a 5/2-competitive strategy also using logarithmic advice for the unrestricted vectors case. These results should be contrasted to the currently best competitive strategy, FIRSTFIT, having competitive ratio 27/10.

Keywords: Bin Packing · Vector Packing · Online Computation · Competitive Analysis · Advice Complexity

1 Introduction

Arguably, the problem of packing items into bins is among the most well-studied in computer science. It asks for the "minimum number of unit sized bins required to pack a set of items, each of at most unit size," and has been shown to be NP-hard [15] but admits a PTAS [12]. It is common to view the bin packing problem through the lens of *online computation*, where the items are delivered one by one and each item has to be packed, either in an existing bin or a new bin, before the next item arrives. The quality of online strategies is measured by their *competitive ratio*, the worst case asymptotic or absolute ratio between the quality of the strategy's solution and that of an optimal one. Currently, the best strategy for online bin packing has asymptotic competitive ratio 1.5815... [16] and it has been shown that no strategy can have asymptotic competitive ratio better than $248/161 = 1.54037...$ [5]. For the absolute competitive ratio the tight bound of 5/3 has been proved [4].

The *vector packing problem* is a natural generalization of bin packing, where each item is a vector from $[0, 1]^D$ and items are to be packed in D-dimensional unit cubes. Approximation algorithms with linear dependency on D exist [10,12]. The online version of vector packing is not as well understood, the FIRSTFIT strategy has been shown to have competitive ratio $D + 7/10$ even for the more general *resource constrained scheduling* problem [14]. Azar *et al.* [2] claim that

© Springer Nature Switzerland AG 2021
T. Calamoneri and F. Corò (Eds.): CIAC 2021, LNCS 12701, pp. 381–393, 2021.
https://doi.org/10.1007/978-3-030-75242-2_27

no online strategy for D-dimensional vector packing can have competitive ratio better than D but offer no proof of this. They show however, that if all the vectors have L_∞-norm at most ϵ^2, there is a $(4/3 + \epsilon)$-competitive algorithm for online two-dimensional vector packing. They also provide a $4/3$ lower bound for arbitrarily small vectors. In general, Galambos *et al.* [13] provide a succinct lower bound for online D-dimensional vector packing that increases with D but remains below 2 for all D. Their result implicitly gives a lower bound of $1.80288\ldots$, for $D = 2$ which is currently the best known. Recently, almost $\Omega(D)$ asymptotic lower bounds have been established for online D-dimensional vector packing for sufficiently large D [3,6,7].

In many cases, the online framework is too restrictive in that it allows an all-powerful adversary to construct the input sequence in the worst possible way for the strategy. To alleviate this, *the advice complexity model* was introduced and has successfully yielded improved competitive ratios for bin packing and similar problems; see [1,8,9,19] for a selection of results. In this model, an oracle that knows both the online strategy and the input sequence provides the strategy with some prearranged information, *the advice*, about the input sequence, thus enabling the strategy to achieve improved competitive ratio.

1.1   Our Results

We consider the general online two-dimensional vector packing problem. We begin by showing a lower bound of $11/5$ for the competitive ratio of any ANY-FIT strategy [17] for the problem. In Sect. 4, we provide a strategy with competitive ratio $\max\{2, 6/(1 + 3\tan(\pi/4 - \gamma/2)) + \epsilon\}$ and logarithmic advice, for any instance where all the input vectors are restricted to have angles in the range $[\pi/4 - \gamma/2, \pi/4 + \gamma/2]$, for $0 \le \gamma < \pi/3$. In Sect. 5, we give a $5/2$-competitive strategy also using logarithmic advice for the unrestricted vectors case. These results should be contrasted to the currently best competitive strategy, FIRST-FIT, where an item is placed in the first bin where it fits, having competitive ratio $27/10$ [14].

2   Preliminaries

We will use two norms in the sequel. For a two-dimensional vector v, the L_1-norm of v is $\|v\|_1 \overset{\text{def}}{=} |v_x| + |v_y|$ and the L_∞-norm (or max-norm) of v is $\|v\|_\infty \overset{\text{def}}{=} \max\{|v_x|, |v_y|\}$, where v_x and v_y are the x- and y-coordinates of v, respectively.

The *online vector packing problem* we consider is, given an input sequence $\sigma = (v_1, v_2, \ldots)$, of two-dimensional vectors $v_i \in [0, 1]^2$, find the minimum number of unit sized square bins that can be packed online with vectors from the input sequence σ. From this problem definition we have that $0 \le v_x \le 1$ and $0 \le v_y \le 1$, i.e., all coordinates are non-negative.

A *packing* is simply a partitioning of the vectors into bins $B_1, B_2, \ldots$ such that for each bin B_j

$$\left\| \sum_{v \in B_j} v \right\|_\infty \leq 1. \tag{1}$$

In the *online* packing variant, the vectors are released from the sequence one by one and a strategy that solves the packing problem must irrevocably assign a vector to a bin before the next vector arrives. The assignment is either to an already open bin, maintaining the feasibility requirement in Inequality (1), or the strategy must open a new bin and assign the vector to this bin.

We measure the quality of an online strategy by its *competitive ratio*, the worst case bound R such that $|A(\sigma)| \leq R \cdot |\mathrm{OPT}(\sigma)| + C$, for every possible input sequence σ, where $A(\sigma)$ is the solution produced by the strategy A on σ, $\mathrm{OPT}(\sigma)$ is a solution on σ for which $|\mathrm{OPT}(\sigma)|$ is minimal, and C is some constant.

In certain situations, the complete lack of information about future input is too restrictive. In a sense, the online strategy plays a game against an all-powerful adversary who can construct the input sequence in the worst possible manner. To alleviate the adversary's advantage, we consider the following *advice-on-tape* model [9]. An *oracle* has knowledge about both the strategy and the full input sequence from the adversary, it writes information on an *advice tape* of unbounded length. The strategy can read bits from the advice tape at any time, before or while the requests are released by the adversary. The *advice complexity* is the number of bits read from the advice tape by the strategy.

We define the *load* of a bin B to be the sum of the L_1-norms of the included vectors, i.e.,

$$\mathrm{ld}(B) \overset{\mathrm{def}}{=} \sum_{v \in B} \|v\|_1 = \sum_{v \in B} |v_x| + |v_y| = \sum_{v \in B} v_x + v_y, \tag{2}$$

since all coordinates are non-negative. The load for the whole request sequence σ is $\mathrm{ld}(\sigma) \overset{\mathrm{def}}{=} \sum_{v \in \sigma} \|v\|_1 = \sum_{v \in \sigma} v_x + v_y$. Since the maximum load in a bin is 2, we immediately have that

$$|\mathrm{OPT}(\sigma)| \geq \lceil \mathrm{ld}(\sigma)/2 \rceil. \tag{3}$$

3   A Lower Bound for Two-Dimensional ANYFIT Strategies

Currently, the best lower bound on the competitive ratio for two-dimensional vector packing is $R \geq 1.80288\ldots$, implicit from the construction by Galambos *et al.* [13]. We show here a lower bound for two-dimensional vector packing valid for the class of ANYFIT strategies. An online strategy A is an ANYFIT strategy, if A does not open a new bin unless the current item released from the input sequence does not fit in any already opened bin [17].

Let $\sigma_1 = (p_i)_{i=1}^n$, $0 < p_i \leq 1$, be an instance of the one-dimensional online bin packing problem for which ANYFIT strategy A has competitive ratio at least $|A(\sigma_1)| \geq \lfloor 17|\text{OPT}(\sigma_1)|/10 \rfloor$, where $\text{OPT}(\sigma_1)$ is an optimal solution. Such sequences σ_1 exist of arbitrary length; see Dósa and Sgall [11] and Johnson [17], and we chose σ_1 so that $|\text{OPT}(\sigma_1)|$ is a multiple of 10.

To construct our lower bound for the two-dimensional case, let $p_{\min} = \min\{p_1, \ldots, p_n\}$, choose $0 < \delta < p_{\min}/2$, and construct a two-dimensional instance σ_2 as follows. The sequence σ_2 has a *prefix* consisting of $|\text{OPT}(\sigma_1)|$ copies of the vector $(0, 1/2)$, followed by a *suffix*, the sequence $(p_1, \delta), (p_2, \delta)$, $\ldots, (p_n, \delta)$. An optimal packing $\text{OPT}(\sigma_2)$ has the same size as $\text{OPT}(\sigma_1)$, since each bin in the optimal solution packs the vectors in the suffix optimally according to the x-coordinate and since we chose $\delta < p_{\min}/2$, the space used on the y-coordinate in each bin is $< 1/2$, so one of the $(0, 1/2)$ vectors can be placed in each bin. Thus, $|\text{OPT}(\sigma_2)| = |\text{OPT}(\sigma_1)|$.

The ANYFIT strategy A, given the vectors in σ_2, will pack the prefix vectors $(0, 1/2)$ pairwise into $\lfloor |\text{OPT}(\sigma_1)|/2 \rfloor$ bins that are full with respect to the y-coordinate. No vector in the suffix can be packed in any of these bins as they all have positive y-coordinate. The suffix is therefore packed by A into at least $\lfloor 17|\text{OPT}(\sigma_1)|/10 \rfloor$ bins, giving us a total of at least $\lfloor |\text{OPT}(\sigma_1)|/2 \rfloor + \lfloor 17|\text{OPT}(\sigma_1)|/10 \rfloor = |\text{OPT}(\sigma_1)|/2 + 17|\text{OPT}(\sigma_1)|/10 = 11|\text{OPT}(\sigma_1)|/5$ bins, since $|\text{OPT}(\sigma_1)|$ is a multiple of 10. We state this as a theorem.

Theorem 1. *Every* ANYFIT *strategy* A *has a lower bound for two-dimensional vector packing of*

$$|A(\sigma)| \geq \frac{11}{5}|\text{OPT}(\sigma)|,$$

for some input sequence σ.

4   A Strategy with Logarithmic Advice for Angle Restricted Vectors

We assume in this case that each vector v in the input sequence σ has angle

$$\arg v \in [\pi/4 - \gamma/2, \pi/4 + \gamma/2], \tag{4}$$

for a given extremal angle γ; see Fig. 1. This set of angles forms a cone issuing from the origin towards the point $(1, 1)$. Let $d = 1 + \tan(\pi/4 - \gamma/2)$, i.e., the abscissa of the line passing through the intersection points of the horizontal and vertical lines through $(1, 1)$.

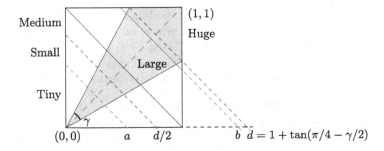

Fig. 1. The partitioning of vectors into five groups.

We follow the exposition in the proof of Theorem 4 in [8] for the one-dimensional case and partition the vectors into five groups. A vector v is

Tiny: if $\|v\|_1 \leq a$, ($a < 1$ is some constant to be established later)
Small: if $a < \|v\|_1 \leq d/2$,
Medium: if $d/2 < \|v\|_1 \leq 1$,
Large: if $1 < \|v\|_1 \leq b$, and ($1 < b < 2$ is some constant to be established later)
Huge: if $b < \|v\|_1 \leq 2$;

see Fig. 1. To ensure that no small or medium vector can be packed together with a huge vector in a bin and no three small or medium vectors can be packed together in a bin, we enforce that $a + b \geq 2$ and $3a \geq 2$, giving us $a \geq 2/3$ and $b \geq 4/3$. Furthermore, $a < d/2$ implies that $d > 4/3$, whereby $\gamma < \pi/3$.

To bound the number of advice bits used, we fix a positive integer parameter k. Each region of large, medium, and small vectors, respectively, is partitioned into k diagonal strips as $a + (i-1)(d/2-a)/k < \|v\|_1 \leq a + i(d/2-a)/k$, for $1 \leq i \leq k$ of the small vectors, $d/2 + (i-1)(1-d/2)/k < \|v\|_1 \leq d/2 + i(1-d/2)/k$, for $1 \leq i \leq k$ of the medium vectors, and $1 + (i-1)(b-1)/k < \|v\|_1 \leq 1 + i(b-1)/k$, for $1 \leq i \leq k$ of the large vectors.

The advice that the strategy obtains is the number of large, medium, and small vectors, respectively in each of the k strips, thus $O(k \log n)$ bits of advice in total.

Our strategy A_γ reads the $3k$ values $L_1, \ldots, L_k$, $M_1, \ldots, M_k$, and $S_1, \ldots, S_k$ corresponding to the number of vectors in each strip, lets $L = \sum_{i=1}^{k} L_i$, $M = \sum_{i=1}^{k} M_i$, $S = \sum_{i=1}^{k} S_i$, and opens $L + M + \lceil S/2 \rceil$ bins, henceforth denoted large, medium and small *critical bins*. It reserves space $1 + i(b-1)/k$ for each of L_i large critical bins and $d/2 + i(1-d/2)/k$ for each of M_i medium critical bins, $1 \leq i \leq k$. We say that a bin has a *virtual load* of $1 + i(b-1)/k$ and $d/2 + i(1-d/2)/k$, respectively. For the small critical bins, we reserve space matching the sum of the upper limit from the lowest non-empty small strip with the upper limit from the highest non-empty small strip, for a pair of vectors that can be matched together, thus reducing the number of vectors in those strips by one each. If the two strips are $a + (i-1)(d/2-a)/k < x + y \leq a + i(d/2-a)/k$ and $a + (j-1)(d/2-a)/k < x + y \leq a + j(d/2-a)/k$ for $i \leq j$, we reserve the virtual load of $2a + (i + j)(d/2 - a)/k$ to the bin. We repeat the process until at most a single small vector remains for which we reserve the virtual load of $a + i(d/2 - a)/k$, if the vector is in the ith strip. The reserved space in a bin is used to pack one large, one medium or two small vectors in the bin when the vector arrives. A_γ then serves each vector v in the sequence σ in order as follows:

- if v is huge, open a new bin and place v in this bin,
- if v is large and lies in strip i, place v in the reserved space of the first unused large critical bin with reserved space $1 + i(b - 1)/k$, reduce the virtual load of the bin to the actual load (by the amount $1 + i(b - 1)/k - \|v\|_1$),
- if v is medium and lies in strip i, place v in the reserved space of the first unused medium critical bin with reserved space $d/2 + i(1 - d/2)/k$, reduce the virtual load of the bin to the actual load (by the amount $d/2 + i(1 - d/2)/k - \|v\|_1$),
- if v is small and lies in strip i, place v in the reserved space of the first small critical bin that contains at most one small vector and has unused reserved space $a + i(d/2 - a)/k$, reduce the virtual load by $a + i(d/2 - a)/k - \|v\|_1$ and, if the bin now contains two small vectors, update the actual load.
- if v is tiny, use the FIRSTFIT strategy and place the vector in the first open bin where it fits based on virtual load (add the current vector to the virtual load, obtaining $\|v\|_1 + x + y$) and if this is not possible, open a new bin and place v in this bin.

Theorem 2. *For any angle $0 \leq \gamma < \pi/3$ and $\epsilon > 0$, the strategy A_γ receives $O(\frac{1}{\epsilon} \log n)$ bits of advice and has cost $c(\gamma, \epsilon)|\text{OPT}(\sigma)| + 1$ for serving any sequence σ of size n, where*

$$c(\gamma, \epsilon) = \max \left\{ 2, \frac{6}{1 + 3\tan(\pi/4 - \gamma/2)} + \epsilon \right\}.$$

Proof. Assume first that our strategy uses $|A_\gamma(\sigma)| = H + L + M + \lceil S/2 \rceil \leq H + L + M + S/2 + 1$ bins, i.e., there is no bin that only contains tiny vectors. The optimal strategy must use at least $|\text{OPT}(\sigma)| \geq H + L$ bins since no two huge or large vectors can be placed in the same bin. If $L \geq M + S$, then the number of bins that our strategy uses is

$$|A_\gamma(\sigma)| \leq H + L + M + S/2 + 1 \leq H + L + M + S + 1 \leq H + 2L + 1$$
$$\leq 2H + 2L + 1 \leq 2|\text{OPT}(\sigma)| + 1. \tag{5}$$

On the other hand, if $L < M + S$, then the optimal strategy can only place one medium or small vector together with a large one in a bin and the remaining medium and small vectors can at best be packed together two and two, the optimal strategy must therefore use at least $|\text{OPT}(\sigma)| \geq H + L + (M + S - L)/2 = H + L/2 + M/2 + S/2$ bins. The number of bins that our strategy then uses is

$$|A_\gamma(\sigma)| \leq H + L + M + S/2 + 1 \leq 2H + L + M + S + 1$$
$$= 2(H + L/2 + M/2 + S/2) + 1 \leq 2|\text{OPT}(\sigma)| + 1. \tag{6}$$

Assume now that our strategy constructs at least one bin with only tiny vectors in it. Any of these vectors did not fit in any of the critical bins initially opened, thus the virtual load of each critical bin is at least $d - a$. Since the difference between virtual load and actual load is at most $1/k$ (we know the number of vectors in each strip), the actual load is at least $d - a - 1/k$. Since the bins with huge vectors also have a load of this magnitude and we can have at most one bin with tiny vectors having less than this load, all but one bin have load at least $d - a - 1/k$. We have, if our strategy uses $|A_\gamma(\sigma)|$ bins, that

$$\mathrm{ld}(\sigma) = \sum_{i=1}^{|A_\gamma(\sigma)|} \mathrm{ld}(B_i) \geq (|A_\gamma(\sigma)| - 1)(d - a - 1/k), \tag{7}$$

so our strategy uses

$$|A_\gamma(\sigma)| \leq \frac{\mathrm{ld}(\sigma)}{d - a - 1/k} + 1 \leq \frac{2|\mathrm{OPT}(\sigma)|}{d - a - 1/k} + 1$$

$$\leq \frac{2|\mathrm{OPT}(\sigma)|}{\tan(\pi/4 - \gamma/2) + 1/3 - 1/k} + 1. \tag{8}$$

bins, by Inequality (3) and since $d = 1 + \tan(\pi/4 - \gamma/2)$ and $a = 2/3$.

Choosing $\epsilon = 8/k$, for $k \geq 8$, we have the competitive ratio as claimed, since $\tan(\pi/4 - \gamma/2) > 1/4$ for every $\gamma \in [0, \pi/3[$.

The strategy receives $3k$ values as advice, the number of vectors in each strip. Each value is encoded with at most $\lceil \log(n+1) \rceil$ bits, where $n = |\sigma|$. Hence, the number of advice bits is at most $3k(\lceil \log(n+1) \rceil) \in O(\frac{1}{\epsilon} \log n)$.   □

5   A Strategy with Logarithmic Advice for General 2D-Vectors

We generalize the approach by Fernandez de la Vega and Lueker [12] for the one-dimensional case and let k be a positive integer. Subdivide the unit square (representing the bins) by a $(k+1) \times (k+1)$ grid with intersection points at $(i/k, j/k)$, for $0 \leq i, j \leq k$. The region $](i-1)/k, i/k] \times](j-1)/k, j/k]$, for $1 \leq i, j \leq k$, is called the (i, j)-box. (Except for the special case when $i = 1$ or $j = 1$, then the $(1,1)$-box is the region $[0, 1/k] \times [0, 1/k]$, the $(1, j)$-box is the region $[0, 1/k] \times](j-1)/k, j/k]$, and the $(i, 1)$-box is the region $](i-1)/k, i/k] \times [0, 1/k]$.) When the specific coordinates of an (i, j)-box are unimportant, we will simply refer to it as a box. A vector v such that $(i-1)/k < v_x \leq i/k$ and $(j-1)/k < v_y \leq j/k$ is said to lie in or be contained in the (i, j)-box. (Again, the special case when $v_x = 0$ or $v_y = 0$, the vector lies in the $(1, j)$-box or the $(i, 1)$-box respectively.) We say that a vector v with $v_x \leq 40/k$ and $v_y \leq 40/k$ is short. All other vectors are long; see Fig. 2(a).

(a) 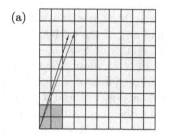   (b)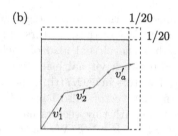

Fig. 2. (a) The partitioning of vectors into boxes ($k = 10$ for illustration). Vectors in green region are short (not to scale for purpose of illustration), a long vector v (blue) and v k-scaled to $\mathrm{sc}_k(v)$ (red). (b) Illustrating the proof of Lemma 1. (Color figure online)

Let $\mathrm{sc}_k(v)$ denote the *k-scaled* vector v, where $\mathrm{sc}_k(v) = (i/k, j/k)$, if v lies in the (i, j)-box. k-Scaling the vectors in σ reduces the types of vectors from possibly $|\sigma|$ to k^2. Disregarding the short vectors, the number of long vectors that can appear in a bin is at most $2\lfloor k/40 \rfloor \leq k/20$, since we can fit at most $\lfloor k/40 \rfloor$ long horizontal vectors and at most $\lfloor k/40 \rfloor$ long vertical vectors in a bin.

Let σ_L be the subsequence of long vectors in σ and let σ_S be the subsequence of short vectors in σ, both dependent on the parameter k. Let $\mathrm{sc}_k(\sigma_L)$ denote the k-scaled vectors in σ_L and let $\mathrm{OPT}(\mathrm{sc}_k(\sigma_L))$ be an optimal solution of the k-scaled long vectors in $\mathrm{sc}_k(\sigma_L)$.

We let the advice given by the oracle be the number of long vectors in each box. (Note that information about short vectors is not provided.) Given the number of long vectors, $n_{i,j}$, in each box, $1 \leq i, j \leq k$, a brute force algorithm can compute an optimal solution $\mathrm{OPT}(\mathrm{sc}_k(\sigma_L))$ in time polynomial in $|\sigma_L| \leq |\sigma| = n$; see the proof of Theorem 3. We let our strategy A_k perform this computation and open a critical bin corresponding to each bin in the solution $\mathrm{OPT}(\mathrm{sc}_k(\sigma_L))$. To each critical bin we reserve space corresponding to the sum of the L_1-norms of the assumed k-scaled vectors placed in it and set the virtual load of the bin to be this value. As the strategy serves requests from the sequence σ, each long vector v contained in an (i, j)-box is placed in the first bin that has remaining space for a k-scaled vector in an (i, j)-box. The virtual load of the bin is reduced by the difference $i/k + j/k - \|v\|_1$. Each short vector that arrives is placed according to the FIRSTFIT rule in the first bin where it fits according to the current virtual load and the virtual load is increased accordingly. If no such bin exists, the strategy opens a new bin and places the short vector there.

We first show a lower bound for our strategy A_k. Assume that k is odd and consider $2s$ copies of the vector $(1/2 - \epsilon, \epsilon)$ and s copies of the vector $(2\epsilon, 1 - 2\epsilon)$, for sufficiently small $\epsilon < 1/3k$. An optimal packing of these vectors uses s bins but the k-scaled vectors become $2s$ copies of $(1/2 + 1/2k, 1/k)$ and s copies of $(1/k, 1)$. No two scaled vectors fit together in a bin, hence an optimal packing of scaled vectors requires $3s$ bins, giving us a competitive ratio of at least 3.

However, when k is even, we can show that the competitive ratio of A_k is $5|\mathrm{OPT}(\sigma)|/2 + 1$ by first proving the following lemma.

Lemma 1. *For $k \geq 100$ and even, $\left|\mathrm{OPT}\big(\mathrm{sc}_k(\sigma_L)\big)\right| \leq \dfrac{5}{2}\left|\mathrm{OPT}(\sigma_L)\right| + 1$.*

Proof. We prove that for any bin B packed with long vectors, the corresponding k-scaled vectors can be packed into at most two bins and one half bin. A *half bin* is a 2-dimensional bin of size $[0, 1/2] \times [0, 1/2]$. Let $Q\big(\mathrm{sc}_k(\sigma_L)\big)$ be such a repacking of the bins in $\mathrm{OPT}(\sigma_L)$. This means that $Q\big(\mathrm{sc}_k(\sigma_L)\big)$ consists of at most $2|\mathrm{OPT}(\sigma_L)|$ bins and $|\mathrm{OPT}(\sigma_L)|$ half bins. Of course, the vectors in any two half bins can be packed together into one bin, giving us a new repacking $R\big(\mathrm{sc}_k(\sigma_L)\big)$ of size.

$$\left|R\big(\mathrm{sc}_k(\sigma_L)\big)\right| \leq 2|\mathrm{OPT}(\sigma_L)| + \lceil |\mathrm{OPT}(\sigma_L)|/2 \rceil \leq 5|\mathrm{OPT}(\sigma_L)|/2 + 1 \quad (9)$$

bins. Since $R\big(\mathrm{sc}_k(\sigma_L)\big)$ is a feasible packing of the long vectors, we have that $\left|\mathrm{OPT}\big(\mathrm{sc}_k(\sigma_L)\big)\right| \leq \left|R\big(\mathrm{sc}_k(\sigma_L)\big)\right|$ and the result as claimed.

Let B be an arbitrary bin and assume that B contains $m \geq 1$ long vectors. Assume the vectors are ordered $v_1, \ldots, v_m$ by decreasing L_1-norms of their k-scaled corresponding vectors, i.e., $\|\mathrm{sc}_k(v_1)\|_1 \geq \|\mathrm{sc}_k(v_2)\|_1 \geq \cdots \geq \|\mathrm{sc}_k(v_m)\|_1$. For ease of notation we use $v_i' \overset{\mathrm{def}}{=} \mathrm{sc}_k(v_i)$, for $1 \leq i \leq m$. Let $x_{i,j}$ and $y_{i,j}$ denote the sum of the x-coordinates and y-coordinates, respectively of $v_i', \ldots, v_j'$, for $i \leq j$, according to our ordering. It is clear that $x_{i,j} \leq 1 + (j - i + 1)/k$ and $y_{i,j} \leq 1 + (j - i + 1)/k$, for any $1 \leq i \leq j \leq m$.

By construction, any single vector v_i' fits in one bin so we can assume that $m \geq 2$. Also, if $m = 2$, then the two vectors can be packed in two separate bins, immediately proving our result, hence we assume $m \geq 3$. Consider the sequence $v_1' + v_2' + \cdots + v_m'$, starting at the origin in the bin; see Fig. 2(b). Fix v_a' to be the first vector that intersects the exterior of the bin (v_a' must exist, otherwise the whole sequence fits in the bin, immediately proving our claim) and assume without loss of generality that it intersects the vertical boundary of the bin. (The other case, where the sequence intersects the horizontal boundary is completely symmetric.) Obviously, $a \geq 2$.

Since the bin B has m long vectors and $m \leq k/20$, we have that $x_{1,m} \leq 1 + m/k \leq 21/20$ and symmetrically $y_{1,m} \leq 1 + m/k \leq 21/20$. Also, by our assumption that v_a' is the first vector intersecting the exterior of the bin, we have $x_{a+1,m} \leq 1/20$; see Fig. 2(b). We make the following case analysis:

if $x_{a,m} \leq 1$ and $y_{a,m} \leq 1$, then we can pack the vectors $v_1', \ldots, v_{a-1}'$ in one bin and the vectors $v_a', \ldots, v_m'$ in a second bin, satisfying our requirement.

if $x_{a,m} \leq 1$ and $y_{a,m} > 1$, we have two further cases:

if there is a vector v_b', $a < b \leq m$, such that $y_{a,b-1} \leq 1$ and $y_{b,m} \leq 1/2$, then we can pack $v_1', \ldots, v_{a-1}'$ in one bin, $v_a', \ldots, v_{b-1}'$ in a second bin, and $v_b', \ldots, v_m'$ in a half bin.

if there is a vector v_b', $a < b \leq m$, such that $y_{a,b-1} \leq 1$, $y_{a,b} > 1$, and $y_{b,m} > 1/2$, then since $y_{a,m} \leq y_{1,m} \leq 21/20$, we have that $y_{b+1,m} =$

$y_{a,m} - y_{a,b} < 21/20 - 1 = 1/20$, if the sequence $v'_{b+1}, \ldots, v'_m$ exists. Thus, $y_{b,b} = y_{b,m} - y_{b+1,m} > 1/2 - 1/20 = 9/20$, whether or not the sequence $v'_{b+1}, \ldots, v'_m$ exists. Since the L_1-norm $\|v'_b\|_1 \geq y_{b,b} > 9/20$, each vector $v'_1, \ldots, v'_{b-1}$ must also have L_1-norm greater than $9/20$, thus $b \leq 4$ in this case. Since $a \geq 2$, it follows that $b \in \{3, 4\}$. Reorder $v'_1, \ldots, v'_{b-1}$ so that $x_{1,1} \geq \cdots \geq x_{b-1,b-1}$.

If $b = 3$, we have three cases.

If $9/20 < y_{3,3} \leq 1/2$, then both $y_{1,1} < 1 + 3/k - 9/20 = 11/20 + 3/k < 12/20 = 3/5$ and $y_{2,2} < 1 + 3/k - 9/20 = 11/20 + 3/k < 12/20 = 3/5$. Since $x_{1,1} > 1/2$ and $x_{2,2} > 1/2$ is not possible, otherwise v_1 and v_2 would not fit together in one bin as they would both have x-coordinate greater than 1. We can pack v'_1 in a bin, v'_2 and $v'_4, \ldots, v'_m$ (if they exist) in a second bin, and v'_3 in a half bin.

If $1/2 < y_{3,3} \leq 19/20$, and again $x_{2,2} \leq 1/2$ so we can pack v'_1 in a bin, v'_2 in a half bin, and v'_3 and $v'_4, \ldots, v'_m$ (if they exist) in a second bin.

If $19/20 < y_{3,3} \leq 1$, and again $x_{2,2} \leq 1/2$. If $x_{1,1} > 19/20$, then $\|v'_2\|_1 < 2 + 6/k - 19/20 - 19/20 = 1/10 + 6/k < 3/20$, a contradiction. Hence, $x_{1,1} \leq 19/20$ and we can pack v'_1 and $v'_4, \ldots, v'_m$ (if they exist) in a bin, v'_2 in a half bin, and v'_3 in a second bin.

If $b = 4$, then since each vector v'_1, v'_2, v'_3, and v'_4 has L_1-norm greater than $9/20$, each of them also has L_1-norm smaller than $2 + 8/k - 3 \cdot 9/20 = 13/20 + 8/k$. Hence, we have $x_{1,1} < 13/20 + 8/k < 14/20 = 7/10$, $x_{2,2} < 7/10$, and $x_{3,3} < 7/10$. We have two cases.

If $9/20 < y_{4,4} \leq 1/2$, then if $x_{1,1} \leq 1/2$, then we can pack v'_1 and v'_2 in a bin, v'_3 and $v'_5, \ldots, v'_m$ (if they exist) in a second bin, and v'_4 in a half bin. If $x_{1,1} > 1/2$ then, since $x_{2,3} = x_{1,3} - x_{1,1} < 1 + 3/k - 1/2 = 1/2 + 3/k < 11/20$, we can pack v'_1 in a bin, v'_2, v'_3, and $v'_5, \ldots, v'_m$ (if they exist) in a second bin, and v'_4 in a half bin.

If $1/2 < y_{4,4} < 13/20 + 8/k$, then each of $y_{1,1} \leq 1/2$, $y_{2,2} \leq 1/2$, and $y_{3,3} \leq 1/2$. If $x_{1,1} \leq 1/2$ then we can pack v'_1 and in a half bin, v'_2 and v'_3 in a bin, and v'_4 and $v'_5, \ldots, v'_m$ (if they exist) in a second bin. If $x_{1,1} > 1/2$ then, both $x_{2,2} \leq 1/2$ and $x_{3,3} \leq 1/2$, so we can pack v'_1 and v'_3 in a bin, v'_2 in a half bin, and v'_4 and $v'_5, \ldots, v'_m$ (if they exist) in a second bin.

finally, if $x_{a,m} > 1$, then since $x_{a+1,m} \leq 1/20$, the L_1-norm $\|v'_a\|_1 \geq x_{a,a} > 19/20$, so each vector $v'_1, \ldots, v'_{a-1}$ must also have L_1-norm greater than $19/20$, thus $a = 2$ in this case, as the maximum sum of L_1-norms of vectors in a bin is 2. Since $x_{2,2} > 19/20$, the value of $x_{1,1} = x_{1,2} - x_{2,2} < 1 + 2/k - 19/20 = 1/20 + 2/k$, whereby $y_{1,1} = \|v'_1\|_1 - x_{1,1} > 19/20 - 1/20 - 2/k = 9/10 - 2/k$. Hence, $y_{3,m} = y_{1,m} - y_{2,2} - y_{1,1} < 21/20 - 0 - 9/10 + 2/k = 3/20 + 2/k < 1/2$. We can therefore pack v'_1 in one bin, v'_2 in a second bin and $v'_3, \ldots, v'_m$ in a half bin.

This completes the case analysis and proves our lemma. $\qquad\square$

We can now prove the main theorem of this section.

Theorem 3. *The strategy A_k receives $O(\log n)$ bits of advice, works in polynomial time, and has cost*

$$\frac{5}{2}|\mathrm{OPT}(\sigma)| + 1,$$

for serving any sequence σ of size n, if $k \geq 640$ is an even constant.

Proof. As in the proof of Theorem 2, assume first that A_k uses $|A_k(\sigma)| = |\mathrm{OPT}(\mathrm{sc}_k(\sigma_L))|$ bins, i.e., there is no bin that only contains short vectors. Consider an optimal solution $\mathrm{OPT}(\sigma)$ and remove all the short vectors from the bins in this solution. This is still a feasible solution for the remaining long vectors, thus $|\mathrm{OPT}(\sigma_L)| \leq |\mathrm{OPT}(\sigma)|$. By Lemma 1, we therefore get

$$|A_k(\sigma)| = |\mathrm{OPT}(\mathrm{sc}_k(\sigma_L))| \leq \frac{5}{2}|\mathrm{OPT}(\sigma_L)| + 1 \leq \frac{5}{2}|\mathrm{OPT}(\sigma)| + 1. \qquad (10)$$

Assume now that A_k constructs a solution $A_k(\sigma)$ having at least one bin with only short vectors in it. Each of these short vectors did not fit in any of the critical bins originally opened, thus the virtual load of each critical bin is greater than $1 - 80/k$, ($80/k$ is the maximum L_1-norm of a short vector). Since a bin can contain at most $k/20$ long vectors, and each long vector is scaled at most $1/k$ in the x-direction and at most $1/k$ in the y-direction, the actual load is greater than $1 - k/20 \cdot (1/k + 1/k) - 80/k = 9/10 - 80/k$. Since the maximum load of a bin is 2, we have in this case, c.f. Inequalities (3) and (7),

$$|A_k(\sigma)| \leq \left\lceil \frac{2|\mathrm{OPT}(\sigma_L)|}{9/10 - 80/k} \right\rceil \leq \left(\frac{20}{9} + \frac{1600}{9k} \right)|\mathrm{OPT}(\sigma_L)| + 1$$

$$\leq \frac{5}{2}|\mathrm{OPT}(\sigma_L)| + 1 \leq \frac{5}{2}|\mathrm{OPT}(\sigma)| + 1, \qquad (11)$$

by choosing $k \geq 640$ and even. The strategy reads at most $k^2 \lceil \log(n+1) \rceil \in O(\log n)$ bits of advice, since k is constant.

It remains to prove that the solution $\mathrm{OPT}(\mathrm{sc}_k(\sigma_L))$ can be computed in polynomial time in $|\sigma_L|$. Let $\mathcal{T}$ be the set of different possible bin types using k-scaled long vectors. Since at most $k/20$ long vectors of k^2 different types can be packed in a bin, we can bound the number of different packing types by

$$|\mathcal{T}| \leq \sum_{l=1}^{k/20} \binom{k^2 + l - 1}{l} \in O\big(k \cdot (k^2)^{k/20}\big) = O\big(k^{1+k/10}\big), \qquad (12)$$

which is constant. We let $t_{i,j}$ be the number of k-scaled long vectors in the (i,j)-box for the bin type $t \in \mathcal{T}$. Given the advice information $n_{1,1}, n_{1,2}, \ldots, n_{k,k}$, the number of long vectors in each box, we can formulate a recurrence for the optimal packing solution as

$$P\big(n_{1,1}, n_{1,2}, \ldots, n_{k,k}\big) =$$
$$\min_{t \in \mathcal{T}} \big\{ P\big(n_{1,1} - t_{1,1}, n_{1,2} - t_{1,2}, \ldots, n_{k,k} - t_{k,k}\big) \big\} + 1, \qquad (13)$$

that we can solve in polynomial time with dynamic programming, since both k and $|\mathcal{T}|$ are constants. $\qquad \square$

When k is even, our analysis is asymptotically tight. Consider the following instance of vectors: s copies of the vector $(1/2 - 2\epsilon, \epsilon)$, s copies of the vector $(1/2 + \epsilon, \epsilon)$, and s copies of the vector $(\epsilon, 1 - 2\epsilon)$, for sufficiently small $\epsilon < 1/3k$. An optimal packing of these vectors uses s bins. The k-scaled vectors become s copies of $(1/2, 1/k)$, $(1/2 + 1/k, 1/k)$ and $(1/k, 1)$, respectively. Only pairs of the first type of vectors fit together in a bin, hence the minimum number of bins required after k-scaling is $\lceil 5s/2 \rceil \geq 5s/2$, giving a lower bound of $5/2$.

6 Combining the Results

Combining our two presented strategies with the result by Angelopoulos *et al.* [1] that achieves a competitive ratio of $1.47012 + \epsilon$ with constant advice (actually $O(\log \epsilon^{-1})$ bits) for the one dimensional case, so that we use this strategy when $\gamma = 0$, strategy A_γ when $0 < \gamma \leq \pi/2 - 2\arctan(7/15)$ and strategy A_k, for $k \geq 640$ and even, when $\gamma > \pi/2 - 2\arctan(7/15)$, we have the following corollary.

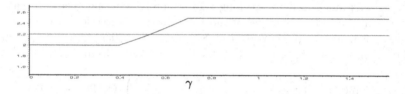

Fig. 3. Plots for worst case competitive ratios for $\gamma \in [0, \pi/2]$ for our combined strategy (red), FIRSTFIT (green), and ANYFIT lower bound (brown). (Color figure online)

Corollary 1. *For any angle* $0 \leq \gamma \leq \pi/2$ *and* $\epsilon > 0$, *the combined strategy described receives* $O(\epsilon^{-1} \log n)$ *bits of advice and has cost* $c(\gamma, \epsilon)|\mathrm{OPT}(\sigma)| + 1$ *for serving any sequence* σ *of size* n, *where*

$$
c(\gamma, \epsilon) = \begin{cases} 1.47012 + \epsilon & \text{for } \gamma = 0, \\ \max\left\{2, \dfrac{6}{1 + 3\tan(\pi/4 - \gamma/2)} + \epsilon\right\} & \text{for } 0 < \gamma \leq \pi/2 - 2\arctan(7/15), \\ 5/2 & \text{for } \pi/2 - 2\arctan(7/15) < \gamma \leq \pi/2. \end{cases}
$$

Figure 3 illustrates the worst case competitive ratio for different values of γ.

7 Conclusions

We consider the online two-dimensional vector packing problem and show a lower bound of $11/5$ for the competitive ratio of any ANYFIT strategy. We also show upper bounds spanning between 2 and $5/2$ depending on the angle restrictions placed on the vectors given logarithmic advice, where the currently best competitive strategy has competitive ratio $27/10$, albeit without using advice.

Interesting open problems include generalizing the lower bound on the competitive ratio to hold for any strategy (without advice) and relating the advice complexity to the competitive ratio, either by giving specific lower bounds on the advice complexity for a given competitive ratio or through some function that relates one with the other.

References

1. Angelopoulos, S., Dürr, C., Kamali, S., Renault, M.P., Rosén, A.: Online bin packing with advice of small size. Theory Comput. Syst. **62**(8), 2006–2034 (2018). https://doi.org/10.1007/s00224-018-9862-5
2. Azar, Y., Cohen, I.R., Fiat, A., Roytman, A.: Packing small vectors. In: Proceedings 27th Annual ACM-SIAM SODA, pp. 1511–1525 (2016)
3. Azar, Y., Cohen, I.R., Kamara, S., Shepherd, B.: Tight bounds for online vector bin packing. In: Proceedings 45th Annual ACM STOC, pp. 961–970 (2013)
4. Balogh, J., Békési, J., Dósa, G., Sgall, J., van Stee, R.: The optimal absolute ratio for online bin packing. In: Proceedings 26th Annual ACM-SIAM SODA, pp. 1425–1438 (2015)
5. Balogh, J., Békési, J., Galambos, G.: New lower bounds for certain classes of bin backing algorithms. Theor. Comput. Sci. **440**, 1–13 (2012)
6. Balogh, J., Epstein, L., Levin, A.: Truly Asymptotic Lower Bounds for Online Vector Bin Packing. arXiv:2008.00811 (2020)
7. Bansal, N., Cohen, I.R.: An Asymptotic Lower Bound for Online Vector Bin Packing. arXiv:2007.15709 (2020)
8. Boyar, J., Kamali, S., Larsen, K.S., López-Ortiz, A.: Online bin packing with advice. Algorithmica **74**(1), 507–527 (2014). https://doi.org/10.1007/s00453-014-9955-8
9. Böckenhauer, H.-J., Komm, D., Královič, R., Královič, R., Mömke, T.: On the advice complexity of online problems. In: Dong, Y., Du, D.-Z., Ibarra, O. (eds.) ISAAC 2009. LNCS, vol. 5878, pp. 331–340. Springer, Heidelberg (2009). https://doi.org/10.1007/978-3-642-10631-6_35
10. Chekuri, C., Khanna, S.: On multidimensional packing problems. SIAM J. Comp. **33**(4), 837–851 (2004)
11. Dósa, G., Sgall, J.: First fit bin packing: a tight analysis. In: Proceedings 30th STACS, pp. 538–549 (2013)
12. de la Vega, W.F., Lueker, G.S.: Bin packing can be solved within $1 + \epsilon$ in linear time. Combinatorica **1**(4), 349–355 (1981). https://doi.org/10.1007/BF02579456
13. Galambos, G., Kellerer, H., Woeginger, G.J.: A lower bound for online vector-packing algorithms. Acta Cybernetica **11**(1–2), 23–34 (1993)
14. Garey, M.R., Graham, R.L., Johnson, D.S., Yao, A.C.-C.: Resource constrained scheduling as generalized bin packing. J. Comb. Theory Ser. A. **21**(3), 257–298 (1976)
15. Garey, M.R., Freeman, R.L., Johnson, D.S.: Computers and Intractability (1979)
16. Heydrich, S., van Stee, R.: Beating the harmonic lower bound for online bin packing. In: Proceedings 43rd ICALP, LIPIcs, vol. 41, pp. 1–14 (2016)
17. Johnson, D.S.: Fast algorithms for bin packing. J. Comput. Syst. Sci. **8**(3), 272–314 (1974)
18. Johnson, D.S., Demers, A., Ullman, J.D., Garey, M.R., Graham, R.L.: Worst-case performance bounds for simple one-dimensional packing algorithms. SIAM J. on Comp. **3**(4), 299–325 (1974)
19. Renault, M.P., Rosén, A., van Stee, R.: Online algorithms with advice for bin packing and scheduling problems. Theor. Comput. Sci. **600**, 155–170 (2015)

Temporal Matching on Geometric Graph Data

Timothe Picavet[1]([✉]), Ngoc-Trung Nguyen[2], and Binh-Minh Bui-Xuan[3]

[1] ENS Lyon, Lyon, France
`timothe.picavet@ens-lyon.fr`
[2] HCM University of Education, Ho Chi Minh City, Vietnam
`trungnn@hcmue.edu.vn`
[3] LIP6 (CNRS – Sorbonne Université), Paris, France
`buixuan@lip6.fr`

Abstract. Temporal graphs are the modeling of pairwise and historical interaction in recordings of a dataset. A temporal matching formalizes the planning of pair working sessions of a required duration. We depict algorithms finding temporal matchings maximizing the total workload, by an exact algorithm and an approximation. The exact algorithm is a dynamic programming solving the general case in $O^*((\gamma+1)^n)$ time, where n is the number of vertices, γ represents the desired duration of each pair working session, and O^* only focuses on exponential factors. When the input data is embedded in an Euclidean space, called geometric data, our approximation is based on a new notion of temporal velocity. We revise a known notion of static density [van Leeuwen, 2009] and result in a polynomial time approximation scheme for temporal geometric graphs of bounded density. We confront our implementations to known opensource implementation (Our source code is available at https://github.com/Talessseed/Temporal-matching-of-historical-and-geometric-graphs).

Keywords: temporal matching · geometric graph · PTAS

1 Introduction

Data collected from automated processes come ordered by the time instants when they are recorded. Graphs in this context appear in several variants: link streams [18], time varying [7], temporal [14] or evolving graphs [6]. These structures occur in the study of transportation timetables [9,15,16], navigation programs [26], email exchanges [17], proximity interactions [27], and many other types of dataset [28]. Therein, a pair working session is a repeated interaction of two vertices over a certain amount of time. Pair working helps in optimizing global parameters such as total fuel consumption when co-sailing with Fello'fly [2], or code reliability when running XP agile projects [1].

Supported by Courtanet – Sorbonne Université convention C19.0665 and ANRT grant 2019.0485.

T. Calamoneri and F. Corò (Eds.): CIAC 2021, LNCS 12701, pp. 394–408, 2021.
https://doi.org/10.1007/978-3-030-75242-2_28

The total workload of pair working is captured in the notion of a temporal matching [4]. Given an integer γ, we define formally problem γ-MATCHING in the subsequent section; informally, it consists in finding a maximum cardinality set of compatible pair working sessions, each to be recorded in at least γ consecutive timestamps in a historical dataset of graphs.

When $\gamma = 1$, the problem can be reduced to (classical) static MATCHING, which consists in computing a maximum independent edge set of a static graph. It can be solved in polynomial time by many well known algorithms [8,13,23], heuristics [11], greedy approximations for large input [31]; as well as in an online algorithmic context [12,22,30]. It is very intriguing to know whether these enthusiastic results extend to the non-static case of γ-MATCHING.

Unfortunately, very little positive results are known for the temporal case. Even when restricting the input to be a path at each instant, one can very naturally obtain a grid-like structure by folding out a temporal graph instance over the time dimension. On these structures, careful polynomial and parameterized reductions allow to obtain very good hardness results, see *e.g.* [3,21] and the extensive bibliography therein. Likewise, γ-MATCHING is NP-hard as soon as $\gamma > 1$ [4], even on very restricted input instances [20]. To the best of our knowledge, most notable positive results for γ-MATCHING are: a fixed parameter tractable (FPT [10]) algorithm parameterized by the matching number of the union graph[1] [20], an implemented kernelization producing quadratic kernels [4,5] (in the sense of FPT algorithms), and an implemented 2-approximation from a greedy approach [5].

Our paper addresses the following question: *Would there be Fast Algorithms Computing γ-MATCHING on Data Recorded from Human Activities? Can they be Implemented?* Human data are not artificial, yet very naturally captured by a geometric graph: an embedding of a vertex set into a Euclidean space, along with a real number representing the threshold below which an edge exists between two vertex-points, see *e.g.* [19]. The formalism is especially useful in transportation and social networks where geometric proximity implies higher probability of successful routing, resp. social relationship [24].

Theoretical Contribution. In order to obtain good runtime, we consider *natural behaviour* of embedded vertex-points. The main crux is to carefully examine a notion of partial derivative, called velocity. This parameter helps ruling out unrealistic leaps of a vertex-point from one recorded instant to another. We revisit, ubiquitously, the parameter control used in [29] which is related to the (static) density of vertex-points. Then, we present and implement a PTAS for temporal geometric graphs of bounded velocity and density.

Numerical Comparison to Previous Works. We compare our result to previously known works. The FPT algorithm given in [20] has not been implemented. In particular, part of this algorithm relies on complex algorithmic results in matroid

---

[1] If a temporal graph is considered to be a sequence $(G_t)_{t \in T}$ of graphs over the same vertex set, then the edge set of the union graph is the union of all edge sets of G_t for all t.

theory. We skip the corresponding analysis. The kernelization implementation [4,5] helps in reducing the input data, but not in solving the reduced instance. On instances where we can afford the runtime, we use it as preprocessing step for the PTAS, exactly the same way done for the greedy implementation [5]. Essentially, the PTAS is compared to the greedy implementation. Our numerical results are in favour of the latter, which finds temporal matching of size ≈10% bigger than the PTAS on generated geometric datasets. Since the theoretical approximation factor of the greedy algorithm is 2 [4], which is much worse than the theoretical ratio of the PTAS on our datasets, these experiments raise the question whether both implementations perform badly, or the greedy approximation factor is near optimal on geometric data.

We devise and implement an optimal solution for the general case of γ-MATCHING terminating in reasonable time on parts of our datasets, that is, in $O^*((\gamma + 1)^n)$ time where n is the number of vertices and O^* only focuses on exponential factors. The PTAS and the greedy experimental approximation ratios are then determined, which average at 1.02-approximation from optimal.

We present in Sect. 2 the formal framework of problem γ-MATCHING. Section 3 presents a PTAS solution for the case of temporal geometric graphs of bounded velocity and density. Section 4 presents a FPT solution for the general case. Due to space restriction, properties marked with ($\star$) are given without proof, and only the essential numerical experiments are summarised in Sect. 5.

2 Pair Working Sessions in Historical Graph Data

Every graph $G = (V, E)$ in this paper is simple, loopless and undirected. We also note $V(G) = V$ and $E(G) = E$. When, and only when, $u \neq v \in V$, we abusively note $uv = vu = \{u, v\} \in \binom{V}{2}$ the *edge* between u and v.

Graph data collected over a duration of time are formalized as a triple $L = (T, V, E)$, called *link stream*, such that $T \subseteq \mathbb{N}$ is an interval, V a finite set of *vertices*, and E a lexicographically ordered subset of $T \times \binom{V}{2}$ called *recorded edges*. We also note $T(L) = T$, $V(L) = V$ and $E(L) = E$.

Pairwise collaborations over a duration are defined as γ-edges, with γ an integer. A γ-*edge* $\Gamma \subseteq E(L)$ is a subset of γ consecutive edges recorded in $E(L)$, namely $\Gamma = \{(i, uv) \in E(L) : t \leq i < t + \gamma\}$ for $t \in T(L)$ and $u \neq v \in V(L)$. We also note such γ-edge $E_\gamma(t, uv)$.

We note $E_\gamma(L)$ the set of all γ-edges of L. Two γ-edges $\Gamma, \Gamma' \in E_\gamma(L)$ are *dependent* if there exist instant i and vertices $u \neq v$, $u \neq w$, such that $(i, uv) \in \Gamma$ and $(i, uw) \in \Gamma'$; the two γ-edges are *independent* otherwise. In planning pair working sessions, a conflict-free planning is called a γ-*matching*, and defined as a set of pairwise independent γ-edges. The following problem is NP-hard for $\gamma > 1$ [4], even on very restricted classes of link streams [20].

Problem γ-MATCHING:
INPUT: a link stream L
OUTPUT: a γ-matching of maximum cardinality in L.

Geometric Model: Let L be a link stream. The subgraph G_t of L induced at time $t \in T(L)$ is defined as $V(G_t) = V(L)$ and $E(G_t) = \{uv : (t, uv) \in E(L)\}$. For $d \in \mathbb{N}$, graph G is a *unit ball graph* if there exists a point set $\{\mathbf{c}_v : v \in V(G)\} \subseteq \mathbb{R}^d$, called set of *centers*, such that $E(G) = \{uv : u \neq v \wedge \|\mathbf{c}_u - \mathbf{c}_v\| \leq 1\}$. Link stream L is a *unit ball stream* if the subgraph of L induced at any time $t \in T(L)$ is a unit ball graph. In this case, we denote the center of vertex v at time t in L as $\mathbf{c}_v(t)$. L has *velocity* ν if $\|\mathbf{c}_v(t+1) - \mathbf{c}_v(t)\| \leq \nu$ for every $t \in T \setminus \{\max(T)\}$ and $v \in V(L)$. We also refer to balls as *intervals* when $d = 1$ and *disks* when $d = 2$.

Line Graph Extrapolation: γ-MATCHING links itself to MAXIMUMINDEPEN-DENTSET on the following input. The *γ-line graph L_γ* of a link stream L is defined as $V(L_\gamma) = E_\gamma(L)$ and $E(L_\gamma) = \{\{\Gamma, \Gamma'\} : \Gamma \text{ and } \Gamma' \text{ are dependent } \gamma\text{-edges}\}$. By definition, solving γ-MATCHING on a link stream is equivalent to solving MAXIMUMINDEPENDENTSET on its γ-line graph.

3 Approximation for Unit Ball Streams

In this section, we use velocity and extend van Leeuwen approximation [29, Theorem 6.3.8, page 74] for MAXIMUMINDEPENDENTSET on unit disk graphs to the γ-line graph of a unit ball stream L. Since the γ-line graph is not necessarily a unit ball graph, our main idea is to examine the middle of the two vertex-points of every γ-edge $\Gamma \in E_\gamma(L)$: the middle point can not vary much because of velocity. Corollary 1 below is crucial to our approach.

For a γ-edge $\Gamma = E_\gamma(t, uv) \in E_\gamma(L)$ between $u \neq v \in V(L)$ starting at time $t \in T(L)$, the *(middle) center* $\mathbf{c}_\Gamma$ is defined as the middle point of the centers of u and v recorded at the starting time t of the γ-edge, $\mathbf{c}_\Gamma = \frac{1}{2} \cdot (\mathbf{c}_u(t) + \mathbf{c}_v(t))$. Using velocity of the centers, which can only move $\gamma - 1$ times while maintaining the existence of Γ, we refer to the *normalized center* of Γ as $\overline{\mathbf{c}_\Gamma} = \frac{1}{1+(\gamma-1)\nu} \cdot \mathbf{c}_\Gamma$.

Proposition 1. *In a unit ball stream, if Γ and Γ' are dependent γ-edges, then* $\|\overline{\mathbf{c}_\Gamma} - \overline{\mathbf{c}_{\Gamma'}}\| \leq 1$.

Proof. Let $\Gamma = E_\gamma(t, uv)$ and $\Gamma' = E_\gamma(t', u'v')$. Because of dependency, we suppose w.l.o.g. that $u = u'$, and $t \leq t' \leq t + \gamma - 1$. We note from Euclidean triangular inequality that $\|\mathbf{c}_\Gamma - \mathbf{c}_{\Gamma'}\| \leq \|\mathbf{c}_\Gamma - \mathbf{c}_u(t)\| + \|\mathbf{c}_u(t) - \mathbf{c}_u(t')\| + \|\mathbf{c}_u(t') - \mathbf{c}_{\Gamma'}\|$. Now, $\|\mathbf{c}_\Gamma - \mathbf{c}_u(t)\| \leq \frac{1}{2}$ because $\|\mathbf{c}_u(t) - \mathbf{c}_v(t)\| \leq 1$. Since $u = u'$, we deduce likewise that $\|\mathbf{c}_u(t') - \mathbf{c}_{\Gamma'}\| \leq \frac{1}{2}$. Finally, $\|\mathbf{c}_u(t) - \mathbf{c}_u(t')\| = \left\| \sum_{i=t}^{t'-1} (\mathbf{c}_u(i+1) - \mathbf{c}_u(i)) \right\| \leq \sum_{i=t}^{t'-1} \|\mathbf{c}_u(i+1) - \mathbf{c}_u(i)\| \leq (t' - t)\nu \leq (\gamma - 1)\nu$. $\square$

Since the normalized centers are uniquely computed from their starting instant, this is also a fast checking method for independent γ-edges. We refer to the unit ball graph having as geometric model the set of normalized centers of all γ-edges of L the *normalized γ-line graph* of L.

Corollary 1. *The normalized γ-line graph of a unit ball stream is a unit ball graph having the γ-line graph as partial subgraph. Any independent set of the γ-line graph is also an independent set of its normalized graph*

We now adapt the notion of density [29] to the normalized γ-line graph of a unit ball stream L of dimension d. Let $A \subseteq E_\gamma(L)$. We refer to the set of all γ-edges of A starting at time $t \in T(L)$ as $A_t = \{E_\gamma(t, uv) \in A : u \neq v \in V(L)\}\}$. The *thickness of* A is the maximum cardinality of such a set, taken over every possible starting time, that is, $\theta(A) = \max\{|A_t| : t \in T(L)\}$.

In the sequel, all cubes are axis-aligned cubes. For a unit d-cube $\mathbf{U} \subseteq \mathbb{R}^d$, we denote $A_{\mathbf{U}} = \{\Gamma \in A : \overline{c_\Gamma} \in \mathbf{U}\}$. The *density of* A is the maximum thickness of such a set, taken over every possible unit d-cube, that is, $\rho(A) = \max\{\theta(A_{\mathbf{U}}) : \mathbf{U} \subseteq \mathbb{R}^d$ a unit d-cube$\}$. The *density of* L is $\rho(L) = \rho(E_\gamma(L))$.

We will describe in Lemma 1 a decrementing process for the Euclidean space dimension. Informally, this is a partial density relaxing the first dimension of the unit d-cubes to infinite unit cuboids. For a unit $(d-1)$-cube $\mathbf{H} \subseteq \mathbb{R}^{(d-1)}$, we denote $A_{\mathbf{H}} = \{\Gamma \in A : \overline{c_\Gamma} \in \mathbb{R} \times \mathbf{H}\}$ (we replace the unit d-cube $\mathbf{U}$ in the definition of density by the infinite unit cuboid $\mathbb{R} \times \mathbf{H}$). The *partial density of A (w.r.t. the first dimension)* is the maximum thickness of such a set, taken over every possible infinite unit cuboids, that is, $\partial\rho(A) = \max\{\theta(A_{\mathbf{H}}) : \mathbf{H} \subseteq \mathbb{R}^{(d-1)}$ a unit $(d-1)$-cube$\}$. We observe when $d = 1$ that the partial density is the thickness.

For a γ-edge $\Gamma \in E_\gamma(L)$, let $\overline{x_\Gamma}$ denote the projection of $\overline{c_\Gamma}$ on the first dimension, that is, $\overline{c_\Gamma} = (\overline{x_\Gamma}, \dots)$. A *decomposition path* X is a set of scalars ordered increasingly, $X = \{x_1, x_2, \dots, x_{|X|}\}$. We define the *incomplete partition of A by X (w.r.t. the first dimension)*, noted $P_X(A) = (P_0(A), P_1(A), \dots, P_{|X|}(A))$, as follows. Firstly, $P_0(A) = \{\Gamma \in A : \overline{x_\Gamma} < x_1 - \frac{1}{2}\}$. For $0 < i < |X|$, $P_i(A) = \{\Gamma \in A : x_i + \frac{1}{2} \leq \overline{x_\Gamma} < x_{i+1} - \frac{1}{2}\}$. Finally, $P_{|X|}(A) = \{\Gamma \in A : x_{|X|} + \frac{1}{2} \leq \overline{x_\Gamma}\}$.

Basically, this is a partition of the Euclidean space into $|X| + 1$ parts called slab decomposition [29]. $P_X(A)$ corresponds to the γ-edges inside each part, while those incident to the boundaries are removed. From Proposition 1, two γ-edges of different parts are independent. Hence, $P_X(A)$ can also be seen as a partial decomposition tree for the pathwidth [25] of the (normalized) γ-line graph of L. In the rest of this section, we describe a way to compute such a set X.

Lemma 1. *Let L be a link stream with density ρ. Let $E_\gamma = E_\gamma(L)$ and $m_\gamma = |E_\gamma|$. Let f_L be a big enough integer, $f_L \geq \rho$. Then, we can compute in time polynomial in m_γ a decomposition path $X = \{x_1, x_2, \dots, x_{|X|}\}$, in such a way that the incomplete partition $P_X(E_\gamma)$ satisfies:*

- $P_0(E_\gamma) = P_{|X|}(E_\gamma) = \emptyset$,
- $\forall 0 < i < |X| - 1, f_L \leq \partial\rho(P_i(E_\gamma)) < f_L + \rho$,
- $0 \leq \partial\rho(P_{|X|-1}(E_\gamma)) < f_L + \rho$,
- $x_{|X|} - x_{|X|-1} \geq \frac{f_L}{\rho}$.

Proof. We would like to scan the γ-edges of E_γ in such a way to only increase the partial density. The main crux is to process greedily along the same x-axis w.r.t. which the partial density is defined: informally, the infinite unit cuboids $\mathbb{R} \times \mathbf{H}$ with $\mathbf{H}$ a unit $(d-1)$-cube can be seen as FIFO strips along this x-axis.

Formally, sort $E_\gamma = \{\Gamma_1, \Gamma_2, \ldots, \Gamma_{m_\gamma}\}$ so that $\overline{x_{\Gamma_1}} \leq \overline{x_{\Gamma_2}} \leq \cdots \leq \overline{x_{\Gamma_{m_\gamma}}}$. In the following, i and P are auxiliary variables containing an index and a set of centers, respectively. Initialize $i \leftarrow 1$, $P \leftarrow \emptyset$, define $x_i = \overline{x_{\Gamma_1}}$ minus one unit, and increment i. For all $\Gamma \in E_\gamma$ in increasing order, if the partial density $\partial\rho(P)$ is strictly less than f_L, add Γ to P along with all other Γ' with $\overline{x_\Gamma} = \overline{x_{\Gamma'}}$, skipping the partial density check for these Γ'. We call this step (**Add**). Else, create a new boundary by defining $x_i = \overline{x_\Gamma}$, emptying P, and incrementing i. At the end of the iteration process, define x_i to be the last seen $\overline{x_\Gamma}$ plus one unit (in order to avoid coinciding with the previous x_i). Finally, increment i again and define x_i to be an arbitrarily big number so that it satisfies the last item of Lemma 1. Partial density checks can be done in polynomial time in m_γ because of Procedure 1 described below. Hence, the overall process is polynomial in m_γ.

All parts $P_i(E_\gamma)$ defined by the computed x_i's have a partial density of at least f_L, except for the first and the last two parts. The only thing left to prove is $\partial\rho(P_i(E_\gamma)) < f_L + \rho$. By contradiction suppose that the partial density exceeds that number on some part P_i. This could only happen after adding a set A of γ-edges in some step (**Add**). Then, we must have $\partial\rho(A) > \rho$ because adding γ-edges along the x-axis can only increase the partial density w.r.t. that axis. Let $\mathbf{H}$ be the unit $(d-1)$-cube such that $\partial\rho(A) = \theta(A_\mathbf{H})$. Consider then the unit cube $\mathbf{U}$ defined by $\mathbf{U} = [\overline{x_\Gamma} - \frac{1}{2}, \overline{x_\Gamma} + \frac{1}{2}] \times \mathbf{H}$, with $\Gamma \in A$. It holds $E_{\gamma\mathbf{U}} \supseteq A_\mathbf{H}$. Hence, $\rho = \rho(E_\gamma) \geq \theta(E_{\gamma\mathbf{U}}) \geq \theta(A_\mathbf{H}) = \partial\rho(A) > \rho$. Contradiction.   □

Procedure 1: Procedure to calculate the density of a set A of γ-edges

---

1 $\rho \leftarrow 0$
   // Each C_i is a normalized center of a γ-edge of A
2 **for** $(C^1, C^2, \ldots, C^d) \in \{\overline{c_\Gamma} : \Gamma \in A\}^d$ **do**
      // We consider the unit hypercube $\mathbf{H}$ with C_i^i as its i-th lowest
      coordinate
3 $\quad$ $\mathbf{H} \leftarrow [C_1^1, C_1^1 + 1] \times \cdots \times [C_d^d, C_d^d + 1]$
4 $\quad$ $\rho \leftarrow \max\{\rho, \theta(A_\mathbf{H})\}$
5 **return** ρ

Due to space restriction, the proof of properties marked with ($\star$) is omitted.

Lemma 2 ($\star$). *Keeping the same notations as in the hypothesis of Lemma 1, we have $x_{i+1} - x_i \geq \frac{f_L}{\rho}$ for any $1 \leq i < |X|$.*

Lemma 3 ($\star$). *Keeping the same notations as in the hypothesis of Lemma 1, let $k \in \mathbb{N}$ be such that $k \leq \left\lfloor \frac{f_L}{\rho} \right\rfloor$. Let $l = |X|$. For $s \in [\![0, k-1]\!]$, we define*

$(P_0(s), P_1(s), \ldots, P_l(s)) = P_{\{x+s:x\in X\}}(A)$. Then, we have $\partial\rho(P_i(s)) < 2f_L$ for $i \in [\![0, l]\!]$ and $s \in [\![0, k-1]\!]$.

Lemma 4 ($\star$). *Keeping the same notations as in the hypothesis of previous Lemma 3, suppose that for every $s \in [\![0, k-1]\!]$ and $i \in [\![0, l]\!]$, $M_i(s)$ is a r-approximation of γ-MATCHING on $P_i(s)$. Let $M(s) = \cup_{i\in[\![0,l]\!]}M_i(s)$. Then, $\gamma MM^{\sim} = \max_{s\in[\![0,k-1]\!]}\{M(s)\}$ is a $r\left(1 - \frac{1}{k}\right)$-approximation of γ-MATCHING on L.*

We will now show an exact algorithm to compute the base case $(d = 1)$ of the approximation. It is also a correct algorithm for arbitrary link streams.

Algorithm 1 (Exact algorithm for the base case of the PTAS, on input an arbitrary link stream).
*On input any link stream L, we note $E_\gamma(t)$ the set of all γ-edges starting at time $t \in T(L)$, $E_\gamma(t) = \{E_\gamma(t, uv) \in E_\gamma(L) : u \neq v \in V(L)\}$. By convention, we note $E_\gamma(t) = \emptyset$ for $t \notin T(L)$. We proceed by dynamic programming and store in $M(t, S_1, S_2, \ldots, S_{\gamma-1})$ a γ-matching $\mathcal{M}$ of maximum cardinality of the restriction of L to time instants between 0 and $t + \gamma - 2$ where we have for $1 \leq i \leq \gamma - 1$ that $\mathcal{M} \cap E_\gamma(t-1+i) = S_i$.
If $\mathcal{T} = S_1 \cup S_2 \cup \cdots \cup S_{\gamma-1}$ is a γ-matching, we have the following formulae:*

- $M(0, S_1, S_2, \ldots, S_{\gamma-1}) = \mathcal{T}$
- $M(t + 1, S_1, S_2, \ldots, S_{\gamma-1}) = S_{\gamma-1}\cup\max_{\substack{S_0\subseteq E_\gamma(t),\\ S_0\cup\mathcal{T} \text{ is a}\\ \gamma-matching}} \{M(t, S_0, S_1, S_2, \ldots, S_{\gamma-2})\}$

If $\mathcal{T} = S_1\cup S_2\cup\cdots\cup S_{\gamma-1}$ is not a γ-matching, we let $M(t, S_1, S_2, \ldots, S_{\gamma-1}) = \emptyset$. After sequentially filling table M by increasing t, we output the value stored in $M(\max(T(L)), \emptyset, \ldots, \emptyset)$. A python implementation is available[2].

Lemma 5. *On input any link stream $L = (T, V, E)$ with γ-edge set $E_\gamma = E_\gamma(L)$, Algorithm 1 computes an optimal solution for γ-MATCHING on L in time $O(t_{max}\gamma^2\theta(E_\gamma)2^{\gamma\theta(E_\gamma)})$, where $t_{max} = \max(T(L))$.*

Proof. We proceed by induction on t. Let $P(\mathcal{M}, t, S_1, S_2, \ldots, S_{\gamma-1})$ denote the fact that $\mathcal{M}$ is a γ-matching of maximum cardinality of the restriction of L to time instants between 0 and $t+\gamma-2$ such that $\mathcal{M} \cap E_\gamma(t-1+i) = S_i$ for every $1 \leq i \leq \gamma - 1$.

Firstly, $M(0, S_1, S_2, \ldots, S_{\gamma-1}) = S_1 \cup S_2 \cup \cdots \cup S_{\gamma-1}$ because $\forall i \in [\![1, \gamma-1]\!]$, $\mathcal{M} \cap E_\gamma(i-1) = S_i$.

Secondly, let $t \in T(L)$, $S_i \subseteq E_\gamma(t+i)$ for each $1 \leq i \leq \gamma - 1$, and suppose that the formula is correct for $t - 1$. For convenience, let $\mathcal{T} = S_1 \cup S_2 \cup \cdots \cup S_{\gamma-1}$. We suppose that $\mathcal{T}$ is a γ-matching. Otherwise, it's trivial. Moreover, let $\mathcal{M}$ be such that $P(\mathcal{M}, t, S_1, S_2, \ldots, S_{\gamma-1})$ is satisfied and $S = \{M(t-1, S_0, S_1, S_2, \ldots, S_{\gamma-2}) : S_0 \subseteq E_\gamma(t-1) \wedge S_0 \cup \mathcal{T} \text{ is a } \gamma - matching\}$. We will show that $|\mathcal{M}| = |M(t, S_1, S_2, \ldots, S_{\gamma-1})|$.

[2] It is implemented in function `base_case` in https://github.com/Talessseed/Temporal-matching-of-historical-and-geometric-graphs/blob/master/approx.py.

$\geq$ We claim that S only contains γ-matchings of the restriction of L to time instants between 0 and $t + \gamma - 3$, where every $\mathcal{M}' = M(t - 1, S_0, S_1, S_2, \ldots, S_{\gamma-2}) \in S$ is a γ-matching such that $\mathcal{M}' \cap E_\gamma(t + i) = S_i$ for every $0 \leq i \leq \gamma - 2$. This is entirely deduced from the induction hypothesis. Thus, because every $\mathcal{M}' \in S$ has only edges living in times between 0 and $t + \gamma - 3$, $M(t, S_1, S_2, \ldots, S_{\gamma-1}) \cap E_\gamma(t + \gamma - 2) = S_{\gamma-1}$. Moreover, $M(t, S_1, S_2, \ldots, S_{\gamma-1})$ is a γ-matching with only edges living in times between 0 and $t + \gamma - 2$. Therefore by definition of $\mathcal{M}$, $|\mathcal{M}| \geq |M(t, S_1, S_2, \ldots, S_{\gamma-1})|$.

$\leq$ Let $S_0 = \mathcal{M} \cap E_\gamma(t - 1)$. We can write $\mathcal{M} = S_{\gamma-1} \cup \mathcal{M}'$ such that $P(\mathcal{M}', t - 1, S_0, S_1, \ldots, S_{\gamma-2})$ is satisfied. Note by induction hypothesis that we have $|\mathcal{M}'| = |M(t-1, S_0, S_1, \ldots, S_{\gamma-2})|$. Hence, it follows from the definition of $M(t, S_1, S_2, \ldots, S_{\gamma-1})$ that $|\mathcal{M}| = |S_{\gamma-1} \cup M(t-1, S_0, S_1, \ldots, S_{\gamma-2})| \leq |M(t, S_1, S_2, \ldots, S_{\gamma-1})|$.

We now address complexity issues. Checking if a set is a γ-matching can be done by Procedure 2. Note that $|\mathcal{M}| \leq \gamma \theta(E_\gamma)$. Hence, Procedure 2 terminates in time $O(\gamma^2 \theta(E_\gamma))$. Algorithm 1 iterates over t and the S_i's. Their number is exactly $t_{max} 2^{\gamma \theta(E_\gamma)}$.

Procedure 2: Procedure to check if $\mathcal{M}$ is a γ-matching

---

```
1  seen ← ∅
2  for Γ = Eγ(t, uv) ∈ M do
3  │   for t' ∈ ⟦t, t + γ − 1⟧ do
4  │   │   if (t', u) ∈ seen ∨ (t', v) ∈ seen then
5  │   │   │   return false
6  │   │   seen ← seen ∪ {(t', u), (t', v)}
7  return true
```

Algorithm 2 (Approximation for γ-MATCHING on unit ball streams).
We keep the same notations as in the hypothesis of Lemma 4. If $d = 1$, we return the output of Algorithm 1. Otherwise, we compute the sets $M_i(s)$ with recursive calls on L but with positions in $\mathbb{R}^{d-1}$: we remove the x dimension by projecting every $\mathbf{c}_\Gamma$ on the hyperplane with equation $(x = 0)$: the input $\mathbf{c}_\Gamma = (x_\Gamma, y_\Gamma, z_\Gamma, \ldots)$ is replaced with $\mathbf{c}_\Gamma \leftarrow (y_\Gamma, z_\Gamma, \ldots)$. We return set $\gamma MM^\sim$ as defined in Lemma 4.

We stress on the use of variable $f_L \geq \rho$. Basically, if we call the approximation algorithm on a link stream with positions in $\mathbb{R}^p$, our algorithm will also use a similar value $f_L \leftarrow f_{p,L}$ with $f_{p,L} \geq \rho$. We define $f_{p,L} = q^{d-p+1} 2^{d-p-1} \frac{\log(m_\gamma)}{\gamma}$ in order to obtain the following result.

Theorem 1. *Algorithm 2 is polynomial in m_γ and is a PTAS for γ-MATCHING on unit ball streams of bounded velocity and density ρ embedded in an d-dimension space. More precisely, for any $q \geq 2\rho\gamma$, a γ-matching with*

approximation ratio $\left(1 - \frac{1}{\left\lceil \frac{q\log(m_\gamma)}{2\rho\gamma} \right\rceil}\right)\left(1 - \frac{1}{\lceil q \rceil}\right)^{d-1}$ *can be computed in time* $O^*(q^d 2^{d-1} m_\gamma{}^{q^d 2^{d-1}})$, *where* O^* *only retains exponential factors.*

Procedure 3: Exact algorithm for the base case of the PTAS ($d = 1$)

---

1 Compute $E_\gamma(1), E_\gamma(2), \ldots, E_\gamma(\max(T(L)))$
2 Initialize M as $\emptyset$ for all elements
3 **for** $S_i \subseteq E_\gamma(i)$ *with* $1 \le i \le \gamma - 1$ **do**
4 $\quad$ **if** $S_1 \cup S_2 \cup \cdots \cup S_{\gamma-1}$ *is a* γ*-matching* **then**
5 $\quad\quad$ $M(0, S_1, S_2, \ldots, S_{\gamma-1}) \leftarrow S_1 \cup S_2 \cup \cdots \cup S_{\gamma-1}$
6 **for** $1 \le t \le \max(T(L)) - \gamma + 1$ **do**
7 $\quad$ **for** $S_i \subseteq E_\gamma(t+i)$ *with* $1 \le i \le \gamma - 1$ **do**
8 $\quad\quad$ $T \leftarrow S_1 \cup S_2 \cup \cdots \cup S_{\gamma-1}$
9 $\quad\quad$ **for** $S' \subseteq E_\gamma(t)$ **do**
10 $\quad\quad\quad$ **if** $S' \cup T$ *is a* γ*-matching* **then**
11 $\quad\quad\quad\quad$ $\mathcal{N} \leftarrow S_{\gamma-1} \cup M(t, S', S_1, ..., S_{\gamma-2})$
12 $\quad\quad\quad\quad$ $M(t+1, S_1, ..., S_{\gamma-1}) \leftarrow \max\{M(t+1, S_1, ..., S_{\gamma-1}), \mathcal{N}\}$
13 $t \leftarrow \max\{-1, \max(T(L)) - \gamma + 1\}$
14 $\mathcal{M} \leftarrow \emptyset$
15 **for** $S_i \subseteq E_\gamma(t+i)$ *with* $1 \le i \le \gamma - 1$ **do**
16 $\quad$ $\mathcal{M} \leftarrow \max\{\mathcal{M}, M(t+1, S_1, S_2, \ldots, S_{\gamma-1})\}$
17 **return** $\mathcal{M}$

Proof. First, we must verify our assumptions on the new transformed vertices each time we do the reduction of dimension. To do so, we need to show that if two γ-edges have their transformed normalized centers at distance strictly greater than 1, they must be independent. We prove this by induction on the dimension p. In dimension d, this is proven following the bounded velocity of the unit ball stream, *cf.* Proposition 1. If the dimension p is strictly smaller than d, we suppose we have proven what we want in dimension $p + 1$. Let Γ and Γ' be two independent γ-edges with normalized centers in dimension $p+1$ $(\overline{x_\Gamma}, \overline{y_\Gamma}, \overline{z_\Gamma}, ...)$ and $(\overline{x_{\Gamma'}}, \overline{y_{\Gamma'}}, \overline{z_{\Gamma'}}, ...)$ respectively. We suppose that $\|(\overline{y_\Gamma}, \overline{z_\Gamma}, ...) - (\overline{y_{\Gamma'}}, \overline{z_{\Gamma'}}, ...)\| > 1$ in $\mathbb{R}^p$. But then, it also holds that $\|(\overline{x_\Gamma}, \overline{y_\Gamma}, \overline{z_\Gamma}, ...) - (\overline{x_{\Gamma'}}, \overline{y_{\Gamma'}}, \overline{z_\Gamma}, ...)\| > 1$ in $\mathbb{R}^{p+1}$. This contradicts the induction hypothesis.

We call the algorithm on a link stream L with positions in $\mathbb{R}^d$ initially. L has m_γ γ-edges. Let ρ_p ($p \le d$) be the maximum density of link stream the algorithm processes with positions in $\mathbb{R}^p$. We claim that $\rho_p \le 2f_{p+1,L} = q^{d-p} 2^{d-p-1} \frac{\log(m_\gamma)}{\gamma}$ for $p < d$. Indeed, if the algorithm processes a link stream with positions in $\mathbb{R}^p$, then for each $M_i(s)$ considered, we have that $\partial\rho(M_i(s)) \le \rho_{p-1}$, because the partial density is actually a density where one dimension is forgotten. We conclude with Lemma 3. The conditions on $f_{p,L}$ are also satisfied for $p < d$. Indeed, $f_{p,L} = 2qf_{p+1,L} \ge q\rho_p \ge \rho_p$.

We first address the approximation ratio. We suppose w.l.o.g. that $\log m_\gamma \ge 1$. Recall in Lemma 3 that k was chosen to satisfy $k \le \left\lfloor \frac{f_L}{\rho} \right\rfloor$. For a link stream

L' with positions in $\mathbb{R}^p$, we choose for the algorithm $k = \left\lfloor \frac{f_{p,L}}{\rho'} \right\rfloor$ where ρ' is the density of L'. Therefore, when the algorithm is called on a link stream with positions in $\mathbb{R}^p$, we have for $p < d$ that $k = \left\lfloor \frac{f_{p,L}}{\rho'} \right\rfloor \geq \left\lfloor \frac{f_{p,L}}{\rho_p} \right\rfloor \geq \left\lfloor \frac{f_{p,L}}{2f_{p+1,L}} \right\rfloor \geq \lfloor q \rfloor$.

Hence, $1 - \frac{1}{k} \geq 1 - \frac{1}{\lfloor q \rfloor} \xrightarrow{q \to +\infty} 1$. Combining this with Lemma 4 implies:

$$\left(1 - \frac{1}{\left\lfloor \frac{f_{d,L}}{\rho} \right\rfloor}\right) \prod_{p=1}^{d-1} \left(1 - \frac{1}{\left\lfloor \frac{f_{p,L}}{\rho_p} \right\rfloor}\right)$$

$$\geq \left(1 - \frac{1}{\left\lfloor \frac{q \log(m_\gamma)}{2\rho\gamma} \right\rfloor}\right) \prod_{p=1}^{d-1} \left(1 - \frac{1}{\lfloor q \rfloor}\right)$$

$$= \left(1 - \frac{1}{\left\lfloor \frac{q \log(m_\gamma)}{2\rho\gamma} \right\rfloor}\right) \left(1 - \frac{1}{\lfloor q \rfloor}\right)^{d-1} \xrightarrow{q \to +\infty} 1 \text{ if } q \geq 2\rho\gamma \text{ and } \log m_\gamma \geq 1.$$

Finally, we show that Algorithm 2 is polynomial when q is fixed. Since the computations when removing a dimension are polynomial in m_γ, it is left to prove that Algorithm 1 also solves the base case in time polynomial in m_γ. We note that $\theta(E_\gamma) \leq \rho_0 \leq q^d 2^{d-1} \frac{\log(m_\gamma)}{\gamma}$, and combine it with Lemma 5 to obtain:

$$O\left(t_{max}\gamma^2\theta(E_\gamma)2^{\gamma\theta(E_\gamma)}\right)$$

$$= O\left(t_{max}\gamma^2 q^d 2^{d-1} \frac{\log(m_\gamma)}{\gamma} 2^{\gamma q^d 2^{d-1} \frac{\log(m_\gamma)}{\gamma}}\right)$$

$$= O\left(t_{max}\gamma q^d 2^{d-1} \log(m_\gamma) 2^{q^d 2^{d-1} \log(m_\gamma)}\right)$$

$$= O\left(t_{max}\gamma q^d 2^{d-1} \log(m_\gamma) m_\gamma^{q^d 2^{d-1}}\right)$$

Notice that we can suppose w.l.o.g that each frame of γ consecutive time instants contains a time-vertex itself contained in a γ-edge. Indeed, if that is not the case, we can delete time instants where no γ-edges exist, without making two γ-edges that were independent dependent. Hence, $t_{max} = O(\gamma m_\gamma)$, implying the algorithm is polynomial in m_γ when q is fixed. Whence, the algorithm is a PTAS of bounded ρ for all q such that $q \geq 2\rho\gamma$. $\qquad\square$

4   Exact Algorithm for Arbitrary Link Streams

For later use in the numerical analysis, we need to find an optimal solution for γ-MATCHING. This section presents an FPT solution for γ-MATCHING parameterized by the vertex number. Without being pushy about the time complexity "in the big O", we are demanding on good runtime performance. Our implementation performs well because with temporal graphs, the vertex number is a very small FPT parameter compared to the more popularly used size of the output.

We shall store in $M(t, A_1, A_2, \ldots, A_\gamma)$, for every $t \in [\![1, t_{max} - \gamma + 1]\!]$, a maximum γ-matching of the restriction of link stream L to time instants between 0 and $t + \gamma - 1$, where we remove all timed edges adjacent to the time-vertices $(t + i - 1, u)$ for all $u \in A_i$. Intuitively, A_i is the set of time vertices at time $t + i - 1$ that are endpoints of already used γ-edges beginning at $t + i - 1$. The recursion for M is:

- $M(-1, A_1, A_2, \ldots, A_\gamma) = \emptyset$
- $M(t, A_1, A_2, \ldots, A_\gamma) = \max\left(\left\{M(t-1, \emptyset, A_1, A_2, \ldots, A_{\gamma-1})\right\} \cup\right.$

$$\left.\left\{\{\Gamma\} \cup M(t, A_1 \cup \{u, v\}, A_2, \ldots, A_\gamma) : \Gamma = E_\gamma(t, u, v) \subseteq L \wedge u, v \notin \bigcup_{i=1}^{\gamma} A_i\right\}\right).$$

Lemma 6 ($\star$). *Keeping the same notations introduced before, $M(t_{max} - \gamma + 1, \emptyset, \ldots, \emptyset)$ is a maximum γ-matching of L.*

Theorem 2 ($\star$). *On any n-vertex, m-recorded-edge link stream L with γ-edge number $m_\gamma = |E_\gamma(L)|$, γ-MATCHING can be solved in time $O(m + n^2 + m_\gamma(\gamma+1)^n)$ by a dynamic programming filling the above described table M. At the end of the process, $M(t_{max} - \gamma + 1, \emptyset, \ldots, \emptyset)$ stores a maximum γ-matching of L, where $t_{max} = \max(T(L))$.*

5 Numerical Analysis

We implement[3] in Python 3 both the PTAS described in Sect. 3 and the DP described in Sect. 4. We compare their numerical results with the JavaScript implementation of the greedy approximation[4] given in [5]. In the latter reference, the authors use an arbitrary total ordering to sort the γ-edge set, then step by step try to add each γ-edge to the matching if it does not conflict with the others that are already present in the matching.

Our experiments are run on a standard laptop with i5 6300HQ 4 cores @2.3 GHz and 8GB memory @2133 MHz. We place automatic timeouts so that the computation stops after spending 100 (one hundred) seconds on an instance. In general, the greedy implementation returns instantaneously, while the PTAS and the DP take much more time. All our computations add up to $\approx$400 h CPU time over the 4 cores. In what follows, we totally skip the discussion about computing time, and solely focus on approximation ratio.

---

[3] Our source code is available at https://github.com/Talessseed/Temporal-matching-of-historical-and-geometric-graphs.

[4] The source code of [5] is available at https://github.com/antoinedimitriroux/Temporal-Matching-in-Link-Streams.

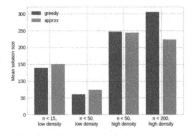
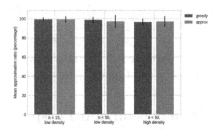

Fig. 1. (Left) Mean of the outputted values of PTAS vs. greedy on 4 generated datasets of unit ball streams; (Right) Mean and standard deviation of the approximation ratio of PTAS vs. greedy, when compared to the optimal values, on ≈90% of the datasets in the left figure. In one of the 4 datasets, we do not have any reliable approximation ratio because the optimal computation runs out of time on most instances.

5.1 PTAS vs. Greedy

Theoretically, the greedy implementation is a 2-approximation [4]. The theoretical approximation ratio of the PTAS is as follows: it is 1.27 for unit interval streams of velocity 5 and density 5; and is 1.38 for unit disk streams of velocity 2 and density 4. Both values are very far from the theoretical ratio of the greedy implementation. Accordingly, we would like to confirm that information on artificially generated data. In the following, we choose the value of q to be the biggest integer from $\left\lceil \frac{2\gamma\rho}{\log(m_\gamma)} \right\rceil$ to $\left\lceil \frac{2\gamma\rho}{\log(m_\gamma)} \right\rceil + 10$ that does not result in a timeout.

Hypothesis 1: PTAS Finds Better Solution than Greedy on Unit Ball Streams of Well Chosen Velocity and Density. Our methodology is to generate four different datasets, *cf.* Fig. 1 (left). We run both implementations on these, and record the outputted γ-matching size.

Discussion: Hypothesis 1 is not Confirmed in our Setting. Our experimental results are in favour of the greedy implementation, which performs ≈10% better than the PTAS, especially when the density is high. This is surprising in the sense that very good conditions for PTAS are met: low dimension of the Euclidean space (good runtime), controlled velocity and density (good approximation ratio), and varying number of vertices (to rule out the help of kernelization [4,5], at least on parts of the dataset). While it does not completely refute *Hypothesis 1* in other settings, our numerical analysis implies at least two possibilities: either both implementations find solutions half the size of the optimal values, or their solutions are near optimal (or they are somewhere in between).

Hypothesis 2: Both Approximation Factors in Previous Experiment are Close to Optimal. We address the dynamic programming explained in Sect. 4. Since the runtime of its implementation is very long, we can only obtain the answer for ≈90% of the datasets described in the above experiment. Our results are presented in Fig. 1 (right).

Discussion: Likely Validation of Hypothesis 2. Our experimental results show that PTAS and greedy find solutions that are more than ≈95% of the optimal values, with a deviation less than ≈5% of the mean, on ≈90% of the unit ball streams presented in the previous experiment. We conclude that Hypothesis 2 seems to be valid on our generated input.

6   Conclusion and Perspectives

We introduce the notion of velocity in a temporal geometric graph. Revisiting van Leeuwen theorem on static geometric graphs of bounded density [29, Theorem 6.3.8, page 74], we extend their PTAS to temporal geometric graphs of bounded density and bounded temporal velocity. Our study case is γ-MATCHING [4], a temporal version of (static) graph matching [13]. Implementation works show that a known greedy implementation [5] finds better approximated solutions by a factor of ≈10%, when compared to the PTAS. Theoretically, the approximation factor is 2 for the greedy algorithm and between 1.27 and 1.38 for the PTAS on our datasets. This raises the question whether the greedy factor 2 is tight on temporal geometric graphs. As a byproduct for obtaining parts of the above numerical analysis, we devise a simple dynamic programming formula solving optimally the general case in FPT time parameterized by the number of vertices. Since vertices are in small number compared to recordings of edges in a temporal graph, our dynamic programming could be of independent practical interest.

Acknowledgements. We are grateful to Hai Bui Xuan for helpful discussion and pointers. We are grateful to the anonymous reviewers for their helpful comments which greatly improved the paper.

References

1. XP. http://www.extremeprogramming.org
2. Airbus Industrie. Fello'fly demonstrator. Dubai Airshow (2019)
3. Akrida, E.C., Mertzios, G.B., Spirakis, P.G., Zamaraev, V.: Temporal vertex cover with a sliding time window. J. Comput. Syst. Sci. **107**, 108–123 (2020)
4. Baste, J., Bui-Xuan, B.-M.: Temporal matching in link stream: kernel and approximation. In: 16th Cologne-Twente Workshop on Graphs and Combinatorial Optimization (2018)
5. Baste, J., Bui-Xuan, B.-M., Roux, A.: Temporal matching. Theor. Comput. Sci. **806**, 184–196 (2020)
6. Bui-Xuan, B.-M., Ferreira, A., Jarry, A.: Computing shortest, fastest, and foremost journeys in dynamic networks. Int. J. Found. Comput. Sci. **14**(2), 267–285 (2003)
7. Casteigts, A., Flocchini, P., Godard, E., Santoro, N., Yamashita, M.: Expressivity of time-varying graphs. In: 19th International Symposium on Fundamentals of Computation Theory, pp. 95–106 (2013)
8. Cygan, M., Gabow, H.N., Sankowski, P.: Algorithmic applications of Baur-Strassen's theorem Shortest cycles, diameter, and matchings. J. ACM. **62**(4), 28:1-28:30 (2015)

9. Dibbelt, J., Pajor, T., Strasser, B., Wagner, D.: Connection scan algorithm. ACM J. Experimental Algorithmics **23** (2018)
10. Downey, R.G., Fellows, M.R.: Parameterized complexity. In: Downey, R.G. (ed.) Monographs in Computer Science. Springer (1999). https://doi.org/10.1007/978-1-4612-0515-9
11. Dufossé, F., Kaya, K., Panagiotas, I., Uçar, B.: Approximation algorithms for maximum matchings in undirected graphs. In: SIAM Workshop on Combinatorial Scientific Computing (2018)
12. Dürr, C., Konrad, C., Renault, M.P.: On the power of advice and randomization for online bipartite matching. In: 24th Annual European Symposium on Algorithms, LIPIcs, vol. 57, pp. 37:1–37:16 (2016)
13. Edmonds, J.: Paths, trees, and flowers. Can. J. Math. **17**, 449–467 (1965)
14. Erlebach, T., Hoffmann, M., Kammer, F.: On temporal graph exploration. In: Halldórsson, M.M., Iwama, K., Kobayashi, N., Speckmann, B. (eds.) ICALP 2015. LNCS, vol. 9134, pp. 444–455. Springer, Heidelberg (2015). https://doi.org/10.1007/978-3-662-47672-7_36
15. Foschini, L., Hershberger, J., Suri, S.: On the complexity of time-dependent shortest paths. Algorithmica **68**(4), 1075–1097 (2014)
16. Kempe, D., Kleinberg, J., Kumar, A.: Connectivity and inference problems for temporal networks. J. Comput. Syst. Sci. **64**(4), 820–842 (2002)
17. Klimt, B., Yang, Y.: Introducing the enron corpus. In: CEAS (2004)
18. Latapy, M., Viard, T., Magnien, C.: Stream graphs and link streams for the modeling of interactions over time. Soc. Network Anal. Min. **8**(1), 1–29 (2018). https://doi.org/10.1007/s13278-018-0537-7
19. McKee, T.A., McMorris, F.R.: Topics in intersection graph theory. SIAM Monographs on Discrete Mathematics and Applications (1999)
20. Mertzios, G.B., Molter, H., Niedermeier, R., Zamaraev, V., Zschoche, P.: Computing maximum matchings in temporal graphs. In: 37th International Symposium on Theoretical Aspects of Computer Science, LIPIcs, vol. 154, pp. 27:1–27:14 (2020)
21. Mertzios, G.B., Spirakis, P.G.: Strong bounds for evolution in networks. J. Comput. Syst. Sci. **97**, 60–82 (2018)
22. Miyazaki, S.: On the advice complexity of online bipartite matching and online stable marriage. Inf. Process. Lett. **114**(12), 714–717 (2014)
23. Mucha, M., Sankowski, P.: Maximum matchings via Gaussian elimination. In: 45th Annual IEEE Symposium on Foundations of Computer Science, pp. 248–255 (2004)
24. Penrose, M.: Random Geometric Graphs. Oxford University Press, Oxford (2003)
25. Robertson, N., Seymour, P.D.: Graph minors. I. Excluding a forest. J. Comb. Theory, Ser. B **35**(1), 39–61 (1983)
26. Ros, F.J., Ruiz, P.M., Stojmenovic, I.: Acknowledgment-based broadcast protocol for reliable and efficient data dissemination in vehicular ad-hoc networks. IEEE Trans. Mobile Comput. **11**(1), 33–46 (2012)
27. Tournoux, P.-U., Leguay, J., Benbadis, F., Conan, V., De Amorim, M.D., Whitbeck, J.: The accordion phenomenon: analysis, characterization, and impact on DTN routing. In: 28th IEEE Conference on Computer Communications (2009)
28. Tsalouchidou, I., Baeza-Yates, R., Bonchi, F., Liao, K., Sellis, T.: Temporal betweenness centrality in dynamic graphs. Int. J. Data Sci. Anal. **9**(3), 257–272 (2019). https://doi.org/10.1007/s41060-019-00189-x
29. van Leeuwen, E.J.: Optimization and Approximation on Systems of Geometric Objects. PhD thesis, Utrecht University (2009)

30. Wang, Y., Wong, S.C.: Two-sided online bipartite matching and vertex cover: beating the greedy algorithm. In: Halldórsson, M.M., Iwama, K., Kobayashi, N., Speckmann, B. (eds.) ICALP 2015. LNCS, vol. 9134, pp. 1070–1081. Springer, Heidelberg (2015). https://doi.org/10.1007/978-3-662-47672-7_87
31. Wøhlk, S., Laporte, G.: Computational comparison of several greedy algorithms for the minimum cost perfect matching problem on large graphs. Comput. Oper. Res. **87**(C), 107–113 (2017)

Author Index

Printed in the United States
by Baker & Taylor Publisher Services